797,885 Books
are available to read at

Forgotten Books

www.ForgottenBooks.com

Forgotten Books' App
Available for mobile, tablet & eReader

ISBN 978-1-330-39681-0
PIBN 10051139

This book is a reproduction of an important historical work. Forgotten Books uses state-of-the-art technology to digitally reconstruct the work, preserving the original format whilst repairing imperfections present in the aged copy. In rare cases, an imperfection in the original, such as a blemish or missing page, may be replicated in our edition. We do, however, repair the vast majority of imperfections successfully; any imperfections that remain are intentionally left to preserve the state of such historical works.

Forgotten Books is a registered trademark of FB &c Ltd.
Copyright © 2017 FB &c Ltd.
FB &c Ltd, Dalton House, 60 Windsor Avenue, London, SW19 2RR.
Company number 08720141. Registered in England and Wales.

For support please visit www.forgottenbooks.com

1 MONTH OF
FREE
READING

at
www.ForgottenBooks.com

By purchasing this book you are eligible for one month membership to ForgottenBooks.com, giving you unlimited access to our entire collection of over 700,000 titles via our web site and mobile apps.

To claim your free month visit: www.forgottenbooks.com/free51139

* Offer is valid for 45 days from date of purchase. Terms and conditions apply.

English
Français
Deutsche
Italiano
Español
Português

www.forgottenbooks.com

Mythology Photography **Fiction** Fishing Christianity **Art** Cooking Essays Buddhism Freemasonry Medicine **Biology** Music **Ancient Egypt** Evolution Carpentry Physics Dance Geology **Mathematics** Fitness Shakespeare **Folklore** Yoga Marketing **Confidence** Immortality Biographies Poetry **Psychology** Witchcraft Electronics Chemistry History **Law** Accounting **Philosophy** Anthropology Alchemy Drama Quantum Mechanics Atheism Sexual Health **Ancient History Entrepreneurship** Languages Sport Paleontology Needlework Islam **Metaphysics** Investment Archaeology Parenting Statistics Criminology **Motivational**

HARDWICKE'S
Science-Gossip:

AN ILLUSTRATED MEDIUM OF INTERCHANGE AND GOSSIP

FOR STUDENTS AND

LOVERS OF NATURE.

EDITED BY

J. E. TAYLOR, Ph.D., F.L.S., F.G.S., F.R.G.S.I., &c.

VOLUME XX.

London:
CHATTO AND WINDUS, PICCADILLY.
1884.

(All rights reserved.)

LONDON:
PRINTED BY WILLIAM CLOWES AND SONS, Limited,
STAMFORD STREET AND CHARING CROSS.

PREFACE.

THE present volume witnesses the Twentieth year of the existence of SCIENCE-GOSSIP—perhaps the fullest and most interesting period in the history of Natural Science. Since this journal came into existence many new methods of research have sprung up, most if not all of which have been duly chronicled in our pages. Never before was the history of discovery so complete as within the last twenty years. Are we assuming too much in stating that, notwithstanding the unpretentious character of our magazine, readers will find it difficult to obtain elsewhere than in the volumes of SCIENCE-GOSSIP, so perfect a scientific chronicle of the period in question?

It is with no small pride the Editor feels that the present volume is the best of the series. Apart from the splendid coloured plates which the enterprise of the publishers has enabled him to present to his readers every month, the Editor was never so ably supported by contributors of such well-known scientific and literary position, as during the past year.

It is equally gratifying to him to know he has the strong support of a large circle of sympathetic readers, who are generously ready to condone faults, and to take such an individual interest in the success of this Magazine.

It is with much pleasure the Editor announces that the programme for 1885 will be in no way behind that of its predecessor. Asking the kindly help of every reader and contributor to enable him to give SCIENCE-GOSSIP a place in every intellectual home in Great Britain, the Editor cordially wishes all, readers and contributors alike, a hearty Christmas Greeting!

LIST OF ILLUSTRATIONS.

AMPHITRITE VENTILABRUM, 181
Amplexus coralloides, 158
Antheridium of Funaria, 84
Aquarium, a New, 8
Archegonium of Funaria, 84
Arenicola piscatorum, 197

BACILLUS OF CHOLERA, 177
Blatta Germanica, 61
Blatta orientalis, 60
Butcher's Broom, 37

CARBONATE OF LIME, 76
Cardiola interrupta, 156
Carcharodon angustidens, 229
Cephalis, 270
Cholera Bacillus, 177
Cladodus mirabilis, 270
Cockroach: Illustrations of the Natural History of:
  Alimentary Canal, 152
  Brain, left half, outer view, 247
  Brain, right half, inner view, 247
  Brain, lobes of, 246
  Brain, side view, 245
  Brain, sections of, 248
  Chylific Stomach of, 158
  Compound Eye, Diagrammatic sections, 251
  Dytiscus-Larva, section through eye of, 250
  Epithelial cell of, 153
  Eye of, 249
  Gizzard of, 152
  Head, back of, 205
  Head, side-view, 205
  Head, top and Front of, 205
  Insect Integument, 250
  Nervous system of female, 244
  Rectum, sections of, 153
  Salivary glands and receptacle, 154
  Skeleton, Outer, 108, 109
  Stomato-gastric nerves, 245
  Thoracic spiracle, abdominal, 207
  Thoracic spiracle, first, 206
  Thoracic spiracle, second, 206
  Tracheal system, dorsal integument removed, 204
  Tracheal system, ventral integument and viscera removed, 205
  Tracheal system, viscera removed, 204
  Vespa, section through eye of, 251
Copodus cornatus, 271
Corax falcatus, 228

Corax pristodontus, 228
Crystals, 207

Deltodus sublævis (lower tooth), 271
Deltodus sublævis (upper tooth), 271
Diagram of teeth of Psammodus, 271
Dodder, 36
Dolomite, 76
Dredge, showing its position on the ground, 201

ELEPHANT, SECTIONS OF TEETH OF, 132
Elm-Leaf, Bifurcation of, 252

FLOWERS OF ORCHID, 20
Fossil Plant, 30
Fossil Pectens, 52, 53
Funaria, antheridium of, 84
Funaria, archegonium of, 84
Funaria, mouth of theca of, 85
Funaria, theca of, 84
Funaria hygrometrica, 85

Galeocerdo latidens, 228
Garlic, 38
Gentiana Charpentieri, 129
Gentiana ciliata, 128
Gentiana gracilis, 129
Gentiana lutea, 128
Geology of Lincolnshire, Illustrations of, 105
Grass Poa, 37
Growing cell, 9

Hemipristis serra, 228
Hoplophora ferruginea, 56, 57
House-Fly, Teeth of, 176, 225
Hybodus grossiconus, 270
Hybodus, dorsal fin-spine, 270
Hybodus Delabechei, 270
Hybodus medius, 270
Hybodus reticulatus, 270

Lamna (Odontaspis), 228
Lamna elegans, 229
Lathræa squamaria, 4, 5
Leaf with Antheridia, 84
Lily of the Valley, Root Action of, 100
Lincolnshire, Illustrations of the Geology of, 105
Lithodendron basaltiforme, 157
Lithostrotion junceum, longitudinal section, 157
Lithostrotion junceum, transverse section, 157

Lonsdalia rugosa, 157
Lophodus mammillaris, 271

MITES, BRITISH FRESH-WATER, 80
Moss, Section of, 84

Notidanus microdon, 229
Notidanus primigenius, 229
Notidanus serratissimus, 229
Notidanus, Upper Tooth of, 229

Ophioglossum vulgatum, var. *ambiguum* 148
Orchid Flowers, 20
Orodus ramosus, 270
Otodus appendiculatus, 228
Otodus macrotus, 228
Oxyrhina macrorhiza, 229
Oxyrhina Mantellii, 228

Pachytheca sphærica, 28, 29
Periplaneta Americana, 61
Petrodus patelliformis, 271
Phyllodoce laminosa, 197
Pleurodus affinis, 271
Pleurogomphus auriculatus, 271
Pontobdella muricata, 197
Portland Spurge, 36
Pristicladodus dentatus, 270
Productus giganteus, 158
Productus punctatus, 158
Productus scabriculus, 157
Psephodus magnus, 271

RAY, DORSAL ASPECT OF, 173
Retepora plebeia, 158
Rotifers, Free-Swimming, 174
Rhynchonella pleurodon, 158

Sabella unispira, 181
Sagitta bipunctata, 197
Sea-Mouse (*Aphrodite aculeata*), 180
Serpula contortuplicata, 180
Sigalian boa, 180
Siphonostoma vestitum, 197
Spinax, Jaws of, 172
Spinax, Outline of, 172
Squills, 38
Stenopora fibrosa, 156

TEREBELLA, 181
Terebratula hastata, 158

YEAST-FUNGI, 12, 13

LIST OF COLOURED PLATES.

Breeze-fly, spiracle of	*To face page* 169	*Lepralia nitida*, polypidom of . . . . *To face page* 193
Cluster cups	,, 145	Limpet, palate of . . . . . ,, 49
Epeira conica, eyes of	,, 25	Locust, pupa of . . . . . ,, 121
Fern, sori of	,, 241	Mallow, anther, pollen and stigma of . ,, 73
Funaria hygrometrica, peristome of . .	,, 97	Mottled Umber Moth, eggs of . . . ,, 265
House-fly, eggs of	,, 217	*Tingis Crassiochari* . . . . . ,, 1

GRAPHIC MICROSCOPY.

By E. T. D.

INTRODUCTION.

IT is proposed, to publish monthly, under the above title, a lithograph, in colour, of a microscopic object, with short description. The main purpose of the series is to place before our subscribers a counterpart or fac-simile, as far as the art of the lithographer can render and elucidate it, of a finished painting from nature, executed with scrupulous exactitude and veracity; both in line and colour, under the most suitable magnifying power, and as a special feature and of equal importance—the best and highest conditions of illumination. The selection of subjects will be made exclusively from objects essentially popular;—easily obtained and prepared, or purchaseable as a " slide."

Each article will embody a note of the conditions necessary to arrange the subject most favourable for good observation, and particularly for drawing.

No. 1.—TINGIS CRASSIOCHARI.

The family Tingidæ is classed in the Heteropterous Section of the Rychota; a subdivision of the Order Hemiptera of Latreille.

There is considerable diversity in the structure of the few groups of which this family is composed, strikingly apparent in the variation of the reticulations of the filmy membranous dilations on each side of the thorax, on the scutellum, as well as on the large elytra, which entirely cover the dorsal surface of the abdomen. In foreign, and in many English specimens these reticulations render them objects of singular and especial beauty.

The number of species is very great. The majority are found in tropical countries, but European specimens disclose markings and colour, under magnification, which vie in splendour with the most gorgeous of the beetle tribes; as in the Orthoptera, the metamorphosis is "imperfect," the young Tingis escapes from the egg, in a form more or less closely approaching that which it is ultimately destined to assume, and in many cases the principal distinction between the larva and perfect insect consists only in the possession of wings (rarely used) by the latter. The larvæ or semi-pupæ, usually of an orange colour, are more convex than when perfectly developed. They are found in the same situations, and often in company with perfect insects. All the species are slow in their motions, and seldom fly; their size, when mature, rarely exceeds an eighth of an inch in length. Insects belonging to this group (commonly known as bugs) do not present any great diversity of habit, or are, to ordinary vision, sufficiently attractive to court or encourage observation; and Tingis, in particular, slowly crawling over, and sucking the juices of plants and fruits, to which it imparts a most offensive odour, might still have lived in comparative obscurity, had not the microscope revealed the matchless beauty of its structure, at once elevating it to the distinction of one of the popular microscopic objects of the day.

Attention was first directed to its peculiar elegance and quaintness by Mr. Richter, in an article published in SCIENCE-GOSSIP, in April 1869, vol. v. page 84. The specimen there described, figured, and provisionally named *Tingis hystricellus*, was one of a few found in Ceylon, by Mr. Staniforth Green, and forwarded by him to Mr. Curties, F.R.M.S. Mr. Richter, in his paper, well described it as "a little insect porcupine."

For secure preservation it should be mounted in balsam, without pressure ; but to obtain fine results for the purpose of drawing, it is better prepared dry, in such a method, as to be capable of being seen under any circumstances—from beneath with the paraboloid, and from above (after removing the covering glass) with reflected light from the side speculum. Under these conditions of double or simultaneous illumination, the finest points of character and especially of colour are revealed.

The preliminary sketch was made with the aid of the camera lucida ; but as the objective used (to bring out the full details) rendered it impossible to get the whole of the subject into one circular field of view, the difficulty was overcome, by shifting the stage adjustments and fitting the parts, for the ultimate drawing from direct observation.

Crouch End.

BLOOD-RED SUNSETS AND THEIR AFTER-GLOW.

ATMOSPHERICAL phenomena appear lately to have been in a very unsettled state. "Nature" has been occupied with letters relating to a "green sun" in India, and the first explanation was that somehow it was due to the Javan earthquake. The last week in November and far into December will long be memorable for the "blood-red sunsets." They reached their maximum of beauty in the Metropolis, where the fog and smoke, unbearable usually, are excellent auxiliaries to sunset effects. All over the country for nearly a month, the most gorgeous sunsets and sunrises were noted, followed by afterglows equal to the most striking displays of the aurora borealis.

Both the scientific journals and public newspapers have contained an unusual amount of correspondence respecting the exceptional brilliance and persistence of the evening after-glow, and the corresponding phenomenon at day dawn.: It is well known that the conditions necessary for the production of this display are most favourable in northerly climates, owing, it is believed, to the suspension of particles of frozen vapour in the air, or other strata of varying density against which the rays of the sun can strike, and from which they can be reflected. The "Lancet" asks whether we may not expect a winter of unusual severity as a possible sequel to such splendid atmospheric displays. Mr. G. J. Symons, the distinguished meteorologist, in a letter to the "Times," suggested that these brilliant afterglows, together with the blue suns and green suns which have been described in various parts of the world, are due to the rise of vapours and volcanic dust from the Java eruptions into the higher regions of the atmosphere, thus affording the necessary conditions. He points out that as early as September 16th Mr. J. P. O'Reilly, of Dublin, called attention to the quantity of gases and vapour emitted during eruptions, to their probable relation to the total quantity of matter emitted, and to their exerting some effect upon the atmosphere. From that time to the present, all kinds of strange and exceptionally brilliant and chromatic effects have occurred in India, Ceylon, the Cape, Venezuela, Barbadoes, New Zealand, Australia, and other places, as well as throughout Great Britain, and Mr. Symons thinks that the enormous discharge of vapours and volcanic matter into the atmosphere from the eruptions in Java may be reasonably held to offer some explanation.

Mrs. Somerville, in her "Physical Geography," shows that the fog and lurid light of 1783 were due to the great eruption of Skaptar, in Iceland. For months after the outbreak the island was obscured by the enormous quantity of fine dust borne aloft in clouds of vapour. This volcanic dust was carried by winds over England and Northern Europe, and caused the atmospheric effects described by Cowper ("Task," book ii.) and by Gilbert White.

Mr. F. A. Rollo-Russell points out that the vesicular nature of pumice, each particle consisting of a small bubble of glass, would allow it, after being shot up by the eruption to an enormous height, to be carried without precipitation to all quarters of the globe, and at the altitude attained it would be far removed from the action of vapour and weather. Nothing like this diffused atmospheric glow after dark, and one or two hours before sunrise, has been observed before, and a singular effect must have its origin in a singular cause. On December 1st and 2nd, the glow, which was of an amber colour, did not become bright until about an hour after sunset, and was partially obscured by clouds. The phenomenon was mistaken for aurora borealis in France and elsewhere, but it yet requires a name of its own.

During the south-westerly gale which raged over England and Wales on December 11th, the blood-red sunset was again splendidly visible in several places, and the after-glow lasted for two hours. The phenomenon appears to have been universal, and was witnessed in both hemispheres, the same general effects being noticeable. Consequently Mr. Symons' explanation, so far, appears the most probable.

Mr. Norman Lockyer, F.R.S., has contributed a very interesting and important paper to the "Times," in which he expresses his belief, that the enormous volumes of fine volcanic dust emitted by the Java earthqukes and eruptions have been carried into the upper air, and that in the course of their progress across and above the earth, the reflection upon them of the sun's setting rays has produced the brilliant phenomena of which so many accounts have been received. There can be no doubt as to the adequacy of the cause thus suggested, when we consider the vast quantities of volcanic dust resulting from the eruptions of August last.

The abnormal sunsets Mr. Lockyer traces directly from the Seychelles to Brazil, not omitting Venezuela and Trinidad, at both which places the sun was observed of a bluish-green between noon and 3 P.M. The line thus recorded is from east to west, but an almost equally definite path of the phenomena is indicated by observations made in places north and south of the scene of the eruptions. Mr. Lockyer's argument is that the phenomena have not only travelled at a definite rate and in definite directions, but have also varied precisely as they might have been expected to vary. At Java the immediate result is darkness, or at any rate a long obscuration of the sun; then as the cloud moves, the pall becomes thinner, the grosser particles have fallen, but the blue and red molecules remain suspended in the upper air, and produce the singular reflected lights that have been lately the wonder of the world. The spectroscope, in fact, has furnished such conclusive evidence on this point that it practically supersedes, while it confirms, the results obtained by the mere record of the phenomena. There can be little doubt, we think, that the volcanic theory, whether true or false, will give a new interest to meteorological studies.

CONCHOLOGICAL NOTES.

HELIX LAMELLATA (SCIENCE-GOSSIP, p. 147). This shell occurs near Huddersfield, which is much farther south than Scarborough. See "Quarterly Journal of Conchology" for May, 1874, where it is described as having been found in 1870.

Helix rufescens.—This species seems to "run out" in the county of Durham. It is found sparingly near Sunderland, and this may be near its northern limit. It does not occur in the list of Middlesbrough-on-Tees shells given by Mr. Hobson in SCIENCE-GOSSIP for July. In the "Zoologist" for June, 1881, Mr. R. M. Christy states that it is very scarce in the immediate neighbourhood of York. The question might be asked: Whether is this species extending northward, retreating southward, or at a standstill? I should like to ascertain if there are any records of its occurrence farther north than Durham.

Helix Cantiana.—This species is exceedingly common in Yorkshire, on the chalk on the east coast, on the red sandstone south of York, and on the magnesian limestone, but seems to thin out rapidly on the carboniferous sandstone westward; very few being found west of a line drawn from Leeds to Sheffield. On August 6th, after a shower, there were thousands of Cantiana in all stages of growth in a lane leading from Pontefract to Ferrybridge, the hedges and ditches, on both sides of the road, being completely lined with them. In this lane there were considerable numbers of *Helix aspersa* and *H. nemoralis*, but no *hortensis*. In one place about ten species of shells were congregated on a large bed of mown and half dead nettles, so thickly that several cracked under the feet at every step. Some conchologists think that the Kentish snail is slowly progressing westward.

Helix aspersa.—Many of the helices climb nearly to the top of the hedges. I noticed *Helix aspersa* perched on the top of a giant cow-parsnip which grew in a garden close to a hedge. The plant was ten feet high.

Helix nemoralis.—On the red sandstone near Milford Junction I found two specimens of white-lipped Nemoralis; the same variety I suppose that was mentioned by Mr. Crowther in January SCIENCE-GOSSIP as occurring near Leeds.

Enemies.—On August 9th I found several specimens of *Sphærium lacustre* which had been taken out of a ditch and on to a high bank and cleared of the animals by water-shrews. I also found several empty Zonites on a dry bank, the shells having been one-fourth eaten away, probably by land-shrews, to get at the animals.

I have taken *Zua lubrica* and *Helix hispida* from the crops of young sparrows, and *Helix caperata* from the crop of the ring-dove. Many beetles kill snails. A friend informs me that he has seen black ants feeding on *Helix Cantiana*. Query: Could ants be made useful in clearing small shells of the animal?

Collecting and packing shells.—Mollusks when confined together rasp the epidermis off each other. This fact is well known to "old hands," but I should like to warn beginners in the study, when sending living shells from one to another through the post, not to pack them all together in a tin canister, but to keep them separate by partitions or in small boxes. *Helix aspersa* should never be associated either with themselves or with other shells.

GEORGE ROBERTS.

HEDGEHOGS.—Can any of the readers of SCIENCE-GOSSIP suggest an explanation of the alleged fact that hedgehogs are proof against the effects of the most virulent poisons? How can the phenomenon be physiologically accounted for?—*Albert Waters.*

GRUBS IN H. CAPERATA.—Whilst cleaning *H. caperata* obtained to-day, I found in several large white grubs, and wherever these grubs occurred the snail was partially eaten. I should say about fifty per cent. of the shells contained these unwelcome intruders. Will some reader tell me what they are? I presume the larvæ of some fly.—*Baker Hudson.*

NOTES ON THE TOOTH-WORT (*LATHRÆA SQUAMARIA*).

EARLY in the present year I procured a copy of J. E. Smith's "Plants of South Kent," published in 1829, to assist me in working out the botany

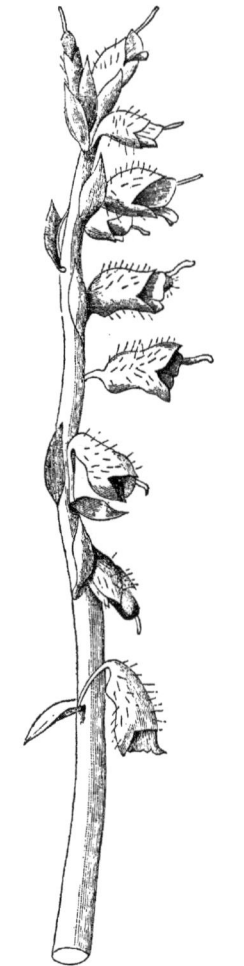

Fig. 1.—*Lathræa squamaria*, as figured in Smith's "Plants of South Kent."

of this district. It has proved of great service, and I am sure that botanists will be glad to learn that most of the rarer plants mentioned by him still occupy the same localities, and in several instances have very considerably extended their borders; for instance, *Orobanche caryophyllacea*, which was confined to one or two localities along the coast, is now found several miles inland, and is comparatively abundant.

It is very pleasing to visit a locality, seeking a plant described as growing there fifty-four years ago, and finding it in luxuriance; on the other hand, it is very disappointing to seek some rarity and find that, through the publication of its locality it has long since been exterminated, as in all probability has been the man orchis from the immediate neighbourhood of Slowling.

The first plant I sought this year was *Lathræa squamaria*, which I found growing in abundance in the locality named by Smith. I gathered some fourteen specimens, and examined many others, seeking anything of different or unusual appearance; but all the plants I saw were of one type. On examination at home, and comparing them with Smith's description and drawing, I found that there was considerable difference; so much so that I had three plants photographed, and made careful drawings of

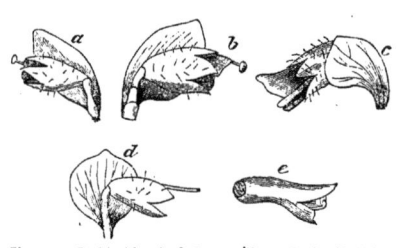

Fig. 2.—*a*, Bud just burst; *b*, flower with corolla about half developed; *c*, flower fully expanded; *d*, bractea and calyx with style projecting, the partially withered corolla being removed.

the flowers in their various stages, together with many analytical and microscopic dissections. Fig. 1 is a copy of *L. squamaria* as figured by Smith. I may here remark that the five plates in his book have been drawn with more than ordinary care, as they are for the express purpose of illustrating some noteworthy character. The description given agrees with the drawing; it is as follows:—

"The plant which I have considered *L. squamaria*, collected upon the first-mentioned locality, presents no slight variation from the characters of *L. squamaria* of 'English Botany,' vol. i., t. 50. With as great variety of habit, for this I presume from the very dissimilar figures to be found in works of accuracy, and a frequently club-shaped and proliferous stem, the Lathræa of Lyminge, and, I suspect, of Hudson in 'Flora Anglica,' and Rudbech, 'Elys.,' vol. ii. p. 234, fig. 17, presents more erect and purplish *pink* flowers, whose upper lip is *entire*, or very slightly cloven, the lower lip involute, the style *scarcely bent*, and *protruded* from the fold of the upper lip. The bracteæ are smooth and lanceolate, the calyx hairy.

The flower purplish-pink, edged with white; occasionally the whole plant is white. The clubbed stems exhibit the true nature of the plant, throwing out from their base squamose offsets, into which also the imperfect flowers above are seen to pass. The herbage is buried about four inches in the loose earth, and bears opposite branches.

"I have seen no figure or dried specimen which satisfactorily explains the difficulty. The figure in 'English Botany' represents the upper lip deeply and acutely cloven, the style bent downwards near the stigma, and hidden. It is sufficient to point out differences, without at once attempting the constitution

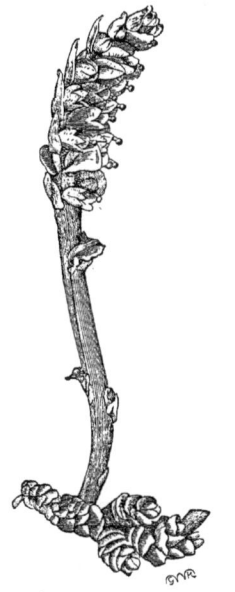

Fig. 3.—Specimen of *Lathræa squamaria* gathered April 1883. [From a photograph.]

of a second species. The plant of Lyminge may stand as *L. squamaria*, β—Lathræa radice squamatâ, bracteis lanceolatis, stylo recto, è labio superiore subintegro, exserto."

From this description and drawing it is evident that fifty-four years ago *L. squamaria*, as found by Smith, was different to its progeny of the present day, which answers the description given in "English Botany," 3rd edition, with these exceptions: the style is described as simple, and curved at the apex; this, however, is not so, what *slight curve* there is in the style is along its entire length. It also describes the style as generally exserted; this is, however, only partially correct, for dividing the life of the flower into four stages, viz. first, the bursting of the bud; second, a stage midway between the bursting of the bud and the fully-expanded flower; third, the fully-expanded flower; and fourth, the flower withering, it will be seen that in the two first stages the style is seen projecting beyond the corolla, while in the fully-expanded flower neither of the organs of generation can be seen; and it is not until the corolla begins to shrivel with decay, that the four stamens with their anthers can be seen. These various stages are shown in the accompanying sketches.

How the difference in the *squamaria* of fifty-four years ago and those of the present day have been brought about, I will not pretend to say; that such marked difference from the normal type existed as to call Smith's attention to it cannot be doubted, he being almost tempted to make a second species of it, but concluded to call it only a variety β.

The accompanying photo proves that the plant has now its normal characters, with broadly ovate bracteas and proliferous stem. Whether these changes mark a period of evolution, or whether the abnormal condition of the plant in Smith's time was only a freak of nature, I must leave for others better able to decide.

Very important lessons may, however, be learnt from the preceding facts, viz. the importance of placing on record careful descriptions, together with drawings of all plants that vary from the normal type, and the exercise of great care and judgment in giving to any plant, even with great differences of character, a new specific name, or even constituting it a new variety.

I do not know if the photo is distinct enough to engrave from; if not, I shall be glad to lend a specimen to any one interested.

W. T. HAYDON.

Dover.

MICROSCOPICAL TECHNOLOGY.

By JOHN ERNEST ADY.

ON THE EXHIBITION OF CANADA BALSAM.

THE following remarks are merely offered to the readers of SCIENCE-GOSSIP as part of the outcome of extended practical research on the subject, and I have essayed to contribute them to these columns, with the hope that they may prove useful to those to whom certain phases in the exhibition of Canada balsam are always looked upon as "bugbears."

I shall confine myself to a consideration of the manner in which sections of tissues, however unmanageable by ordinary methods, on account of their inherent physical properties, may be coerced, as it were, to yield to the desires of the manipulator. But before I proceed further, I would like to satisfy my readers as to the *raison d'être* of this communication.

There are several text-books on the microscope, and on microscopical technology; most of these works are bulky and expensive, and even if they are within the easy reach of many, they are too formidable to be "read, marked, learned, and inwardly digested." Space will not permit me to criticise these stupendous volumes, more than to state, that they do not contain what I am about to explain. But there are other smaller works, inexpensive, and full of practical hints; they are greedily read by every tyro microscopist, and to them do we owe the main impetus which has of late been given to popular microscopical research, by bringing the subject within the easy grasp of those who study in their leisure to derive pleasure. In these little books the harassing details of their bulkier brethren are avoided, and only practical items are noted. As I must not prolong this paper by criticising the directions given in Mr. Marsh's little volume, or in Dr. Heneage Gibbes' practical work, I shall merely refer the reader to one or two things in those treatises, concerning the difficulties of mounting objects for the microscope in Canada balsam, and then describe a method which, if followed in its integrity, will dispense with all those difficulties and a great deal of unnecessary trouble as well.

Thus, Dr. Gibbes, in speaking of mounting large sections,[*] finds that the cover glass on to which he has succeeded in transferring the section is smeared all over the top with clove oil, and has thus to remain for some time ere it can be cleansed for examination. Mr. Marsh devotes some pages[†] to the description of apparatus for drying slides after they have been mounted, and in a footnote [‡] gives a "wrinkle," which he ascribes to Mr. Kay, for the prevention of air bubbles in slides mounted in viscous balsam, with a few remarks to advise the use of viscous balsam for thick tissues, because the prepared mobile balsam is apt to evaporate and leave vacuoles behind. I have referred to these two handy volumes because they are so well known, and I am aware that the plans inculcated in them are those in general use, and in almost every text-book, whilst very few persons know how to mount objects in Canada balsam and avoid the defects which have been alluded to above. That all such defects can most readily be obviated, even by the most inexperienced worker, it shall be my present endeavour to show.

As success in mounting objects depends upon a certain amount, the larger the better, of methodicity and cleanliness, I shall preface these remarks with a few maxims, and describe the mounting process progressively by means of an example.

---

[*] "Practical Histology and Pathology," by Heneage Gibbes, M.D., 2nd ed., H. K. Lewis, Gower Street, London, 1883, pp. 54.
[†] "Microscopical Section Cutting," by Sylvester Marsh, L.R.C.P.E., 2nd ed., J. & A. Churchill, New Burlington Street, London, 1882, pp. 115-119.
[‡] *Op. cit.*, p. 109.

I. Before commencing your work always see that your table, windows, and in fact everything in the room, are quite clean and free from dust.

II. See that your microscope, spirit lamp or Bunsen burner, bottle of balsam, dipping-rod, forceps, knife, lifter, etc., are all in good condition, and arranged on the table in the most convenient positions.

III. Have ready by your side a box containing a sufficiency of cleaned cover glasses and glass slips, so as to avoid having to clean anything whilst at work.

I shall now describe the way in which any thin section, however large and prone to curl up, may be successfully mounted.

1st. Remove the section to a dish[*] containing some clean filtered methylated spirit, [†] and allow it to remain in this for about an hour, so as to thoroughly dehydrate it. The fastidious worker may re-transfer it to absolute alcohol for another hour, but this is quite unnecessary.

2nd. Pour a little oil of cloves into another palette. Drain as much spirit as possible off the section, and allow it to float on to the surface of the oil. In the course of a short time, which varies from the fraction of a minute to a few minutes with the nature of the section, the specimen will sink to the bottom of the oil, and thus show that it has been thoroughly permeated; it may now be removed, but not necessarily, as prolonged soaking in the clove oil will not damage it much. Care must be taken, however, to keep the palette covered, to prevent the admission of foreign particles, of dust, hairs, &c.

3rd. Procure a clean glass slip, breathe on it gently, and to the surface of condensed vapour apply a clean cover glass; the latter will be found to adhere to the glass slip sufficiently to prevent its falling off during the subsequent processes. On to the centre of the cover glass place a drop of benzoled balsam; lay the section in this drop, and examine it under a low power of the microscope (two-inch objective with A eye-piece). If any foreign particles have crept in they must be removed with a needle.[‡] Air-bubbles may be disregarded.

Should a very large section be floated on to the cover glass from the oil of cloves, drain off the superfluous oil and place the cover on a clean glass slip with the section uppermost and exposed, and cover the section with benzoled balsam. Care must be

---

[*] A useful kind of dish may be procured from any artists' colourman in the form of a circular china palette about four inches in diameter. The palettes are usually sold in nests of six, and may be used as staining and clearing troughs, and for a variety of other purposes.
[†] Methylated spirit should be perfectly clear and colourless. It should register 63 (or at least 59) above proof on the alcoholometer.
[‡] It is of the utmost importance in the production of neatly-mounted sides, that the worker should be able to use a needle to remove dust particles from the preparation under a compound microscope. Perseverance for a couple of hours at a slide full of such particles will enable any ordinarily neat-handed person to succeed in this. A simple microscope may be used by those who are not endowed with a small stock of patience; but I recommend the mastery of this feat with the compound microscope, as experience has proved it to be the most satisfactory method.

taken in this, as in the previous case, not to permit the balsam to overflow the edges of the cover glass; otherwise, the cover will become firmly cemented to the slip on which it is temporarily placed, and the future processes rendered impossible or difficult, as the cover glass bearing the section will have to be detached by heat or otherwise from the supporting slip. I may add that none but a careless and slovenly worker need fear the overflow of balsam spoken of. The section itself ought to be completely covered with balsam.

4th. Lay the glass slip with the cover-glass upon which the balsamed preparation lies exposed, either in a special dust-proof box (an ordinary object cabinet with trays will do), or under a glass shade, for about twelve hours. If left thus exposed for much less than twelve hours, e.g. for two or three hours only, the balsam will not be reduced to the right degree of consistency through the evaporation of the solvent (benzol in this case). If exposed for much over twelve hours, e.g. for twenty hours, the evaporation will have progressed too far, but this may be rectified by the addition of another small drop, and the exposure continued for an hour or two more. If examined after about the twelfth hour under the microscope, it will be observed that all air bubbles have disappeared.

5th. Gently warm a clean glass slip, place it upon the mounting tile* so as to show its centre, upon which a drop of benzoled balsam should be placed; now remove the cover glass with its supported preparation from the slip upon which it lay exposed (a pair of blunt-edged smooth-tipped forceps will be found useful here), pass it quickly over, not into, the flame of a spirit lamp or Bunsen burner, and apply it slantingly, section downwards this time, to the warmed slip with its central drop of balsam. Squeeze out the superfluous balsam, and lay the now completed slide aside for a couple of hours. The squeezed out balsam, if there is much of it, may now be scraped off with a small knife and the rest of the balsam cleared off with a rag dipped in methylated spirit. If it so happens that the cover-glass has not sufficiently set after being laid aside for a couple of hours, if left for double that time, it may be handled with impunity, and thoroughly cleansed.

Of course, the section may be placed upon a clean glass slip, balsamed, exposed, and then covered with a cover-glass, to the under side of which a drop of balsam has been applied; but the above plan will be found the most convenient in the mounting of a large number of sections—*experientia docet*.†

If air-bubbles arise through the application of too much new balsam after exposure, they will invariably disappear in the course of a few hours. If, however, as sometimes though rarely happens, air-bubbles get entangled in very reticular tissues and wear an obstinately persistent aspect, the cover-glass should be removed at once, a fresh drop of balsam applied, exposure for two or three hours repeated, or for such time as it may take to dispel the atmospheric demons, another small drop of fresh balsam applied, and the preparation covered and cleaned in the manner directed above.

In the case of all thick tissues, such as sections of decalcified bone, teeth, cartilage, the rhizomes and leaf-stalks of ferns, &c., it will be found, that what I shall term *the exposure method*, is sufficient to harden the balsam to such an extent, that their curling tendencies are subdued or altogether checked by that process. But to make success doubly sure, such refractory sections may be flattened during the clearing process. As soon as they sink to the bottom of the oil of cloves, they are to be removed to a clean glass slip with their adhering pools of the oil, covered with another similar slip and the two held together with an elastic band or piece of string, and laid aside for a few hours (four or five). On removal, and during the exposure method which follows, they may recur, but their limit of elasticity has been greatly diminished by the continued pressure, and they will be found to be quite manageable.

Whilst mounting very thick sections, in cells, or otherwise, the benzoled balsam will evaporate during the exposure process, and should be replenished at the end of the twelfth hour, exposed for another three hours, re-added to with a small drop of fresh balsam, and covered. Vacuolar spaces cannot possibly appear after these precautions.

One of the most important things in balsam mounting is the preparation of balsam of suitable consistency. To secure a good result, some viscid balsam should be placed to harden in a sand or water bath. An old glue pot provided with a cover will do. The hardening of the balsam must be conducted gradually, to avoid burning the material. It should be tested from time to time, by the removal of a small quantity upon a rod. As soon as the material removed hardens into an almost brittle mass upon cooling, the still liquid hot balsam in its pot should be taken off the flame, and an equivalent bulk of pure benzol added to it. A little stirring with a glass rod will accelerate the solution of the hot balsam, and it should now be filtered through a cone of fine filter-paper. The filtrate should be of the consistency of thick syrup. If too thick, a little more benzol should be added and the process continued in a warm place, in an oven or in front of a fire, preferably the former. If too mobile, exposure to the air under a bell glass will reduce it to the right degree of viscosity in a few hours; refiltration in a warm place will effect the same end in a few minutes.

---

* A mounting tile of white pottery-ware about 6×8 inches in surface, with lines painted upon it, to indicate the centres of slides, 3×1, and 3×1½, is a useful adjunct to the microscopist. A piece of white cardboard similarly marked may be substituted for such a tile.
† I choose to curtail the proverb, "*Experientia docet stultos.*"

Canada balsam, thus prepared, may be variously applied; but other methods are modifications rather than essentially distinct from the processes described above. I hope in future papers to treat explicitly of some of these. In every case I shall make it a rule to explain *only what I have verified by repeated experiment*, so that these notes may be of essentially practical utility to my readers.

A NEW AQUARIUM.

ONE of the chief drawbacks to private marine aquaria, or hatching tanks, is the elaborate and costly machinery necessary to aerate the water: and even when such aeration is, or may be, carried on by

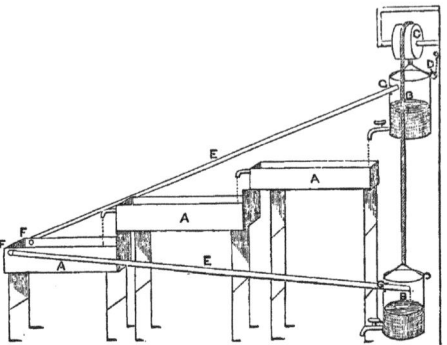

Fig. 4.—A, Series of hatching tanks; B, two dark reservoirs working by a rope over C; c, a wheel; D, hooks to hold up the full reservoirs; E, overflow pipe from bottom tank to the empty reservoir; F, flexible joints to E; G, hole in reservoir in which overflow pipe slides easily.

a gentle circulation, promoted by a fall of water from a higher level, the labour of getting the water to such higher level constantly, is a source of trouble, and therefore soon left undone, to the destruction of the living contents of the tank.

In the annexed sketch I have given an idea for producing a flow of water through a series of shallow tanks, which may be adapted to a small or large undertaking as desired. The principle is, that whilst one reservoir is discharging its contents into the top tank, the bottom tank is overflowing into the other, which, when full, can by a contrivance so simple that I have not thought it necessary to show it in the sketch, be pulled up to discharge its contents, in turn, into the top tank.

Now we will suppose that the reservoirs have a capacity of five gals., and that the full one has been hauled up to its position and secured there. By turning on the tap, which must of course be on flexible joints to prevent fouling the edge of the tank in its descent, a small stream of water will fall into the top tank, and overflow into the next and so on, till it reaches the empty reservoir. When the upper reservoir is empty the lower one is full, the circulation ceases, and therefore no waste of water takes place.

When the time occupied by the discharge of the full reservoir is known, it is a very simple matter for anyone near the tank room to go at stated intervals and reverse the gear, an operation occupying about half a minute.

Of course if the tanks be wide ones, two wheels, at a corresponding width apart, will be necessary.

The advantages of this plan are as follows:—

Its comparative cheapness of material.

The ease and rapidity with which the level of the reserve tank can be altered.

That a certain quantity of water is always in the dark, and therefore the "dark reservoir" is thus obtained.

That when the reservoirs are large, and the stream small, thus occupying some time to discharge, an almost constant stream can be maintained if some one near at hand can be relied upon to devote 30 seconds, say four times a day, to reversing the gear.

That it is not likely to get out of order, if well made in the first place.

I shall be very pleased to give anyone desiring it, any further information, and I feel sure that this simple and labour-saving method will commend itself to those who wish to keep a marine tank or study the minute organisms of the sea in their metamorphoses.

I may mention that the tanks should be covered with sheets of glass to keep off dust, but having a round hole for the overflow water to drop through. A series of tanks like this, rigged up in a cool room, say an out-house for instance, will, with care, do wonders.

EDWARD LOVETT.

Croydon.

A GROWING-CELL FOR MINUTE ORGANISMS.

THE number and variety of "growing-cells" from which microscopical workers may choose is large, but it now seems necessary to increase it by one more. Desiring a slide of the kind for the study of minute animal organisms, capable of use with immersion objectives, I have hit upon the following arrangements, which I hope the microscopical readers of SCIENCE-GOSSIP will find as useful as I have.

The materials needed to build this little device are an ordinary slip, a thin cover ⅛ inch or less in diameter, two glass rings, a thin square, and a drop or two of Canada balsam. Every microscopist has all these at his command, except, perhaps, the glass rings, which, on this side the Atlantic, are sold for more than their weight in gold; consequently, those familiar with a "small dodge" buy not, but make. And these "home-brewed" affairs are quite as useful, if not so ornamental, as those cut with a diamond.

With whom originated the idea of making rings by punching out the centre of a thin cover I do not know. Beale refers to it in his "How to Work with the Microscope," and my impression is that its mention may be found in microscopical literature of a date earlier than that of his book. However, for cells of moderate depth they are superior to tin, zinc, or vulcanite.

In a brass plate ¼ inch or less in thickness, drill

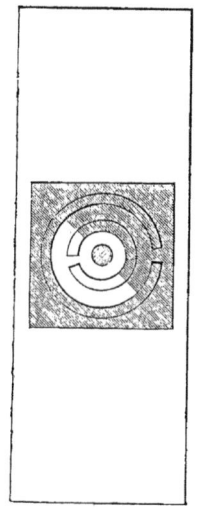

Fig. 5.—Growing-cell.

holes to correspond with the openings in the rings of the various sizes desired, and with Canada balsam cement thin covers over these holes; place the plate on the kitchen stove for a few minutes, or until the balsam will cool hard; when entirely cold boldly thrust an iron nail through the glass. If the cementing has been properly done the fractures will not extend beyond the edge of the hole in the plate. The two secrets of success are to have the balsam reach quite to the edge of the aperture in the brass, and to be cold. If the balsam flows over toward the centre of the cover the fractures made by the nail will extend irregularly to the edge of the cement, and the ring can then be completed by grinding the glass away to the brass, giving the nail a rotation in addition to its downward movement. Do not draw the nail upward. A gentle heat will loosen the cover, and a slight rubbing with gasoline will clean it. Gasoline I have found to be a kind of universal solvent for cements microscopical, and wonderfully impressive.

In the "American Monthly Microscopical Journal," June 1882, Mr. C. H. Kain describes a similar plan, using in his method sheet wax instead of Canada balsam. With wax I have always failed, smashing covers till on the verge of thin-glass bankruptcy.

Having punched the covers, to make the growing cell cement the small disc in the centre of a slip, take a ring with a quarter inch aperture, break a little piece from one side, and fasten this broken circular band about the central circle, as in fig. 5. From another ring with a ⅜ or larger aperture break a piece as before, and cement about the inner ring, so that its broken part shall be opposite the unbroken curve of the former, and the cell is made.

To use, place on the central disc a small drop of the water containing the organisms to be kept alive, and over it arrange a large square cover, taking pains to prevent the water from overflowing into the inner annular space. With a camel's hair pencil carefully, and in small quantities, add fresh water at the top or side of the square, until the space covered by the latter and bounded by the outer ring is filled. It will be found that this water will flow between the square and upper surface of the exterior ring, will enter through the break in the latter, partially filling the outer annular space, and by capillary attraction will occupy a part of the vacancy between the cover and the interior ring, as shown by the diagonal lines in the diagram, fig. 5, but unless too much water is used, or is supplied in too great quantities at a time, it will not pass the opening in the inner ring, thus leaving an abundance of air to supply the animal life under observation. The imprisoned air at once becomes saturated with moisture, as evidenced by the fogginess of the cover; the central drop cannot evaporate, and the external water will not come in contact with it if care is taken in filling and in adding that lost by evaporation. When not in use, the slide is placed across a small vessel of water, a double and twisted thread arranged in contact with the edge of the square cover, and the whole left for another examination at some future time.

DR. ALFRED C. STOKES.

Trenton, New Jersey, U.S.A.

CORMORANT IN WORCESTERSHIRE.—In October, 1882, a fine specimen of the above was shot on the large reservoir between Cofton Hackett and Alvechurch. What I think to be a still more unusual visitor to the Midlands is the Manx Shearwater, a specimen of which was, in September, 1873, caught in a hedge by some terriers when I was out walking at Trygull, near Wolverhampton. Though much exhausted, the bird seemed otherwise uninjured.—*K. D., Cofton Hackett.*

DARWIN'S OPINIONS ON INSTINCT.

A POSTHUMOUS essay on "Instinct," by the late Charles Darwin, was read on December 6th, at the Linnean Society, before a very large and distinguished audience of Fellows. The paper, which treated of the instincts of animals, and the bearing of the subject on the theory of natural selection, was originally written for the "Origin of Species," but never published. It discussed the migration of birds and mammals, and after narrating a great series of curious facts, concluded with the inference that though there were many aspects of the question which admitted of no immediate explanation, the migratory instinct was inherited from ancestors who had to compass (for the sake of food or other causes) long distances, when the conditions of land and water were different from what they are at present. He then considered how the more remarkable migrations could possibly have originated. Take the case of a bird being driven each year, by cold or want of food, slowly to travel northward, as is the case with some birds; and in time we may well believe that this compulsory travelling would become instinctive, as with the sheep in Spain. Now, during the long course of ages, let valleys become converted into estuaries, and then into wider and wider arms of the sea; and still he could well believe that the impulse which leads the pinioned goose to scramble northward would lead our bird over the trackless waters; and that, by the aid of the unknown power by which many animals (and savage men) can retain a true course, it would safely cross the sea now covering the submerged path of its ancient land journey. Animals on oceanic islands, and other localities where they have never met with man or beasts of prey, are devoid of fear. This instinctive dread they subsequently acquire, for their own preservation, and transmit it to their descendants. At the Galapagos Islands Mr. Darwin pushed a hawk off a tree with the muzzle of his gun, and the little bird drank water out of a vessel which he held in his hand. But this tameness is not general, but special towards man; for at the Falklands the geese build on the outlying islands on account of the foxes. These wolf-like foxes were here as fearless of man as were the birds, and the sailors in Byron's voyage, mistaking their curiosity for fierceness, ran into the water to avoid them. In all old civilised countries the wariness and fear of even young foxes and wolves are well known. At the Galapagos Islands the great land lizards (Amblyrhynchus) were extremely tame, so that Mr. Darwin could pull them by the tail; whereas in other parts of the world large lizards are wary enough. The aquatic lizard of the same genus lives on the coast, is adapted to swim and dive perfectly, and feeds on submerged algæ; no doubt it must be exposed to danger from the sharks, and consequently, though quite tame on the land, he could not drive them into the water; and when he threw them in they always swam directly back to the shore. Animals' feigning death seemed to Mr. Darwin a remarkable instinct, but he considered that there was much exaggeration on the subject. It struck him as a strange coincidence that the insects should have come to exactly simulate the state which they took when dead. Hence he carefully noted the simulated positions of seventeen kinds of insects (including an Iulus, spider, and Oniscus) belonging to the most distinct genera, both poor and first-rate shammers; afterwards he procured naturally dead specimens of some of these insects, others he killed with camphor by an easy slow death. The result was that in no one instance was the attitude exactly the same, and in several instances the attitude of the feigners and of the really dead were as unlike as they possibly could be. Bird-nesting and the habitations of other animals were next discussed, the general conclusion being that though there are various adaptations of inherited instincts to suit varying circumstances, yet that these variations all tend to preserve the species in the struggle for existence, by conducing to the "survival of the fittest." Although he did not doubt that intelligence and experience often come into play in the nidification of birds, yet both often fail; a jackdaw has been seen trying in vain to get a stick through a turret window, and had not sense to draw it in lengthways; Gilbert White describes some martins which year after year built their nests on an exposed wall, and year after year they were washed down. The *Furnarius cunicularius* in S. America makes a deep burrow in mud-banks for its nest; and he saw these little birds vainly burrowing numerous holes through mud-walls, over which they were constantly flitting, without thus perceiving that the walls were not nearly thick enough for their nests. After an exhaustive account of various traits of instinct, and difficulties in the way of his theory, explaining all of them, the paper closed with the following general conclusion:—"We have chiefly considered the instinct of animals under the point of view whether it is possible that they could have been acquired through the means indicated on our theory, or whether, even if the simpler ones could have been thus acquired, others are so complex and wonderful that they must have been specially endowed, and thus overthrow the theory. Bearing in mind the facts given on the acquirement, through the selection of self-originating tricks or modification of instinct, or through training and habit, aided in some slight degree by imitation, of hereditary actions and dispositions in our domesticated animals, and their parallelism (subject to having less time) to the instincts of animals in a state of nature; bearing in mind that in a state of nature instincts do certainly vary in some slight degree; bearing in mind how very generally we find in allied but distinct animals a gradation in the more complex instincts, which show that it is at least possible that a complex instinct might have been acquired by successive steps; and which, moreover, generally indicate, according to our theory, the actual steps by which the instinct has been acquired, inasmuch as we suppose allied instincts to have branched off at different stages of descent from a common ancestor, and therefore to have retained, more or less unaltered, the instincts of the several lineal ancestral forms of any one species; bearing all this in mind, together with the certainty that instincts are as important to an animal as their generally correlated structures, and that in the struggle for life under changing conditions, slight modifications of instinct could hardly fail occasionally to be profitable to individuals, I can see no overwhelming difficulty on our theory. Even in the most marvellous instinct known—that of the cells of the hive-bee—we have seen how a simple instinctive action may lead to results which fill the mind with astonishment. Moreover, it seems to me that the very general fact of the gradation of complexity of instincts within the limits of the same group of animals, and likewise the fact of two allied species, placed in two distant parts of the world, and surrounded by wholly different conditions of life, still having very much in common in their instincts, supports our theory of descent, for they are explained by it; whereas if we look at each instinct as specially endowed, we can only say that it is so. The imperfections and mistakes of instinct, on our theory, cease to be surprising; indeed, it would be wonderful that far more numerous and flagrant cases could not be detected, if it were not that a species, which has failed to become modified and so far perfected in its instincts that it could continue struggling with the co-inhabitants of the same region would simply add one more to the myriads which have become extinct. It may not be logical, but to my imagination it is far more satisfactory to look at the young cuckoo ejecting its foster-brothers, ants making slaves, the larvæ of the ichneumidæ feeding within the live bodies of their prey, cats playing with mice, otters and cormorants with living fish, not as instincts specially given by the Creator, but as very small parts of one general law leading to the advancement of all organic bodies—Multiply, vary; let the strongest live and the weakest die."

THE STOAT IN JERSEY.—Having just read in SCIENCE-GOSSIP for July, 1883, that, "Some authors seem to speak as though the stoat were never white in England, but only farther north," I thought it might interest some of your readers to know that about seven or eight years ago my father shot one in Jersey. It was seen running along a clay bank just behind our house, and was shot from the staircase-window; the distance being so short the skin was unfortunately much spoilt, and therefore was not kept. The animal was perfectly white, with the exception of a black tip to its tail. We sometimes get strange visitors to the island, even a hoopoe having been shot here about twenty-five years ago.—*J. J. B.*

NOTES ON YEAST-FUNGI.*

THE Saccharomycetes, or Yeast-Fungi, are unicellular plants, which multiply themselves by budding, and reproduce themselves by endogenous spores. They live singly or united in bud-colonies, chiefly in saccharine solutions, where they excite alcoholic fermentation.

In most of the Saccharomycetes the cells are round, oval, or elliptic; seldom are they elongated into cylindrical tubes, which are divided by transverse partitions, and may be regarded as the first indication of the formation of hyphæ, i.e. of a mycelium. For the purpose of multiplication the cell forms an outgrowth, which is filled with a portion of the contents of the mother-cell, gradually assumes the form and size of the latter, and separates itself from it by a wall. Both cells can in like manner produce fresh daughter-cells, which often remain for a considerable time united with one another, and on separation continue to grow independently.

The formation of spores succeeds most easily on a moist solid substratum. Typically the whole cell contents divide themselves into 2-4 roundish portions, or contract into a single spherical body. The portions of the contents surround themselves each with a membrane, and so produce the spores, which can bud like the vegetative cells.

To the Yeast-Fungi (in the narrower sense) belongs the capacity of decomposing the sugar of a fluid into alcohol and carbonic acid, i.e. of exciting alcoholic fermentation.

The carbonic acid comes off in rapid streams of bubbles, while the alcohol, as well as certain subordinate constituents of sugar, remains behind.

The fermentation proceeds most energetically with restricted access of air; but, if the air is excluded for a long time, the yeast-cells perish.

SACCHAROMYCETES.

The same is true of the Saccharomycetes, especially in a botanical aspect, as of the Schizomycetes. Just as in the latter case, so also in this, is it necessary to impose a limit upon the accepted species, and only those founded by trustworthy investigators can be considered. Of course there remain even then many doubtful points; for the majority of the now accepted species of Saccharomycetes may be only various forms of one and the same species, which have become differentiated by changed conditions of growth.

XVI. SACCHAROMYCES, Meyen. Unicellular fungi, with vegetative increase by budding, and reproduction by spores, which, for the most part, arise by subdivision of the contents of the mother-cell.

* [Translated from Dr. Winter's edition of the "Kryptogamen-Flora," with additions.]

A.—Species not producing a mycelium.

70. *S. cerevisiæ*, Meyen.
 Torula cerevisiæ, Turpin.
 Cryptococcus fermentum, Kützing.
 Cryptococcus cerevisiæ, Kützing.
 Hormiscium cerevisiæ, Bail.
 [*Saccharomyces minor*, Engel?]

Cells mostly round or oval, 8–9 μ long, isolated or united in small colonies. Spore-forming cells isolated,

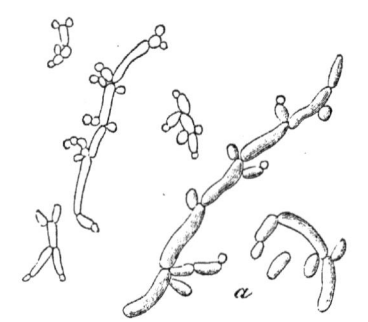

Fig. 6.—*S. pastorianus*; *a*, the same more highly magnified. (After Pasteur.)

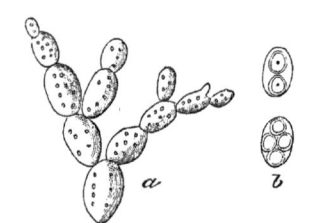

Fig. 7.—*S. cerevisiæ*; *a*, a bud-colony; *b*, two spore-forming cells. (After Winter.)

Fig. 8.—"High yeast," *S. cerevisiæ*; *a*, the same, budding actively. (After Pasteur.)

11–14 μ long; spores mostly three or four together in each mother-cell, 4–5 μ in diameter.

In beer, in both high and low fermentation.

The true beer-ferment is found in the various sorts of beer, in both modes of fermentation; it is cultivated on a large scale, and then yields the German yeast, a mass which consists of yeast cells and water.

[There are two races of this species, "high" yeast and "low" yeast. The cells of "low" yeast are slightly smaller, and more oval in shape, than those of "high" yeast, and in budding produce less ramifica-

tions, so that there is an absence of the globular clusters which are so striking a feature in the development of "high" yeast, when examined at an early stage of growth. "Low" yeast never rises to the surface of the fermenting fluid, which is thus left clear, but it produces, in the opinion of Englishmen at least, an inferior beer. With high yeast, the newly-formed cells rise to the surface as the fermentation proceeds, and there form large foam-like masses. It is doubtful whether the names "high" and "low" arose from

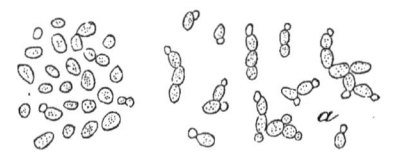

Fig. 9.—"Low yeast," *S. cerevisiæ*; *a*, the same, budding actively. (After Pasteur.)

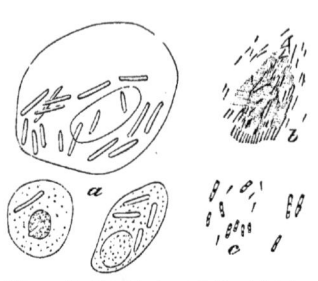

Fig. 10.—*Bacillus lepræ*; *a*, cells from tubercles, fresh; *b*, a "brown element" coloured with methyl-violet, from a tubercle treated with osmic acid; *c*, bacilli, with spores. (*a* and *b*, after Hansen; *c*, after Neisser.)

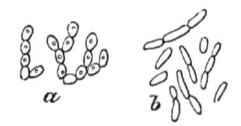

Fig. 11.—*S. mycoderma*, budding; *a* is the *Hormiscum vini* of Bonorden.

these different positions of the yeast, or from the difference in the temperatures at which they work. High yeast ferments at a temperature between 16° C. and 20° C., while low yeast is usually employed at a temperature of from 6° C. to 8° C., and rarely more than 10° C. In Pasteur's (from a morphological point of view) confused "Études sur la Bière," these are considered as distinct species, but this position is untenable.

The wildest possible theories have been started to explain the origin of this ferment. These are of two kinds; the first attempted to derive it by a sort of spontaneous generation, the second held that it was merely a state of other fungi, such as *Penicillium*, *Bacteria*, etc. These mistaken ideas were all due to a forgetfulness of the minuteness and omnipresence of the spores of these fungi, and also perhaps to the fact that as it appears, other fungi, such as Mucor, can act the part of a ferment under certain conditions. It may be regarded as certain that most of the so-called "trans-

Producing spontaneous fermentation in must; [this is the ordinary ferment of wine.]

72. *S. conglomeratus*, Reess.

Cells almost round, 5–6 μ in diameter, united in clusters, which consist of the numerous cells produced by budding from one or a few mother-cells. Spore-forming cells often united in twos, or with a vegetative cell; spores 2–4 in each mother-cell.

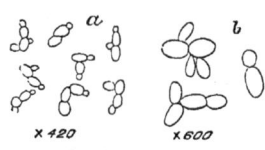

Fig. 12.—*S. ellipsoideus*; *b*, the same more highly magnified.

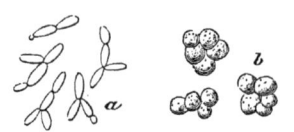

Fig. 15.—*a*, *S. exiguus*; *b*, *S. conglomeratus*; × 600.

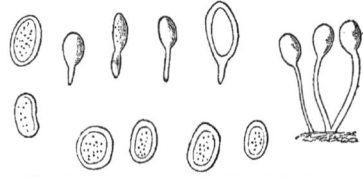

Fig. 16.—*S. coprogenus*. (After Saccardo.) × 500.

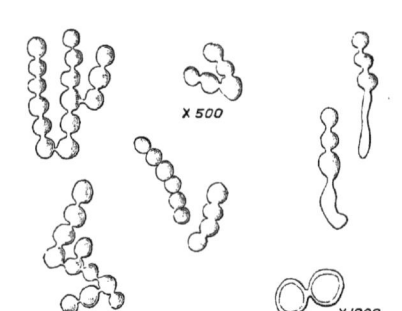

Fig. 13.—*S. sphæricus*. (After Saccardo.)

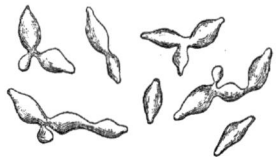

Fig. 17.—*S. apiculatus*. × about 500.

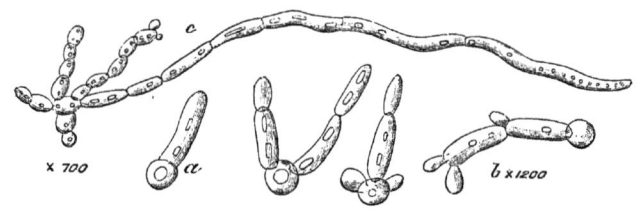

Fig. 14.—*S. albicans*; *a*, beginning of growth; *b*, farther advanced; *c*, formation of mycelium. (After Grawitz.)

formations" to which we are alluding had their origin merely in the carelessness of the experimenter, or in the inadequacy of the means adopted for securing a pure cultivation. The Saccharomycetes are as truly autonomous as any other fungi.—TR.]

71. *S. ellipsoideus*, Reess.

Cells elliptic, mostly 6 μ long, isolated or united in little branched colonies. Spore-forming cells mostly isolated; spores 2–4 together in each mother-cell, 3–3½ μ in diameter.

In wine at the beginning of the fermentation, and on decaying grapes.

73. *S. exiguus*, Reess.

Cells conical or top-shaped, about 5 μ long, reaching 2·5 μ in thickness, united in sparingly branched colonies. Spore-forming cells isolated, each with 2–3 spores, which lie in a row.

In the after fermentation of beer.

74. *S. Pastorianus*, Reess.

Cells roundish-oval or elongated-clavate, of varied

size. Colonies branched, consisting of primary clavate cells, 18–22 μ long, which produce secondary roundish or oval daughter-cells, 5–6 μ long. Spore-forming cells roundish or oval; spores from 2 to 4 together, 2 μ in diameter.

In the after fermentation of wine, and fruit-wine, or spontaneously fermenting beer. [The "caseous ferment" of Pasteur; may be obtained sometimes in English yeast.]

75. *S. apiculatus*, Reess.
Carpozyma apiculatum, Engel.

Cells lemon-shaped, shortly apiculate at each end, 6–8 μ long, 2–3 μ broad, sometimes slightly elongated; daughter-cells arising only from the ends of the mother-cells; for the most part soon isolated, rarely united in small, scarcely branched colonies. Spores unknown.

In the principal fermentation of wine, and in other spontaneous fermentations. [On all kinds of fruit, stone-fruits, etc., in must, and in certain kinds of beer.]

76. *S. sphæricus*, Saccardo, "Fungi Italici," fig. 76.

Cells of various forms; the basal ones (of a colony) oblong or cylindrical, 10–15 μ long, 5 μ thick; the others round, 5–6 μ in diameter, united in bent, branched, often clustered families. Spore formation unknown.

On the fermenting juice of *Lycopersicum esculentum*, the tomato.

[Saccardo, who regards this as a Hyphomycete of low organisation, says (Michelia, i. p. 90):— "Occurring in minute, flatly-convex, gregarious and confluent, dirty-white heaps; conidia perfectly spherical, 5–6 μ in diameter, collected in variously curved, branched and often clustered chains, separating with difficulty, hyaline, usually supported on oblong or subcylindrical bases, 10–15 μ × 5 μ." There is a strong likeness between this and *Hormiscium album*, Bonorden, except in habitat.]

77. *S. glutinis* (Fresenius), Cohn.
Cryptococcus glutinis, Fresenius.

Cells round, oval, oblong, elliptic to shortly cylindrical, 5–11 μ long, about 4 μ broad, isolated or united in twos, seldom more together. Cell-membrane and contents colourless, when fresh; but, when moistened again after drying, [with a faintly reddish central nucleus. Spore formation unknown.

On starch-paste, slices of potato, etc., forming rose-coloured, slimy spots, which have at first a diameter of ½–1 millimeter, but by degrees spread and become confluent in patches of as much as one centimeter broad.

The colouring matter is unchanged by acids and alkalies.

B.—Species producing a mycelium.

78. *S. Mycoderma*, Reess.
Mycoderma cerevisiæ, and *M. vini*, Desmazières.

Hormiscium vini, and *H. cerevisiæ*, Bonorden.

Cells oval, elliptic or cylindrical, about 6–7 μ long, 2–3 μ thick, united in richly branched colonies. The cells are often elongated, so as to resemble a mycelium. Spore-forming cells as much as 20 μ long; spores 1–4 in each mother-cell.

On fermented fluids, sauer-kraut, juices of fruit, etc., forming on beer and wine the so-called "mould."

This and the following species reach in their development the highest rank among the Saccharomycetes. The cells often form, especially in watery fluids, long tubes, which are divided by transverse partitions, and fall into single pieces at those points. These bud, in their turn, in the same manner.

While the true yeast-fungi grow submerged in the higher layers of the fluid, and there excite active alcoholic fermentation, the "mould" grows on the surface, without exciting fermentation. When artificially forced to grow submerged, of course a little alcohol is produced, but the fungus soon perishes.

Although the growth of the layer of "mould" goes hand in hand with the souring of the wine or beer, yet the Saccharomyces is not the cause of the latter. The formation of vinegar from alcohol is produced rather by other fungi, whose systematic position is still undetermined. According to some, it is a species of Vibrio (Spirillum), which causes this decomposition.

79. *S. albicans* (Robin), Reess.
Oidium albicans, Robin.

Cells partly round, partly oval, oblong or cylindrical, 3·5–5 μ thick; the round ones 4 μ in diameter, the cylindrical ones 10 to 20 times as long as thick. Bud-colonies mostly consisting of rows of cylindrical cells, from the ends of which spring rows of oval or round cells. Spores formed singly in roundish cells.

On the mucous membrane of the mouth, especially of infants, forming the disease known as aphtha or thrush. Also in animals.

This fungus appears in the form of larger or smaller greyish-white heaps, which nevertheless do not consist exclusively of the Saccharomyces, but also contain Schizomycetes, and the mycelia of moulds. When cultivated, the fungus forms long-jointed, richly-branched threads; at the upper end of each articulation there is usually a crown or bundle of shorter cells, which are oval or round in form, and bud in their turn. In other cases, all the cells of a bud-colony remain short, and assume a rounded form. This fungus excites alcoholic fermentation only in a small degree.

According to Grawitz (Virchow's "Annalen für Path. Anat. und Phys.," vol. lxx. p. 557), *S. albicans* is identical with *S. Mycoderma*.

C.—Doubtful species.

80. *S. guttulatus* (Robin).
Cryptococcus guttulatus, Robin.

Cells elliptic or elongated-ovate, 15–24 μ long, 5–8 μ thick, brown, opaque, with 2–4 colourless drops, isolated or from two to five together. Spore formation unknown.

In the œsophagus and intestines of mammals, birds, and reptiles.

[81. *S. coprogenus*, Saccardo et Speggazini, "Fungi Italici," fig. 911.·

Effused, superficial, rather compact, dirty rose colour; conidia ovoid and then globose, 12–14 μ long, 10–11 μ broad, forming very short chains or solitary, often provided with a tail-like appendage (? from germination), clouded within, when in clusters pale rose-coloured, hyaline ("Michelia," ii. p. 287).

On fermenting human ordure, where it forms a somewhat waxy layer, almost like a Corticium. This also is considered by Saccardo to be a Hyphomycete. —TR.]

W. B. GROVE, B.A.

SCIENCE-GOSSIP.

THE venerable Swedish palæontologist and antiquary, Professor Nilsson, has just died, aged ninety-six years. He is best known in this country, perhaps, by the translation of his work on "The Primitive Inhabitants of Scandinavia," published in 1868, and edited by Sir John Lubbock.

DR. ANDREW WILSON has been delivering four popular lectures on anatomy and physiology in Princes' Hall, Piccadilly.

A VERY interesting example (illustrated) of the mutual spiral growth of two carrots is given in the *Gardener's Chronicle* for December 8th.

PROF. ARCH. GEIKIE is contributing a series of papers to *Nature* on "The Origin of Coral Reefs."

AN International Exhibition of Health and Education is intended to be held in 1884, in the buildings recently occupied by the Fisheries Exhibition.

THE works in connection with the new Bridge across the Forth are lighted at night by electricity, numerous arc as well as incandescent lights being employed, so that the operations can go on by night as well as by day.

M. DE LESSEPS recently made a communication before the Paris Academy of Sciences on the propagation across the Indian and Atlantic Oceans of the great earthquake wave caused by the recent disturbances in Java. The engineers engaged on the Panama Inter-oceanic Canal at Colon made some observations upon it, from which it appears that the wave made its way in about thirty hours from Java, round the Cape of Good Hope to the east coast of Central America.

A REMARKABLE whirlwind occurred on the 17th of November, in Somersetshire, about mid-day. In the village of Brympton, three miles from Yeovil; trees were uprooted by it, and in a neighbouring village it unroofed several houses. It necessarily occurred over only a limited area, but its track was plainly marked by uprooted trees, unroofed houses and devastated hay-ricks.

THE GEOLOGICAL MAGAZINE for December last contains a capitally written memoir of the eminent geologist, Joachim Barrande, recently deceased, and is accompanied by an excellent portrait.

THE PRACTICAL NATURALIST is now incorporated with "The Naturalists' World," under which title this excellent monthly will henceforth appear.

WE have received the second number of our new contemporary ["The Science Monthly" (David Bogue). It is well turned out, and contains some excellent articles on a variety of scientific subjects. The portrait of Sir John Lubbock is even better than that of Sir George Biddell Airy which appeared in the first number.

THE LIVERPOOL GEOLOGICAL ASSOCIATION have just issued their annual Report for 1883. It contains the Rules, list of members, &c., and outline of papers read and excursions made during the past year.

THE latest official report of the Imperial German Post-office states that at the end of October the telephone was fully in operation in thirty-six cities and towns.

A MEETING was held in the Royal Society rooms on the 7th December, Professor Huxley in the chair, when it was resolved that a memorial to the late William Spottiswoode, President of the Royal Society, &c., should be formed, and that it should take the form of an endowment for a pension in perpetuity to be called the "William Spottiswoode Memorial Pension," the proceeds to be devoted to an incapacitated printer or widow.

MR. WILLIAM WESLEY'S "Natural History and Scientific Book Circular," is now in the thirteenth year of publication, and No. 58 has just appeared. We commend it to all scientific book buyers.

THE Scientific Expedition sanctioned by the Geographical Society, for the purpose of exploring New Guinea, under the command of Mr. Wilfred Powell, will shortly leave England.

AT a recent meeting of the British Archæological Association, Mr. Worthington Smith, F.L.S., produced a fine Palæolithic Flint implement which had been found in Clerkenwell.

THE Victoria Government have asked the distinguished botanist, Baron von Mueller, to report on a new kind of prairie grass which is spreading in the Colac district, and which is said to bind the sand together, and to furnish good food for cattle, although the rabbits won't eat it.

WHITTINGTON'S days are not quite over. An advertisement has just appeared in the Adelaide newspapers inviting a supply of one thousand cats at ninepence each, for the purpose of putting down the rabbits on a certain estate.

A CAPITAL chemical laboratory, with convenience for fifteen working students, has just been fitted up with all accessories, stoves, &c., in connection with the Ipswich Museum, and presented to the town by Mr. Alderman Packard, J.P.

THE "Popular Science News" says that the most costly pharmaceutical preparation in the market at present is the Ergotine prepared by the manufacturing chemists, Gehe & Co. of Dresden. It contains the active principle of ergot of rye, and costs ten pounds a gram, or about 21,000 dollars a pound.

HERR PALISA, of the Vienna Observatory, has just chronicled the discovery of a new minor planet, at 1h. 20m. (Greenwich mean time), when it was in Right Ascension 3h. 19min. 15sec., and North Polar Declination 74° 7' 43". It is of the 12th magnitude, and therefore is only visible with a large telescope.

WE hear from Victoria that the supplementary estimates recently laid on the table of the Assembly included the sum of £1000, for scientific exploration in New Guinea. Cannot the English and Australian explorers arrange for some kind of joint action?

THE largest locomotive in the world is being constructed at Sacramento for the Central Pacific Railway. It will weigh 73 tons, and have 5 pairs of driving-wheels. The tender alone will weigh 25 tons, and the total length of engine and tender will be 65 feet.

A MEETING has been held in Chester for the purpose of establishing a museum, to be a centre for scientific information for Cheshire and North Wales. The Duke of Westminster gives the ground for the site, and £4000 towards the building fund.

THE Anniversary of the Royal Society was held on the 30th of November, when the President (Professor Huxley) delivered his anniversary address.—The following were chosen Council and officers for the year ensuing: President, Professor T. H. Huxley; Treasurer, Dr. J. Evans; Secretaries, Professor G. G. Stokes and Professor M. Foster; Foreign Secretary, Professor A. W. Williamson; other members of the Council, Captain W. de W. Abney, Professor W. G. Adams, the Duke of Argyll, J. G. Baker, Dr. T. L. Brunton, W. H. M. Christie (Astronomer Royal), Warren De La Rue, Sir F. J. O. Evans, Professor G. C. Foster, F. Galton, J. W. L. Glaisher, Sir W. W. Gull, Bart., Dr. H. Müller, Professor J. Prestwich, Professor O. Reynolds, and O. Salvin.

MR. CHANDLER, of the Harvard Observatory, has been carefully noticing the rapid changes in the brightness of Pon's comet, especially at the latter end of last September and during October. So rapid an increase and diminution of the light of a comet is an unusual phenomenon, and Mr. Chandler thinks that phases of this kind may be characteristic of the comet's mode of light development. Schiaparelli, the distinguished astronomer, has written concerning the same phenomena. On September 22nd, the comet was faint and diffuse, whereas on the following night it had greatly increased in brightness. This comet attains its maximum brightness on January 14th. Before then it will be visible with the naked eye.

ECONOMIC botanists can no longer say seaweeds are of little importance, for the last new thing is that they are being employed in the manufacture of "Gooseberry Jelly!" The jelly is obtained from sea-weeds, coloured by some substance, and flavoured by acetic ether, tartaric acid, small quantities of benzoic, succinic, and œnanthic acids, and aldehyde.

TARCHNOFF has discovered that the white of eggs of those birds whose young are born unfeathered differs from ordinary albumen, its most striking peculiarity being that it remains transparent after coagulation by heat.

THE Russian scientific expedition at the mouth of the Lena, which has just entered upon the second winter of its sojourn in its frost-bound region, is reported to be in excellent health and spirits. Last winter the cold seldom reached 40 deg. centigrade till January, but during January and February it was seldom less. The fluctuations of the magnetic needle were so great that the instruments were not capable of measuring them.

DR. LE CONTE, the distinguished American coleopterist, has just died, in his fifty-ninth year.

Two Fellows of the Royal Society, both distinguished medical men, Professor William Bowman, and Professor Lister, have recently had the honour of baronetcy conferred upon them.

M. DAUBRÉE, the distinguished French geologist, from the mineralogical analysis of the ashes collected in Batavia during the recent Javanese volcanic disturbances, thinks it highly probable that the surface water penetrated deeply into the underground cavities, and there becoming super-heated, formed the chief agency in such volcanic eruptions as those of Krakatoa and Ischia.

WE regret to hear that Professor Owen's health has obliged him to resign his post as Superintendent of the Natural History Department of the British Museum.

UNDER the title of "The Botanic Stand," Mr. David Bogue has published a brochure by Thomas Twining, price 6d., in which is set forth the easy way in which the natural orders of plants might be learned by means of living specimens arranged in stands.

WE have received a copy of Mr. T. M. Reade's "Traverse of the Yorkshire Drift," originally read before the Liverpool Geological Society. The "Traverse" extends from Ribblesdale across the Pennine Chain, and the Vale of York to Scarborough.

WHAT can be done by the humblest workers has recently been demonstrated in the case of Messrs. Robert Law and James Horsfall, who for three years carefully and tediously explored the surface of the millstone grit moors between Todmorden and Marsden, for neolithic flint implements. They found no fewer than eighty-one, including stone hammers, arrows, scrapers, lances, &c. The results have been communicated in a paper (illustrated) read before the Geological and Polytechnic Society of Yorkshire.

A REALLY scientific article on "Cremation," by Mr. W. H. France, appeared in the last number of "The Midland Naturalist."

A CAPITAL paper on "The Geology of Central Australia" appeared in the November number of "The American Naturalist," by Edward B. Sanger, who has explored the district. He shows how the continent of Australia has been gradually built up.

MR. THOMAS BOLTON, the well-known and enterprising provider of microscopic material, has sent us a copy of his "Popular Account of the Fish's Nest." It is illustrated by sketches showing the development of the stickleback from its earliest stage in the egg, and is accompanied by a notice of the anatomy of the fish by Mr. J. E. Ady.

DR. LANG's "Butterflies of Europe" has just reached the issue of its sixteenth part, and is devoted to the various species of the genus Erebia, all of which are illustrated by the exquisite coloured drawings which have already rendered this work noteworthy.

PART 5 of Dr. Greene's "Parrots in Captivity" has been issued by Messrs. Geo. Bell & Sons. It describes Barraband's parakeet, the red-winged parakeet, and the Turquoisine, each of which is illustrated by a beautiful coloured plate.

AT the meeting of the American National Academy of Science recently held in New Haven for a few days' session, the most important paper appears to have been read by Dr. Graham Bell, upon the "Formation of a Deaf Variety of the Human Race," in which he spoke against isolating deaf-mutes from ordinary children, and the greater benefit which would ensue from their being educated together, and taught to speak by lip-signs, instead of the old sign language.

MICROSCOPY.

PHOTOGRAPHING MICROSCOPIC OBJECTS.—At a recent meeting of the Academy of Medicine in Ireland, Dr. Dickenson read a note on the "Art of Photographing Microscopic Objects," exhibiting a number of specimens produced by himself and the apparatus used for the purpose. His apparatus consists of three parts—the first, an inexpensive magic lantern, illuminated by a triplex petroleum lamp with the ordinary combination of lenses, and an extra tube with a small bull's-eye condenser; the second, a microscope, placed horizontally without the eyepiece; and the third, a frame to hold the glass screen for focussing the image, and to receive the sensitised plate when photographing. The period of exposure was from eighteen seconds to two hours. Dr. Hayes congratulated Dr. Dickenson on the great simplicity of the apparatus, and on the results which it produced. He had no doubt that if the electric light could be applied to it there would be nothing to compare with it. Its use with a light intense and strongly actinic would be a great advantage; for the illumination could be cut down by diaphragms, and the sharpness and depth of the image increased. With the electric light he had got pictures up to 300 diameters taken in four seconds, and with the oxyhydrogen lamp pictures could be obtained in from ten to twenty seconds.

MR. HUGH POWELL.—We regret having to announce the death of this well-known microscopist (of the well-known firm of Powell & Lealand), at the advanced age of eighty-five. How long he has been before the world in connection with microscopy is shown by the fact that in the year 1834 he was awarded a silver medal by the Society of Arts, for a stage for the microscope. Lately he has been chiefly distinguished for his "fiftieth," and other high powers.

DRAWING FROM THE MICROSCOPE.—E. T. D., in SCIENCE-GOSSIP for November, in his article, makes a difficulty in the use of the neutral tint reflector where none really exists. He has but to turn his slide over, i.e. cover downwards, on the stage to make his outlines, and then put his slide right way up when he fills in his detail freehand. He will thus have identically the same presentation of his subject in each case. I do not agree with his remarks deprecatory of the use of neutral tint reflectors. I take it the first requisite of a microscopical drawing is exactness and truth. Beauty is a secondary thing. An artist is generally the worst possible delineator of microscopical objects. He wants to make a picture, and sacrifices nature on the altar of art. May I add that a thin cover makes as good a reflector as any neutral tint, and costs nothing. The reflection is quite vivid enough for anything in the way of outline drawing, and the pencil is seen through it without difficulty.—*Edwin Holmes.*

STUDIES IN MICROSCOPIC SCIENCE.—Edited by A. C. Cole, F.R.M.S.—We very much regret that in the paragraph in our last issue on "A New Morphological Institution," we inadvertently over-stated the relationships which existed between Mr. A. C. Cole and Mr. J. E. Ady, in the first volume of this admirable work, which was so well done, and so thoroughly successful from all points, literary, scientific, and artistic, that we deem it a great pity any difference of opinion should exist among those who contributed to so great a success. We inadvertently stated that Mr. Ady was the *actual* and Mr. A. C. Cole the *nominal* editor; whereas the legal status of the relationship was that Mr. Ady edited the "Studies," under "the advice and direction" of Mr. Cole. Mr. Cole's name appeared all through as editor, and the weekly parts of the 2nd vol. now issuing, also bear his name as editor. The preface to the first volume states that all the articles except three were written by Mr. J. E. Ady. This much, however, appears certain, that to Mr. Cole belongs the credit and honour of bringing out the "Studies." The second volume is issuing its weekly parts in quite as attractive a manner, and, as regards the coloured plates, the drawings are very much improved, from an artistic point of view, as we should expect from seeing the initials of the same artist who is engaged on our own coloured plate. We have only space to notic Nos. 7 and 8, dealing with Epithelium, and accompanied by a beautiful coloured plate showing the various kinds; squamous, from the tongue; columnar, from the intestine; and ciliated, from the fauces. No. 8 deals with the "Cell as an Individual," and is illustrated by an exquisite coloured drawing of *Micrasterias denticulata*, in different stages of cell development. In addition to the above, we are pleased also to call attention to No. III. of "Popular Microscopical Studies," which treats upon the human scalp, its hair follicles, glands, and tissues. This is accompanied by what we consider the best coloured plate Mr. Cole has ever turned out. All the parts above mentioned (including the latter) were accompanied by beautifully mounted slides of the objects described, so that the student is doubly benefited.

MICROSCOPY SAMPLE SLIDES.—We have received two of the slides prepared by Messrs. Ady & Hensoldt, as advertised in our columns. One of them is a double object, containing both a longitudinal and a transverse section of the compact tissue of the middle of shaft of the human humerus, mounted in gum and Canada balsam; and the other a section of the Eozoonal white Serpentine, recently discovered by Dr. Heddle in Sutherlandshire. These specimens approach the best style of mounting we have yet seen, and if the new Morphological Laboratory continues to send forth slides of this character it cannot fail soon to command general attention.

ZOOLOGY.

ASSOCIATION OF HELIX NEMORALIS AND H. HORTENSIS.—Having read Mr. Crowther's communication hereon, I have thought the following might be of interest to him and others among your readers. During 1882 I took *H. nemoralis* abundantly near Stokesley, where it was associated with *hortensis*, which, however, was not so common. I also took a single specimen of *hybrida*. This summer I have taken, at Redcar, var. *roseo-labiata* of Nemoralis (?) These specimens (I took two) are smaller than Nemoralis usually is at the place I mention and are in fact more like Hortensis in size, though much more solid than that species, which does not occur, so far as my experience goes, within eight miles of Redcar. In 1882 I turned out a canister load of Hortensis, near the place where the var. *roseo-labiata* were found, with a view to colonisation, but soon lost sight of them, and have not seen them since. I must therefore query, are the shells I have taken hybrids? In September this year I found Nemoralis and Hortensis living together near Durham, and at once made a search for Hybrida, which search was rewarded. Off one bunch of nettles, about two yards square, I took seven of the purply-brown shells with the flesh-coloured lip, and both Nemoralis and Hortensis were present. My experience with Hybrida is that I have never found it apart from Hortensis, and in all cases Nemoralis was also present. For Mr. Gann's information, I may say that I have raised in confinement from parents taken in copulation broods of both Nemoralis and Hortensis, which I am keeping with a view to experiments. I find that a glass-covered box, moderately deep, and containing a depth of soil of about five inches, which is planted with nettles, plantain and coltsfoot, suits my prisoners very well. I have some at present in confinement, which were placed in the box in May 1882, and are in possession of excellent health. I have raised a fine lot of *Limax maximus* in confinement this year.—*Baker Hudson.*

"YOUTH" NATURALISTS' SOCIETY.—This society has just been formed with a view of promoting a love of Natural History among the youths of the

United Kingdom, by the establishment of branches in all our large cities and towns. The originator of the scheme was the Editor of "Youth," who acts as President, and the idea has been warmly taken up by not a few eminent naturalists, who have already added their names to the honorary membership list. The subscription for honorary members is 2s. 6d. and for ordinary 1s. per annum. Further particulars may be had on application to the Chairman of Council—Alfred J. Weyman, Esq. F.R.Hist.S., etc. *Keppoch House, Glasgow*, or J. W. Williams, Esq., B.Sc., 100, *Albert Street, Regent's Park, W.*

APANTELES GLOMERATUS.—As a rule, very few caterpillars of the great white cabbage butterfly escape being eaten up alive by the larvæ of this ichneumon fly, but last autumn I scarcely saw one caterpillar attacked by them, and the fact struck me as somewhat remarkable and worthy of record.— *Albert H. Waters, B.A., Cambridge.*

BUTTERFLIES AT CAMBRIDGE.—I noticed last year that *Colias edusa* had begun to put in an appearance. I had not seen it flying here for the last six years. I also observed that *Pyrameis Cardui* was tolerably abundant, *Io Urticæ* and *Atalanta* have been very plentiful, as have also the common whites; but *Janira, Hyperanthus, Rhamni,* and *Cardamines* seem to have been scarcer than they ordinarily are. *Tithonus* and the common blue seemed, however, to have been pretty plentiful.— *Albert H. Waters, B.A., Cambridge.*

THE SALMON DISEASE.—Mr. H. Marshall Ward, Fellow of Christ's College, Cambridge, has made some interesting experiments in the reproduction of the fungus of the salmon disease—*Saprolegnia ferax*. The disease first appeared in 1877 in certain rivers flowing into the Solway Firth. Since then it has extended rapidly and widely, and in 1880 it appeared in North Wales. Salmon affected by the disease show signs of languor, feed badly, and when severely affected, die. The fungus is reproduced very rapidly, and may be cultivated on the bodies of ordinary flies. Contact of the diseased salmon with a dead fly in fresh water, for twenty-four hours or less, results in infection of the latter, and very fine silky filaments are soon observed to shoot forth in all directions from the body of the fly into the water. If proper precautions are taken, the silky filaments soon form multitudes of reproductive bodies by which other flies may be infected. Mr. Ward recently described his experiments in detail to the members of the Manchester Literary and Philosophical Society, and said that the results he had obtained confirmed those published by Professor Huxley.

SPOTTED REDSHANK AT LYNN.—Four beautiful specimens of the spotted redshank (*Totanus fuscus*) were shot on the mud-banks at Lynn on October 28th last. The birds were in the winter plumage. Two were adults, and two immature, males and females of each. The latter sex was the largest; the adult female weighed eight ounces, and the male seven. Adult specimens are not often obtained in Norfolk, but individuals in the immature plumage sometimes occur in the autumn.—*E. W. Gunn, jun., St. Giles Street, Norwich.*

THE CULTIVATION OF THE HERRING.—Some very interesting experiments were made by the Investigation Committee of the Fishery Board for Scotland, during the recent summer, in connection with the early history of the herring. The ova of that fish were successfully artificially impregnated and developed. The first experiment was made on August 7th with spawn from herring which had been several hours out of the water, but these results were unsatisfactory. Afterwards ripe fish were obtained, the roe and the milt were pressed out on separate slips of glass; and then placed in specially-designed carrying boxes, and conveyed to a laboratory, where they were transferred to hatching boxes, through which a constant current of water flowed. In from three to five days well-formed active little embryos were visible through the transparent egg-membrane, and in ten days successfully hatched fry were obtained from the artificially impregnated ova. It was found that the chief feature in their successful development was having an abundant supply of pure sea water, at an equable temperature. The committee report that from the experience gained last autumn, they are now able to hatch immense numbers of herring; each herring produces from 30,000 to 50,000 eggs, but they are so small that 20,000 are only a layer thick, and can be placed on a square foot of glass. From 1000 herrings it would be possible to obtain about thirty millions of fry, and this in from ten to fifteen days. It is well known that following herrings is an abundance of cod, and other food fishes. Hence the annual introduction of some millions of young herring into British waters might serve to attract numerous large food fishes to our shores. What is true of the herring is equally so for other fishes, such as the sole and the turbot (which have recently been declining in numbers). These are less migratory than the herring, and the Committee of the Fishery Board think they might be manipulated in much the same way as trout and salmon, if we only knew their habits.

GEESE MIGRATING.—A large flock of geese passed over Norwich city about 2 P.M. on 29th November. The direction of flight was from north-west to south-east. As they passed overhead some of the birds were distinctly recognised as the Canada (*Anser Canadensis*). Several persons counted the flock, and reported it to consist of some eighty-three or eighty-four individuals.—*E. W. Gunn, jun., St. Giles Street, Norwich.*

BOTANY.

ORCHID FLOWERS.—Lately whilst examining a large series of orchids, I have been struck with the fact that several of our British species run gradually in a series, the one into the other; notably is this the case with *Orchis purpurea*, Huds., *O. militaris*, L., and *O. Simia*, Lam. I herewith give a sketch of three specimens of the lips of the above species to show this gradual transition. Hooker, in the "Student's Flora," makes *O. Simia*, Lam., into a subspecies of *O. militaris*. I have cultivated all three

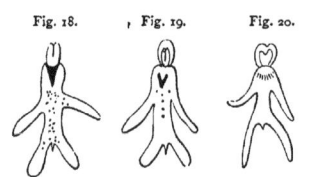

Fig. 18.   Fig. 19.   Fig. 20.

Fig. 18.—*Orchis purpurea*.   Fig. 19.—*Orchis militaris*.
Fig. 120.—*Orchis Simia*.

species, growing them in strong loamy soil, and I will defy anyone to point out any strong distinctive character of permanent value whereby they may be distinguished as species. After careful comparison, I am compelled to come to the conclusion, that the only marks are as follows: *Orchis purpurea*, Huds. Lip rough with tufts of red hairs. This is the *O. fusca*, Jacq., and *O. militaris*, Sm., in Eng. Bot. 16. Fig. 18. *O. militaris*, L. Sepals acuminate. Lip rough with scattered red hairs. E. B. S. 2675. Fig. 19, *O. Simia*, Lam. Lip generally quite smooth. It is the *O. tephrosanthos* of E. B. 1873, fig. 20.—*James F. Robinson*.

CLOVER IN NEW ZEALAND.—It has been stated in all books dealing with the fertilisation of flowers, that the red clover does not bear seed in New Zealand, owing to the absence of humble bees. Mr. J. B. Armstrong, of the Christchurch Botanic Garden, has recently contradicted this statement, and shown that there are four varieties of red clover in New Zealand, all of which produce seeds of good germinating power. One variety is partly self fertile and partly self sterile. The produce of those which have been grown in the colony for several generations tend almost invariably to become self-fertilising. Mr. Armstrong thinks there is every reason to believe that the red clover is there becoming modified in its structure, so as to admit the visits of insects not known to visit it in England, and that such modification tends to render the plant self-fertilising, but at the same time enables it to be improved in constitutional vigour by occasional inter-crossing.

REPRODUCTION OF THE ZYGNEMACEÆ.—A paper on this subject has just been read before the Linnean Society by Mr. A. W. Bennett. Twenty-five years ago Dr. Bary, and since then Wittrock, have instanced what they thought were sexual differences between the conjugating cells. Mr. Bennett has directed his investigations chiefly to the genera Spirogyra and Zygnema, and from these he supports the opinions of the above-mentioned botanists. He finds there is an appreciable difference of length and diameter in the conjugating cells, that deemed the female being the larger. He also finds that the protoplasmic contents pass only in one direction, and change first commences in the chlorophyll bands of the supposed male cells, with accompanying contraction of the protoplasmic material.

THE FLORA OF SOCOTRA.—Professor Bayley-Balfour, who was sent out purposely to study the flora of this island, in a lecture upon it, shows that the flora of a continental island, such as Socotra, is in the main interesting in connection with the geographical distribution of plants, and the working out of the history of their migrations over the face of the globe. But in the flora of Socotra there are a number of special features in individual plants well deserving of attention. To summarise them, we may say the flora of Socotra is that of a continental island, and presents features of great antiquity. Relative proportion of orders to genera, and of these to species, is large. There are few annuals. It possesses much individuality, and further exhibits three distinct elements; (*a*) of a dry parched region ; (*b*) of a moister tropical region ; (*c*) of a cooler and more temperate region. Its affinities are essentially tropical, African, and Asian, but the African element predominates, and in the African element we find in great force the features of the flora of the mountainous region of Abyssinia, West Tropical Africa, and South Africa, and also of Madagascar. This element of the flora, too, is that of the higher regions of the island. The flora of the dry region is the typical Arabo-Sahara. The flora of the moister tropical region is that of the old world tropics generally. There are only a few Indian and American types in Socotra.

A NOVEMBER NOSEGAY IN WORCESTERSHIRE.—Groundsel, nipplewort, thistle, black knapweed, buttercup, white dead nettle, charlock, sowthistle, bramble, herb Robert, harebell, trailing dog-rose, feverfew, chickweed, gorse, wild strawberry, wood sage, tiny tare, bitter vetch, black medick, red clover, hawkweed, potentilla, milfoil, fool's parsley, wood avens, ragwort, red campion, black berry, devil's-bit, scabious, foxglove, dock, cross-leaved heath, honeysuckle, broom, daisy, pimpernel.—*K. D., Cofton Hackett*.

GEOLOGY, &c.

THE MAMMOTHS AT GUILDFORD. — There has just been an exhumation of the great teeth, tusks, and portions of the skulls of these marvellous elephants, in the railway cutting a few hundred yards from the spot where I am writing. Excitement has given rise to the most wonderful stories, and one good woman has been dreadfully shocked to think that these prodigious beings were exterminated by man, and not all drowned by the deluge. There is some little geological interest attached to the matter. When Mr. Godwin Austen came across mammoth remains in the Peas Marsh by the banks of the Wey, the occurrence seemed natural enough; this time the existence of an elephant bed is not so easy to account for. As is known, the plastic clay overlaps in a ribbon the northern slope of the Downs. East of Guildford this ribbon consists of mustard-coloured sand traversed by a layer of bluish, red mottled brick clay, evidently incipient clay-slate, averaging ten feet in thickness. In this clay are vast heaps of shells here and there, as if washed together. As you approach the town this seam of brick clay thins out and leaves the sand reposing in hollows on the chalk, which rises in little waves. The sand-bed here is composed of elements which bespeak rest and disorder. First there are noticed traversing it, rice-like layers of minute chalk-pebbles, and others of small flint pebbles, with here and there a bit of iron-stone, perchance derived from the greensand. There are likewise beds of angular flints and shingle patches. All this would seem to indicate that the arenaceous accumulation had taken place in tranquil water, on the shore of the tertiary sea that drowned the north of France and Belgium. Otherwise the deposition of the sand has not so tranquilly proceeded; for the layers are sometimes tilted up at an angle, and great detached masses of chalk plainly suggest that the earthquake heaved the while. Now I think that when it is borne in mind that there is no perceptible demarcation in the sand-bed from the plastic clay onwards, as seen in section; and save that oysters disappear, its consistence is uniform, and when it is likewise taken into consideration that the mammoth remains were dug up forty-two feet deep in the sand, and consequently near the level of the plastic clay, some confusion in geological ideas will naturally arise. My friend, Mr. Kidd, and another gentleman who appears to have given attention to local geology, suggested to me last spring that the Wey may anciently have flowed though this sand-bed, which now is raised upwards of two hundred feet on the flank of the Downs. We all know that elephant-beds do occur at intervals along the slope of the English and Continental chalk hills, but no one seems to have furnished an adequate reason for their existence, while those who revert to a deluge to explain the chaos on a hill-side, seem to ignore the earthquakes that were invoked to raise the hills.—*A. H. Swinton, Guildford.*

THE BATTLE OF THE ATOLL.*

Is there a moral from the coral
    Reefs and Atolls we can quote?
Were they stable or unable
    To withstand the seas that smote
At their base and living face
    With oft-repeated shock?
Did they shiver, did they quiver?
    No, the seas they mock!
In the battle of the Atoll
    Which the little creatures wage
In a strife for their life,
    They tame the breakers' rage
As on they rush as if to crush
    The fabric at one bound,
But Polypifer built it stiffer
    And made it safe and sound.

Moral.

So Darwin's theory builded up with care,
Facts piled on facts and nothing left to chance,
Good Master Geikie pray beware! beware!
Your trivial efforts cannot but enhance.
The minute labours of the master mind,
That first the facts explained, the truth divined
Will most assuredly win in this small battle
And breast your breakers like the living Atoll.

A. Conifer.

SUPPOSED ANIMAL TRACKS.—At a recent meeting of the Geological Society, Professor Hughes made some observations on certain pits in the district about Cambridge which are filled with the fine mud produced in washing out the phosphatic nodules from the "Cambridge Greensand"—a seam at the base of the Chalk Marl. As the water gradually dries up, a surface of extremely fine calcareous mud is exposed. This deposit is often very finely laminated, and occasionally among the laminæ old surfaces can be discovered, which, after having been exposed for some time to the air, had been covered up by a fresh inflow of watery mud into the pit. The author described the character of the cracks made in the process of drying, and the results produced when these were filled up. He also described the tracks made by various insects, indicating how these were modified by the degree of softness of the mud, and pointed out the differences in the tracks produced by insects with legs and elytra, and by Annelids, such as earthworms. The marks made by various worms and larvæ which burrow in the mud were also de-

---

* See 'Origin of Coral Reefs,' "Nature," Nov. 3 and Dec. 6.

scribed. Marks resembling those called Nereites and Myrianites are produced by a variety of animals. The groups of ice-spicules which are formed during a frosty night also leave their impress on the mud. The author concluded by expressing the opinion that Cruziana, Nereites, Crossopodia, and Palæchorda were mere tracks, not marine vegetation, as has been suggested in the case of the first, or, in the second, the impression of the actual body of ciliated worms.

NOTES AND QUERIES.

A BEE'S LEG.—There are bee's legs, and bee's legs. Many years ago, I went with my dear old friend P. S. Mitchell to call on R. Beck, who had just brought out his twentieth of an inch object glass. We had a most pleasant time and saw things which made us feel how small is man, how marvellous the works of the Creator, whose ways are beyond our ken. I, as a country bumpkin, then and there submitted to Mr. Beck's inspection a bee's leg. "Oh," he said, "I have seen lots of them, nothing new in that." "Look," I said, "ah, this is something new." In the slide went, under a lovely 2-inch object glass, microscope binocular. "Before you use the microscope," I said, "look at it again." Beck : "Dear me, it's shot with gold, never saw a bee's leg like it before, they are black." He took a long look, and said, "The most wonderful thing ; why, there are rows of combs, one rank behind the other, just like the tortoise-shell ones ladies used to wear when I was a boy." "Look again."—"Ah, I see, sir, muscles connected with these combs." "Yes," I said, "they may be erected vertically when the bees are getting pollen, and by the action of those muscles packed close together when the insect is fully laden for home and hive." At the side between the slightly concave part of the thigh, where these combs are, and the convex part, runs a long dark pocket. If the bee had been a Highlander instead of a Southron, I should have said this is the sheath for his skean-dhu, slit from end to end by a microtome, nothing was found in it. These bees' legs are rare ; you may pick up fifty dead bees, and only find a small percentage of these golden thighs.—*A. H. B.*

MIND IN THE LOWER ANIMALS.—Dr. Quin Keegan appears to lay down as a premise what he professes to seek as a conclusion. "There is no such thing as mind in the lower animal" is his proposition, and the facts must bend to suit. No one asserts that a dog can reason like a doctor, but then all human beings are not gifted to that extent. His test-question is, "Are they capable of understanding these words as signs of abstract or generalised thought." He mistakes in saying he has "demonstrated" they do not. He has only "asserted" this—a very different thing. But he admits the possession of mind among the lower animals in his very efforts to deny it. He asks, "Is not the human general term associated in the animal mind" "with some individual impression." Clearly he admits animal intelligence, for there is an idea and the association of ideas, but he will not call it mind. As he inquires further about my dog, I beg to inform him that if I was at home when he was fed by some one else, he never attempted to delude me into giving him a second dinner. "Well," he says, "the probability is that the dog merely intended to convey that he would have no objection to a second dinner." Then in Dr. Keegan's opinion he did intend to convey some idea. He had built up a conclusion from a fact, and he proceeded to act upon it. Dr. Keegan's explanation of the facts implies the possession of mind—that is, reasoning power to a certain extent—just as much as my explanation ; I think the difference betwixt us is only that Dr. Keegan objects to the use of the word "mind." We are all agreed that human intelligence is limited, and animal much more so, and some of us call the more extended mind, and deny it to the less ; others apply it to both. If Dr. Keegan will come to the inquiry not bound hand and foot to a preconceived notion, he will admit that though a dog may be incapable of a sneer, or a horse of solving a quadratic equation, or determining the solar parallax, still both are possessed of what ordinary men mean by the word mind.—*Edwin Holmes.*

KEEPING SERPULÆ.—I have a large shell in my marine aquarium with a great many serpulæ on it, but one by one all the serpulæ have died. Can anyone tell me the probable cause of this ? The other animals are well and happy, so there is nothing wrong with the water. Do they require feeding ? I don't see how it is possible to feed them. Their fans seem to dwindle away until nothing is left of them, when the worm itself dies and has to be removed. I always thought that they were very hardy creatures. Any hints will greatly oblige.—*R. A. R. Bennett.*

THE STORM-GLASS.—Would some reader kindly inform me whether the following recipe for the "Chemical Barometer" or "Storm-Glass" usually sold in opticians' shops is correct. "Put into a tube 2½ drachms camphor, and 11 drachms spirits of wine. When the camphor is dissolved, which it will readily do by agitation, add the following : water 9 drachms, nitrate of potash, 38 grains, muriate of ammonia, 38 grains. Dissolve in water, prior to mixing with the camphorated spirit, then shake the whole together. Cork the bottle well and wax top, but put a little hole in top or cover the top with skin. The above forms an excellent indicator of the change in the weather." I have made several glasses according to the above recipe, but they have all had a heavy white appearance, owing, I think, to the excess of camphor, very different from the fleecy grey matter composing the optician's glass. This thick white matter solidifies, and does not change its position. The optician's storm-glass is sealed at the top with no air-hole. The above leads me to conclude the recipe I have followed is incorrect. Would you kindly supply me with the proper one, and oblige by inserting the recipe in your answers to correspondents in your valuable periodical.—*John H. Milne.*

GOLD-FISH KEEPING.—In reply to Mr. Eaton, I offer my own experience. Having tried the usual way with globe, and changing the water twice a week, without success, I obtained a glass aquarium, 12 inches high, 10 inches wide, and 20 inches long. The bottom is covered with rough sand and small pebbles. A rockery of nice-looking stones is placed at one end, sloping towards the glass. From between the stones, and planted in a small flower-pot saucer amongst some well-washed yellow loam, grow various water-plants. At the other end, and in each corner, is a piece of stone covered with water-moss. When filled up with river water, and having stood undisturbed for some days to allow the water to clear and the plants to take root, the fish may be introduced along with a variety of pond-life, which helps to give the aquarium a natural appearance. The nearer we can copy nature in the aquarium, the more successful are we likely to become. The main object is to have a proper balance between plant and animal life. The plants supply oxygen to the fish, while the latter in return nourish the plants with their dung and carbonic gas. Place the aquarium in a window with a westerly exposure, as too much sunlight is objectionable. A few trumpet snails look well, and serve as scavengers to clear off the vegetation which gathers on the glass and stones. Their spawn, which is a jelly-like substance, adheres to the glass, and is eaten by the fish. Arrowroot biscuit is a good food, and may be given in very small quantities at a time, not more than the size of a pea for three fish; if more be given it is not eaten, and sours the water. Small worms and a fly occasionally are dainty morsels. Select the smallest fish, and four will be quite sufficient for the space I have described. I kept minnows amongst the gold-fish; and out of five gold-fish, two have lived with me over three years. I change the water only twice in the year. If Mr. Easton is a microscopist, he will find this aquarium teeming with an endless variety of aquatic life.—*G. M., Brechin, N. B.*

GOLD-FISH IN GLASS AQUARIA.—There is no difficulty in keeping gold-fishes in glass aquaria that I am aware of, providing they are not overcrowded. Mr. Easton should have his tank or vase (he does not say which) as broad as possible in proportion to its depth. Cover the bottom two inches or more in well-washed sand, plant some vallisneria and anacharis in the centre, arrange some rockwork, beneath which the fish can retreat (otherwise they are apt to worry themselves), fill the aquarium up with water from a brook or river, and do not be constantly changing it. Let the aquarium stand three or four days until the water is perfectly clear, before the fish are introduced, and let these be of small size and few in number. Be careful never to let the direct rays of the sun fall on the glass. By attending to these directions, I have no doubt Mr. Easton will succeed in keeping his fish a long time. I take it for granted he knows how to feed them, so I have said nothing about this.—*Albert Waters, Cambridge.*

MEDICINAL PLANTS.—My query as to medicinal plants has received due attention in the November number of SCIENCE-GOSSIP, pp. 262–263, from three correspondents, and for which I thank them. Their explanation does not elucidate matters, as, for instance, "St. Peter's wort" is referred to different plants by the three correspondents. Bishop's-weed, again, is stated to be a name used for plants with different medicinal qualities. It would not be safe therefore to make a decoction from Robinson's Herbal unless one were sure of the plants.—*S. A. B.*

LOCAL NATURALISTS.—Any of your readers visiting Australia and wishing for conchological, geological, botanical specimens or birds of Victoria, Ent., I should be pleased to see them at 191 Swanston Street, Melbourne, and would be most happy to show them the best localities for collecting, especially tertiary fossils.—*J. F. Bailey, Melbourne, Victoria, Australia.*

QUERY AS TO STRANGE BIRDS.—On the 17th of last month I visited the famous "Warden Landslip" in the Isle of Sheppey, and when near the edge of the cliffs was surprised at seeing a small and very brilliantly-coloured bird flying very close to me. It settled in a tree, but on my attempting very cautiously to approach it, flew off inland. It was not so large as a robin, of a brilliant scarlet, approaching crimson on the head, and shading into orange on the coverts. The wings and tail were brown. Neither my friend, who was accompanying me, nor myself, knew of any English bird answering to this description, and we came to the conclusion that it was a foreigner. Can any reader tell me what this is most likely to have been?—*W. H. Summers.*

WOOD PIGEONS AND OWL.—I am pleased to testify to the truth of the species mentioned in SCIENCE-GOSSIP, feeding on birds. In September a friend of mine at the Werribee on several occasions missed his chickens, and as it appeared to be at dusk they disappeared, he this evening kept watch, and was astonished to see a large bird swoop down into his yard and seize a chick. He struck at the bird and brought it down, breaking its leg. He killed it and sent it to me for preserving; it duly came to hand, when I was astonished to find it to be a beautiful specimen of *Strix flammea*. Reading the notes in SCIENCE-GOSSIP, I venture to hope this may prove of interest to your many readers. I exhibited the specimen last night at our Victoria Field Naturalists' Club, and the facts above mentioned created considerable discussion among the members. Some of our members who have observed the habits of this bird (very scarce out here) are very decided to the fact that it will attack anything when hard up for a feed.—*J. F. Bailey, Melbourne, Victoria, Australia.*

SOAP-BUBBLES.—On pp. 80–81 of "On the Various Forces of Nature," by M. Faraday, edited by William Crookes, F.C.S., there is a description of an experiment with a soap-bubble, whereby, by introducing one end of a glass tube into the bubble, it may be shown "that it has the power of contracting so powerfully as to force enough air through the tube to blow out a light," and an illustration of the experiment is given on p. 82 (fig. 22). Would some one of your numerous readers kindly inform me how the experiment should be conducted? I can of course blow a soap-bubble on a plate, and insert one end of a glass tube as directed, but my bubbles are so obstinate they will not contract. Why is this?—*H. J. G.*

NOTICES TO CORRESPONDENTS.

TO CORRESPONDENTS AND EXCHANGERS.—As we now publish SCIENCE-GOSSIP earlier than heretofore, we cannot possibly insert in the following number any communications which reach us later than the 8th of the previous month.

TO ANONYMOUS QUERISTS.—We receive so many queries which do not bear the writers' names that we are forced to adhere to our rule of not noticing them.

TO DEALERS AND OTHERS.—We are always glad to treat dealers in natural history objects on the same fair and general ground as amateurs, in so far as the "exchanges" offered are fair exchanges. But it is evident that, when their offers are simply disguised advertisements, for the purpose of evading the cost of advertising, an advantage is taken of our *gratuitous* insertion of "exchanges" which cannot be tolerated.

WE request that all exchanges may be signed with name (or initials) and full address at the end.

T. BOYLE.—White varieties of blue flowers are not uncommon. In every instance they may be regarded as reversions to an ancient style of colouring.
C. F. OAKLEY.—The "Marine Objects" sent are not animal, although they much resemble Bryozoa at first sight. They are undoubtedly some kind of Algoid growth, but we have been unable to detect the species.
J. W. R. AND OTHERS.—Write to Mr. Thomas Bolton, 57 Newhall Street, Birmingham, who, we have no doubt, will be able to supply you with aquarium objects, or tell you where you can procure them.

C. STICKLAND.—Dr. M. C. Cooke's "Handbook of British Fungi," 2 vols., published by Macmillan, contains a full description of all known species, and also their classification.

R. TODD (Wigan).—A capital little book has just been published, price 2s. (Thomas Laurie, 31 Paternoster Row) which would meet your requirements—"Chemical Analysis for Schools and Science Classes," by A. H. Scott-White, B.Sc., &c.

J. F. M.—It is intended to publish "Our Common British Fossils, and Where to Find Them" early in 1884. Many thanks.

"OUIDA."—We fear many of our novelists are better acquainted with the names of our great scientists than their discoveries—hence the confusion.

R. A. BILHALD.—The micro-fungus on the leaf of Sowthistle (*Sonchus arvensis*) is the *Coleosporium sonchi-arvensis*.

ROBERT F.—Townsend's admirable "Flora of Hampshire," recently issued, would supply you with all the information you seek.

FRED LEE.—Your specimens are (1) the Sea Fir Coralline (*Sertularia abietina*); (2) Sickle Coralline (*Hydrallmannia falcata*); (3) Bottle-brush Coralline (*Thuiaria theria*); (4) the Sea-Mat (*Flustra foliacea*), a polyzoan; (5) on edge of seaweed, *Laomedea dichotoma*. See Taylor's "Half Hours at Sea Side," price 4s., chapter on 'Corallines.'

J. P. (Exeter).—The polished specimen contains a fossil (very common in the Devonian Limestone, of the neighbourhood of Torquay) formerly set down as a coral, but now generally regarded as a Calcareous sponge, *Stromatopora concentrica*.

K. A. D. (Redditch).—Your leaves are the nurseryman's variety, or a laciniated form of the Maple.

EXCHANGES.

WELL-mounted objects for leaves of *Drosera rotundifolia* (sundew), also spicules or infusorial earths, quantity preferred.—R. M., 59 Hind St., Poplar, London, E.

WANTED, good section cutter, also back vols. of SCIENCE-GOSSIP, in exchange for Valentine's knife, and first-class micro slides.—C. S. Boutfell, 3 Chestnut Villas, Forest Gate, London.

OFFERED, skins of magpie, rook, starling, robin, greenfinch, sky-lark, blue tit, wren, swallow, hedge sparrow, &c., for eggs of hawk, tits, seacoast birds, &c.—Albert Newton, 24 Rycroft Place, Ashton-on-Lyne.

OFFERED, post-tertiary and raised beach shells, *Lucinia borealis*, *Cardium Norvegicum*, *Astarte compressa*, *Tapes virgineus*, *Mya truncata*, *Tectura virginea*, *Puncturella Noachina*, *Trochus tumidus*, *Trophon clathratus*, *Venus fasciata*, &c. Wanted species of the British genera, Defrancia, Cyliclina, Scaphander, Modiolaria, Fusus, Odostomia, Isocardia, Pandora, Poromya, Neæra, Thracia, Pinna.—J. Smith, Kilwinning, Ayrshire.

WANTED, succinea from all parts of England, Scotland, and Ireland. Will give in exchange land and freshwater shells, either British or European.—J. Fitzgerald, 10 West Terrace, Folkestone.

WELL-mounted slides of sections of *Trichomane radicans*, *Hymenophyllum demissum*, with other British ferns; also slides of section hop bine, mostly double stained, for other well-mounted slides, parasites preferred.—A Norris, Church Road, Urmston.

Unio tumidus, *U. pictorum*, *A. anatina*, *A. anatina*, var. *ventricosa*, from Grand Junction Canal, either alive or cleaned out shells. Many common shells wanted, especially marine. Mrs. S., 21 Loudon Road, Brentford, Middlesex.

A SPLENDID collection of Swiss, Pyrenean, and Mediterranean plants, in the most perfect condition as to dying, &c., and all correctly named. Price 6d. each specimen. Address Dr. B., care of Editor of SCIENCE-GOSSIP.

FORAMINIFEROUS sand from the Levant, and from West Indies; wanted in exchange other micro-material, and well-mounted slides.—J. L. Smithett, 45 Highbury Hill, London, N.

FORAMINIFEROUS sand from Mediterranean Sea (being the washings of sponge), will be forwarded to any reader sending me a stamped and directed envelope for same.—F. A. A. Skuse, 143 Stepney Green, E.

I WANT to correspond with conchologists (home or abroad) with the view of exchanging marine shells, of which I have a large number of species.—C. Jefferys, Langharne, Carmarthenshire.

SCIENCE-GOSSIP for 1881-3, what offers? books or otherwise.—E. H. Smith, 5 Hurley Road, Lower Kennington Lane, London, S.E.

FORAMINIFERA, selected and well mounted in exchange for foraminiferous material, parasites, &c.—A. C. Tipple, 35 Alexander Road, Upper Holloway, N.

WANTED, Stylopæ, various species. Well-prepared slides of rare Acari in exchange.—H. E. Freeman, 60 Plimsoll Road, Finsbury Park, N.

ONE dozen good micro slides in exchange for others. Lists exchanged. Unaccepted offers not answered.—F. R. Rowley, 31 St. Stephen's Road, Highfields, Leicester.

WANTED, books on bees, old or modern. Will exchange books of general literature, natural history and sporting, or will purchase them.—W. T. Cooper, 16 Earl's Court Road, Kensington, W.

WHAT offers for vol. xvi. 1880 SCIENCE-GOSSIP?—W. E. Collinge, 68 Springfield Place, Leeds.

DUPLICATES. *Limnæa stagnalis* and *P. corneus*, very fine; *P. complanatus*, *P. spirorbis*, *P. carinatus*, *Physa Hypnorum*, *L. palustris*, *L. peregra*, and *Sphærium corneum*. Desiderata.—W. Hewett, 26 Clarence Street, York.

FOSSILS, from Greensand and a few Lias to exchange for other fossils, books on scientific subjects, or fossil cabinet.—J. A, Floyd, 4 Springfield Villas, Bury-St.-Edmunds.

FILMY ferns of New Zealand for the filmy ferns of other countries.—T. Rogers, 27 Oldham Road, Manchester.

WANTED, a thoroughly good two-inch objective. Will give good one-inch in exchange. Also wanted Kent's Infusoria for cash.—E. B. L. Brayley, 13 Burlington Road, Clifton, Bristol.

RARE British plants sent in exchange for local and common British moths—one large exchange preferred.—Wm. R. Hayward, Wingfield House, Birdhurst Road, Croydon.

LIVING specimens of freshwater shells suitable for aquariums, *L. stagnalis* and *Planorbis corneus*. Would be glad for offers of British birds' eggs, side blown, or land and freshwater shells.—Robert Barker, 11 Fowend Street, Groves, York.

WANTED, Bell's stalk-eyed, Spence Bates' sessile-eyed Crustacea, and Baird's Entomostraca. A valuable assortment of Crustacea, Echinodermata, Mollusca, Fishes, Rock specimens or embryological microscopic preparations offered in exchange.—Edward Lovett, 43 Clyde Road, Addiscombe, Croydon.

WANTED, British marine algæ, mosses and lichens, mounted. Will give land and freshwater shells.—C. T. Musson, 1 Clinton Terrace, Derby Road, Nottingham.

WANTED, micro material for first class mounted specimen of diatom insects, anatomical, pathological, and other specimens.—J. Noakes, Clairville, Archway Road, Upper Holloway, N.

TOBACCO, leaf well mounted, in exchange for other slide of interest.—E. C. Stedman, 115 Stepney Green, London, E.

FOR sand from Mediterranean Sea containing minute shell please send stamped envelope to E. C. Stedman, 115 Stepney Green, London, E.

WHAT offers in micro aparatus for Carpenter's "Microscope" and "Manual of Physiology," both as good as new?—G. E. Ward, Wallwood Nursery, Leytonstone.

BOOKS, ETC., RECEIVED.

"The Pedigrees of Plants." By Grant Allen. London: Longmans & Co.
"Injurious, &c., Insects of New York." Albany, Weed & Co.
"The New Principia." By Newton Crosland. London: Trubner & Co.
"The Parallel Roads of Glen Roy." By James Macfadzean. Edinburgh: Menzies & Co.
"Solar Physics." By A. H. Swinton. Londou: W. H. Allen & Co.
"Journal of the Royal Microscopical Society." December.
"Perennial Plants." (Parts.)
"Studies in Microscopical Science," edited by A. C. Cole.
"The Methods of Microscopical Research," edited by A. C. Cole.
"Journal of Conchology."
"Land and Water."
"The Science Monthly."
"Midland Naturalist."
"The Inventor's Record."
"The Truth Seeker."
"Ben Brierley's Journal."
"Cricket."
"The Medical Student." (New York.)
"Natural History Notes."
"Science."4
"American Naturalist."
"The Microscope."
"Canadian Naturalist."
"American Monthly Microscopical Journal."
"Popular Science News."
"The Botanical Gazette."
"Revue de Botanique."
"La Feuille des Jeunes Naturalistes."
"Le Monde de la Science."
"Ciel et Terre."
"Cosmos: les Mondes."
&c. &c. &c.

COMMUNICATIONS RECEIVED UP TO 10TH ULT. FROM :—H. G. G.—C. H. R.—J. F. R.—E. T. D.—J. G.—A. H. B.—G. B.—W. L. B.—J. D.—H. W. K.—S. A. B.—Dr. P. Q. K.—W. T. H.—W. F. H.—W. L. W.—C. R.—A. M. P.—A. H.—C. H. W.—H. C. W.—C. E. F.—W. H. S.—E. T. B.—L. M.—B. H.—H. G. S.—J. W.—A. H. W.—J. E. A.—Dr. A. C. S.—H. W.—R. M.—Dr. J. F.—W. B. G.—A. N.—C. S. B.—J. S.—E. L.—W. B. G.—W. H. B.—J. B.—J. F. G.—A. N.—J. R. D.—J. F. B.—B. H.—A. O.—H. J. G.—T. M. R.—E. L.—E. B. L. B.—W. M.—E. H.—W. R. H.—R. B.—C. S.—W. H.—E. J. G.—W. M. C. P. N.—J. S. K.—E. C. S.—J. W.—C. T. M, &c.

GRAPHIC MICROSCOPY.

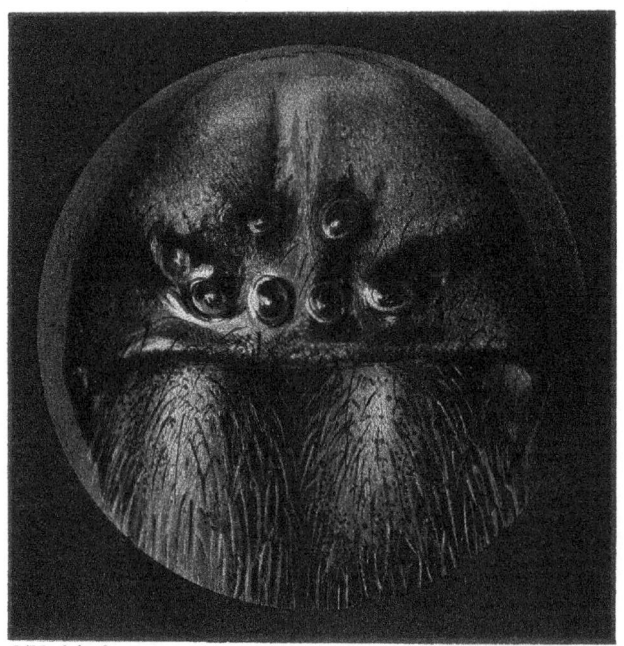

E.T.D. del.ad nat.   Vincent Brooks,Day & Son, Lith.

EYES OF EPËIRA CONICA.
× 30

GRAPHIC MICROSCOPY.

BY E. T. D.

NO. II.—EYES OF EPEÏRA CONICA.

THE Arachnida class is one so elevated above the Articulata, as to constitute the highest condition of the Invertebrata. It has a wide range, extending from the microscopic Tardigrada (water bears), the parasitie Demodex, the Acari, and spiders; reaching the organised Mygale and scorpions.

The Araneidæ, or Spiders proper, form one Order of this class, separated into two tribes: 1. Octonoculina (eight eyes); 2. Senoculina (six eyes), divided into families, genera, and species.

Epeïra conica, the subject of the plate, is a species of the Family Epeïridæ of the Tribe Octonoculina, and fairly represents the popular idea of the entire Order, as to them especially belong the webs, so well known, formed by precise geometrical rule radiating from a centre, and connected at regular intervals by threads.

Interesting as it would be, space does not admit touching on the general anatomy of the spider, but it may be comprehensively stated, that the differences of structure between spiders and insects, consist in the continuation of head and chest in one mass, no articulation of the body into rings or segments, the number of legs, the absence of antennæ, the structure of the mouth and palpi, and particularly the presence of none but simple eyes; these points at once elevate and characterise the spider above the insect.

The eyes are placed in front of the cephalothorax, or forehead in English species; they are invariably six or eight, varying much in size, in the manner of grouping and position, conditions depending entirely on the habits of the species, and especially affected by the necessity of catching prey, under the many diversities of locality in which they are found.

In spiders, the eyes or ocelli are grouped in the middle of the anterior part of the cephalothorax; the position of the four median ones, as shown, is generally in a square with two pairs outside; any variety in the arrangement depends on the habits of the species, and constitutes an important characteristic feature in their classification. Amongst many specialities, those which hide in obscurity have the eyes aggregated in a close group in the middle of the forehead (Clubiona), those which inhabit short tubes terminated by a long web have the eyes separated and spread over the front (Philodromus); those which lie in ambush, in and under leaves, and in fissures in walls, have the eyes on a prominence, permitting a slight divergence (Theridion), and in the wandering, hunting, and ground spiders, the eyes are large and grand beyond description, studding in an even row the front margin of the cephalothorax. In a well-known large species with zebra-like markings (*Salticus scenicus*), which by sudden leaps, seizes its prey; on the edge of the forehead are four "oculars," touching each other in a parallel line, two in the middle, of such dimensions and startling appearance, that when arranged under reflected illumination, carefully focussed from the side speculum, they present a sight when the words "ghastly" and "beautiful" may be appropriately associated, for they gleam with weird-like cruelty. These eyes have been minutely investigated, and present a type of structure so far superior to those of insects, as to admit of comparison of similar parts in vertebrate animals; a cornea, a globular lens, aqueous and vitreous humours, a retina and choroid; all are found in nearly the same positions, so that the sense of vision in the Araneidæ must be, as many must have noticed, extremely quick and keen; and like the eyes of cats, they are said to gleam in the dark.

The object is sufficiently popular to be easily obtained, and under good illumination singularly attractive; it may be quickly arranged by gumming the head on a cork slide, as soon after death as possible, and immersed in a deep cell, in balsam, without pressure it may be permanently preserved.

It is possible to paint this portrait from life; a cone of pasted paper rather larger than the specimen, with the apex cut off, makes a suitable receptacle. A vigorous spider will soon project its head through the aperture. When in this position it should be blocked behind with cotton-wool slightly wetted, the arrangement is gummed to a slip apex upwards, and a drawing may then be carried forward with leisure for a considerable time. Spiders are very tenacious of life; Mr. Blackwell, the greatest authority on the subject, kept one alive, without food, for nineteen months, fifteen days, a result necessarily depending on the gastric cœca and stores of fat which contributed to the powers of endurance of such a prolonged fast.

Many insects for the observation of facial movements can be arranged in the same way, and such front views admit of interesting and extended study, the action of the antennæ, palpi, and various organs of the mouth may be watched, and curious effects may be produced by the excitation of saccharine, or nitrogenous juices, administered from the tip of a sable pencil. Mr. W. J. Slack has devised something similar as a permanent accessory; a "tubular livebox" (See "Journal Microscopical Society," Dec. 1883, p. 906), by which insects may be watched imbibing syrups smeared on the under side of a covering glass.

The drawing represents a type of objects for the microscope obtainable without professional preparation, or purchase, and eminently consonant with the admiration and skill of an artist.
Crouch End.

OBSERVATIONS ON DOMED NESTS.

ANY ONE who has given any attention to the nidification of birds must have been struck with the great diversity of form, and style of architecture of nests. For within our own limited avifauna we find all kinds of nests, ranging between a cavity in the sand to the open, hemispherical, decorated nest of the chaffinch, and to the closed and elaborately-formed domicile of the long-tailed titmouse. No two birds, no matter how nearly allied in structure they may be, build exactly alike. And yet all these diverse nests are constructed for one identical purpose—the rearing of the young. It would be hopeless to attempt to inquire into the circumstances which originally caused these diversities. The primary cause or causes which gave rise to certain habits may have been swept away thousands of years ago, still we have the effects in the continuation of the habits, with more or less variation. We may assume that each variation that we see, has been brought about for the benefit and perpetuation of the species, though a particular form of nest, or colour of an egg, may not be of the same use to the bird now as it was formerly. In long-past ages certain birds might have built peculiar nests to guard against the attacks of a certain specific enemy, but should circumstances have arisen to cause a decline, or the extinction of that enemy, the particular form of nest would not then be an imperative necessity to the bird. It is impossible to say how many ages a bird would go on guarding and defending itself against an enemy that had ceased to exist, but we may conjecture that the withdrawal or removal of a certain specific, destructive enemy would, sooner or later, cause some alteration in the habits of the bird.

Those birds that exist amid a host of enemies are more intelligent than those that have few enemies, hence new habits would sooner arise, and new and more advantageous forms of nests be necessitated and constructed among the former class than the latter. The rook, the sparrow, and the wren, living in genial, cultivated, or partly wooded regions, have a far greater number of enemies than the grouse or the ptarmigan on the mountain tops.

With us in Britain the domed nest would seem to be the most convenient kind of nest, and the most conducive to the well-being of the species which adopt this style of architecture; for the dom$_e$ is a protection against rain and wind, is an effectual screen and concealment against egg-destroying birds, and also insures uniform warmth to the eggs and nestlings. And these advantages seem to be apparent, for although we have only ten species that construct domed nests, including the house-martin, these species are numerous in individuals, and general in distribution.

It is impossible to say what first occasioned the construction of domed nests, but we may safely assume that no species would jump all at once from the construction of an open nest to that of a covered one. We find that with the small birds—birds that cannot defend themselves by main force against their enemies, concealment is one of the chief things aimed at in nest-building, and we may surmise that this object alone would induce, firstly, a placing of the nest amid dense herbage or foliage, then a covering of the eggs, and then an arching of the nest. But it must not be overlooked that the arched nest is not without disadvantages, for the sitting bird is unable to see the approach of an enemy, and the hole in the side is just large enough to admit the paw of a cat or the head of a snake, and thus facilitate as it were the capture of the occupant or occupants.

All the dome-builders lay conspicuous eggs, and this might be one reason for the necessity of a screen. It is true that many of the small birds that build open nests lay conspicuous eggs, but none of their eggs are perfectly white. Those birds that lay conspicuous eggs in an open nest find it necessary to sit upon the eggs or young to hide them by their dull-coloured upper plumage, and thus elude the eye of the passing jay or hawk.

Instances of the observation of nests in a state of transition between the open and the covered condition have been rare, but one such example is supplied by an American species. That acute ornithologist, Wilson, describes the nest of the towhe bunting as being "formed outwardly of leaves and pieces of grape vine bark, and the inside of fine stalks of dried grass, the cavity completely sunk beneath the surface of the ground, and sometimes half-covered above with dry grass and hay." And he states further that "the bird is remarkable for the cunning with which he conceals his nest." Mudie informs us that sometimes the robin covers its eggs, and occasionally constructs an arch of leaves where concealing foliage is wanting, under which it passes to its nest; he also states that the goldcrest builds an open nest in pine forests, but a domed one in other districts, where presumably enemies are more numerous. He further remarks that the long-tailed titmouse does not uniformly confine itself to one style of nest, but often constructs two entrances or means of exit instead of one, the nest having a backdoor for escape. ("Feathered Tribes of the British Islands.") These are deviations, but deviations may not always indicate progression, as many circumstances, the withdrawal or destruction of some peculiar enemy, for instance, might occasion a reversion to some former and simpler type of nest. Possibly the habits of one portion of a species might undergo change in a particular region, such change being induced by peculiar circumstances. It has long since been pointed out that birds inhabiting pine forests are different in structure from birds of the same species inhabiting oaken or beechen forests.

Nearly all birds that have acquired a habit of building in darkened places lay conspicuous eggs. Many birds build deceptive nests; many produce eggs that assimilate in colour with the lining of the nest; many sit upon the eggs to guard them; and some cover their eggs, but none of these means of concealment and security seem to be equal to the domed nest. Mr. Wallace, as quoted by Darwin, gives his opinion that domed nests were originally built to hide birds of brilliant plumage while sitting. In Britain this theory would scarcely hold good now, as none of our dome-builders possess conspicuous plumage* though we cannot tell what kind of plumage they might have possessed when they first began to build domed nests.†

Of fifty of our smallest insessorial birds (omitting very rare species) that build ordinary exposed nests twenty-five are uncommon; of fifteen that build in darkened places six are uncommon; of ten that build domed nests none can be called uncommon: so that it would appear that birds that build in darkened places have advantages over those which build an ordinary hemispherical nest in exposed places, but those which build domed nests have still greater advantages, as, by the aid of their domed nest and general mode of nidification, every species is able to keep up its numbers.

No classification can be grounded on architecture, or on the colour of eggs. Of the ten birds above mentioned the "three willow wrens" have a close kinship, but the others differ much, being included in widely separated genera; some have long tails, some have short tails, some lay white eggs, some lay spotted eggs, some build on or near the ground, some at a considerable height, the only thing that is common to them all is the vaulted nest. Yet one cannot but imagine that these birds, along with others which may have withdrawn from the British Isles, have been originally subject to common, but, perhaps, peculiar influences, to determine them to take action in one respect nearly all alike.

However near two species may be, or appear to be in structure, the marked differences we find in architecture and in the colour of eggs indicate that a long time has elapsed since the two diverged from a common parent. On the other hand slighter differences in nests and eggs almost certainly indicate that a shorter time has elapsed since the period of divergence. For instance the three willow-wrens are near together, they are similar in size and plumage, they build domed nests and lay spotted eggs, and the two sparrows are similar in plumage, build domed nests and lay spotted eggs. From these circumstances we may infer that a shorter period has elapsed since the three willow-wrens or the two sparrows began to diverge, than that which has elapsed since the divergence of two species like the swallow and the house-martin, or the thrush and the blackbird; birds which though much alike in structure build very different nests and lay very different eggs.

In regard to superiority there are, in my mind, four grades of birds: the water birds, the wading birds, the scratching birds, and the perching birds, all succeeding each other in intelligence. So far as British birds go, our ten dome-builders stand in the front, as being more artful as designers, adapters, and executors than the rest of their kin. If all the nests of British birds could be arranged in a chain the simple cavity in the sand would stand at one end, and the chef-d'œuvre of the long-tailed titmouse at the other.

GEORGE ROBERTS.

Lofthouse.

---

* The magpie excepted. This bird preys on eggs of other birds, but it has sufficient intelligence to erect a screen to protect its own.
† It is interesting to notice here that the American quail that lays white eggs has a domed nest, but our quail that lays spotted or clouded eggs has an open nest.

FOSSIL PLANTS IN THE SILURIAN FORMATION NEAR CARDIFF.

By W. H. Harris.

ABOUT two miles from Cardiff, just on the borders of Monmouthshire, lies the little village of Rumney, with the river from which it is named winding around the foot of the hill, and in its course discovering the best exposure of the Silurian beds obtainable in this neighbourhood. In the hillside is a quarry which is occasionally worked, and furnishes a very compact stone of a fine grit character, while the overlying beds are of the usual friable nature and break up more readily. Beautiful examples of ripple marking were to be seen at one time in some of the beds that had been exposed to the action of the atmosphere, but they have been removed, owing to the recent working of the quarry.

In these soft beds are frequently found black patches of carbonised vegetable material, usually of a lustreless and powdery character, while occasionally other patches are bright and compact, being in fact converted into pure coal, but which break up into small fragments having a cubical fracture, and like ordinary coal is worthless for any practical purpose under the microscope, perfect mineralisation being entirely destructive to all structure. Occasionally these patches bear some fanciful resemblance to a leaf or frond, but usually they are mere films of an irregular size and shape.

Associated with these remains, in some beds, are numbers of the little spherical bodies ranging in size from a dust shot to a small pea, which are known by the name of *Pachytheca sphærica*, Hooker.

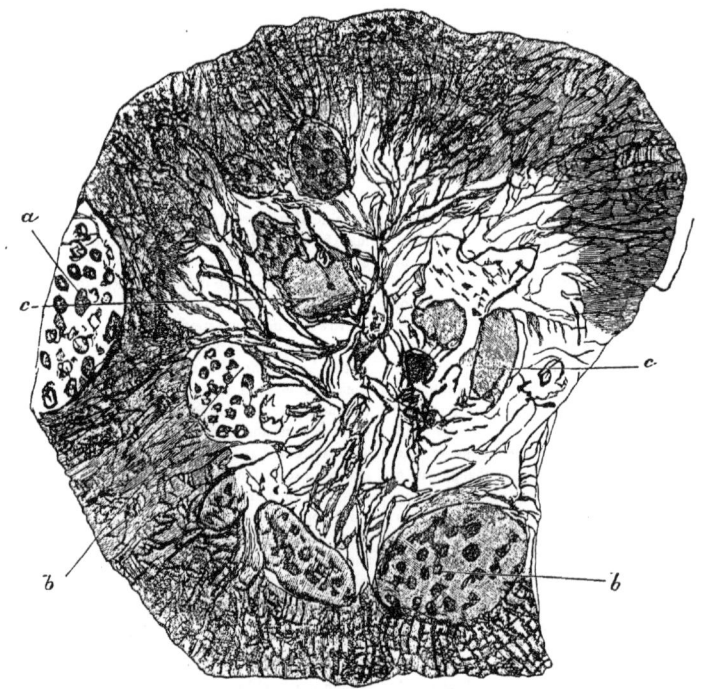

Fig. 21.—*Pachytheca sphærica*, Hooker. × 30 diam. *a*, supposed point of attachment to parent plant; *b*, spherical bodies, probably spores; *c*, cavities containing fine particles of the enclosing rock.

These are well preserved externally, but to obtain the interior detail is a much more difficult matter.

Some years since, in company with my friend Mr. Storrie, the present curator of the Cardiff Museum, we were fortunate enough to discover a fragment of this carbonised material, which at once gave indication of being useful in giving an insight into the nature of the material above described. On carefully sectioning the fragment (for such it was), we were rewarded with a slide or two of what is possibly the oldest vegetable matter, showing the structure perfectly well preserved, yet discovered, and although it cannot claim to be ranked among the beautiful, yet it compensates for this by its rarity and venerable age.

It has lain in my cabinet for some years, in the hope that I might make further discoveries from the same source, but after many long searches and subsequent labour in preparing the material I have never done more than verify the typical slide I am about to describe. That it is the type of the prevailing vegetation of the period I have no doubt, as I have frequently found traces of the same structure, but always in such a bad state of preservation that without the knowledge obtained from this identical slide it would have been difficult to make out its detail at all.

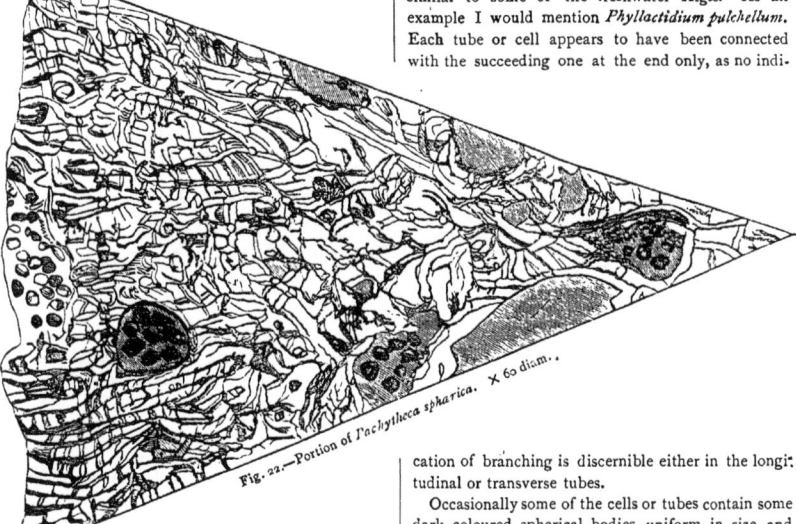

Fig. 22.—Portion of *Pachytheca sphærica*. × 60 diam.

Description.—The piece of fossil vegetable, as mounted, measures two tenths of an inch long by just one-tenth in its widest part; it is somewhat angular in shape, and, for the reasons presently to be stated, appears to be an almost transverse section of the original plant. In colour it resembles a piece of seasoned oak. It was exceedingly friable and is traversed by many fine fractures. The preservative material has doubtless been protoxide of iron, but I cannot help thinking the perfect preservation of the structure has been in a great measure due to accidental protection from the great pressure these plants must have suffered, as in one part a perfectly structureless portion appears, although from the size of the cells and the general surroundings, the structure should not have been absent. I have found the same occur in sections of Sigillaria from our coal measures, and this induces me to think this fragment received its greatest pressure at this point, and thus comparatively relieved the immediately surrounding parts from the pressure necessary to destroy structure.

It appears to consist entirely of a vast number of simple tubular cells, having a very thick wall which by its slightly darker colour clearly defines the boundary of each. They are mostly exhibited as cut obliquely transverse, but the few that are discovered having a longitudinal section measure about $\frac{1}{100}$ of an inch in length, by $\frac{1}{1000}$ in breadth, they terminate abruptly, and appear to be separated from the succeeding cell by an exceedingly thin division, similar to some of the freshwater Algæ. As an example I would mention *Phyllactidium pulchellum*. Each tube or cell appears to have been connected with the succeeding one at the end only, as no indication of branching is discernible either in the longitudinal or transverse tubes.

Occasionally some of the cells or tubes contain some dark coloured spherical bodies, uniform in size, and apparently arranged in a double row. I have counted twenty in a cell $\frac{3}{1000}$ by $\frac{1}{1000}$. Sections of these bodies are seen in other cells, but their minuteness defies the limit of my instrument to detect structure in any. I look upon these as being the germs of future plants, had not mineralisation interfered with their development.

In *Pachytheca sphærica* we have a much more complicated structure, but still bearing a strong resemblance to that of the plant above described. The tubular structure is exhibited here, with the same absence of constriction at the ends of the cells, but instead of being connected simply with each other by their ends they branch off at right angles in all directions, leaving small interstices between.

Towards the boundary of the sphere these tubes become compacted closer together, and probably gave a tolerable degree of solidity to the organism. Ramifications of the same arrangement form the centre of the object, but much more loose in texture. In some cases cavities of an irregular shape appear filled only with the fine particles forming the rock, and which polarise feebly. Interspersed throughout the object are a large number of spherical bodies, varying in size from ·001 downwards. In some cases they are quite transparent, in others, opaque, being filled with iron oxide, while in others a mineral which under polarised light behaves like glauconite fills the cavity.

Remarks.—It would be difficult from the evidence afforded by these sections to assign the correct habitat of this plant, whether terrestrial or aquatic ; in itself it is a mere fragment, and if of terrestrial origin may give us an insight into only a small part of the plant, from which it was derived ; but if this were the case, I am disposed to think other portions would have been met with ere this, as the rocks have not been neglected from which this specimen was procured.

If, on the other hand, it was of aquatic origin, we are more likely to have the entire detail of its structure present, seeing that most Algal forms are simple in their structure. In *Pachytheca sphærica* we find the same general character, the elaboration of branching cells or tubes being doubtless to give consistency to the covering of the contained spores.

If *P. sphærica* is the fruit of a plant, as doubtless it is, and the bodies described in the first section are really germs or spores, I should hesitate to believe they were different parts of one plant. Pachytheca extends higher in the series of beds than the one from which my first described section was taken, and this would agree with the fact that they are more specially organised.

The remains having been deposited in shallow water, and the general simplicity of structure, inclines me to think they are both of algal nature, but that in

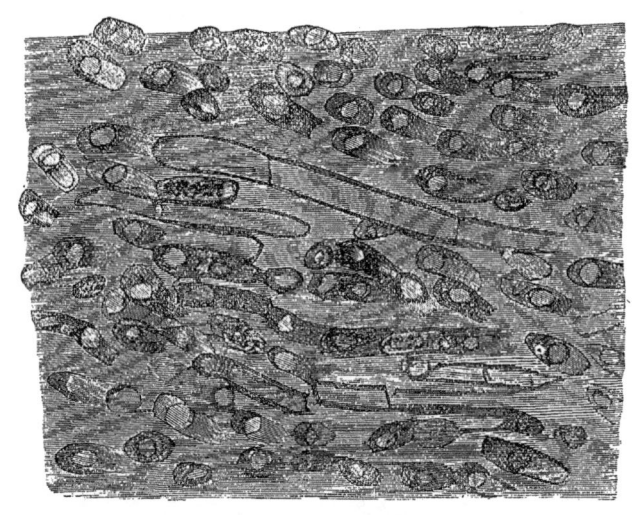

Fig. 23.—Fossil plant from the Silurian at Rumney, near Cardiff. × 140 diam.

this unknown and as yet unnamed specimen we have one of the pioneers of the vegetation that now so abundantly flourishes on land and in the seas.

Cardiff.

A WATER-SPOUT.—On the 17th November, my wife, and son (who has been round the world four times) saw near the Flat Holmes, Burnham, a waterspout, or rather, I should say, a huge column of black cloud revolving rapidly. It tore up the water in a most astounding way. Its course was from northwest to north-east, then to south-east and south-southeast. They saw it a little before twelve. Its course was so rapid that my son said it went faster than the Flying Dutchman on the Great Western Railway. It reached Yeovil, Dorset, a little after twelve, and the damage done there and elsewhere was enormous. The roar was deafening, as a man said, like a lot of express trains running through a tunnel. Soon after a terrific squall came up from the west and north, with tremendous rain and hail. On November 19th a heavy thunderstorm with vivid lightning passed over the Mendips and beyond them, travelling from north to east, and from east to south-south-east at night.—*A. H. B.*

RECOLLECTIONS OF AUSTRALIAN ENTOMOLOGY.

By W. T. Greene, M.D., F.Z.S., Author of "Parrots in Captivity," &c. &c.

IF Australia has occasionally been spoken of in a more or less disparaging manner, because of the paucity in actual numbers as well as in species, of its Mammalia, and in a less degree of its birds, the same cannot be said of its insects; whose tribes, flying, crawling, and swimming, absolutely defy computation; as anyone who has ever resided in, or travelled for a couple of days through, the "bush," is but too well aware.

I say "too well," because they are unpleasant, some of these insects—exceedingly so, in fact, to every sense. Some are positively terrifying, so strange and weird, not to say unnatural, are their forms, while others are dangerous in the highest degree, bearing, as they do, almost certain destruction to their adversaries in their heads or tails. Others, again, are interesting from their habits, others from the periodicity that marks their appearance and disappearance; others from their peculiar forms, which simulate twigs, grass, or leaves; others from the gorgeous livery they wear, and others again from all the above peculiarities combined.

I propose, in this article, to consider briefly a few of the more remarkable species of insects with which I became acquainted during a sojourn of some years in the great southern land, beginning with that widely diffused persecutor of "new chums," the mosquito.

I remember some years ago, reading a story that went the round of the papers at the time, to the effect that a colonist sent one of these pests home in a letter to a friend in England, and that the friend while reading the missive, which informed him of the nature of its enclosure, felt a sudden sharp prick on the back of his hand, and glancing at the injured part, perceived a small golden fly that immediately flew away and escaped.

It is unnecessary for me to dilate upon the improbability of the story. Of course, it was utterly impossible for the insect—a particularly fragile creature—to have survived the pressure to which it must necessarily have been subjected in the mail-bag, and unfortunately for the narrative, the mosquito, far from being a "golden fly," is very plainly dressed, being, in point of fact, neither more nor less than a first cousin to the well-known gnat, so familiar of an evening to the rambler through the green lanes of old England, where it may be seen disporting itself in myriads, beneath the shade of the overhanging trees and hedges; the only difference being that our native mosquito seldom bites, while its Australian congener is one of the most bloodthirsty little abominations in existence.

During the great heat of the day, at our antipodes, the mosquito wisely keeps himself under cover; but once the sun has disappeared beneath the horizon, out he comes from his retirement with keenest appetite, and pounces with unerring aim upon his prey. You may shut up your tent as closely as you please, you cannot keep him out; you may hang mosquito curtains round your bed and fancy yourself free from his attacks; pooh! he laughs at your vain precautions, and no sooner have you extinguished your candle and settled yourself down—as you think—for a comfortable sleep, than "bizz!" the awe-inspiring sound is heard in painful proximity to your ear, and presently a sharp prick, possibly on the end of your nose, announces that war has actually begun. You have nothing for it but to relight your candle, and hunt your foe to death; unless you prefer allowing him to sate himself uninterrupted with your blood; which done he will retire to rest, and be found next morning clinging to your curtain, a bloated little vampire, too heavy to fly, when he will fall an easy prey to your avenging fingers. There is one drawback, however, to this course of proceeding; the longer the mosquito sucks, the larger and more painful will be the tumour that arises round the puncture he has made; so, as I remarked before, you must declare war, and war to the knife at once, with your tiny but implacable foe.

The mosquito neither bites during the middle of the day, nor the middle of the night, but just before and after sunrise and sunset he is on the alert, and positively ubiquitous. The deepest shaft at Ballarat or the closest room in Melbourne are alike familiar with his presence; the margins of rivers and creeks, and clumps of bush, fifty or more miles distant from water, he frequents them all alike; town and country are the same to him; mountain or valley, wooded plain, or table-land, he has no more predilection for the one than the other; nay, he has even been met with ten miles out at sea! If he cannot get at you by any other means, he will be down upon you through the chimney; and if that is stuffed up, which can scarcely be done without placing you in some danger of suffocation, it will go hard with him if he cannot find an entrance somewhere, for a pin-hole will afford him ample scope for ingress where there is English blood to be sucked, or French either, for that matter, for he is not hard to please.

The "old hands" declare that he only bites "new chums," as they term the recent arrivals in the colonies. All I can say is that he feasted upon me as eagerly at the close as he did at the beginning of a six years' residence; but I suppose that lapse of time did not entitle me to consider myself an old hand; the mosquitoes at least, did not seem to think it did.

Another almost insupportable pest are the flies, which are so numerous and troublesome that one could imagine the fourth plague of the old Egyptians to be endemic in Australia. Be that as it may, these abominable insects, which vary in size from the tiniest little blue midge imaginable, celebrated for the pertinacity with which it insists upon getting into the corners of your eyes, down to the enormous red, bloated meat-flies, which oblige you to keep the closest watch over your provisions, are almost as bad as, if not worse than, the mosquitoes. It is next to impossible to keep anything out of their reach; I have seen mutton spoiled by them in less than a quarter an hour after the sheep had been killed, and even hard salt junk does not escape their assaults, for they are viviparous, and the maggots begin to feed immediately after being laid, and grow with amazing rapidity.

The intermediate kinds, or sizes, of flies are not particularly different from those we are accustomed to here at home; they are just as inquisitive, familiar, and just as annoying as their European relatives, with whom I really think, they not infrequently intermarry; for in the ship that carried me to England from Melbourne, we had their delightful company all the way, though where they came from no one could make out. They were cunning too, and not to be lured by any bribe of peppered sugar to their destruction, but stuck to us to the last, though they retired from observation during the cold weather at the Horn, to re-appear some weeks before we cast anchor in the Mersey.

It is a curious fact that the farther you go into the bush, the more numerous do the meat-flies become, while they are comparatively scarce in Melbourne; but it is just the reverse with the house-flies, whose name, in the town, is legion, whilst "up the country" they are found in moderate numbers only, and in some places not at all.

Australia does not possess many butterflies. A few grey, and brown, insignificant looking little beings, were the only representatives of that class of insects which I chanced to become acquainted with in the bush; but it has many varieties of moths.

Wonderful creatures are some of these, which, at rest, so exactly resemble a withered leaf, that you would never suppose them to be anything else, unless you chanced to see them move, which they are careful not to do while you are looking on. The only thing that betrays them is the phosphorescent glare of their eyes, which shine even in the daytime, like little carriage lamps, and are positively, tiny meteors in the dark. Another species, a tremendous brown fellow, is very nearly as large as a sparrow, and comes against the window at night with a thump that is almost alarming.

I recollect once meeting with the pupa of one of these giant moths, as it was working its way out of the ground, preparatory to casting off its chrysalis shell, and completing its metamorphosis. I mistook it at first sight, for the cone of some species of pine, and under that impression stooped to pick it up, wondering where such a thing could have come from, in that land of gum-trees and acacias; it was fully five inches in length, and thick in proportion. The moment it felt my hand touching it, it wriggled back into its hole, greatly to my astonishment; however, I proceeded to dig it out with my knife, and must have injured it in doing so, for it bled a great deal after I got it out, a colourless, ichory kind of blood, and never came to anything. I fell in with plenty of the creatures afterwards, and was told by a Cornish acquaintance that they were "Buskum Sneevers," or some such name, which I had never heard of before, and of which I doubt whether I caught the true pronunciation.

Australia possesses several kinds of native bees; and it is a curious fact that these laborious and useful insects—I am now speaking of the European variety— seldom succeed well in that country. They either fly away into some unknown region, or if they remain with their owners, refuse to work. The reason of this strange conduct appears to be that the climate is so fine, flowers so plentiful all the year round, and so large a quantity of "manna" is secreted by several kinds of eucalyptus trees, that they grow lazy. "Why should we toil when we can live comfortably without fatiguing ourselves?" seems to be their mode of reasoning; whether or not, they act as if it were, and lay up no provision for the winter that never comes. Perhaps they are led astray by the bad example of the native bees, which are thorough vagabonds, destitute of sting, leading an erratic, merry life, flitting from flower to flower, and from sweet to sweet, all the day long, taking no thought for the morrow, like the human aborigines of their native land; though unlike them, they have a fixed dwelling-place to which they resort at night.

From bees to wasps, the transition is natural and easy; some of the latter are tremendous fellows—one, especially, a handsome blue insect, with great gauzy wings, is quite two inches in length, and carries a sting a quarter of an inch long, with which it is said to attack small birds.

I was once the spectator of a strange combat between one of these monsters and a large tarantula, which terminated tragically for the latter. The tarantula was quietly walking down the smooth bole of a large gum-tree, probably on the look out for prey. I had had my eye on him for some time, and was meditating an attack, for I had no desire to see him about my premises, when I suddenly beheld him drop, as if he had been shot, or galvanized; I expected he would have fallen to the ground, but he did not: he had thrown out a thread, and swung on it, at about a foot below the spot where I had seen him throw himself down. At the same instant something whizzed past me, and flew straight at the tarantula, which received it in a close embrace that lasted for a second or two; then, releasing its hold, the great spider permitted a blue wasp to escape; it returned to the charge again, however, almost directly, and was again embraced by the tarantula, and again released as before. This was repeated several times, and at last the spider fell to the ground dead, slain by the more potent venom of its adversary, which seemed to have received no injury during the encounter, in spite of the tarantula's powerful arms and formidable jaws.

(*To be continued.*)

A STRANGE VISITOR.

ONE evening, only a few weeks ago, I was resting on the sofa in my drawing-room, thinking of nothing in particular, when suddenly a dim winged shadow seemed to emerge from the wall, and flit in front of my eyes. I am not superstitious; but I confess that I was startled. A London drawing-room in October, at nine o'clock in the evening, with lamp and fire lighted, was the last place in which one would expect to see a ghost; and yet the winged-shape was decidedly ghostly. Had my eyes deceived me? I thought so at first, and rubbed them vigorously. But no—there it was again, the strange winged shadow. And now I looked more attentively, and, as I watched, the dim moving shadow rose higher in the room, and began flitting round the ceiling. Its flight was wonderfully graceful, and almost perfectly noiseless, which added to its uncanny and ghost-like character. And now I began to suspect that my visitor was only a visitor from the other world, in the sense that October was late in the year, and a London drawing-room a strange place for its appearance. I began to suspect that I had seen similar ghosts before in the other world of June lanes, and August meadows, flitting about cheerily in company with ghost-moths, and watchman-beetles. In other words, I surmised that my visitor was a bat.

While I had been pondering on his nature, my ghost had continued flitting round the room, occasionally pausing to rest on the cornice. Birds not uncommonly enter rooms, and I was not sure at first whether my visitor might not be a belated bird. But, when I listened carefully, a peculiar faint shiver or rustle that accompanied the creature's flight showed me that it could not be a bird. A bird's wings have not this weird, crape-like sound. Moreover, a bird entrapped in a room is always wild with fright, and dashes itself against the walls and windows. My little visitant was not frightened at all; and his flight was more like a large moth's than a bird's, and the skilful way in which he avoided the furniture in the room showed that he could see well in a dim light. So he was undoubtedly a bat.

The next point was how to catch him. After some thought, I rummaged out an old butterfly net, and watching my opportunity, when he was perched quietly on the cornice, and calling to my aid old butterfly-catching experience, I succeeded in enclosing the curious little ghost-creature in the net. He did not struggle much; he took the thing very philosophically. And now I had an opportunity of examining him more closely. Yes: he was a bat. And what a marvellous animal a bat is: one of the most marvellous animals that exist, I should think. His nose and countenance were like those of a little pig (his nose was pink), or, perhaps still more, like those of a pug dog; his wings were stretched from leg to leg, and were weird and leathern, and as I said above, rustled half metallically when he moved; his ears* were nearly as long as his little dun mouse-like body, and were very like those of a rabbit; he had small black eyes, and his fussy gait when walking was the funniest thing in the world.

Some of these particulars, of course, I noted later on; for, as far as my story goes, I have left him still in the butterfly net. To take up the thread of the narrative—it now became a question what was to be done with the captured ghost. After mature consideration, and, bearing in mind the accommodation which I had found most suitable for some short-tailed field-mice many years ago, I sent for a tray and a common fire-guard, placed the fire-guard over the tray, filled up the ends with two halves of a broken ship-board, tied carefully to the wires of the fire-guard; and there was a cage complete. And thoroughly secure it looked.

The bat soon evinced his appreciation of the measures that were being taken for his comfort and safety. He hooked himself up by his curious little claws to the wires of the fire-guard and scrambled all along the top and sides, head downwards, with the most reckless indifference to possible congestion of the brain. And the most charming point about this very charming and beautiful little creature (for the vulgar ideas of the ugliness and spitefulness of the bat are the very reverse of the truth) was his complete fearlessness. This is what makes the bat such a delightful pet. For most wild creatures—birds for instance—require an elaborate process of training before they become tame or happy in captivity, and some remain more or less untamed to the very last. But this bat was tame from the first. He took food from my hand immediately after I caught him, and had evidently had no personal experience (nor any inherited experiences, which raises a number of curious and interesting scientific questions) of the ways of men with bats.

Having provided lodging for him, the next question was that of his board and maintenance. It became necessary to provide supper for the bat.

---

* He turned out to be a long-eared bat (*Plecotus communis*), much the prettiest species.

Raw meat, chopped up small, seemed the most suitable thing for an experiment. And this I tried. As just stated, he took it without the slightest hesitation or timidity. And after he had amply supped, first taking care to provide him with a saucer of water, and looking once more to the security of the fastenings of his prison, I left him for the night.

But in the morning he was gone. Not one bolt or bar of his prison had been disturbed. Everything was left as I had left it overnight—with the exception of the meat, that is to say, for the rascal had taken care to finish his supper before he departed. And then, ghost or bat—he had vanished without paying his bill, and without leaving the slightest trace of the manner of his disappearance! The ghost-theory began to look up again, and I felt uncomfortable. His cage had been left in the drawing-room all night, and unless he had flown up the chimney he must still—barring the ghost-theory—be somewhere in the room. The most careful and patient search, however, failed to reveal him. The bat had fled, and though not actually moved to tears, I confess that it was with some tribulation, and the faint semblance of a heart-ache, that I felt that in all probability I should never set eyes on my little pet again, or only perhaps discover him after the lapse of years (as I once discovered a newt that had escaped from my aquarium) dead, stiff, and cold, a wretched and hapless little corpse, jammed tight behind the drawing-room piano.

However, though search throughout the day proved fruitless, when evening came on a sort of splash of brown mud was discernible on the white cornice of the drawing-room ceiling. This was the bat who had emerged from whatever hole or crevice he had been concealing himself in during the day, and was now waiting—doubtless expectant of another supper. The proceedings of the preceding evening were then again repeated—except that, he being again captured in the butterfly net, he was this time consigned to a safer prison in the shape of a wire meat guard (one of those used in the summer to keep off flies) placed over the tray. There he has remained safely incarcerated till the present date. In fact, Victor Hugo's fine line with reference to the escape of Marshal Bazaine is strictly applicable to him :—

'*Et qui donc maintenant dit qu'il s'est évadé?*'

He seems perfectly happy in captivity, and eats bread and milk out of my hand. Raw meat he will not touch, nor underdone meat. He has decided tastes of his own, preferring mutton to beef, and highly relishing chicken.

Contrary to expectation, he does not seem to care for flies. He darts at them, seizes them and shakes them as a dog does a bone—but then abandons them, and goes back to his bread and milk. As a rule he sleeps during the day, generally waking up in a very lively and hungry condition about eight o'clock in the evening. His sleeping position is always head downwards, hanging suspended from the wire roof of his cage. Bats, I believe, generally die in this head-downward position.

Altogether, I can recommend all persons—especially those of æsthetic tastes—in search of a new pet, to capture a long-eared bat, and tame him. It is because mice are so common, and bats so uncommon, as pets—and because a bat, if the fact were only more usually known, is an infinitely prettier and quainter and more interesting creature than a mouse—that I have written this little account of my "strange visitor,"—in the hope of making more widely known the name by which we have elected to designate the curious little living lump of fur and claws and skin and leather—the name of " Tommy the Bat."

GEORGE BARLOW.

ON THE DISTRIBUTION OF PLANTS WITH REGARD TO THE ROCKS BENEATH.

WHEN some years ago, I first commenced to make observations on this subject, in my own neighbourhood, Godalming, I was under the impression that mineral, or, perhaps, I should say chemical, composition of the rocks, would entirely influence the distribution of our flora. On the whole this seems to be the general rule; there are however some facts which seem to prove that there are other causes which influence distribution. In this paper, I must of course entirely make use of local examples, but I wish it to be understood that I send these remarks to your columns, not so much as a record of local facts, as to give a stimulus to inquiry in other districts. First, I may mention the beech (*Fagus sylvatica*). This is abundant on the chalk, especially along the escarpment, whilst upon the lower greensand it is well-nigh unknown, until you come to the sandstone beds, which are nearly, I think, devoid of calcareous matter; these beds form the escarpment of the lower greensand, where the beech is particularly abundant,[*] take Hascombe Beech for example. The absence of this tree from the intervening area, is, to my thinking, somewhat remarkable, seeing that a good part of that area is covered by beds of a local limestone called bargate, forming the top of the Hythe group. This bargate stone area has not a few chalk-loving plants, take for example the spindel, the cornel, and more sparingly clematis vitalba, and *Viburnum lantana*, and many other chalk or limestone plants, which I have never seen along the escarpment of the lower greensand. The whitebeam also frequents the escarpment of the chalk and lower green-

---

[*] I believe it will be found that the same distribution of the beech, appertains on the south side of the Weald, as on the north, where all the aspects are of course reversed, but I shall be glad to have this corroborated.

sand, while it avoids the intermediate bargate. I am inclined to think that these trees, the whitebeam and beech therefore, are more influenced by drainage than by soil. The beds forming both the escarpments having a sharp dip, a perfect drainage is secured. This however is a mere surmise, as one would think that the hargate stone with all its open joints would furnish a tract equally well drained. Turning now to the juniper (*Juniperus communis*), I find that this tree (excepting a few stunted specimens on the chalk) appears to be confined to the ironsands of the Folkestone beds, and more than this I believe they only grow where these beds lie rather thinly on the bargate, as at Munstead, and Shackleford. These beds, I should observe, are very frequently false bedded, the bed lines running at high angles, and are supposed by some geologists to be of eolian origin. When therefore they rest on bargate beds, with its open joints, a most perfect drainage is secured. Professor Wyville Thompson mentions a species of juniper as growing on eolian beds of modern date in the Bermudas. In the Farnham area, where a large tract of the Folkestone beds occur, very few, if any junipers are to be seen. Here, however, I believe the beds to have some clayey sand belonging to the Sandgate beds intervening between them and the bargate below, thus rendering the drainage less perfect, and the same facts I believe hold good with the Woolmer Forest district. In conclusion it seems probable that some plants are more influenced by the drainage of the soil in the selection of their habitat, than by its chemical composition.

I trust you may deem these suggestions worthy of insertion in SCIENCE-GOSSIP.

H. W. KIDD.

P.S.—Since penning the foregoing, it has struck me as worthy of remark, that the common elm which abounds on the calcareous bargate series, and also on the chalk without the weald, is absent alike from the escarpment of the lower greensand and the escarpment of the chalk.—H. W. KIDD.

DRIED FLOWERS WITH COLOURS.—I shall feel very thankful for a list of such ordinary garden or greenhouse flowers as will retain their colour in a dried state. I specially desire blue and red, and other bright colours, of form and size suitable for mounting for Christmas cards and other decoration. I understand there are many secrets in the success of some in the drying of their flowers, and any hints which will help me in preserving the natural tints, or at least sufficient brightness for my purpose, will be most gratefully accepted. I should also be glad of a list of wild flowers which retain a bright colour in drying. Forget-me-not would be exceedingly useful, but that it sadly fades after a short time. —*T. McGann.*

CHRISTMAS NOTES.

THE weather this Christmas-tide has been unusually mild. In the course of a hasty stroll or so at Great Marlow, Bucks, I observed the following—

1. Plants in flower:

a. Plants which flower the whole season.

Profusely: *Lamium album, Senecio vulgaris* (α), *Veronica agrestis* var. *polita, Ulex Europæus* (β), *Lamium purpureum, Stellaria media.*

b. Plants which do not as a rule flower the whole season.

Profusely: *Galium saxatile, Capsella bursa-pastoris, Stellaria graminea, Pimpinella saxifraga, Bellis perennis, Viola tricolor* (seen by a friend).

Occasionally: *Ranunculus repens* (γ), *Scabiosa arvensis, Geum urbanum,* ; *Veronica Chamædrys, Erigeron acris, Taraxacum dens-leonis, Senecio sylvaticus, Heracleum sphondylium, Crepis virens.*

Singly: *Rubus fruticosus* var. *carpinifolius* (δ).

2. Plants leafing profusely:

a. Last season's stems (perennials) — *Lonicera periclymenum* (ε), *Galium mollugo, Asperula odorata.*

b. New stems (annuals)—*Fragraria vesca, Galium aparine.*

3. Birds: *Troglodytes vulgaris, Parus major, Passer domesticus, Accentor modularis, Fringilla cœlebs, Pyrrhula vulgaris, Motacilla alba, Vanellus cristatus, Corvus frugilegus, Turdus musicus, Turdus merula, Picus viridis* (seen by a friend).

4. Other signs of activity:

Culex pipiens, Ichneumon sp., Limax cinereus.

A beetle, a spider, a few diptera, and a lepidopterous larva on food-plant.

(α) Amongst other localities, on the top of a brick wall 8 feet high,

(β) Occurs on Marlow Common, a clay-capped elevation, reappearing in a sandy cutting between Cookham and Maidenhead. Similarly with *Tanacetum vulgare* and *Lychnis diurna* (not, however, now in flower), which reappear to the east of Burnham Beeches. None of these spp. are to be found on chalk itself, on which this district is mainly situate.

(γ) 4-sepalled, 6-petalled.

(δ) This flowering stem bore two monstrous leaf-forms out of three, in one the lateral pinnæ being represented merely by obscure lobes adherent to the terminal, and in the other one lateral pinna coalescing with the terminal, exhibiting a deep incision on its margin.

(ε) On one stem, where a last season's leaf was still attached, the development of the new bud in that spot was arrested. The opposite bud, however, sprouted undiminished.

From (γ) (δ) and (ε) it would almost appear that in abnormal seasons of development plants have a tendency to monstrous growth. I append a sketch or two.

E. G. HARMER.

A BOTANICAL RAMBLE ROUND WEYMOUTH AND IN THE CHANNEL ISLANDS.

STARTING from Weymouth on a fine summer's morning last July, I took the road for Portland, via Wyke Regis. It was a glorious hot summer's day, with just enough of the fresh sea breeze to make it pleasant. Crossing the fields towards the bridge, I came upon *Triticum acutum*; after leaving the mainland, just on the other side of the bridge, I found *Euphorbia Portlandica*, *Eryngium maritimum*, *Trifolium scabrum*, *Sclerochloa loliacea*, and *Convolvulus soldanella*. Farther on, on that wonderful bank of pebbles, yclept Chesil Bank, which stretches away through the water for six or seven miles to the westward, I found the true samphire Crithmum, not growing in its usual position on the dizzy cliffs, but among the pebbles. Casting my eyes along the ridge, I noticed large dark green patches in the distance, and on going to look at them, I was rewarded by finding that they consisted entirely of the rare *Lathyrus maritimus*, then in full flower. Returning towards Portland, I came across a patch of *Sueda fruticosa*, also *Festuca unighumis*, *Diplotaxis muralis*, the small *Lepturus filiformis*, and *Helminthia echoides*, all near the railway. On Portland Island I did not stay very long, but found a quantity of *Centranthus ruber* growing in the quarries. Returning to Weymouth, I found *Lathyrus nissolia* and *aphaca*, *Trifolium maritimum*, and *Vicia bithynica* on the shore to the west of the bridge. Next day I set out for the chalk cliffs some miles to the east of Weymouth, hoping to find something worth having; about three miles out of Weymouth I found the banks perfectly blue with *Echium vulgare*. Not far from a coast-guard station, about six miles on, I found a large patch of *Silybum Marianum*, also, close to, *Reseda lutea* and *luteola* growing together. Leaving the coast, I struck across country inland, and came across *Campanula trachelium* growing in a hedge, and *Habenaria chlorantha* in a small copse. That night I took the boat at 11.30 (a most inconvenient time; I cannot think why they should start at such an unearthly hour) for Guernsey. About 7.30 A.M., we steamed into the rocky harbour of St. Peter Port. After breakfast I started to explore the north of the island; leaving St. Peter Port, on the north road I soon came to St. Sampson, the port from which the granite quarried in the island, is shipped. Numerous round towers, barracks and look-out stations, are scattered round the coasts, reminding one of the times when a French invasion was contemplated, in fact, I noticed several marked with the date of 1789, or a bit later on. I went to L'Ancresse Bay and found quantities of *Lagurus ovatus*, a common grass on the sandy bays of both Guernsey and Jersey. *Oxalis corniculata* and *Polycarpon tetraphyllum* were not uncommon, growing on the walls, especially the latter. On the rocks at Paradis, I found *Inula crithmoides* and *Lotus hispidus*, but could not find the rare *Cicendia pusilla*, said to grow there. Next day I took the south coast; above

Fig. 24.—Portland Spurge (*Euphorbia Portlandica*).

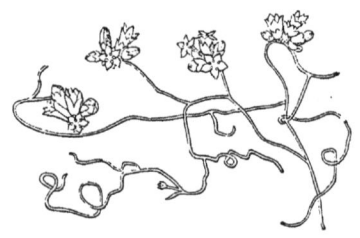

Fig. 25.—Dodder (*Cuscuta Epithymum*).

the lovely Fermain Bay, the best bathing place on the island, I found *Anchusa sempervirens*. On the rocks below Jerbourg grew *Statice occidentalis*;

all about the cliffs in great profusion the sea radish showed its curious pods and yellow flowers, while among the bushes the prickly plants of the butcher's broom hid themselves away. The south coast of Guernsey is by far the finest; it consists of bold rocky bays and inlets, where the sea rushes in among the rocks, dashing up its white foam high into the air, even on the calmest day. All the furze bushes along this coast are covered with a thick red web of *Cuscuta Epithymum*. On the ivy-covered banks and than English, in fact, in winter you seldom see an English coin. They have a copper coinage of their own, but use French gold and silver. In Jersey, English gold and silver only are used. On the third morning I took passage on the good ship 'Cygnus' for Jersey, and after a three hours' passage landed at St. Heliers. Walking along the sands of St. Aubin's Bay, I came across *Allium sphærocephalum*, *Allium vineale*, var. *capsuliferina*, *Echium plantagineum*, *Onothera odorata*, *Alyssum maritimum*, and *Bromus*

Fig. 26.—Grass Poa (*Lathyrus Nissolia*).

Fig. 27.—Butcher's Broom (*Ruscus aculeatus*).

walls in the narrow shady lanes may be found *Orobanche Hederæ*. The west coast was my last day's work; after crossing the island from St. Peter Port, came to Grand Cobo, where I found *Cyperus longus*, *Orobanche amethystea*, *Tamarix Anglica*, *Lavatera arborea*, and, about Vazon Bay, *Centaurea aspera*, *Juncus acutus*, *Schœnus nigricans*, *Bromus mollis*, var. β. *glabrescens*, and *Festuca uniglumis*. Guernsey seems to have more of the French element in it than Jersey, for you see notices up in French all about the island, and there is much more French money used *maximus*; *Senebiera didyma* grew about St. Aubin's in great profusion, and very fine. I walked on to St. Peter's, and by the roadside found *Asplenium lanceolatum* and *Sedum sexangulare*, also *Verbascum nigrum* in a field. In a small valley running down to St. Ouen's Bay, the small *Sibthorpia Europæa* crept over the damp shady banks. In the evening I arrived at the best hunting ground in Jersey for rare plants, St. Ouen's Bay; here were the hoary sinuate-leaved sea stock, *Mathiola sinuata* and *Armeria plantaginea* in great quantity. That night I stayed at a very comfortable hotel at St. Brelade's Bay, which I beg to recommend to any one who thinks of botanising in this part of the island, as a very good place for headquarters. Next morning I set off again for St. Ouen's Bay, and on the way made a good find in *Helianthemum guttatum*; on a bank I found a single specimen of *Orobanche cærulea*, while in a sandy lane *Antirrhinum Orontium* was discovered. On reaching the bay I began to use my eyes to look for the good *Epipactis palustris*, and, I believe, though I could not find any, *Orchis laxiflora*. Another day took me by the Jersey Eastern Railway to Gorey, near which is the fine old castle of Mont Orgueil; it is in a pretty good state of preservation, many of the rooms being still habitable. Among the crevices of the stones grows *Erodium maritimum*, and the walls are covered with parietaria.

Near Mont Orgueil is Ann Port, a small fishing

Fig. 28.—Garlic (*Allium vineale*).

Fig. 29.—Squills (*Scilla autumnalis*).

things which I knew were to be had there, and was soon rewarded by dropping on *Orobanche amethystea*, *Silene conica*, *Kœleria cristata* var. β. *gracilis*, *Crambe maritima*, *Diotis maritima*, *Sinapis incana*, *Raphanus maritimus*, and *Juncus acutus*. On this bay, a little distance from the sea, is a large pond surrounded by marshy ground; this, too, is a good hunting ground for the botanist. Here may be found *Scirpus pungens*, *Cladium Mariscus*, *Cyperus longus*, *Hypericum elodes*, hamlet; on a heathy hill above it I was lucky enough to make a rare find, viz. *Hypericum linariifolium*, with it grew *Scilla autumnalis*. This being my last day in the Channel Islands, I crossed back that evening to England, well satisfied with my botanical ramble in those delightful little islands.

A. E. L.

Liverpool.

SCIENCE-GOSSIP.

AT the last meeting of the Cryptogamic Society of Scotland, Mr. J. King gave an account of the mould producing the potato disease, and expressed his belief that the constitution of the potato plant had become enfeebled through the methods of cultivation it had experienced, so that it could not withstand the attacks of the parasitic fungus so well. The disease is unknown among the potatoes cultivated in Chili, but some grown in this country by Mr. King were not quite proof against it.

THE pastor and members of the Free Christian Church, Shrewsbury, have just put up a mural tablet with the following inscription:—"To the memory of Charles Robert Darwin, author of 'The Origin of Species,' born in Shrewsbury, February 12th, 1809. In early life a member of, and a constant worshipper in this church. Died April 19th, 1882."

PROFESSOR J. W. H. TRAIL describes in the last number of "The Scottish Naturalist" several species of leaf-parasites which are new or rare in Britain. They include *Doassansia alismates*, *Entyloma calendulæ*, *Entyloma canescens*, and *Protomyces rhizobius*.

MR. C. B. PLOWRIGHT, from actual observation, records the fact that the squirrel is a fungus-eating animal.

ALL the five older planets were visible during some part of the evening, in the first week of January.

A NEW book is announced by the editor of SCIENCE-GOSSIP, under the title of "The Sagacity and Morality of Plants" (illustrated), to be published by Messrs. Chatto & Windus shortly.

A MEETING and conversazione of the National Association of Science and Art teachers was held in Manchester on December 22nd. All kinds of books, models, diagrams, and other apparatus connected with science teaching were exhibited. Professor Roscoe was in the chair. Such an association is much required among science teachers, who are just now rather too much under the imperious command of the South Kensington Bureaucracy.

AMONG the lectures to be delivered at the Royal Institution before Easter, are the following:—"The Origin of the Scenery of the British Isles," by Professor Geikie; "Animal Heat; its Origin, Distribution, and Regulation," by Professor McKendrick; "The older Electricity; its Phenomena and Investigation," and on "Rainbows," by Professor Tyndall; "The Building of the Alps," by Professor Bonney; "The Darwinian Theory of Instinct," by Dr. G. J. Romanes; "Theory of Magnetism," by Professor Hughes; "The two Manners of Motion of Water," by Professor Osborne Reynolds.

THE population of China has recently been estimated at 250,000,000.

WE are sorry to have to record the death of the well-known American microscopist, Mr. Robert B. Tolles, of Boston. In the United States he was distinguished as the maker of the highest power microscopes. His telescopes were almost equally famous for their perfection.

WE are pleased to observe that an old and valued contributor to our columns, Professor Sollas, late of the Bristol Museum, has been appointed to the Professorship of Geology in Trinity College, Dublin.

ATTENTION has been recently drawn to the fact that in the Japanese seas numerous fish are very poisonous if eaten, and it has been suggested that all the ships going there should be provided with descriptive representations of the poisonous kinds.

"A WHITE ELEPHANT" has long done duty as a figure of speech. Now we may have the opportunity of seeing a real one in the flesh, for one has been purchased in British Burmah for Mr. Barnum, and is now on view in London.

WE are sorry to find that Mr. R. A. Proctor, the energetic editor of our contemporary, "Knowledge," is obliged to take two months' mental rest, by medical advice. Mr. Proctor did not give himself time to get over the shock he received in the railway accident last July.

THE planting of our waste grounds with suitable trees for timber is advocated in "Woods and Forests." The writer states there is hardly a vacant corner or heathy waste which will not produce valuable crops of timber, of one kind or another

THE "Report of the Local Scientific Societies' Committee," consisting of Mr. Francis Galton (chairman), the Rev. Dr. Crosskey, Mr. C. E. De Rance, Mr. H. G. Fordham (secretary), Mr. John Hopkinson, Mr. R. Meldola, Mr. A. Ramsay, Professor Sollas, Mr. G. J. Symons, and Mr. W. Whitaker, has been published. It gives a list of all the local scientific societies which publish "Proceedings," &c., their headquarters, number of members, name of secretary, &c.

UNDER the title of "The Medical Annual," a year-book for the study table of the medical practitioner, has just appeared, edited by Mr. Percy R. Wilde, M.B. There can be no doubt this book meets a long-felt want. Its contents are exceedingly well planned, and include a concise summary of the principal hints and facts in medicine, surgery, and therapeutics, which have appeared within the last twelve months in no fewer than one hundred and fifty British and foreign medical journals. Dr. Wilde has done his work well, and we have no doubt will reap his reward. The book is published by Henry Kimpton, 82 High Holborn.

Mr. J. B. Sutton has recently shown that the common opinion that monkeys in confinement generally die of tuberculosis is not correct.

Mr. R. M. Christy, in a communication to "Nature," points out that the vast region known as Manitoba and North-West Territories, comprising about three million square miles, is remarkable for the total absence of earthworms. He thinks this is due to the prairie fires which annually sweep over great portions of the country.

Professor Milne-Edwards has read a Report before the Paris Academy of Science, on the "Talisman" Expedition made last year, to dredge the coasts of Senegal, the Cape Verde, Canary, and Azores Islands. They found a marvellous abundance of marine life, and at depths as great as 1900 metres numerous fishes (many of them possessing phosphorescent spots) were brought up. Among other rarities the soft echinoderm Calveria was dredged alive. Species of Arctic mollusca and crustacea were also obtained from the deeper parts of the sea-bed. Magnificent sponges allied to *Euplectella suberea* were discovered, together with Holtenia, &c. In the deep waters of the Cape Verde Archipelago life displayed surprising energy. A large number of new species was discovered.

Professor McIntosh, whose labours in cataloguing the marine fauna off the east coast of Scotland are so well-known to naturalists, has made arrangements to proceed once a fortnight for the trawling-grounds off the east coast, in order to investigate the fishing-grounds and their inhabitants.

Hops have at length been successfully cultivated in the Province of Wellington, New Zealand. Singularly enough, the chief want experienced now is that of poles, and trees will have to be grown for the purpose!

Every British naturalist will be delighted to hear that natural science has been honoured in the person of our distinguished scientific veteran, Professor Owen, who, on his retirement from the post of Superintendent of the British Museum, has been made a K.C.B.

Sir John Lubbock, in an article written for "Nature," thinks that the reason why we know so little about the mental condition of animals is because hitherto we have tried to teach them rather than to learn from them. He then narrates some very clever experiments he has made with his dog. He had some cardboards, with the word "Food," &c., legibly printed. The card labelled "Food" was placed in the saucer containing eatables, and plain cards on saucers that were empty. The dog soon learned to distinguish between the two, and when he wanted food learned to bring the card with the word printed on it.

A correspondent writes in reply to Sir John Lubbock's experiments, suggesting that it would be simpler to commence with drawings on the cards instead of words.

Dr. David Shier has published some very interesting notes on *Trichonema columnæ* in the "Transactions of the Devonshire Association for the Advancement of Science, Literature, and Art." This plant is one of the few southern kinds which reach their northernmost limit of distribution along the southern shores of England. It is the smallest representative of the order Iridiæ, and was found near the Warren, near Exmouth.

On January 21st Dr. J. E. Taylor, F.G.S., &c., Editor of Science-Gossip, delivered a lecture before the Hitchin Natural History Society on "The History and Origin of Earthquakes and Volcanoes."

The brilliant after-glows have continued more or less throughout January. Perhaps the most beautiful was that on January 11th, which filled the entire western sky with crimson light. At the same time in the east the full moon was rising. The fore-glows of the sunrises have hardly been less noticeable on many occasions.

We much regret having to record the death of Mr. C. W. Merrifield, F.R.S., at the comparatively early age of fifty-five. The deceased gentleman was an authority on most matters relating to naval architecture, wave-motion, atmospheric resistance to projectiles, &c., and on one occasion was President of the Mechanical Section at the British Association Meeting.

Earthquake shocks of some violence have been reported lately from Bucharest, various parts of France, and the Hautes-Pyrénées.

Achard's continuous electric brake has been successfully worked in competition with those of Westinghouse and others.

A very interesting discourse was given by the Rev. W. S. Green before the Royal Geographical Society on January 7th, on the subject of the Southern Alps of New Zealand, which the reverend gentleman has recently explored. Mount Cook 12,362 feet, is the highest point. The south-western portion of New Zealand much resembles Norway in its numerous fiords, but is more picturesque. In latitudes equal to Florence, in the old world, glaciers descend to within a few hundred feet of the sea-level.

A good "find" of palæolithic and neolithic implements is reported from Maidenhead. The latter were obtained from the bed of the river Thames.

"Sunlight" (published by Simpkin, Marshall, & Co.) is a new competitor in the world of popular science.

WE have received a copy of the Annual Report (with list of members, &c.) of the Manchester Microscopical Society, which shows that its affairs are in every way prospering.

THE Report of the Lambeth Field Club and Scientific Society contains the President's Address, and the Conchologist's Report of the land and freshwater shells of the London district.

THE GARDENER'S CHRONICLE of January 12th, figures and describes a very remarkable "sport" in a specimen of the cristatum of the Hart's-tongue fern; the upper surface of the frond is thickly studded with young plants, produced in linear groups along the lines the sori would follow on the lower surface.

MR. GLADSTONE in his speech to the Hawarden farmers the other day, urged the advantages of fruit culture—a subject he first took up two years ago. He showed how sadly we have fallen off from our ancient agricultural customs. It should be remembered, however, that it would not be sufficient merely to plant orchards all over the country, unless we also extended apiculture as well. The flowers of our orchards are hardly half fertilised as it is, owing to the comparative scarcity of bees. In ancient times, when sugar was scarce and dear, people were forced to keep bees for the sake of their honey, and then apples, pears, and plums were abundant.

MICROSCOPY.

GLASS CELLS.—Many years ago I described in this journal my method of making glass cells, by which plan I could perforate not only thin glass of various thicknesses, but also glass slips $\frac{1}{12}$ inch thick, the perforations in thin glass varying from $\frac{1}{8}$ to $\frac{5}{8}$, and those in the slips from $\frac{1}{4}$ to $\frac{5}{8}$ in diameter. The apparatus required are, as many brass plates 3 × 1 and $\frac{1}{16}$ inch thick as perforations, and a small steel pointed hammer. I have 1 dozen plates, 3 with $\frac{1}{4}$ in. aperture, 3 with $\frac{1}{2}$ in., 3 with $\frac{5}{8}$ in., 1 $\frac{5}{8}$ in., 1 oval, and 1 square; if the $\frac{1}{4}$ inch is wanted, I heat the 3 plates with that aperture; when hot enough to melt shellac, I select a thick piece and smear the margins of the holes with it, and place the covers upon them, and press them down on the brass; when cold the centres can be knocked out, reheat the plates, push off the glass rings into methylated spirit, apply fresh lac, and proceed as before; there is no difficulty in making 40 cells in an hour. To perforate thick glass slips, I place the slip on the turn-table and make a ring the size required with a cutting diamond. I now smear the central square inch of the brass plate with the lac, the slip (which should be previously heated) must be placed at right angles to the brass plate with the ring directly over the hole, a few scratches made with the diamond within the ring facilitates the removal of the centre, which should be knocked out with the little hammer. These perforated slips are very useful for objects which require examination of both surfaces. In the last part of the "Quekett" Journal Mr. Whitwell directs gum arabic to be used for cementing the covers previous to perforating them, and in the January part of SCIENCE-GOSSIP Dr. A. Stokes proposes Canada balsam. I think they will find shellac preferable for that purpose.—*F. Kitton*.

DRAWING WITH THE MICROSCOPE.—Mr. Holmes' suggestion of placing the slide with the cover downwards must have been made on the spur of the moment, or he would have remembered that the upper and under surfaces of an object are not as a rule alike; a further objection is that all powers exceeding $\frac{1}{10}$ could not work through an ordinary slide. Having used the Wollaston camera constantly for many years, I give it the preference over all others I have tried, the neutral tint-reflector included. The reversal of the image, and a certain haziness of outline, renders the latter objectionable, and at best it is a cheap substitute for the more costly camera lucida. In the December number of the J. R. M. S. a new form of camera is described by the inventor, Dr. Hugh Schroeder, which appears to possess many advantages over any of the cameras I have had the opportunity of trying, of which the non-limitation of the field and the object being seen reflected on the paper without bisecting the lens of the eye, as in the Wollaston camera, are not the least. The inventor speaks highly of its performance; judging from the accuracy required in its construction, I fear it is somewhat costly.—*F. K.*

QUERY.—Is *Tingis crassichari* the same as *T. crassicornis*? My specimen appears to be identical with the coloured drawing of Mr. Hudson, and is so named.—*F. K.*

MESSRS. COLES' SERIAL "STUDIES."—Nos. 9 and 10 of the well-known "Studies in Microscopical Science," deal, the first with "Cartilage" (illustrated by a beautiful coloured plate of the transverse section of hyaline cartilage from the human trachea ×250), whilst the latter has an introductory chapter to the "Morphology of Tissues." The slides sent out with these parts fully maintain their high character, No. 10 being illustrated by a stained section of the pileus of *Agaricus campestris*. No. 4 of the "Popular Microscopical Studies," also edited by Mr. A. C. Cole, F.R.M.S., deals with the Ovary of the Poppy, illustrated by a coloured plate of a transverse section of the unfertilised ovary of *Papaver rhœas* × 50. The slide which accompanied this part is one of the best Mr. Cole has distributed.

"PETROGRAPHICAL STUDIES."—We have received No. 1 of Messrs. Ady & Hensoldt's new publication bearing the above title. It deals with the specimen of Calciferous Serpentine sent out as a slide, to which reference was made in our last number, commonly called eozoonal. A brief, but exceedingly clear abstract is given of the opinions of Carpenter and others as to the organic character of Eozoon, on the one hand, and of Möbius on the other. The Sutherlandshire Eozoon lends considerable evidence to the mineralogical theory of Professors King, Rowney, and Möbius. The sketches accompanying the part are very carefully drawn and executed.

SLIDES ILLUSTRATIVE OF MARINE ZOOLOGY.—If there is "no royal road to knowledge," it cannot be denied that the old road has been rendered very much smoother and easier for the modern student to travel upon. In the matter of marine zoology, for instance, the possibility of obtaining such slides as Sinel & Co. are sending out from their natural history depot at St. Heliers, Jersey, for microscopical examination, saves an enormous amount of time and trouble. The latest of these slides, mounted in a medium, and after a manner which Sinel & Co. alone possess, include the zœa of crab (*Pisa tetraodon*), two days old, which admirably exhibits the long-jointed, lobster-like body, subsequently abbreviated into the "apron." Other slides are that of young fishes (*Gobius niger*), one day old, of another interesting crustacean (*Hippolyte varians*), and an exquisite young starlet (*Asteria gibbosa*), so mounted that both surfaces, upper and lower, can be equally well studied.

HARDENING ANIMAL TISSUES.—I should be very glad if any reader would inform me how to successfully harden animal tissues which have been injected. I have tried spirit, as Dr. Marsh recommends in his little book, but cannot succeed to my satisfaction.—*W. H. P.*

THE POSTAL MICROSCOPICAL SOCIETY.—Part 9 of vol. iii. of the "Journal" of this society has appeared under the able editorship of Mr. Alfred Allen. It gives the presidential address, papers on "Living Bacilli in the Cells of Vallisneridæ," by Dr. T. S. Ralph ; "The Foraminiferæ of Galway," by Messrs. F. P. Balkwill and F. W. Millatt ; "Solorina Saccata," by Arthur J. Doherty ; "Thymol as a Polariscope Object," by Dr. Ralph ; " Half an Hour at the Microscope with Mr. Tuffen West ;" Selected Notes ; Reviews, Current Notes, &c.

PRESERVATION OF SOFT TISSUES.—"Science" reports that at a recent meeting of the Philadelphia Academy, Dr. Benjamin Sharp called attention to Professor Semper's mode of preparing dried specimens of soft animals, and exhibited a couple of snails as illustrations of the admirable results of the process. The tissues are first hardened by being steeped in chromic acid, which is afterwards thoroughly washed out in water. The specimen is then allowed to remain in absolute alcohol until the water is perfectly extracted, when it is placed in turpentine for three or four days. It may then be dried and mounted. Specimens prepared in this way retain their characters in a very satisfactory degree, and are strong and flexible, the example shown resembling kid. If the surface be treated, after drying, with a solution of sugar and glycerine, the natural colours will be restored ; but the specimens must then be kept in hermetically-sealed glass cases to preserve them from the dust. The objection to this mode of treating large specimens is the expense of absolute alcohol ; otherwise there is no reason why the largest animals should not be preserved by this process.

"THE JOURNAL OF THE ROYAL MICROSCOPICAL SOCIETY."—The December issue of this compact and encyclopædic scientific journal, besides the usual "Summary of Current Researches," contains the following papers : "On Some New Cladocera of the English Lakes," by Conrad Beck ; "On an Improved Method of Preparing Embryological and other Delicate Organisms for Microscopical Examination," by Edward Lovett ; "The Relation of Aperture and Power in the Microscope," by Professor E. Abbe ; "On a New Camera Lucida," by Dr. Hugo Schröder ; and "On Optical Tube Length ; an Unconsidered Element in the Theory of the Microscope," by Frank Crisp, Hon. Sec.

CARLISLE MICROSCOPICAL SOCIETY.—This newly-founded society is in the full flush of active enjoyable work. The session for 1883-84, according to the programme forwarded to us, has numerous papers of great interest to be brought before it, among which are the following : "The Salmon Disease," by the vice-president (Dr. Lediard) ; "Structural Botany," by the hon. sec. (Mr. A. Barnes-Moss) ; "Animal Tissues," by Dr. Maclaren ; "Fertilisation of Flowers," by Mr. R. A. Allison ; "The Microscope in Manufactures," by the president (Mr. C. S. Hall) ; "Adulteration of Food," by Mr. W. Parker ; "Micro-Photography," by Mr. J. Forsyth. A communication from Dr. W. B. Carpenter, the eminent physiologist and microscopist, was presented. The communication is printed in the syllabus of this promising society.

ZOOLOGY.

WHALES AND THEIR ORIGIN.—The President of the Zoological Society, Professor W. H. Fowler, LL.D., F.R.S., on the 20th inst., delivered at the London Institution a profusely illustrated and in all respects most interesting lecture on "Whales," in which he took a survey of the existing and fossil species, with a view to some sort of solution of the obscure but important problem of their origin. It needed no great foresight, he said, to forecast their more or less speedy improvement out of existence. But what was their probable origin ? In the first place, the evidence was absolutely conclusive that they were not originally aquatic, but sprung from land mammals of the placental division, animals with a hairy covering, and with sense organs, especially that of smell, adapted for living on land; animals, moreover, with four completely developed pairs of limbs on the type of the higher vertebrates, and not that of fishes. Their now simple homodont and monophyodont teeth had evidently degraded from a more perfect type. But the great difficulty was in determining the particular group of mammals whence the whale family (Cetacea) arose. One of the methods by which a land mammal might have changed into an aquatic one, was clearly shown in still surviving stages among the Carnivores. The seals were obviously modifications of the land Carnivores, the sea-lions and sea-bears being curiously intermediate. Many naturalists had been tempted to deem the whales a still further stage of a like modification. But there was a fatal objection to this view, as was shown. It was far more reasonable to regard whales as derived from animals with large tails, which were used in swimming, to such effect at last that the hind limbs were no longer needed, and so at length disappeared. The powerful tail with side flanges of skin of the *Pteroneura Sandbachii*, an American species of otter, or the beaver's tail, might give some idea of a primitive Cetacean. As pointed out long ago by Hunter, there are many points in the structural organisation of the Cetacean viscera far more like those of the Ungulates than the Carnivores, such as the complex stomach, simple liver, respiratory organs, and especially the reproductive organs and structures relating to the development of the young. Though there was, perhaps, generally more error than truth in popular ideas in natural history, the lecturer said he could not help thinking some insight had been shown in the common names attached to the most familiar Cetaceans by those enjoying the best opportunities of knowing its nature. The names were the "Sea Hog," "Sea Pig," or "Herring Hog," of our fishermen, the equivalent "Meerschwein" of the Germans, corrupted into the French "Marsouin," just as we had shortened the French "Porcpoisson" into "porpoise." The difficulty that might be suggested in the derivation of the Cetaceans or whale family from the Ungulates arising from the latter being mostly vegetable feeders, was not great, as the earliest Ungulates were most likely omnivorous, like their progeny, the pigs, now, and the aquatic branch might easily have gradually become more and more piscivorous. The audience might picture to themselves some primitive, generalised, marsh-hunting animals, with scant hair, like the modern hippopotamus, but with broad swimming tails and short limbs, omnivorous in their mode of feeding, probably combining water-plants with mussels, worms, and fresh-water crustaceans, gradually becoming more and more adapted to fill the void place ready for them on the water side of the borderland on which they dwelt, and so by degrees becoming modified into dolphin-like creatures inhabiting lakes and rivers, and at last finding their way into the ocean. There the disappearance of the huge Enaliosaurians, the Ichthyosaurians, and the Plesiosaurians, which formerly played the part the Cetaceans did now, had left them ample scope. Favoured by various conditions of temperature and climate, wealth of food supply, almost complete immunity from deadly enemies, and wide watery fields to roam in, they had undergone the various modifications traceable in the evolution of the Cetacean species, existing and fossil, and by slow degrees grew to that colossal size, which, as they had seen, was not always an attribute of the whale family.

STRIX BRACHYOTUS.—An adult female specimen of the short-eared owl (*Strix brachyotus*) was shot near Norwich on January 1st, 1884. It was a very fine bird, and above the usual average; the length being from tip of beak to end of tail 15½ ins., and the fully extended wings measured 42¼ ins. Upon dissecting it, I found it had taken a three-quarter grown rat (*Mus decumanus*), the skull of which was almost entire, as also were the hind legs. I saw also the remains of several mice.—*E. W. Gunn, jun., St. Giles Street, Norwich.*

ADDENDA TO "THE MOLLUSCA OF MARGATE."[*]—Since writing the above list I have referred some of my small and doubtful specimens to Dr. J. Gwyn Jeffreys, who has kindly named them for me. I have also made two additional excursions to Shellness. The result is that the following species may be added: (Conchifera) *Nucula nitida*; now and then at Shellness, but not nearly so abundant as *N. nucleus*. *Montacuta bidentata*; common in shell-sand at Margate. *Venus verrucosa*; single valves only, and much waterworn, at Shellness. *Tellina donacina*; a single valve at Shellness. (Gasteropoda) *Capulus Hungaricus*; a broken specimen at Shellness. *Trochus montacuti*; I found two of this uncommon species at Shellness. *Lacuna divaricata* and *L. puteolus*; frequently in shell-sand from Margate. *Rissoa costata* and *R. semistriata*; ditto. *R. cancellata*; a single specimen from Shellness. *Aclis unica*; one only from Margate. *Odostomia acuta*, *O. unidentata*, *O. plicata*, and *O. indistincta*; all from Margate. *Eulima polita*; Margate and Shellness, somewhat frequently. *Cerithiopsis tubercularis*; common all round the coast with *Cerithium reversum*. *Pleurotoma attenuata* and *P. lævigata*; Margate, both very scarce. *Philine catena*; two specimens of this pretty little shell from Margate. (Cephalopoda) *Loligo vulgaris*; occasionally thrown up at Margate after stormy weather. *Sepia officinalis*;

---
[*] SCIENCE-GOSSIP, Sept. 1883.

I picked up a shell of this species at Shellness, but they are not so common as might be expected. Of estuarine species I have lately met with *Melampus bidentatus* in the rejectamenta of the Stour, and for *Assiminea Grayana*, I have a new locality to record, namely Sandwich, where it occurs alive and in abundance, on the banks of the dykes. These dykes are slightly brackish, but they cannot be much so, as I noticed a freshwater beetle in them. I shall be happy to send specimens to anyone interested in this little mollusc.—*Sydney C. Cockerell, Glen Druid, Chislehurst.*

BOTANY.

NEW FACTS CONCERNING THE FLOW OF PROTOPLASM.—A thorough revolution will shortly take place in the views hitherto held as to the manner in which fluids passed from cell to cell in plants. The phenomenon has hitherto been ascribed to Osmosis, but W. Gardiner in a paper just read before the Royal Society shows that the protoplasm in many cases actually passes from cell to cell through minute openings which exist for the purpose. Mr. Gardiner, however, was antedated by Mr. Thomas Hick, B.Sc., who read a paper at the Southport Meeting of the British Association (an abstract of which appeared in our December number) on "Protoplasmic Continuity in the Floridæ."

LATHRÆA SQUAMARIA. — I read Mr. Haydon's note on this parasitical species with much interest. Several years since I studied its habit, &c., very closely in Shropshire, and I came to the conclusion that it differed considerably when found on the poplar roots, to the one I collected from the hazel-root. Smith's figure is evidently taken from an old plant bearing seed, whilst Mr. Haydon's photograph is a very young plant.—*James F. Robinson.*

WATSON'S "TOPOGRAPHICAL BOTANY."—I lately purchased the new edition of H. C. Watson's "Topographical Botany," and am greatly astonished at finding such cosmopolitan plants as the nettle, dandelion, buttercup, daisy, and primrose, recorded as not indigenous to the county of Wigton, as given under the authority of Mr. Balfour. I expressed my surprise to a friend from over the border, who was equally astonished with myself, and in order to test the statement he wrote to one of his clan in Wigton, who replies that such plants as buttercups and daisies are exceedingly common in that county. If such is the case, is it to be considered that these plants have taken possession of that county since the publishing of the first edition of the above-named valuable work? An authority such as that of Watson or Balfour cannot for one moment be doubted. Would botanists actually resident in the county of Wigton kindly communicate?—*T. H.*

CHRISTMAS FLOWERS, 1883.—Between Christmas Day and New Year's Day I have seen the following wild-flowers: *Ranunculus hederaceus, R. ficaria, R. repens, Fumaria officinalis, Cardamine sylvatica, Sinapis arvensis, Lepidium Smithii, Viola arvensis, Polygala vulgaris* (both blue and white), *Stellaria media, Cerastium vulgatum, Spergula arvensis, Hypericum humifusum, Geranium molle, Ulex Europæus, U. Gallii, Rubus discolor, Potentilla fragariastrum, Daucus carota, Sherardia arvensis, Valerianella olitoria, Knautia arvensis, Petasites fragrans, Bellis perennis, Chrysanthemum leucanthemum, C. segetum, Senecio vulgaris, S. jacobæa, Lapsana communis, Hypochæris radicata, Leontodon taraxacum, Sonchus oleraceus, Crepis virens, Jasione montana, Erica cinerea, Scrophularia nodosa, Primula vulgaris, Veronica agrestis, V. polita, V. serpyllifolia, V. arvensis, Galeopsis tetrahit, Stachys arvensis, Teucrium scorodonium, Euphorbia peplus.* If the Christmas of 1881 deserved to be called mild, when I sent you from this district a list of twenty-two wild-flowers found in the week succeeding Christmas Day, what is to be said of the corresponding list this winter, including forty-five species? I doubt, however, whether many parts of the United Kingdom can boast as many flowers in mid-winter as we have now in Wexford. In the garden and shrubberies during the same period were violets, primroses, pansies, wall-flowers, stocks, blue gentians, yellow calceolarias, a fuchsia, a veronica, an early rhododendron, snap-dragons, mignonette, monthly roses, Michaelmas daisies, common double daisies, Japanese primroses, hydrangeas, laurustinus, arbutus, and lemon-scented thyme. Both song-thrush and missel-thrush were in full chorus through the week. Yesterday (January 2nd), I saw in flower, *Ranunculus acris, R. flammula,* and *Potentilla tormentilla,* and to-day, *Veronica Chamædrys* and *Senecio aquaticus.* Most of these had evidently been some days in flower; but of all my instances of the "mildness of the season," the flowering of *Veronica Chamædrys* seems to me the most astonishing. My list now amounts to fifty species, found within a few days on either side of New Year's Day. As a correspondent of the "Standard" wrote a while ago from Naples stating that the planet Venus, during one of the recent sunsets shone like an emerald in the rosy sky. I may say that the same phenomenon was witnessed here, by me and at least four others, on the evening of December 30th, 1883. —*C. B. Moffat, Ballyhyland, Enniscorthy.*

LATE OCCURRENCE OF THE HOUSE-MARTIN.— A solitary specimen of the house-martin (*Hirundo urbica*) was seen flying at 1 P.M. (November 30th) on Earlham Road, Norwich.—*E. W. Gunn, jun., St. Giles Street, Norwich.*

GEOLOGY, &c.

GEOLOGY IN LIVERPOOL.—The "Stony Science" has always flourished in this city, doubtless owing to the presence of such active and well-known geologists as Messrs. Morton, Ricketts, T. M. Reade, Roberts, &c. The Proceedings of the Liverpool Geological Society, part 5, vol. iv., in addition to the President's Address, dealing with the post-tertiary changes of level around the coasts of England and Wales, contains papers (chiefly local, or dealing with the geology of the neighbourhood) by Messrs. G. H. Morton and T. M. Reade. The Transactions of the Liverpool Geological Association is published, price five shillings, in a neat volume of 157 pp., with coloured sections, &c. It contains a vast deal of local geological matter, although many of its papers go farther afield for material. The chief contributors are Messrs. Bramall (President), F. P. Marratt, Beasley, Brennan, Hall, Logeman, T. M. Reade, Fox, Miles, George, &c.

PECULIAR BLOWHOLE, NEAR BIDEFORD.—On the western bank of the river Torridge, between Bideford and Appledore, there is, just below a rocky cliff, a narrow strip of shaly beach from which, when the tide flows over it, a great number of bubbles burst forth until the water in several places appears to be boiling. I have not been able to ascertain accurately the composition of the gas in the bubbles, nor what quantity of it was given off, but apparently it was ordinary air and came out at the rate of about five cubic feet per minute. At spring tides the water covers the beach to a depth of three or four feet, and the bubbling will then go on for an hour or more. The only explanation that occurs to me is that there must be some cave with a floor below high water mark, and with one opening through which the rising tide enters, and thus forces out the air above it through another opening so placed in a vertical face of rock that sand and mud cannot be carried in with the receding tide. The tidal water is often heavily charged with mud, and it is certainly remarkable that the cave has not been filled up yet. Apparently this action has been going on for centuries, as there is no historical record of any change in the relative level of land and sea in this neighbourhood. Could any of your readers tell me of similar phenomena?—*Herbert G. Spearing.*

THE GEOLOGISTS' ASSOCIATION.—The last number of the "Proceedings" contains the following papers: "On the Drift Deposits at Hunstanton, Norfolk," by B. B. Woodward; "On Some of the Optical Characters of Minerals," by Professor G. S. Boulger; "On the Geology of Hunstanton," by W. Whitaker; "On the Bagshot Series of the London Basin," by the Rev. A. Irving; "On Probable Glacial Deposits at Ealing," by J. Allen Browne.

THE DROITWICH BRINE SPRINGS AND SALIFEROUS MARLS.—A paper has just been read on this important subject by Mr. C. Parkinson, F.G.S. The author referred to the effects of the pumping of brine from beneath Droitwich in producing insecurity in the buildings, and proceeded to discuss the possible source of the brine-water system. He referred to the probable existence of extensive beds of rock-salt, lower than the present brine-cavities, towards the north-east of Droitwich—a conclusion which receives support from the deeper borings carried on at the Stoke Works. Full details of these and other recent borings were given by the author. In the discussion which followed, Captain Douglas Galton said that for every ton of salt at Droitwich 900 gallons of water were pumped, a quantity sufficient to exhaust the rainfall of about six square miles, and that whilst formerly the Droitwich brine-springs overflowed at the surface, the brine is now pumped up 200 feet. He remarked on the difference of level between the Droitwich and Stoke deposits—the highest bed of rock-salt at Droitwich being about 120 feet below mean sea-level and the lowest 170 feet; whilst at Stoke the first bed, which is very thin, is at 170 feet below sea-level, and the lowest yet reached at 300 feet. He suggested the existence of a great fault between Stoke and Droitwich. He thought the question of the existence of deeper supplies of brine at Droitwich a doubtful one.

NOTES AND QUERIES.

HEDGEHOGS.—It would be well for R. T. V. S. W. to acquire at any rate some knowledge of Natural History before sending queries to a scientific magazine. The first one is a fine example of the "scientific method of generalising from observed fact," but it would be well not to forget that generalisations may come under the denomination "rash," especially when founded on insufficient evidence. Moreover "verification" is an important part of this (the scientific) method. I have observed mice not infrequently in the dwellings of *Homo sapiens*; do these ferocious animals devour him? In this and the other matters I would suggest observation, but let R. T. V. S. W. be very careful in making deductions. Moreover, a reference to "Common Objects of the Country" (price only a shilling), in future, would doubtless save this querist much exercise of mind and ingenuity.—*J. R. D.*

EARLY EMERGENCE OF INSECTS (SCIENCE-GOSSIP, xix. p. 280). It is not unusual for the imagines of *Attacus Pernyi* to emerge from the pupæ the same year. This bombyx is a native of North China, where it is double-brooded. This species passes through its metamorphoses more rapidly under natural conditions than with us; there is therefore ample time for the development of a second brood, which in this country would hardly be possible. The following notes, relative to *A. Pernyi*, extracted from my diary may possibly interest Mr. W. Finch. On July 31st, 1875, I received from a friend six pupæ of *A. Pernyi*. The first, a female, emerged on

August 23rd, the next day (24th) a male came out, these at once paired, and on the 25th the female deposited a large quantity of eggs, and three more moths came out. Another also emerged on the 27th. The eggs deposited on the 25th hatched on September 12th, and the young larvæ were supplied with branches of oak, on which they fed well for a time, but did not increase in size very rapidly; most of them passed the second moult, but by November 10th all had died. In 1876 I fed up a quantity of larvæ of *A. Pernyi*, the pupæ from these I placed in a cool cellar to prevent their emerging, but in spite of this an imago appeared on September 22nd, and another on October 9th, the remainder did not produce imagines until the following spring. *Attacus Cynthia* also with me, have occasionally reached the perfect state the same year.—*Robert Laddiman, Hellesdon Road, Norwich.*

BLOOD PRODIGY.—In regard to my communication to SCIENCE-GOSSIP on this subject a relative of mine writes me : "I am much obliged to you for sending me the bramble leaves with the fungi on them. I took them to the Curator of our Institute, who has been giving his attention to microscopic fungi, and he and I come to the conclusion that it must be *Lecythea ruborum*, belonging to the order Puccinæi, another member of which is *Puccinea graminis*, which gives rise to wheat mildew, so that your conjecture was in a right direction. 'Smut' is caused by Ustilago, also belonging to a family of the same order Puccinæi." Shakespeare makes Quintus exclaim when Martius falls into the pit where Bassianus is lying murdered, it will be remembered :

"What art thou fallen? What subtle hole is this,
Whose mouth is cover'd with rude-growing briars ;
Upon whose leaves are drops of new-shed blood,
As fresh as morning's dew distill'd on flowers?"

But since I cannot vouch for every conceit in Shakespeare, I must in justice say that the umber-crimson drops in point are somewhat more suggestive of the liquified blood of some old saint than morning's dew. The artist whose duty it is to paint the tragic stains at Holyrood and Rothesay might take copy. —*A. H. Swinton.*

THE COMMON HEDGEHOG (*Erinaceus Europæus*). —A correspondent in the November number of SCIENCE-GOSSIP, asks for information about the habits of the hedgehog. Possibly the following facts may be of interest to him. Hedgehogs eat insects (especially beetles), snails, slugs, earth-worms, frogs, lizards, snakes, fallen fruit, and small animals, such as mice. I do not think they will eat rabbits—unless they are very small. In captivity they relish small pieces of raw meat. The celerity with which they will clear a kitchen of cockroaches and crickets is well known. I am not able to state with certainty their length of life, possibly about ten or twelve years. They are nocturnal animals and slumber in the daytime. A tame hedgehog will remain persistently curled up and somnolent as long as the sun is above the horizon, giving his possessor the idea that he is an animal of exceptional lethargic disposition. But see him at night, take a candle into the kitchen, during the small hours and see him racing about after the blattidæ and gryllidæ. During the cold months of winter when insect life is for the most part dead or dormant, the hedgehog retires to some snug hollow —frequently a deserted rabbit's burrow, and there curls himself up and sleeps (or hybernates) until warm weather returns. The number of young brought forth by the female is few in number (usually but two, sometimes four). At first the infant hedgehog is white and devoid of the prickly coat so characteristic of adult individuals. It is a singular fact about the hedgehog that you cannot poison it. Strychnine, arsenic, or prussic acid have no visible effect whatever upon it. Moreover, it is indifferent to the bite of the most poisonous snake. Adders are, indeed, a favourite diet with it, and it attacks and devours them, heedless of their venomous fangs. On this account it is a very useful animal, checking the increase of what might become dangerous pests.— *Albert H. Waters, B.A., Cambridge.*

CAUSE OF GOLD-FISH DYING.—In answer to Mr. Easton's query, as to "how it is that gold fishes in a glass aquarium die so soon?" There are probably several reasons for this. He says that he has tried both rain and town water, and that he changes it every three days. Now, I have had some gold fishes for the last two years, and they have always had town water, that is, water which is supplied from the waterworks. I have also in the same tank two chub, these I have had for *five years*. When first I had them they were no more than two inches long, now they are about seven inches long, and very tame, taking flies from the hand. So you will see that it is not the town water which has killed the fish ; and rain water would not be likely to do so, as it is their natural element. With regard to the frequent changing of the water, that may perhaps be partly the reason ; these fish are very delicate, and too much handling, or moving from one vessel to the other, will often cause death. My tank is capable of containing about ten or twelve gallons of water, and in it are the two chub mentioned, three gold fish and three small minnows ; which are quite enough. Perhaps Mr. Easton overstocks his aquarium? I change the water in my aquarium once a week in the summer, and once a month in the winter. I have no aquatic plants or mollusc in it, nothing but the fish, and I never have any trouble with it. Bread should not be put into an aquarium as food for the fish, as it will pollute the water and kill the fish. A little vermicelli is all they require for food, say,—once or twice a week. Then again, some of these gold fish are obtained from water or gas-works, where they are bred in *warm* or *tepid* water ; to place such fish in *cold* water would be certain death. My fish are coldwater bred and therefore are kept in cold water. Fish should not be kept in a room, where gas is burnt, as it is very unhealthy for them. Hoping Mr. Easton will have better success, I remain, &c., *W. Finch, jun., Nottingham.*

WHITE STOAT IN ENGLAND.—A friend of mine saw one—quite white, except for black tip to tail— a few weeks since (December) at Clevedon, Somerset. —*C. Jeffreys, Langharne.*

SOAP-BUBBLES.—The reason why H. J. G. has failed in his experiment is fairly obvious. In introducing the glass tube he has forgotten that the film of the bubble will adhere to its edges and cover its opening. In proof of this let a glass tube be introduced into the bubble in the manner performed by H. J. G., and blow through the tube, when a small bubble will be formed inside the large one, *i.e.*, the film over the opening of the tube will be distended into a spherical shape. This difficulty will be avoided if the bubble be blown by means of the glass tube, and the mouth then removed ; the air will be forced back through the tube, and the bubble will certainly contract, though not, in general, with such force as to blow out a light.—*G. H. Bryan.*

UNSEASONABLE FOLIAGE.—With reference to a paragraph under the above heading, in the December number of SCIENCE-GOSSIP, it may interest your readers to know that a similar phenomenon occurred in the south of Devon. A severe gale on September 1st withered the foliage of many trees, especially the horse-chestnuts. In the end of September fresh leaves appeared on several of these trees, and on one in a garden in Adelaide Terrace, Exmouth, several spikes of blossom opened fully. These were, however, small and poor compared with the usual spring flowers. At the same time, in the same garden, a good deal of apple and pear blossom appeared, and in an adjoining garden there was also a laburnum in full flower. Both these gardens were much exposed to the gale and suffered severely. I may also mention that several swallows (house-martins) were seen at Exmouth on November 11th, shortly before eleven o'clock A.M., flying about or sitting on the telegraph-wire. They had been seen near the same spot (a favourite haunt) once or twice in October, but have not been observed since.—*E. S., Exmouth.*

THE HOLLY.—Many years ago when curate of Checkendon, Oxon, my old squire, the late Adam Duff, drew my attention to a curious fact about the holly. In the woods there it attains the dimensions of a tree. Well, as high as cattle can possibly reach the leaves are armed with sharp prickles, above this they are quite smooth and without this armature. This is a curious fact. I looked over Selby, but he does not allude to it in any way. —*A. H. B.*

[Southey alludes to the fact in his poem on the Holly.—ED. S.-G.]

CONCHOLOGICAL NOTES.—Referring to Mr. George Roberts' query respecting ants, permit me to say that I have frequently availed myself of their services in clearing small and delicate shells of the animal. I have placed the shells in a cardboard tray, under a bell-glass, near an ant's nest in my garden, and the result has been most satisfactory.—*J. W. Cundall, Redland, Bristol.*

NOTICES TO CORRESPONDENTS.

TO CORRESPONDENTS AND EXCHANGERS.—As we now publish SCIENCE-GOSSIP earlier than heretofore, we cannot possibly insert in the following number any communications which reach us later than the 8th of the previous month.

TO ANONYMOUS QUERISTS.—We receive so many queries which do not bear the writers' names that we are forced to adhere to our rule of not noticing them.

TO DEALERS AND OTHERS.—We are always glad to treat dealers in natural history objects on the same fair and general ground as amateurs, in so far as the "exchanges" offered are fair exchanges. But it is evident that, when their offers are simply disguised advertisements, for the purpose of evading the cost of advertising, an advantage is taken of our *gratuitous* insertion of "exchanges" which cannot be tolerated.

WE request that all exchanges may be signed with name (or initials) and full address at the end.

J. F. (Manchester.)—Many thanks for your friendly hints and congratulations.

GEORGE E. EAST.—"The Transactions of the Liverpool Geological Association," vol. iii., Session 1882–3, is published at five shillings, and may be had of Mr. Henry Young, 12 South Castle Street, Liverpool.

M. L. W.—Write to the secretary of the Geological and Polytechnic Society of Yorkshire. No doubt he will be able to supply you with a copy of Messrs. Law and Horsfall's paper.

CHEMICUS.—Your formula was correctly expressed.

G. SMITH (Dudley).—Your box was duly received, but not the slightest trace of any insects was within.

J. FLEMING AND OTHERS.—Accept our best thanks for congratulations on the artistic character of our first coloured plate. It is pleasing to find our efforts to improve SCIENCE-GOSSIP meeting with so much success.

S. B. AXFORD.—Raphides may be easily obtained from the onion, Turkey rhubarb, pine-apple (*Scilla maritima*), the common dock, nettles, leaves of iris and most species of Lily, sepals of geraniums. They may be found in most monocotyledonous plants. See articles on the subject in vol. of SCIENCE-GOSSIP for 1873.

R. A. B. (Kingston).—Grove's "Characeæ" is published by Newman, price 2s. The best work on the British Hepaticæ is that by Dr. Carrington, published by Messrs. W. H. Allen & Co., in coloured parts, price 3s. 6d. each. A cheap work on the Liverworts, &c. (with numerous wood-cut illustrations) was written many years ago by Dr. M. C. Cooke, and published by Hardwicke. Apply for it at Messrs. Allen & Co., 13 Waterloo Place, London.

C. H. WADDEL (Warrenpoint).—Your fungus usually attacks oak wood. It is *Xylostroma giganteum*—sometimes known as "oak leather." Try a strong solution of carbolic acid for it. We warrant that if properly applied you will not be troubled with the fungus again. You can obtain SCIENCE-GOSSIP direct from our publishers by sending postal order for 5s., which covers the cost of one magazine and postage for one year.

GEORGE TIMMINS (Troy, N. Y.).—As you will see by referring to Dr. Stoke's article in our last number, gasoline is a universal solvent for microscopical cements.

W. P. B.—You will find a complete account of the life-history of the caterpillar which destroys the turnips in Miss Ormerod's "Manual of Economical Entomology," and also in her "Reports," for the last two or three years.

A. OGILVY.—Please send us another supply of diseased leaves of Gloxinia.

A. H. B. (San Francisco).—A slight reference to the Infusorian genus Ceratium may be found in the "Micrographical Dictionary," but the best and fullest occurs in Saville Kent's "Manual of the Infusoria." As far as we have been able to warrant them your slides are correctly named. We will report as to the supposed lichen in next number.

M. ROUX wishes to know where he can procure a "Mechanical finger to pick up diatoms." Perhaps some of our readers can tell him?

EXCHANGES.

BETWEEN two and three thousand continental Phanerogams in exchange for recent works on cryptogamic botany or microscopy.—W. B. Waterfall, 9 Redland Grove, Bristol.

WANTED, seeds of British wild flowers, especially woodland species. Exchange seeds of New Zealand plants (mostly shrubs, trees, and alpines), or spores of ferns; or if means of transit offer, dried flowering plants or ferns (named), insects, &c.—T. P. Arnold, Boys' High School, Christchurch, New Zealand.

SCIENCE-GOSSIP, unbound, for 1877, 1881–3 inclusive, with a few odd numbers. What offers?—George Pirie, Carron Terrace, Stonehaven, N.B.

WANTED, quantity of pretty shells, small flat kinds preferred; also corals and other things suitable for shell baskets, &c., exchange for marine algæ, first-class micro-slides, unmounted material, ferns, &c.—T. M'Gann, Burren, Co. Clare.

WANTED to correspond with collectors of spiders, in order to purchase or exchange specimens.—Jno. Rhodes, 360 Blackburn Road, Accrington.

STAMP album, containing upwards of 400 miscellaneous foreign stamps in tolerably good condition; in exchange for micro-slide, books, or bees.—W. T. C., 16 Earl's Court Road, Kensington, W.

WANTED, perfect fossil shells, all formations. Recent British shells, marine, land, and freshwater, offered in exchange.—C. Jefferys, Langharne, Carmarthenshire.

WANTED, exotic butterflies, set or unset. British, European, and American Lepidoptera offered in exchange.—A. H. Shepherd, 4 Cathcart Street, Kentish Town, London.

WILL exchange collection of British birds' eggs (120 specimens 60 species), collection of rocks, fossils, &c., skulls, and few odd natural history specimens for fishing tackle, creel, &c., or fishing books.—F. J. Corkett, High Street, Winslow, Bucks.

ADAMS, G., "Essays on the Microscope," containing a general history of insects, with frontispiece and 32 folio plates. 2nd ed., 4to., half calf, 1798. By F. Kanmacher, F.L.S. What offers? Local floras preferred.—G. H. Knowles, 4 Carfax Square, Clapham Park Road, S.W.

CASSELL'S new edition of "European Butterflies and Moths" By W. F. Kirby. Complete in 61 parts, for instruments for stuffing birds.—Hugh Fleming, 40 Thomson Street, Aberdeen.

WILL send four specimens of *Hydra viridis*, some budding, on the receipt of a good mounted object. Wanted a first-class ⅛ inch objective, also Foster's "Text-book of Physiology" and Geikie's "Text-book of Geology," last editions.—T. W. Lockwood, Lobley Street, Heckmondwike, Yorkshire.

SCIENCE-GOSSIP, Vols. 1, 3, 4, 5, 6, 7, 8, 13, 14, complete. Vol. 2 wanting No. 13, vol. 9 wanting three numbers, vol. 10 wanting eight numbers, and vol. 11 wanting eleven numbers. Desiderata, side parabolic reflector, good lamp, 4th O. G., by good maker, and camera lucida.—R. C. P., The Robin's Nest, Blackburn.

WANTED, back volumes of SCIENCE-GOSSIP, "Botanical Register," "Botanical Magazine," rare British and foreign shells, especially Helices, good micro-slides. Will give in exchange rare and choice lilies and other hardy bulbs, also many rare alpine and hardy flowers. Address H. V. R., Post Office, Horncastle, Lincolnshire.

SIX fancy mice of three different colours, in exchange for other pets or natural history books.—F. H. Parrott, Walton House, Aylesbury.

MICROSCOPIC preparations of marine embryology, &c. Also rare shells. Crustacea or echinoderms offered for books of scientific reference or fossil crustacea.—Edward Lovett, 43 Clyde Road, Croydon.

A SPLENDID collection of Swiss, Pyrenean, and Mediterranean plants, in the most perfect condition as to drying, &c., and all correctly named. Price 6d. each specimen. Address Dr. B., care of Editor of SCIENCE-GOSSIP.

BEAUTIFUL sections of Devonian corals for the microscope, or polished slabs of the corals from Devon. In exchange for the following British marine shells, will take any or a dozen of any one sort as follows : *Pholadidea papyracea, Sphæria Binghami, Neaera costellata, N. cuspidata, N. abbreviata, Pandora obtusa, Lyonsia Norvegica, Thracia distorta, T. convexa, T. pubescens, T. villiosulca, Solecurtis coarctatus, S. candidus, Donax politus, Psamobia costulata, P. tellinella, Astarte elliptica, A. compressa, A. crebricostata, Cytherea chione, Crenella nigra, C. discors, C. marmorata, C. costulata, C. rhombea, Arca trigonia, A. nodulosa, Lama hians, Pecten niveus.*—A. J. R. Sclater, 23 Bank Street, Teignmouth, Devon.

LAND, freshwater, and marine shells, including *H. cartusiana, Ach. acicula, Lim. glutinosa, Claus. biplicata* (foreign), *Ass. Grayana* (from Sandwich, new locality), *Mya truncata*, and *Nucula nucleus.* Northern marine shells especially wanted. —S. C. Cockerell, Glen Druid, Chislehurst, Kent.

FOR any one species in the above list, send box and return postage to S. G. C., Glen Druid, Chislehurst.

WANTED, Hobkirk's "Synopsis of British Mosses," latest edition. State price, &c., to F. J. George, Chorley, Lancashire.

VIVIANITE, Chrome Garnet, Zincite, Franklinite, Calamine, Zircon, Sodalite, Cancrinite, Precious Tourmaline, Lepidolite, Chalcophanite, &c., to exchange for British lead minerals. Fuller lists sent to those wishing to exchange.—W. F. Ferrier, P. O. Box 377, Montreal, Canada.

A SPLENDID specimen of *A. atropos*, female; well set. What offers? Communicate first.—J. Boggust, Alton, Hants.

WELL-MOUNTED slides of the following: Spores of Lycopodium, Spores of *Filix-mas, Puccinia malvacearum, Puccinia arundinacea*, tobacco leaf, scales of bream, and foraminifera (opaque mountings) from the following localities, Mediterranean Sea, West Indies, Singapore, Japan, Orkney and Sheppy, and will send any on receipt of a good slide (diatoms preferred) in exchange.—F. A. A. Skuse, 143 Stepney Green, London, E.

WANTED, foraminiferous sand from all parts of the world, also dredging, deposits of diatoms, infusorial earths, &c. Will send in return other foraminiferous sand, &c.—F. A. A. Skuse, 143 Stepney Green, London, E., England.

WANTED, to exchange foraminifera in slides or material for other foraminifera.—J. H. Harvey, St. John's College, Cambridge.

SEND stamped envelope to address as below for foraminifera sand containing microscopic shells from Mediterranean Sea. Well-mounted slides of tobacco leaf in exchange for other slides of interest.—E. C. Stedman, 115 Stepney Green, London, E.

"NATURE," Vols. 1 to 10, bound, clean, and in best condition, for a few skeletons and preserved specimens of British freshwater fishes, such as would suit a boys' school museum.—George Fyfe, Jedburgh, N.B.

MOUNTED stained pig lice in exchange for rock sections or other slides of interest, except spicules, foraminifera, and diatoms; a lot of wood sections in exchange also.—S. R. Hallam, 22 High Street, Burton-on-Trent, Staffordshire.

PREPARING ROCK SECTIONS.—Can any reader give me any help in mounting rock sections and grinding them for microscope.—S. R. Hallam, 22 High Street, Burton-on-Trent, Staffordshire.

CONCHOLOGY, *Unio margaritifer, tumidus, pictorum, Dreseina polymorpha*; any number of above duplicates given in exchange for common marine specimens.—Holstead, Lees Street, Lodge Road, Birmingham.

DUPLICATES: *Rhamni, paphia*, Cardamines, Cardui, Atalanta. Desiderata: Sinapis, Semele, Rubi, Betulæ, Iris, Argiolas, *C. album*, Villica, Napta, Promissa, Sponsa, Præcox, Prodromalia, larvæ of Villica, Monacha, &c.—J. Bates, 10 Orchard Terrace, Wellingborough.

DUPLICATES: *Limnæa glabra, L. palustris, L. peregra, Planorbis spirorbis (carinatus), P. complanatus, Physa hypnorum, Limnæa stagnalis*, var. *fragilis*, and *Planorbis corneus*, fine; locality, Strensall Common, near York. Desiderata : British marine and land and freshwater shells.—W. Hewett, 26 Clarence Street, York.

Limnæa glabra at the head of my duplicates. The other duplicates were *Limnæa stagnalis* and *Planorbis corneus*, very fine, *Limnæa palustris* and *peregra, Planorbis, Spirorbis, P. complanatus* and *Carinatus* (*Physa hypnorum*). Desiderata very numerous.—W. Hewett, 26 Clarence Street, York.

STARFISHES.—Fine specimens of *Solaster endeca, Luidia fragilissima, Goniaster equestris, Ophiocoma granulata*, &c., for other British starfishes, fishes, or crustacea.—George Sim, 20 King Street, Aberdeen.

DUPLICATES: *Bulimus miltocheilus*, several species of *F. cyclostomidæ* and *F. helicidæ*. Wanted, recent Brachiopoda, Pteropoda, Chamidæ; good well-mounted microscopical slides. —J. E. Linter, Colony House, Grosvenor Road, Twickenham.

A FEW specimens of *Cheilanthes Californica, C. Clevelandii, C. Fendleri*, from Southern California, for rare British or other foreign ferns or shells.—J. Edward Reed, Santa Clara, Santa Clara Co., California.

OFFERED, the following Diatomaceæ unmounted but prepared, *Himantidium pectinale*, Black Moss (Scotland), *Schizonema obtusum, Gomphonema geminatum* (Richmond, Virginia), *Tabularia flocculosa* (Leghorn, Italy), Cherryfield (Maine); wanted unmounted or mounted animal objects.—H. T., South Cross, Musbury, Axminster, Devon.

LITTLE auk nicely mounted on rock-work under glass-shade, also great auk egg (cast), and some Indian eggs to exchange for eggs, English or Foreign.—G. A. Widdas, Bond Street, Leeds.

BOOKS, ETC., RECEIVED.

"The English Flower Garden." By W. Robinson. London John Murray.
"The Chemical Effect of the Spectrum." By Dr. J. M. Eder. London : Harrison & Sons.
"Studies in Microscopical Science," edited by A. C. Cole.
"The Methods of Microscopical Research," edited by A. C. Cole.
"The Antiquary."
"Journal of Conchology."
"The Journal of Microscopy."
"Land and Water."
"The Science Monthly."
"Midland Naturalist."
"The Inventor's Record."
"The Naturalist's World."
"Ben Brierley's Journal."
"Sunlight."
"The Medical Student." (New York.)
"Natural History Notes."
"Science."
"American Naturalist."
"Medico-Legal Journal." (New York.)
"Canadian Naturalist."
"American Monthly Microscopical Journal."
"Popular Science News."
"The Botanical Gazette."
"The Arcadian Scientist"
"The Ornithologist and Oologist."
"Revue de Botanique."
"La Feuille des Jeunes Naturalistes."
"Le Monde de la Science."
"Ciel et Terre."
"Cosmos: les Mondes."
&c.   &c.   &c.

COMMUNICATIONS RECEIVED UP TO 11TH ULT. FROM :— A. H. S.—A. E. L.—T. L.—A. O.—J. W.—E. F. B.—C. H. W.—W. T. H.—H. C. W.—G. B.—T. W. B.—W. L. B.—T. D.—C. B. M.—J. H. B.—C. S. H.—W. H.—G. T.—A. N.—H. W.—S. W. B.—W. B. W.—T. P. A.—Dr. P. Q. K.—T. D. A. C.—J. E. A.—G. W. B.—S. A. B.—A. H. B.—T. H. F. K.—J. F.—D. M.—F. A. A. S.—C. B. M.—C. J. S.—M. L. W.—J. B.—G. R.—W. F. F.—G. E. E., jun.—F. J. G.—E. G. H.—H. T. M.—C. H. O. C.—J. H. H.—F. H. P.—J. W. C.—E. L.—T. M. C.—C. G. P.—F. R.—W. F.—A. J. R. S.—Dr. D. S.—S. C. C.—J. E. L.—E. S.—A. H. B.—H. Y. R.—R. C. P.—T. W. L.—H. F.—J. H. K.—F. T. C.—A. H. S.—J. E. R.—G. H. B.—W. H. P.—E. W. G.—W. T. C.—J. R.—C. J. S.—S. C. C.—E. S. S.—S. B. A.—J. H. M.—T. W. H.—G. P.—J. B.—S. R. H.—G. S.—W. H.—M. A. H.—V. G.—C. B.—H. M., jun., &c.

GRAPHIC MICROSCOPY

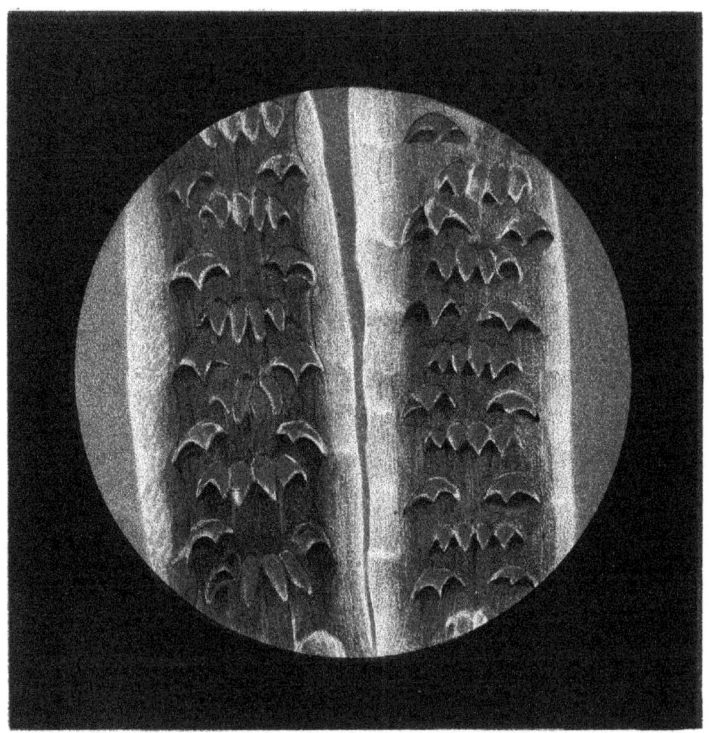

PALATE OF LIMPET.
× 50

GRAPHIC MICROSCOPY.

By E. T. D.

No. III.—Palate of Limpet.

IN reference to this subject, the words of Mr. Gosse may be quoted, "Who that looks at the weather-worn cone of the limpet, as he adheres sluggishly to the rock, between tide-levels, would suspect that he carries coiled up in his throat, a tongue twice as long as his shell? and that this tongue is armed with thousands of crystal teeth, all arranged with the most consummate art, in a pattern of perfect regularity!"

The tongue, or palate of the limpet (*Patella vulgata*) is a typical representative of the contrivance, or cutting instrument, attached to the muscular cavity which contains the oral apparatus of the gasteropod mollusks. It is a long, or ribbon-like tooth-bearing membrane, spread open upon the floor of the mouth, forming nearly a flat surface; and then assuming a more or less lengthened tubular form, the interior studded with transverse rows of teeth arranged upon flattened plates; when dissected and opened, it merges into the form of a consecutive or continuous band (the drawing represents two portions of the same band), and is in the nature of a rasp or file, admirably adapted, first to graze upon, and then to divide, bruise, and engulf vegetable or other nutriment; as the teeth, or spines, are arranged points downward on the flexible cartilaginous strap; they not only collect aliment, but assist in propelling it into the œsophagus; the wear and tear are made good by constant growth and new development. As the anterior prickles are worn away and absorbed, another portion of the tongue is brought forward to supply its place, and that there may be no deficiency in its length, the apex, the point where the continual growth and addition are going on is soft and vascular; each principal tooth sometimes has a basal plate of its own, in other instances, one plate carries several.

So strangely diversified is the character of the lingual ribbon in the gasteropods, as showing generic character, that the classification of the Mollusca has been attempted on the basis of its structural arrangement and disposition. In the large garden slug (*Limax maximus*) there are as many as one hundred and fifty rows of teeth, and the rows so closely packed, that the aggregate number of teeth is said to amount to over twenty-six thousand. In the marine gasteropods, particularly the limpet, we find the teeth larger, and the tongue so long, that it is even folded up in the abdominal cavity. It has been urged by Dr. Gray that these structures might be an important guide to the natural affinities of the species, genera, and families of the group, since diversities so strongly marked must necessarily affect the habits, form, and character of the animal. This view of the subject is full of interest, and a systematic examination opens out to the student of microscopy, a most genial employment.

The actual operation of the action of the tongue, in newly-hatched individuals, may be observed on the stage of the microscope, and is conveniently seen in an ordinary freshwater aquarium containing pond snails in various stages of growth; the introduction of mollusks, in such arrangements, is in some degree necessary to preserve an equilibrium of health, as their presence suppresses the overgrowth of minute confervæ. And when these plants are attached to the side of the tank, the snails may be found continually browsing; with an ordinary lens the rolling action of the palate can be plainly seen, rasping, or rather mowing the minute vegetation, in regular and consecutive swathes, and leaving behind the marks of progress. A graphic passage in Woodward's "Mollusca," evidently written from direct observation, may be cited. "The upper lip with its mandible is raised, the lower lip expands; the tongue is protruded, and applied to the surface for an instant, and then withdrawn; its teeth glitter like glass-paper, and in the pond snail it is so flexible that frequently it will catch against projecting points, and be drawn out of the shape slightly, as it vibrates over the surface."

In the limpet, the food is grasped by the lips, drawn forward and retained by the prickly tongue, and simultaneously pressed against the upper horny jaw, by which means a portion is bitten off; the detached morsel is then pressed along the tongue, torn rasped down in its progress, and forced onwards by peristaltic motion; thus the mass is made to enter the gullet.

The gasteropods are not entirely vegetable feeders. The popular whelk and snail will eat flesh of all kinds, and the slug has been observed browsing on an individual of its own species, accidentally crushed and scarcely dead, and they feed eagerly on earthworms. Of the aquatic tribes, as may be seen in aquaria, the food of the Limnei is frequently animal matter, which makes them deserve the name of "scavengers of the waters." In the absence of other nourishment they will even devour each other.

The tongues, or palates, form beautiful microscopic objects, not very difficult to procure or prepare, and they may be arranged to meet all conditions of illumination (for the purpose of the drawing the paraboloid was used), but as an opaque object, with reflected light, structure and colour are seen to equal advantage; in balsam, with polarised light, the most gorgeous effects may be produced.

ERRATUM.—In No. 1 of this series, the title of the Plate and paper should have been *Tingis crassicornis*. Crouch End.

NEW EVIDENCES OF PRIMEVAL MAN.

A WEEK or two ago, when lecturing to the Hitchin Natural History Society, I took advantage of the opportunity to visit the remarkable brick-pits in the neighbourhood. I had paid one of them a hasty visit about three years before, when palæolithic flint implements were first found there; and on this later occasion I had the privilege of being accompanied by Mr. William Hill, the President of the Society, who pointed out all the details of the surface deposits.

The latter are exceedingly interesting, and the implements are found under conditions quite different to any I have heard of before. In this county Palæolithic weapons have been met with chiefly, if not only, in valley gravels, and in such ancient cave breccias as those of Kent's Hole, Torquay.

In the neighbourhood of Hitchin, however, they are found in deposits of quite another character. One of the best places for them is an extensive pit worked for the brick-earth. The deposit here was evidently formed on the bottoms of small lakes, for about six feet of the lower beds at one end of a pit I visited are of a rich cream colour, and full of the remains of Bithynia and other fresh-water shells. Indeed it is a kind of fresh-water marl, formed chiefly by their partial decomposition. The upper beds are of a darker colour, and appear to me to have been more arenaceous, so that the surface water has percolated through them, and dissolved and carried away the lime, leaving them of a different colour to the unaltered beds beneath. At the other end of the pit is quite a different deposit, more or less stratified, and lying at an angle of about twenty degrees. It is a mass of rubbish, and has quite a morainic appearance.

It appeared to me that into one end of this lake a small glacier must have found its way, and there deposited this rubbly material, whilst in the stiller parts the mollusca lived and died, and their shells accumulated to form the marl. The country round about is of such a physical character as would easily allow of these conditions taking place at the close of the Glacial Period. The original surface must have been very undulating, so that the drainage would flow into the hollows. Masses of boulder-clay occur here and there in the neighbourhood of the marl-pits at a higher level, but they are of more ancient date.

The flint implements appear to be tolerably numerous, and all of them are beautifully chipped and worked. I looked over a number of them in the cabinets of local collectors, and was particularly pleased with those I saw in the collections of Messrs. Hill and Ransom. In another pit (where I saw no traces of freshwater marl, though this may be because it is worked at a higher level, and has not been carried so low down), not only are numerous palæolithic weapons found, but abundance of chips and splinters, and even hand-hammers formed of flint nodules, all of which bear plain evidence on their surfaces of having been used to detach the flakes.

The discovery of these implements under these interesting conditions, was due, in the first instance, I believe, to one of the workmen in the brick-pits, who was reading an article in the "Leisure Hour" about Primitive Man, in which some of the flint implements were figured. He immediately recognised them, and actually had one in his possession at the time, for he appears to have been of an observing and collecting turn of mind. He soon found others after that, and now they are turning up every week.

The discovery of palæolithic implements in freshwater deposits which bear evidences of glacial conditions is a novel addition to our knowledge of the subject.

J. E. TAYLOR.

DARWIN AT HOME.

A WRITER in the "American Naturalist," for January, gives the following interesting account of some incidents in the life of our celebrated naturalist which he managed to pick up :—

In a recent visit to England, the writer strolled into the village of Down in Kent, and talked with some of the villagers in regard to Mr. Darwin, whose beautiful home is just outside the town.

Some of this talk, although in itself idle and valueless, may have an interest to readers, as showing how a great man looks towards his smaller neighbours. The landlord of the "George Inn" said that "All the people wished to have Mr. Darwin buried in Down, but the Government would not let them. It would have helped the place so much. It would have brought hosts of people down to see his grave. Especially it would have helped the hotel business, which is pretty dull in winter time." "Mr. Darwin was a very fine-looking man. He had a high forehead, and wore a long beard. Still, if you had met him on the street, perhaps you would not have taken much notice of him, unless you knew that he was a clever man." "Sir John Lubbock (Darwin's friend and near neighbour) is a very clever man, too, but not so clever nor so remarkable-looking as Mr. Darwin. He is very fond of hants (ants) and plants and things." At Keston, three miles from Down, the landlady of the "Greyhound" had never heard of Mr. Darwin until after his death. There was then considerable talk about his being buried in Westminster, but nothing was said of him before. Several persons had considerable to say of Mr. Darwin's extensive and judicious charity to the poor. To Mr. Parslow, for many years his personal servant, Mr. Darwin gave a life pension of £50, and the rent of the handsome "Home Cottage" in Down. During the time of a water famine in that region, he used to ride about on horseback to see who needed water, and had it brought to them at his own expense from the stream at St. Mary's Cray. "He was," said Mr. Parslow, "a very social, nice sort of a gentleman, very joking and jolly indeed ; a good husband and a good father, and a most excellent master. Even his footmen used to stay with him as long as five years. They would rather stay with him than take a higher salary somewhere else. The cook came there while young, and stayed there till his death, nearly thirty years later." Mrs. Darwin is a pleasant lady, a year older than her husband. Their boys are all jolly, nice young fellows. All have turned out so well, not one of them rackety, you know. Seven children out of the ten are now living. George Darwin is now a professor in Oxford. He was a barrister at first; had his wig and gown and all, but had to give it up on account of bad health. He would have made a hornament to the profession. Francis Darwin is a doctor, and used to work with his father in the greenhouse. He is soon to marry a lady who lectures on Botany in Oxford. For the first twenty years after Mr. Darwin's return from South America, his health was very bad —much more than later. He had a stomach disease which resulted from sea-sickness while on the voyage around the world. Mr. Parslow learned the water-cure treatment, and treated Mr. Darwin in that system, for a long time, giving much relief. Mr. Darwin used to do his own writing, but had copyists to get his work ready for the printer. He was always an early man. He used to get up at half-past six. He used to bathe, and then go out for a walk all around the place. Then Parslow used to get breakfast for him before the rest of the family came down. He used to eat rapidly, then went to his study and wrote till after the rest had breakfasted. Then Mrs. Darwin came in, and he used to lie half an hour on the sofa while she or some one else read to him. Then he wrote till noon, then went out for an hour to walk. He used to walk all around the place. Later in life he had a cab, and used to ride on horseback. Then after lunch at one, he used to write awhile. Afterwards he and Mrs. Darwin used to go to the bedroom, where he lay on a sofa and often smoked a cigarette while she read to him. After this he used to walk till dinner-time at five. Before the family grew up, they used to dine early, at half-past one, and had a meat tea at half-past six. Sometimes there were eighteen or twenty young Darwins of different families in the house. Four-in-hand coaches of young Darwins used sometimes to come down from London. Mr. Darwin liked children. They didn't disturb him in the least. There were sometimes twenty or thirty pairs of little shoes to be cleaned of a morning, but there were always plenty of servants to do this. The gardener used to bring plants into his room often of a morning, and he used to tie bits of cotton on them, and try to make them do things. He used to try all sorts of seeds. He would sow them in pots in his study.

There were a quantity of people in Westminster Abbey when he was buried. Mr. Parslow and the cook were among the chief mourners, and sat in the Jerusalem Chamber. The whole church was as full of people as they could stand. There was a great disappointment in Down that he was not buried there. He loved the place, and we think that he would rather have rested there had he been consulted.

STORM GLASS.—In reply to J. H. Milne's query in January number, I find on reference to "Gardening Illustrated," vol. i. p. 104, that the bottle, or chemical barometer "should be corked very loosely, or better tie over the orifice a piece of linen or cotton cloth, and place the storm-glass in a good light out of the sunshine, where it can be observed without handling."—*S. A. B.*

THE FOSSIL PECTENS OF THE UPPER GREENSAND AND CHLORITIC MARL OF THE ISLE OF WIGHT.

By C. PARKINSON, F.G.S.

SOME of the drawings accompanying this paper are from specimens found by me, most of which are now included in the collections of the Geological Department of the Natural History Museum, South Kensington. During the last few years I have spent a great deal of time and trouble in cataloguing the fossil of upper greensand, chloritic (or glaucanitic) marl, and chalk marl, at the south side of the Isle of Wight, taking pains, at the same time, to separate the fossils of each zone; as a result of this care in referring every fossil to its own horizon, we find near to Ventnor railway station; the result corresponds altogether with my previous measurement in the same spot. At the same time it is only fair to state that in different parts of the undercliff the chloritic marl varies in thickness, as it apparently fills up depressions in the stratum below. I am certain, however, that six feet is an average thickness, and that it may be divided into two divisions; $3\frac{1}{2}$ feet fossiliferous, with base of hard phosphatic nodules and crushed *Pecten asper*; $2\frac{1}{2}$ compact, darker grains, and few fossils.

Mr. Etheridge thought that *Pecten asper* was more properly an upper greensand form, and very unusual in the chloritic marl. In the five years' experience I have had in the Isle of Wight, I have never been able to find a single specimen of *Pecten asper* below the phosphate nodules referred to above. In that

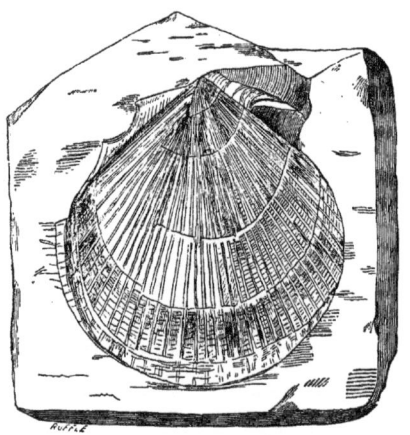

Fig. 30.—*Pecten interstriatus.*

Fig. 31.—*Pecten asper.*

evidence of no less than three successive faunæ between the chloritic marl and the base of the upper greensand. In a paper communicated to the Geological Society in March, 1881, I gave the measurements of these rocks as found at St. Lawrence and Ventnor, I. W., at the same time pointing out how remarkably certain zones might be distinguished and determined by the careful observation of the palæontological remains. In the short discussion which followed, some astonishment was expressed at my measurement of the chloritic marl, and at the position I had given *Pecten asper*, while an opinion was also expressed that such widely distributed genera as Pecten and Lima were not of themselves sufficient to form a guide in separating the zones of life in the greensands.

Since that time I have had an opportunity of remeasuring the chloritic marl in a section of a quarry band, a few inches only in thickness, many specimens may be found, all more or less broken, a fact which would seem to indicate some violent and sudden action by which the nodules were formed and the molluscs destroyed. Fig. 31 is a specimen I found *in situ* in the nodule band; the denticulated ribs are here visible, being worn away from the larger specimen. It is most certainly a characteristic species (though possibly derived) of the chloritic marl.

Pecten Beaveri (Sow.) also occurs plentifully in the marl; the valves of specimens are different from chloritic and chalk marl respectively.

Pecten orbicularis (Sow.) is a species common to several zones in the upper greensand and marls. Below the chloritic marl there lies twenty-four feet of alternate bands of greensand and hard blue chalk.

P. interstriatus (Seym.) in the second band of

coarse greensand, occurring, as far as I have been able to discover, in this horizon only. The drawing represents the left valve, although the characteristic details of the ribs, so well shown in D'Orbigny, are not visible in this specimen.

P. inæquivalvus is one of the most beautifully preserved fossils of the upper greensand. I have found it only in a zone which lies some four feet below that of *Pecten interstriatus* (this species closely resembles *P. elongatus*, Lam., figured in D'Orbigny).

Pecten elongatus is apparently a young shell, possibly the earlier stage of interstriatus, as Mr. Etheridge has kindly suggested to me. I have, however, found many specimens of the same size, in a zone distinct from that of *P. interstriatus*; they occur on account of the deep indented divisions of the lower valve, whereas they usually are flat (*vide* figure in D'Orbigny of *Janira digitalis*), corresponding rather closely with my specimen.

A specimen of another pecten I showed to Mr. A. J. Jukes-Browne, F.G.S., and he expressed an opinion that the species was possibly undescribed. I think, however, that it may be referred to *P. interstriatus* (Leym.), showing the characteristics of the left valve, which apparently differs considerably from the right valve. This specimen was moreover from the same zone as fig. 30.

It may be unsafe to state absolutely that these pectenidæ actually afford means of grouping successive zones of the upper greensand in their relative

Fig. 32.—*Pecten Beaveri.*

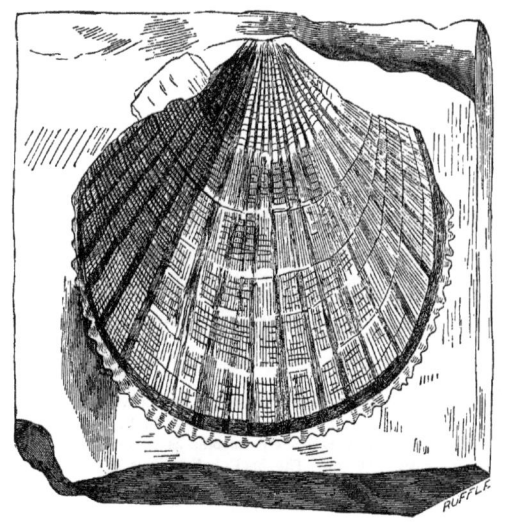

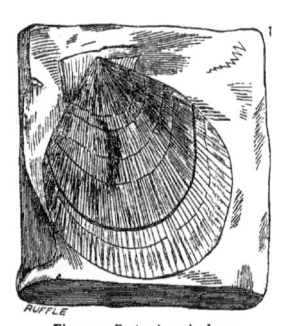

Fig. 33.—*Pecten inequivalvus.*

only in the Malon rock, in the four foot band of Ibbeston.

Lima, sp. This fossil might either be referred to the genus Pecten, although from the unequal form it has more the characteristic of a lima. When a perfect specimen is discovered it may correspond with *L. Gallieunei* (Derb.), as it has exactly the same number of strong ribs as that species figured in D'Orbigny. It is confined in the upper greensand to a blue chert band, below the zone of *P. inæquivalvus*. This chert is so hard that specimens are extremely difficult to extract; this lima took nearly a week to develope after the piece of chert had been broken off the exposed surface of rock.

I found a peculiar variety of *P. 5-costatus* (Sow.), differing from any specimen I had previously found, order, but I cannot help thinking they do indicate some true succession of fauna. *P. asper, P. interstriatus, P. inæquivalvus,* and Lima, sp., I have never been able to find, except in the zones named—except in cases at St. Lawrence, where strata are inverted through the catastrophe which formed the beautiful scenery of the undercliff.

It happened one day that I was examining a fragment of plocamium under the microscope, in search of some hydroid zoophyte, when I noticed a minute pecten, hardly bigger than a large pin's head, disporting itself in the water in the full vigour of life, a fit object for comparison with the representations of the same genus from the rags and cherts of the greensands. With a rapid movement, caused by the expansion and contraction of the valves, this tiny bivalve shot across the salt water trough, playing in and out of the red fronds of plocamium. It clung to the frond by means of the hinges, while the valves continued to open and shut with great rapidity, as the hidden mollusc protruded a series of filaments or thread-like processes which correspond in number to the ribs of the external shell. It has even been said that the extremity of each filament consists of an organ of vision, by means of which the creature surveys its surroundings. Whether this be so, or whether the processes are for the purpose of catching food, the restless movement is most interesting for observation, affording some insight, not only into the life of the recent species, but also into the life-history of the greensand genera. With their restless expansion and contraction they lived in the ocean as it then existed, each successive species surviving so long as the necessary conditions of life were maintained, and each in turn giving place to the developments more adapted to the ever-changing ocean bed or littoral zones.

NOTES ON NEW BOOKS.

THE ENGLISH FLOWER-GARDEN, by W. Robinson (London : John Murray). The author's name is well known, both as a botanist and horticulturist ; and it is only such a man who ought to undertake a book of this kind. Within the last few years the garden has become something more than a mere pleasure-ground, enjoyable and healthful as it is in that respect—it has developed into a scientific observation-ground and even a laboratory. The smallest garden may have collected within its narrow limits, plants from all parts of the world, representing the most widely separated of orders. Since biological botany came to the front, the deviation of every exotic species from British types has obtained a fresh significance. Mr. Robinson has arranged his work in alphabetical order—the most convenient for his readers he could have devised. The illustrations are numerous, and for the most part very effective. The author tells us, "The whole aim of the book is to make the flower-garden a reflex of the world of beautiful plant-life, instead of the poor formal array it has long been." Mr. Robinson has called in to his aid the best writers and contributors to his own journal, " The Garden," and has thus produced an encyclopædic work of upwards of 424 pp. and 274 plates of flowers, &c. In addition to the alphabetical description of garden-plants, there is an introduction extending over 124 pages, dealing with such subjects as " Examples from English Gardens," " Hardy Plants, and the Modes of Arranging Them," "Hardy Flowers, Bulbs, &c.," "Spring Flowers," "Alpine, Bog, &c., Plants," "The Garden of Sweet-smelling Flowers," "The Garden of Beautiful Form," "The Wild Garden," "Roses," &c. Altogether a most useful and readable, and thoroughly profitable book has thus grown together, which will meet and satisfy a long-felt want.

Flowers and their Pedigrees, by Grant Allen (London : Longmans & Co.). Under this attractive and suggestive title the author has collected various essays and articles which have appeared in magazines lately. Mr. Grant Allen plays "ducks and drakes" with the conservatism of academic botany. Many of the extreme supporters of the latter seem to think that plants existed for the purpose of being technically described, and that a man who offended in the least nomenclature matter was guilty of an unpardonable sin, notwithstanding his intimate knowledge of plant life. Mr. Grant Allen has made himself sympathetically familiar with the life and habits of the commonest plants. We often read of naturalists who have so thoroughly identified themselves with the animal world, that the latter has responded in unison. Nathaniel Hawthorne beautifully indicated this kind of sympathy between animals and men in his character of "Donelli." But we never knew any naturalist who seemed to similarly identify himself with plants. All of us love flowers, for the sake of the pleasure they give us—but to love them for their own sakes, in spite of unattractiveness, of lack of colour and perfume—to read off in dwarfed stems, degenerated stamens, aborted petals, modified and altered leaves, botanical details concerning fruit and seeds, the genealogy of the plant and the numerous life-changes the individuals have passed through since the original species attained its individual development geological ages ago—nobody has done this like Grant Allen. And, even if the task had been achieved, none will deny that it has never been done in English so graceful and exquisitely smooth, as in the works of our author. The book before us is richly suggestive. A few errors of fact, and possibly also of logic, cannot detract from the merit of this little work. Mr. Allen offers it as the first instalment of a work he hopes some day more fully to carry out—a Functional Companion to the British Flora. We sincerely hope he may live, not only long enough to carry out his hopes, but also to enjoy the honour which rightly belongs to a man devoted to such work. The eight chapters of *Flowers and their Pedigrees* are among the brightest and best things in our English botanical literature.

The Poet's Birds, by Phil. Robinson (London : Chatto & Windus). Ornithologists and zoologists surely cannot have had this remarkable book brought vividly before their notice, or we should have heard more of it from their quarters. It deals with no fewer than 90 different kinds of birds, mostly British species, about which our poets have written, or to which references are made in their works. No fewer than 80 of our British poets are quoted. Alas ! their poetical references and descriptions are not only seldom true, but as a rule the very reverse of truth. All this Mr. Phil. Robinson (to use his own penname) shows up in the raciest of English, and with the driest of humour and satire. Nothing like this book has appeared since Butler's *Erewhon*. There is hardly a paragraph which does not force a laugh or an exclamation.

The Parallel Roads of Glen Roy, by James Macfadyean (Edinburgh : John Menzies & Co.). The old geological bone of contention, which is regarded by most geologists as finally settled, is brought up again by Mr. Macfadyean in connection with the Deluge. A good deal of geographical research is displayed by the author, particularly in the chapter on "The Cosmical Change of Level that followed the Deluge," and he evidently connects the events of the glacial period with that fact in the Hebrew Cosmogony.

Solar Physics, by A. H. Swinton (London : W. H. Allen & Co.). The writer of this work is well known to the readers of SCIENCE-GOSSIP, as a frequent and welcome contributor. His scheme in this little book is to present a handy epitome of the years, or universal almanack, that may be consulted by every one. It is in reality an almanack of the Christian Era, and contains a prediction of the weather, disasters by sea and rain, shipwrecks and river floods, prognostications of the harvest, havoc by vermin and infection, famines and panics, electrical disturbances, calamities by earthquakes and volcanic eruptions—a record of the past, and glimpse into the future,—based on solar physics. Mr. Swinton has expended a vast amount of industry on his book, and the pages devoted to "The Sun Cycles," wherein are arranged, in tabular form, the occurrences of earthquakes and volcanic outbursts, &c., with observed sun-spot years, is both interesting and suggestive. Altogether this is a remarkable work, in which both zoological and physiographical students will find much to ponder over.

The Chemical Effect of the Spectrum, by Dr. J. M. Eder (London : Harrison & Sons). This little book is translated and edited by Captain Abney—a circumstance that will be quite sufficient to lovers of photography and physics to recommend it to their notice. It has been reprinted from the "Photographic Journal" of 1881 and 1882.

Where did Life Begin? by G. Hilton Scribner (New York : Charles Scribner's Sons). This short monograph is a brief enquiry as to the probable place of beginning, and the natural courses of migration therefrom, of the flora and fauna of our planet. The author sets forth his reasons for thinking that life first commenced and spread from the polar regions, where the earth would be first cooled. If we mistake not, Professor Thistleton Dyer elaborated a similar view some years ago. Mr. Scribner's short essay is written in a most vigorous and fervid style, as the following concluding sentence will show : "Thus the Arctic zone, which was earliest in cooling down to the first and highest heat degree in the great life-gamut, was also first to become fertile, first to bear life, and first to send forth her progeny over the earth. So, too, in obedience to the universal order of things, she was first to reach maturity, first to pass all the subdivisions of life-bearing climate, and finally the lowest heat degree in the great life-range, and so the first to reach sterility, old age, degeneration and death. And now, cold and lifeless, wrapped in a snowy winding-sheet, the once fair mother of us all rests in the frozen embrace of an ice-bound and everlasting sepulchre."

A Tour in the United States and Canada by Thomas Greenwood (London : L. Upcott Gill). Mr. Greenwood gives us a very lively account of a run out and home again in six weeks. America is becoming a recreation ground, as well as an emigration settlement, and to all who meditate a run across the Atlantic we recommend Mr. Greenwood's little book.

Botanical Micro-Chemistry, by V. A. Poulsen (London : Trübner & Co. ; Boston : S. E. Cassino & Co.). This most valuable handbook to all engaged in histological work, owes its appearance in our English language to Professor W. Trelease, of the University of Wisconsin, who became acquainted with the German edition in 1881, and felt what a valuable work required translation. Mr. Trelease has accordingly translated it, and enriched it with numerous notes, the result of discoveries subsequent to the original publication of the book, which has received the honour of translation from the Danish original, into German, French, Italian ; and now, thanks to Professor Trelease, into English also. The first part of the contents deals with microchemical reagents, and their application, mounting media, cements, &c., whilst the second part is engaged with vegetable substances, and the means of recognising them. We strongly recommend the little volume to all our microscopical readers.

Energy in Nature, by W. Lant Carpenter (London : Cassell & Co.). This attractive little book contains the substance of a course of six lectures upon the energies of Nature, and their mutual relations, delivered by the author under the auspices of the Gilchrist Educational Trust. They deal with the following subjects : Matter and Motion, Force and Energy, Heat and Form of Energy, Chemical Attraction (especially Combustion), Electricity and Chemical Action, Magnetism and Electricity, and Energy in Organic Nature. The illustrations are numerous, and of a superior artistic character, the style is terse but plain, [and the author's allusions always felicitous. A much-needed want is admirably met by this little work. The last chapter on "Energy in Organic Nature" is one we recommend all naturalists to read.

Whence; What, Where? by James R. Nichols, M.D. (Boston : A. Williams & Co. ; London : Trübner & Co.). This is the fourth edition of a very thoughtful and suggestive book, written by the Editor of "The Popular Science News" (formerly known as "The Boston Journal of Chemistry"). It covers (as its title indicates) a good deal of ground of a theological character. Many of the subjects have been discussed times out of mind, but Dr. Nichols brings to their consideration singularly original methods of contemplation, and his familiarity with the latest views of science enables him to discuss abstract questions with fulness and breadth.

A Handbook to the Fernery and Aquarium, by J. H. Martin and James Weston (London : T. Fisher Unwin). A cheap, handy, well-written, and very useful little book, which would have been better if it had been fuller and more extensive.

The Organs of Speech, by George Hermann von Meyer (London : Kegan Paul, Trench & Co.). This is the latest issued volume of the celebrated "International Scientific Series." The author is one of the most distinguished and widely-known of continental scientists, and is Professor of Anatomy at the University of Zurich. The present work is something more than an ordinary anatomical handbook of the subjects. It specially deals with the organs of speech with a view to assisting the philologist in obtaining a knowledge of the laws which govern the transformation of the elements of speech in the formation of dialects and derivative languages. The author gives a most important chapter on all possible articulate sounds, and constructs a system upon them, which cannot fail to influence future writers who take up the subject of comparative philology. All who are interested in vocal sounds, whether of music or speech, have here a clearly written and lucid, as well as a thoroughly exhaustive manual.

Chemical Analysis for Schools and Science Classes, by A. H. Scott-White, B.Sc., &c., (London : Thomas Laurie). Chemical students are perhaps better supplied with cheap manuals and handbooks than any other class. But we have not seen any which comes up to the present little volume, either in method of treatment or usefulness. Mr. Scott-White has started a new line of departure from the old and hackneyed methods of teaching chemistry. Having had considerable laboratory practice and demonstrative work, he knows exactly what practical students require. By the aid of this book any beginner can work out by himself all the fundamental principles of modern chemistry ; and even advanced students will find its succinct and tabulated arrangement of reactions of the various groups, examinations, tables, of the highest value as working references.

Bee-Keeping, by Alfred Rusbridge (London : E. W. Allen). This is a cheap and useful manual, very practical in all its directions respecting every department of bee-keeping, and well illustrated. We are always glad to welcome any book which helps on or encourages apiculture.

Manual of Taxidermy, by C. J. Maynard (London Trübner & Co. ; Boston : L. E. Cassino & Co.). Works on animal preserving have not been uncommon of late years, but this little work is the most thorough and practical of any we have yet seen. The chapters are devoted to collecting, skinning birds, making skins, mounting birds, making stands, collecting mammals, making skins of mammals, mounting mammals, mounting reptiles, batrachians, fishes, &c. The illustrations are very useful and effective.

HOPLOPHORA FERRUGINEA.

MR. C. F. GEORGE, who, in 1877, first drew attention in your Journal to the occurrence in England of the above singular mite, and probably others of your readers, will be interested to learn that I.

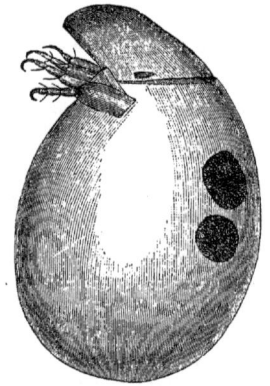

Fig. 34.—Side view of *Hoplophora ferruginea*.

have come across it under somewhat singular circumstances. A considerable amount of excitement and alarm has been occasioned by the finding of the vine-pest (*Phylloxera vastatrix*) at no great distance from our locality. Many gardeners in our more immediate neighbourhood, who had been completely baffled with their vines, at once concluded that here was a solution of their difficulties. Quite a number of them dug up their vine roots, and submitted them to me for examination, with the result of the phylloxera being found on some roots in great numbers. One of these gardeners, whose vines had undoubtedly been attacked by this insect, on a subsequent occasion, brought more root, on which I was not able to discover any ; but from a root which he said had been dug up several feet, I noticed a small shiny insect, which, upon microscopical examination, I had no hesitation in pronouncing to be the above pretty and interesting mite. I have made two sketches, side and ventral views, copies of which I send for reproduction in SCIENCE-GOSSIP, if you think it advisable. I can confirm most of the description given by the above authority. My specimens, however, had eyes situated on the side of

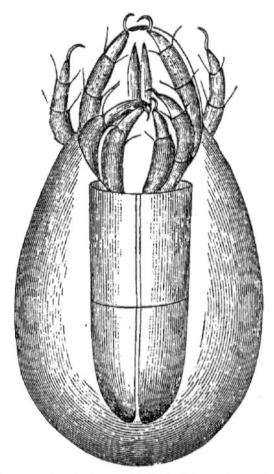

Fig. 35.—Ventral view of *Hoplophora ferruginea*.

the head, close to its juncture with the abdomen. Also I was not able to discover moveable plates or shield on the under side of the head, although the abdominal ones were very distinct. Mr. George, at the time, was not sure of the species, and it is possible, of course, that my specimen and his are specifically different. I notice that the last edition of the "Micro. Dictionary," has eliminated the statement that these insects are "not British."

J. E. LORD.
Rawtenstall.

THE ORIGIN OF DOUBLE FLOWERS.

THIS is a subject which has much exercised the minds of botanists and scientific horticulturists, not because it is more obscure than that of many other abnormal forms, but because the beauty of double flowers attracts attention to them. It has been a favourite notion that hybridism or crossing has had something to do with their production. As in many cases double flowers do not contain pollen, it is obvious that they cannot supply the means for crossing other flowers; but then their barrenness might be regarded as presumptive evidence of hybrid origin; it being believed that hybrid plants were commonly sterile. The theories of scientific botanists could throw no light upon a subject which it was the interest of scientific seedsmen to keep dark, the production of seeds which would give origin to plants with double flowers being a valuable professional secret. Now that the experiments of Dr. Darwin have shown that the seeds of flowers that have been self-fertilised or fertilised illegitimately give origin to plants in many respects resembling hybrids, it may be worth while to see if this discovery, along with other facts and theories connected with the doctrine of evolution, may not help us to form a notion approximately true as to the origin of double flowers.

By double flowers I mean flowers with an increased number of petals or petaloid organs. These are to be carefully distinguished from synanthic flowers in which two or more flowers are united collaterally, or it may be rather that the parts which might form two or more flowers are blended in one. In these there is no transformation of one organ into another, nor any impairment of reproductive energy; the stamens contain abundance of good pollen, the pistil has no lack of ovules, and the fruit when ripe contains its full number of seeds capable of growing into strong healthy plants. Besides which synanthy does not commonly affect all the flowers of a plant, but only those in such positions as we might expect to find the largest flowers in, the summit of a plant with terminal flowers, or the lower part of a raceme. Multiplication of petals on the contrary usually affects all the flowers of a plant producing them, and they seem to be formed at the expense of the stamens, for even if there be stamens in such a flower, they are usually deformed or imperfect. Thus it seems that a vigorous plant of a favoured race with no set limits to its power to reproduce itself by seed will be likely to bear synanthic flowers. A plant equally vigorous, but with a rule made absolute forbidding the banns of marriage, will bear double flowers. Such a rule may be expressed in the words of Darwin, that "Nature abhors perpetual self-fertilisation." Several years after having stated this as an axiom, Darwin laid especial emphasis upon the word "perpetual," admitting that nature may tolerate, or even favour self-fertilisation for a long while, but not for ever.

Now let us see if we can find anything in the history of flowering plants as they exist in nature to warrant the assumption that a course of self-fertilisation tends to change stamens into petals or make them petaloid. After all that has been written about evolution during the last twenty years there are still some who think it quite as likely that orchis and camera were created as they now exist, as that they should have been derived from petaloid endogens with six stamens, four or five of which have become petaloid or otherwise very different from their original state. These persons, however, are not in a majority, for in works written years before the "Origin of Species" disturbed the philosophic mind, we find the orders of *Marantaceæ, Zingiberaceæ* and *Orchidaceæ* as having stamens transformed into the appearance of petals. Now it may be interesting to consider how, when, or why, did such a change take place. If the ancestors of such plants were anemophilous, as it is likely that they were, suppose an individual to appear, with pollen grains coherent in masses, it is obvious that they could not be transported by the wind so as to reach the stigmas of other flowers. Now if such a phenomenon came to pass before that winged insects were attracted by flowers, as indeed it may have happened when no winged insects existed, then such flowers must have either been fertilised with their own pollen, or not fertilised at all. Under such circumstances it is not unlikely that the ancestors of orchids were subjected to a long course of self-fertilisation, during which most of the stamens became sterile and petaloid, and even the stigmas could not escape the brand of sterility that was stealing over the flowers when some winged insects crossed their path to extinction and saved them from it. If in the meantime any flowers had appeared with all their organs barren and petaloid, as in the double stocks of our time, the plants bearing such flowers would perish, and their race become extinct, leaving no record of their having lived. So we do not find many double flowers in a state of nature. But we do find plants whose flowers are said to be habitually self-fertilised. What shall we say of them? In Darwin's book on the effects of "Cross and Self-fertilisation," he has recorded experiments which he made on plants for several generations, extending over a period of eleven years, a time long enough to afford valuable results, but not decisive as to the ultimate consequences that would follow in such a length of time as must have elapsed since the species originated. We find, as might have been expected, that the effects of self-fertilisation are by no means uniform on the offspring. Some of them seem to have been very diminutive, and were thrown away by the experimenter. It is by no means certain that such plants would be thrown away by nature. Some of them would be passed over by the sickle or the scythe that had cut down taller plants, or might grow by the wayside and the borders of fields where their very insignificance might save them from destruction. So we find a plant with prostrate stem, as *Malva rotundifolia*, and such other plants as *Vicia hirsuta* and *Trifolium procumbens*, with flowers reduced and purposely rendered inconspicuous, so that they are now but little visited by insects. It is not among such plants as these that double flowers occur. Their smallness hinders them effectually from crowding out other plants and filling the earth with vegetation. But among the plants on which Darwin made experiments, three cases occurred of plants varying in such a manner as to be more fertile with their own pollen than they originally were, namely, with Mimulus. Ipomœa, and Nicotiana. One plant of Ipomœa he also mentions which he called Hero, from its exceeding in height and size plants from seeds that were cross-fertilised.

Such plants as these were probably the ancestors from which plants now under cultivation have been for the most part derived; Nature tolerating the self-fertilisation of flowers on plants that are protected by art from competition with others that are cross-fertilised. Thus we can understand that cultivated plants would be destroyed if neglected, by reason of the delicacy of their constitution as compared with others growing wild. But even among cultivated plants Nature draws the line somewhere beyond which self-fertilisation cannot go. Stamens become sterile and assume the form of petals. Even during the eleven years of Darwin's experiments he found some flowers with stamens partially deformed and petaloid on plants that were the offspring of self-fertilisation. Sometimes the change comes on suddenly, but perhaps more often gradually, under circumstances not generally known. In an appendix to Dr. Masters' "Vegetable Teratology," several facts are given bearing on the subject. One is of a plant of *Camellia Japonica* at Vienna, from which seeds were saved, the flowers having been fertilised with their own pollen. All the plants raised from these seeds bore double flowers. This result was attributed by the parties concerned to geographical and climatal conditions, which might indeed have had something to do with it, but it is at least as likely that the progenitors of these plants had already gone through a course of self-fertilisation which had reached the limit Nature would permit.

A double variety of *Primula Sinensis* is also mentioned as having been raised at Southampton by seedsmen or florists, who say that to obtain double varieties the raiser fertilises certain fine and striking single flowers with the pollen of other equally fine single blooms, and the desired result is obtained. It is admitted, however, that there is a reservation of some important item which is kept as a professional secret. Now Darwin tells us that flowers fertilised by the pollen of other flowers on the same plants yield seeds which give origin to plants having no advantage over those raised from the seeds of self-fertilised flowers; so that it is quite possible that the whole secret may be revealed by adding four short words, thus, after the words single blooms read "on the same plant." This I write hypothetically, as an inference deduced from the premises hereinbefore stated, for nobody has told me anything on the subject. When I once asked a seedsman by what means double varieties of the Japan pink were obtained, he told me by hybridising with another species of dianthus. It might be that somebody had told him thus, and that he believed it. I did not. Rhododendrons are said to be hybridised, but are not commonly double-flowered, nor are hybrid gladioli. Of pelargoniums, the one most commonly seen with double flowers is *P. zonale*, which in that condition does not show any signs of hybridism. So that it seems as if a variety of considerations point to the conclusion at which I have arrived, that double flowers are the consequence of a course of self-fertilisation under conditions favourable to the vital energy of vegetative growth.

JOHN. GIBBS.

THE NATURAL HISTORY OF THE COCKROACH (*PERIPLANETA ORIENTALIS*).

By Professor L. C. MIALL AND ALFRED DENNY.

[The present is the first of a short series of sketches which will deal with the natural history, structure, and physiology of one of our commonest insects.]

THE cockroach or black-beetle is, to our sorrow, very widely distributed in England, as in other commercial countries. It is not a native of Europe, nor does any member of the genus, as now limited by naturalists, belong strictly to the European fauna.

The species of cockroach which the English naturalist may expect to meet with are enumerated and shortly characterised in the following table.

Order ORTHOPTERA.

Section *Cursoria* (One family only).

Family *Blattina*. Legs adapted for running only. Abdomen usually flattened. Wing-cases usually leathery, opaque, overlapping (if well developed) when at rest. Head vertical, retractile beneath the pronotum. Eyes large, ocelli rudimentary, antennæ long and slender.

Group 1. Both sexes wingless (*Polyzosteria*).

Group 2. Males winged, females wingless (*Perisphæria, Heterogamia*).

Group 3. Both sexes with more or less developed wings (about 10 genera).

Genus *Blatta*. A pulvillus between the claws of the feet. The seventh sternum of the abdomen entire; no externally visible sub-anal styles in the male.

B. Lapponica. Smaller than the common cockroach. The wing-cases and wings long in both sexes.

This is the insect of which Linnæus tells, that in company with *Silpha Lapponica* it has been known to devour in one day the whole stock of dried but unsalted fish of a Lapp village. Out of Lapland it has no home among men, but frequents thickets over a great part of Europe.

B. ericetorum. This, and some other species of which we have no critical knowledge, have been described as natives of England. They frequent the sea-shore, where the larvæ may be taken under stones, or marshes and woods not far from the coast.

B. Germanica. The wing-cases and wings are well developed in both sexes. Two longitudinal stripes on the pronotum or first dorsal plate of the thorax are the most certain mark of this species, which is considerably smaller than our common cockroach. There is little doubt that *Germanica* is a native, not of Germany, as the name implies, but of Asia and the extreme east of Europe. To the Swede Linnæus, who named the species, it was truly the German cockroach, but Germany is now known to have received the unwelcome guest from Russia, where, especially in Asiatic Russia, it occurs wild, feeding upon the leaves of the birch. The Russians, it is true, disclaim the credit of originating this pest, and to many Russian peasants the insect is known as the Prussian cockroach (Fischer de Waldheim), tradition affirming that the Russian soldiers brought it back in their knapsacks from the seven years' war.

Genus *Periplaneta*. Readily distinguished from Blatta by the divided seventh abdominal sternum of the female and the sub-anal styles of the male.

P. orientalis. The wing-cases of the male reach the 5th abdominal segment, and the wings the 4th abdominal; in the female the wing-cases reach only to a little beyond the middle of the last thoracic segment, while there are no free wings. This species is native to tropical Asia,* and long ago made its way by the old trade-routes to the Mediterranean countries. At the end of the sixteenth century it appears to have got access to England and Holland,† and has gradually spread thence to every part of the world.

P. Americana. The wings and wing-cases of this species, which is much larger than *orientalis* or *Germanica*, extend to the end of the abdomen in both sexes. It belongs to tropical America, but occurs sporadically in all countries which trade with America.

An Australian species also (*P. Australasiæ*) has been observed beyond its native limits in Sweden (De Borck, "Skandinaviens rätvingade insekter Nat. Hist." I. i. 35) and in Florida (Scudder, "Proc. Boston Soc. N. H.," vol. xix. p. 94). In Florida it is said to be the torment of housekeepers.

Panchlora Maderæ is said by Stephens to be

---

* Linnæus was certainly mistaken in his remark (Syst. Nat.) that this species is native to America, and introduced to the east—"Habitat in America: hospitatur in Oriehte."
† This is to be inferred from Moufet's "Insectorum Theatrum" (1634), in which he speaks of the Blattæ as occurring in wine-cellars, flour-mills, &c., in England. It is hard to determine in all cases of what insects he is speaking, since one of his rude wood-cuts of a "Blatta" is plainly *Blaps mortisaga*; another is, however, recognisable as the female of *P. orientalis*, a third more doubtfully, as the male of the same species. He tells how Sir Francis Drake took the ship Philip (of Spain?) laden with spices, and found a great multitude of winged Blattæ on board, "which were a little larger, softer and darker than ours." Perhaps these belonged to the American species, but the description is obscure. Swammerdam was also acquainted with our cockroach as an inhabitant of Holland early in the seventeenth century (Bibl. Nat. p. 92).

occasionally seen in London, and *Blabera gigantea*, the drummer of the West Indies, has often been found alive in ships in the London Docks.

Blatta Germanica, Periplaneta orientalis, and *P. Americana*, are so similar in habits and mode of life as to be interchangeable, and each is known to maintain itself in particular houses or towns within the territory of another species, though usually without spreading.

Orientalis is, for example, the common cockroach of England, but *Germanica* frequently gets a settlement and remains long in the same quarters. H. C. R., in SCIENCE-GOSSIP for 1868, p. 15, speaks of it as swarming in an hotel near Covent Garden, where it can be traced back as far as 1857. In Leeds one baker's shop is infested by this species; it is believed to have been brought by soldiers to the barracks after the Crimean war, and to have been carried to the baker's in bread-baskets. We have met with no instance in which it has continued to gain ground at the expense of *orientalis*. *Americana* also seems well established in particular houses or districts in England. H. C. R. (*loc. cit.*) mentions warehouses near the Thames, Red Lion and Bloomsbury Squares, and the Zoological Gardens, Regent's Park. It frequents one single warehouse in Bradford, and is similarly local in other towns with foreign trade.

Many cases are recorded in which *Germanica* has been replaced by *orientalis*, as in parts of Russia and Western Germany, but detailed and authenticated accounts are still desired. On the whole *orientalis* seems to be dominant over both *Germanica* and *Americana*.

The slow spread of the cockroaches in Europe is noteworthy, not as exceptional among invading species, but as one more illustration of the length of time requisite for changes of the equilibrium of nature. It took two centuries from the first introduction of *orientalis* into England for it to spread far from London. Gilbert White, writing, as it would appear, at some date before 1790, speaks of the appearance of "an unusual insect," which proved to be the cockroach, at Selborne, and says: "How long they have abounded in England I cannot say; but have never observed them in my house till lately."* It is probable that many English villages are still clear of the pest. The house-cricket, which the cockroaches seem destined to supplant, still dwells in our houses, often side by side with its rival, sharing the same warm crannies and the same food. The other imported species, though there is reason to suppose that

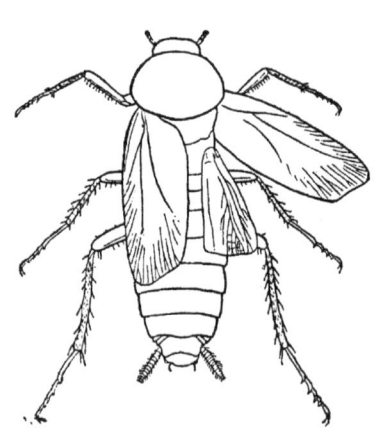

Fig. 36.—*Blatta (Periplaneta) orientalis*, male. Twice natural size.

Fig. 37.—*Blatta (Periplaneta) orientalis*, female. Twice natural size.

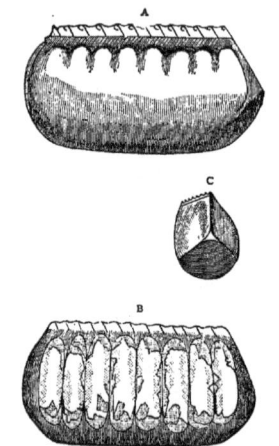

Fig. 38.—Capsule of Cockroach. A, external view; B, opened; C, end view.

they cannot permanently withstand *orientalis*, are by no means beaten out of the field; they retreat slowly where they retreat at all, and display inferiority chiefly in this, that in countries where both are found, they do not spread, while their competitor does. It may yet require some centuries to settle the petty wars of the cockroaches.

It is also worth notice that in this, as in most other cases, the causes of such dominance over the rest as one species enjoys are very hard to discover. We cannot explain what peculiarities enable cockroaches to invade ground thoroughly occupied by the house-cricket, an insect of quite similar mode of life: and it is equally hard to account for the superiority of *orientalis* over the other species. It is neither the largest nor the smallest; it is not perceptibly more.

* Bell's edition, vol. i. p. 454.

prolific, or more voracious, or fonder of warmth, or swifter than its rivals, nor is it easy to see how the one conspicuous structural difference, viz., the rudimentary state of the wings of the female, can greatly favour *orientalis*. Some slight advantage cycads, paper, woollen clothes, sugar, cheese, bread, flesh, fish, leather, the dead bodies of their own species, all are greedily consumed. Cucumber, too, they will eat, though it disagrees with them horribly.

In the matter of temperature they are less easy to

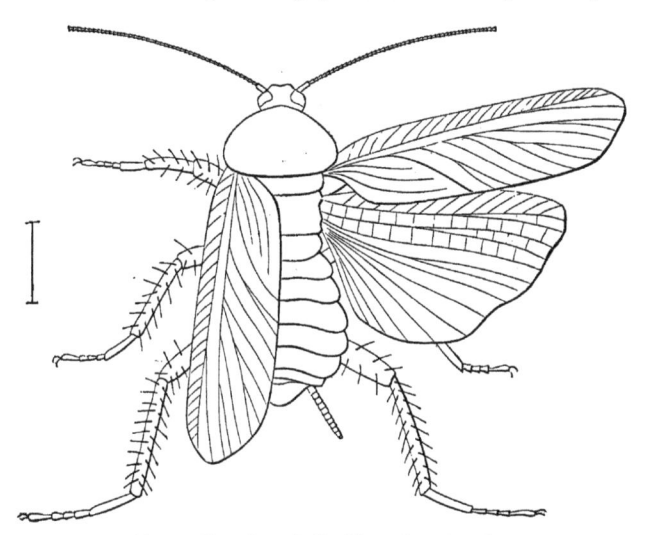

Fig. 39.—*Blatta Germanica* (female), × 4 times natural size.

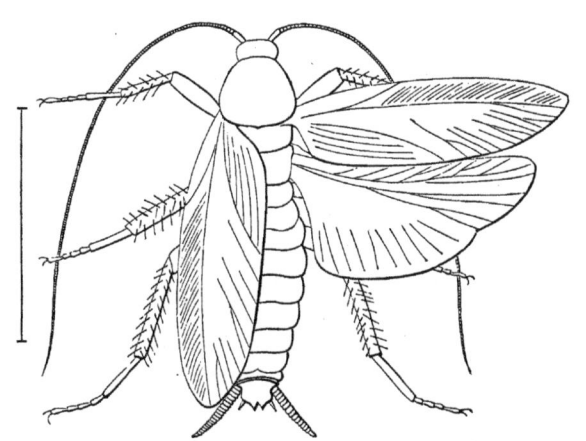

Fig. 40.—*Periplaneta Americana* (male), × 1¼ times natural size.

seems to lie in characteristics too subtle for our detection or comprehension.

As to the food of cockroaches, we can hardly except any animal or vegetable substance from the long list of their depredations. Bark, leaves, the pith of living please. They are extremely fond of warmth, lurking in nooks near the oven, and abounding in bakehouses, distilleries, and all kinds of factories which provide a steady heat together with a supply of something eatable. Cold is the only check, and an unwarmed room during an English winter is more than they can endure. They are strictly nocturnal, and shun the light, although when long unmolested they become bolder.

The cockroach belongs to a miscellaneous group of animals, which may be described as in various degrees parasitic upon men. These are all in a vague sense domestic species, but have not, like the ox, sheep, goat, or pig, been forcibly reduced to servitude; they have rather attached themselves to man in various degrees of intimacy. The dog has slowly won his place as our companion; the cat is tolerated and even caressed, but her attachment is to the dwelling and not to us; the jackal and rat are scavengers and thieves; the weasel, jackdaw, and magpie, are wild species, which show a slight preference for the neighbourhood of man. All of these, except the cat, which holds a very peculiar place, possess in a considerable degree qualities which bring success in the great competitive examination. They are not eminently specialised, their diet is mixed, their range as natural species is wide. Apart from man, they would have become numerous and strong, but those qualities which fit them so well to shift for themselves, have had full play in the dwellings of a wealthy and careless host. Of these domestic parasites at least two are insects, the house-fly and the cockroach; and the cockroach in particular is eminent in its peculiar sphere of activity. The successful competition of cockroaches with other insects under natural conditions is sufficiently proved by the fact that about nine hundred species have already been described,* while their rapid multiplication and almost world-wide dissemination in the dwellings of man is an equally striking proof of their versatility and readiness to adapt themselves to artificial circumstances. In numerical frequency they probably exceed all domestic animals of larger size, while in geographical range the five species, *Lapponica*, *Germanica*, *orientalis*, *Americana*, and *Australasiæ*, are together comparable to the dog or pig, which have been multiplied and transported by man for his own purposes, and which cover the habitable globe.

The cockroach is historically one of the most ancient, and structurally one of the most primitive, of our surviving insects. Its immense antiquity is shown by the fact that so many cockroaches have been found in the coal measures, where nearly sixty species have been met with.† The absence of well-defined stages of growth, such as the soft-bodied larva or inactive pupa, the little-specialised wings and jaws, the simple structure of the thorax, the jointed appendages carried on the end of the abdomen, and the unconcentrated nervous system, are marks of the most primitive insect-types. The order (*Orthoptera*) which includes the cockroaches, is undeniably the least specialised among winged insects at least, and within this order, which comprises the termites or white ants, the leaf and stick insects, the crickets, locusts and grasshoppers, the earwigs, the ephemeræ, and the dragon-flies, none are more simple in structure, or reach farther back in the geological record than the cockroaches. The wingless Thysanura are even more generalised, but their geological history is illegible.

The eggs of the cockroach are laid sixteen together in a large horny capsule. This capsule is oval, with roundish ends, and has a longitudinal serrated ridge, which is uppermost while in position within the body of the female. The capsule is formed by the secretion of a "colleterial" gland, poured out upon the inner surface of a chamber (vulva) into which the oviducts lead. The secretion is at first fluid and white, but hardens and turns brown on exposure to the air. In this way a sort of mould of the vulva is formed, which is hollow, and opens forwards towards the outlet of the common oviduct. Eggs are now passed one by one into the capsule, and as it becomes full, its length is gradually increased by fresh additions, while the first-formed portion begins to protrude from the body of the female. When sixteen eggs have descended, the capsule is closed in front, and after an interval of time for hardening, is dropped in a warm and sheltered crevice. In *Periplaneta orientalis* it measures about ·45 in. by ·25 in. The ova develope within the capsule, and when ready to escape are of elongate-oval shape, resembling mummies in their wrappings. Eight embryos in one row face eight others on the opposite side, being alternated for close packing. Their ventral surfaces, which are afterwards turned towards the ground, are opposed, and their rounded dorsal surfaces are turned towards the wall of the capsule; their heads are all directed towards the serrated edge. The ripe embryos are said by Westwood to discharge a fluid (saliva?) which softens the cement along the dorsal edge, and enables them to escape from their prison. In *Blatta Germanica* the female is believed to help in the process of extrication.* The larvæ are at first white, with black eyes, but soon darken; they run about with great activity, feeding upon any starchy food which they can find; they love warmth even more than their parents, and often huddle together in snug and protected corners.

Cornelius, in his very interesting "Beiträge zur nähern Kenntniss von *Periplaneta orientalis*" (1853), gives the following account of the moults of the cockroach. The first change of skin occurs immediately after escape from the egg-capsule, the second four weeks later, the third at the end of the first year, and each succeeding moult after a year's interval.

---

* British Museum Catalogue of Blattariæ (1868) and supplement (1869).
† Scudder on Palæozoic Cockroaches, Mem. Boston Soc. Nat. Hist. vol. iii.

* "Modern Classification of Insects," vol. i. pp. 421–1.

At the sixth moult the insect becomes a pupa, and at the seventh (being now five years old*) it assumes the form of the perfect insect. The following moults are annual, and like fertilisation and oviposition, take place in the summer months only. He tells us further that the ova require about a year for their development. These statements are partly based upon observation of captive cockroaches, and are the best accessible, but they require confirmation by independent observers, especially as they altogether differ from Hummel's account of the life-history of *Blatta Germanica*, and are at variance with the popular belief that new generations of the cockroach are produced with great rapidity. The wings and wing-cases appear first in the pupa-stage, but are then rudimentary, and constitute a mere sculpturing of the dorsal plates of the thoracic rings; the white ocelli internal to the antennary sockets appear first in the pupa. The insect is active in all its stages, and is therefore with other Orthoptera described as undergoing "incomplete metamorphosis." After each moult it is for a few hours nearly pure white. Of the duration of life in this species we have no certain information, and there is great difficulty in procuring any.

We have before us a long list of parasites which infest the cockroach. There is a conferva, several infusoria, nematoid worms (one of which migrates to and fro between the rat and the cockroach), as well as hymenopterous and coleopterous insects. The cockroach has a still longer array of foes, which includes monkeys, hedgehogs, cats, rats, birds, chamæleons, and wasps, but no single friend, unless those are reckoned as friends which are the foes of its foes.

A few lines must be added upon the popular and scientific names of this insect. Etymologists have found it hard to explain the common English name, which seems to be related to *cock* and *roach*, but has really nothing to do with either. The lexicographers usually hold their peace about it, or give derivations which are absurd. Mr. James M. Miall informs us that "*Cockroach* can be traced to the Spanish *cucarácha*, a diminutive form of *cuco* or *coco* (Lat. *coccum*, a berry). *Cucarácha* is used also of the woodlouse, which, when rolled up, resembles a berry. The termination -*ácha* (Ital. -*accio* -*accia*) signifies *mean* or *contemptible*. The word perhaps reached us through the French; at least *coqueraches* has some currency (see for example Tylor's 'Anahuac,' p. 325)." The German word *Schabe*, often turned into *Schwabe*, means perhaps *Suabian*, as Moufet, quoting Cordus, seems to explain ("Insectorum Theatrum," p. 138). *Franzose* is another German word for the insect, applied specially to *Blatta Germanica*, and both *Schabe* and *Franzose* would be thus interpreted as geographical descriptions, implying some popular theory as to the native country of the cockroach. *Kakerlac*, much used in France and French-speaking colonies, is a Dutch word of unknown signification. The name *Blatta* was applied by the ancients to quite different insects, of which Virgil and Pliny make mention; *Periplaneta* is a modern generic term, coined by Burmeister.

Of the uses to which cockroaches have been put we have little to say. They constitute a popular remedy for dropsy in Russia, and both cockroach-tea and cockroach-pills are known in the medical practice of Philadelphia.

---

* Cornelius says four, but this conflicts with his own account.

SCIENCE-GOSSIP.

A WRITER who signed himself C., recently gave an account in "St. James's Gazette" of an "experiment with Hachish" upon himself. His narrative of the visions he saw exceeds what Alexandre Dumas or any other writer ever declared of this Indian drug. He was under its operation for more than four hours, but woke none the worse for the experiment, unless the tendency to eat a great supper at midnight is to be regarded as a misfortune.

OUR readers will remember the interesting incident of small jelly-fishes appearing in the freshwater tanks of the Horticultural Society a few years ago. They excited much attention, and turned out to be new species, but the most singular fact about them was that they should assume freshwater conditions, instead of marine. Dr. Bohn, a Prussian geographer, travelling in Africa, has just sent word he has discovered freshwater medusæ in Lake Tanganika. They possess a broad, umbrella-shaped disk, and have numerous long and prehensile tentacles.

EXPERIMENTS have been going on on one of the shorter American railways, to see whether the wire-fences on either side the line cannot be used for telegraphing purposes, the wires at the level crossings being run under, so as to make them continuous. It is stated that the experiments prove the possibility of the thing being done.

M. TREPIED, of Algiers, has spectroscopically examined the Pons-Brook comet, and found its spectrum to consist of two bands in the green of different degrees of brilliancy, and a third faint band in the blue. He states that these bands are practically identical with those seen in the spectrum of an alcohol flame.

WE are exceedingly sorry to have to record the death of an old contributor to our Botanical columns. Mr. Thomas Brittain, the well-known Lancashire botanist, died at Urmston, near Manchester, at the end of January. He was President of the Manchester Microscopical Society. A fine, genial old man, beloved and respected by all, has passed away.

A FRENCH electrician has invented a new sounding lead, which tells the exact moment of its reaching the bottom by means of an electric alarm bell.

IN a paper read before the Penzance Natural History Society, on "Wild Flowers at the end of the year," by Mr. Samuel Tait, the writer shows that at Christmas, 1882, there were no fewer than 111 species in bloom at Madron, near Penzance.

IT is with much pleasure we notice the appointment of Professor Flower, F.R.S., the distinguished osteologist and comparative anatomist, to the position of Superintendent of the British Museum, vacant by the retirement of Professor Owen.

RECENTLY, near Letterkenny, co. Donegal, Domhnal Kinahan shot a young Greenland falcon. These birds are very rare visitors, only about six being recorded during the last century.

AT the meeting of the Royal Hist. and Arch. Ass. of Ireland, in Ballymena, G. H. Kinahan, M.R.I.A., &c., called attention to flint implements picked up in the co. Donegal. He finds these occur in different localities near the luscas, or cave dwellings, of the primitive inhabitants; they being exposed when tilling the ground. The luscas are of peculiar types, being long, narrow and low. Allied to them are fosliac or flag dwelling-places on the surface of the ground. These are also long and narrow, having at one or both ends, "standing stones" or gallâns, if a ridge pole rested on them to hold up a roof of either rushes, grass or sods; some, however, were covered by flags; the roofs appear to have been made of flags when they were found of sufficient size.

PROFESSOR TAIT has been lecturing on thunderstorms, with a view to showing the amount of energy involved in changes then produced. He proved that, to evaporate a tenth of an inch of water on a square foot of ground, required a power equal to one horse for half an hour, so that to condense one-tenth of an inch of water on a square mile would require one million horses working for the same time. In this way there is little difficulty in understanding how it is that hurricanes and typhoons can be produced by the amount of energy in the heat formed out of that small quantity of water in condensing from the vaporous into the liquid form. Speaking of the three forms under which lightning is usually manifested—forked, sheet, and globular—Professor Tait said that forked lightning was of the same kind that we obtained from an electrical machine. A brilliant flash of lightning lasts only the one-millionth part of a second. "Sheet" lightning is merely the illumination of the clouds and vapour in the atmosphere by the forked kind. Of "globe" lightning very little is at present known. Professor Tait thinks that atmospherical electricity is caused by the friction of the water-vapour molecules with those of the air.

THE sixth annual reception of the New York Microscopical Society, was held on February 1st, in the Lyric Hall. The retiring President, Mr. B. Braman, delivered an address on "The Microscope in Art," and the programme of objects exhibited was unusually interesting.

PROFESSOR MILNE, of Japan, has just published in "Nature" an important practical paper on "Earthquakes and Buildings." The hints contained in it ought to be studied by all who are concerned with architecture in countries affected by earthquake disturbances. Mr. Milne recommends, that a house built in such places to be aseismic (that is unaffected by earthquakes) should be a low frame building, with iron roof and chimneys supported by a number of slightly concave surfaces resting on segments of stone, or metal spheres, the latter being in connection with the ground. The streets ought to be wide in earthquake towns, and open spaces left for refuges. Chimneys with heavy tops, like heavy roofs, should be avoided. The pitch of the roofs must not be too great. If it is necessary to have substantial buildings, their upper portion ought to be as light as is consistent with the requisite strength. The city of Manilla has been re-erected much upon this plan, and now presents a singular appearance of light roofs rising from old foundations.

A FRENCH scientist, M. Duchatre, has given an account of some experiments he has been making with seeds. Everybody is aware of the influence which direct sunlight has upon the growth and development of young plants. M. Duchatre has been experimenting upon the germination of seeds with moonlight instead of sunlight. He subjected the seedlings of lentils, vetches, &c., to its influence. When the seeds had sprouted he put them in a dark place, and kept them there for a time, so that their stalks grew slender and of a yellowish-white colour. Afterwards, on three nights, when there was clear moonlight, he exposed them to its influence for six hours each night. He found that the stalks at once became seleniotropic, that is, they turned towards and followed the moonlight, just as many plants, such as the sun-flower, are heliotropic, or turn towards and follow the progress of the sun through the heavens. From the very first the stalks of the plants began to bend so that they constantly presented themselves and their budding leaves towards the moonlight, or rather towards the moon.

THE Javanese volcanic eruption will perhaps not be required to account for all the dust thrown into the atmosphere, and which is believed to be the efficient cause of the recent beautiful sunsets, for news has just reached us of a great volcanic eruption in Alaska. A new submarine volcano was formed there last summer. It burst out from the bottom of the Behring Sea, and has already formed an island from 800 to 1200 feet high. This volcano is still in active operation, and our atmosphere has very likely received a contribution of volcanic dust from it, as well as from Krakatoa.

DR. CLEVENGER writes in the "American Naturalist" to show the disadvantages of the upright position in man. He shows that nothing but the original "all fours" position, from which he sprang, will account for the occurrence of valves in the intercostal veins, &c., the absence of valves from parts where they are needed, such as the venæ cavæ, &c., and the exposed and dangerous position of the femoral artery. All of these facts are fraught with a certain amount of danger to man in the upright position; but they are beautifully adapted to the quadrupedal. With so many drawbacks to the upright position, it is singular that man ever assumed it. But perhaps he thought it worth risking a little to obtain it!

ANOTHER discovery of coal has been made in the Canadian North-West. It is a six-feet seam of hard coal, and has been found in the Cascade Range of the Rocky Mountains, near the line of the Canadian Pacific Railway.

In the February number of the "Gentleman's Magazine," Mr. Mathieu Williams has some very interesting "Science Notes" relating to the Wonderful Twilights and their Causes, and he there discusses the several theories advanced as to their causes, but personally he appears to advocate that of meteoric dust.

MESSRS. CHRISTY & CO. are issuing, in parts, accounts of "New Commercial Plants and Drugs." No. 7 is just out, dealing with the Calisaya Verde, the Pahu Cabbage, the Siam Benzoin Tree, the Menthol plant, the New Fibre plant, &c., as well as giving the latest particulars as to results obtained from new drugs, &c.

THE Geologists' Association held their annual Conversazione at University College on the 1st of February, when short addresses were given by the following geologists: on "Pre-historic Man, and recently extinct Mammalia," by Dr. Henry Woodward, F.R.S.; on "Fossil Plants from various Formations," by Mr. William Carruthers, F.R.S.; on "Rocks and Rock Sections," by Mr. Hudleston, and on "The Volcanic Eruption at Krakatoa," by Professor Judd.

MR. A. D. MICHAEL, in a paper read before the Linnean Society, concludes, from a careful series of experiments and observations, that the "Hypopi" are not adult animals, but only a stage, or heteromorphous nymphæ, of Tyroglyphus and allied genera. All individuals do not become " Hypopi," the latter stage occurs during the second nymphal skin-casting. It seems a provision for the distribution of the species, irrespective of adverse conditions. "Hypopi" are not truly parasitic.

Dr. M. C. COOKE has recently shown from microscopical examination that *Sphæria pocula* is a hymenomycetal, and not an ascomycetal fungus, being allied to the genus Polyporus or Porothelium.

AN interesting paper by Mr. J. Gilbert Bowick has come under our notice. The writer thinks the peculiar richness and melody of the Italian voice is due to meteorological causes—in fact to the abundance of hydrogen peroxide, or volcanic ammonia, present in the atmosphere of Italy.

DR. CARTER MOFFAT, who has been for long resident in Italy, has, as the result of a series of experiments, constructed an instrument which he calls the "Ammoniaphone," by whose means a person may inspire air charged with an artificial mixture of the above-named gases.

EXPERIMENTS with the Ammoniaphone appear to have proved satisfactory. Dr. Moffat's own voice has much improved, and is now a light tenor of large compass. Similar results with other people are also detailed. The use of hydrogen peroxide in throat or lung affections appears to be of no recent date, and the ammoniaphone promises to be useful in such cases.

IN a late number of the "Botanische Zeitung," Büsgen gives an account of some experiments he has been performing at Strassburg for two seasons in the feeding of *Drosera rotundifolia*. The $_{re}e^{su}{}_{lt}s$ confirm the conclusions of Darwin and others; and the plants fed with animal matter through their leaves were stronger and more vigorous in every way, than those that were not thus fed, but equally favoured in every other respect.

MR. HALSTED in "SCIENCE," describes and figures a "*Combination walnut*," being a nut which is covered partly by a walnut hull, and partly by a shellbark hull, as if Carya and Juglans had been cross-fertilised. Within the hull it seems that the nut was entirely walnut. The specimen is worth a careful examination.

PROFESSOR BESSEY, in the "American Naturalist" for December, describes a new species of insect-destroying fungus, under the name *Entomophthara Calopteni*. It occurs as a clay-coloured mass in the body cavity and femora of the common locust (*Caloptenis differentialis*).

MR. RUSKIN has been lecturing at the Royal Institution on "The Storm-cloud of the Nineteenth Century." He thinks this form of cloud is peculiar to our time, and has been sent as a punishment for our sins! If so, the greatest sinners ought to suffer most by it; but do they?

WE learn the Philadelphia Academy of Sciences is building up a very fine herbarium, claiming to possess probably one-half the known species of plants. The growth has been very rapid for some years, the past year showing an addition of 2,868 species. The species are all poisoned, labelled, and systematically arranged, and this great work is being done gratuitously, by the persistent labours of Mr. J. H. Redfield, assisted by other botanists.

DR. H. W. CROSSKEY's paper in the Birmingham Philosophical Society's Proceedings, on "The Grooved Blocks and Boulder-clays of Rowley Hill," and on other unmarked boulders, is accompanied by some of the most beautiful lithographs we have seen illustrating scientific papers. Dr. Crosskey gives sketches of ice-marked boulders from Switzerland, to illustrate those from the above neighbourhood.

THE storm of Saturday night, January 26th, was, for the short time it lasted, one of the most violent experienced for some time. The wind-pressure in the north-east of Scotland was for a few seconds, thirty-five pounds on the square foot. The barometer settled to a little over twenty-seven degrees, and the difference in barometrical pressure between Scotland and France at the same time was two inches.

MICROSCOPY.

ON GUM STYRAX AS A MEDIUM FOR MOUNTING DIATOMS.—In the "Bulletin de la Soc. Belge de Microscopie," No. ix., 1883, Dr. Henri Van-Heurck describes two resins he had discovered in his search for a fluid to take the place of monobromide of naphthaline; this medium, in spite of its high refractive index (1·65), is on account of the difficulty of permanently securing it in the cell, and its disagreeable odour, but rarely employed. The two resins are the products of *Liquidambar orientale* and *L. styraciflua*; the former plant grows in Syria, the latter in North America. The resin yielded by the Syrian plant is prescribed in the British Pharmacopœia under the name of Gum Styrax, and in the drug trade is known as "strained gum styrax." It has the colour of the old-fashioned black treacle, but is of greater consistency; a temperature of 212° renders it fluid. In its commercial state it is unfit for microscopic purposes, first from its impurities (probably owing to the rough method employed in making it; the stems are cut in small pieces and boiled, when the gum rises to the surface and is skimmed off), and second, from its thickness. It is therefore necessary that it should be dissolved in one of the following menstrua: chloroform, benzole, ether, a mixture of benzole and absolute alcohol. When the resin is dissolved it must be filtered, and it is then ready for use; the solution should be of the colour of brown sherry, and the consistency of limpid olive oil. Its consistency can of course be increased by evaporating a portion of the benzole, and the whole of the latter should be eliminated before placing the cover on the slip; its refractive index is then 1·63, very nearly that of monobromide of naphthaline. The American liquid-amber is prescribed in the American "Pharmacopœia," but seems to be unknown in Europe. It would, if obtainable, be preferable to gum styrax, as its colour is a pale yellow. The colour of the styrax is practically of little consequence, as the film between the cover and slip is very thin, and does not show any appreciable amount of colour when placed under the microscope. I have, during the past four or five months, used this medium for various diatomaceæ; among others *Vanheurkia rhomboides*, *Amphipleura Lindheimeri*, *Pleurosigma angulatum*, *P. littorale*, *Navicula cuspidata*; the transverse striæ on *P. littorale*, and the longitudinal on *N. cuspidata*, are much more sharply defined, and the striæ on all of them are more easily resolved than when mounted in Canada balsam. The most striking difference between gum styrax and Canada balsam is displayed by *Polymyxus coronalis*; in balsam the valves are perfectly hyaline, and the rays and puncta almost invisible; in gum styrax the valves are light brown, and the markings easily resolved. Heliopelta, as might be expected, does not exhibit more structural detail, but every line and dot are more distinct than when it is balsam mounted. Several of the Aulisci are also much improved when mounted in this medium. I cannot say much of its merits as a medium for mounting other microscopic objects. I have tried it for thin wood-sections, hairs, chalk foraminifera, and a few butterfly scales, all of which show better than they do in balsam. The colour of styrax becomes objectionable when a thick layer is necessary. Dr. Van Heurck directs that the commercial gum styrax should be exposed in thin layers to the light and air for several weeks, to eliminate the moisture contained in it previous to dissolving it. I have not found this necessary with my sample. The gum storax, one of the ingredients of incense, is useless.—*Fred. Kitton, Hon. F.R.M.S.*

PRESERVATION OF PROTOZOA AND SMALL LARVÆ. —Hermann Fol recommends an alcoholic solution of ferric perchloride to kill small animals without injury to the tissues. It is diluted with water down to two per cent. and then poured into the vessel holding the animals. These then sink to the bottom. The water is poured off, and seventy per cent. alcohol substituted. Change the alcohol and add to the second dose of it a few drops of sulphuric acid; otherwise the iron may remain in the tissues, and cause them to overstain with colouring reagents. The alcoholic washing should be thorough. Even larger animals (medusæ, dolium, &c.) may be perfectly preserved by this method. The tissues may be subsequently stained by adding a few drops of gallic acid (one per cent. solution) to the alcohol containing the specimens. The nuclei are stained dark, the protoplasm light brown, in twenty-four hours.

PRESERVING IN GLYCERINE.—I put some freshwater entomostraca in glycerine to preserve them. Within five hours I found that the glycerine had apparently shrivelled them. Is this on account of the glycerine being impure, as I was under the impression that glycerine was an excellent preservative without altering the form of the specimen to be preserved.—*G. F. B.*

ENOCK'S ENTOMOLOGICAL SLIDES.—Some time ago we called attention to the head of a spider, showing the opal-like eyes prepared by Mr. Enock. It is with much pleasure we note a specimen of the wood ant (*Formica rufa*), prepared expressly to show the internal structure and arrangements of the muscles. When examined with proper illumination (with blue selenite) the effect is extremely beautiful, every muscle being distinctly seen. A word of praise is further due to Mr. Enock for the manner in which the labels scientifically describe the object.

"PETROGRAPHICAL STUDIES."—We have received two exquisitely prepared slides illustrative of this series, now being issued by Messrs. Ady & Hensoldt. One is a section of Nephelinite, from the Odenwald, perhaps the most remarkable of all volcanic rocks, for its beautiful appearance under the polariscope. The other is a section of Amazon stone (a green felspar) from America. Both slides are accompanied with terse, but accurately written, descriptions. A new and valuable feature in this work is the series of hand-tinted plates illustrating the essays, which enable a student easily to verify the various mineral constituents in each specimen.

ZOOLOGICAL SLIDES.—Mr. E. Ward, F.R.M.S., of Manchester, has just issued two very neatly mounted and highly interesting slides, one of the hydroid zoophyte, *Halecium halecinum*; and the other of a polyzoan, *Bugula plumosa*. Both are mounted with their tentacles expanded so naturally (the hydrozoan especially), that we feel there must have been an understood arrangement between the mounter and his objects !

COLE'S "STUDIES."—The high activity displayed by Mr. A. C. Cole, in issuing his weekly microscopical serials, accompanied with beautiful slides illustrating each, is certainly remarkable. Part v. of "The Methods of Microscopical Research," deals with section cutting, and descriptions of microtomes. No. v. of "Popular Microscopical Studies," gives a technically written, but a very clear description of "A Grain of Wheat." The coloured plate (by E. T. D.) accompanying this part is a marvel of microscopical drawing; and the stained section of the slide will teach a young botanist more in half an hour than book-study would in a day. Nos. 11 and 12 of the older "Studies in Microscopical Science" deal with "Areolar Tissue," and "Morphology of Tissues," with illustrations of the types of simple tissues, and a beautiful coloured plate (to illustrate the slide sent out) of the prothallus of fern × 250. The essays in Animal Histology in the "Studies" are now written by Dr. Fearnley, and those on Botanical Histology by Mr. David Houston, F.L.S.

ZOOLOGY.

MOLLUSCA NEAR MARGATE.—I can add the following land and freshwater species to my brother's list in the September number of SCIENCE-GOSSIP (p. 208). *Pisidium fontinale*; Ebbsfleet. *Neritina fluviatilis*; dead shells washed up at Margate. *Physa fontinalis*; St. Nicholas marsh, &c. *Limnaea truncatula*; Minster. *Limax flavus*; Margate. *L. marginatus*; common at Margate. *L. agrestis*; Ramsgate, &c., common. *Vitrina pellucida*, *Zonites cellarius*, *Z. nitidulus*. *Helix aspersa*, *H. nemoralis*, var. *libellula*, and var. *castanea*; Minster. *H. arbustorum*; dead shells in rejectamenta of River Stour. *H. Cantiana*, var. *galloprovincialis*; Sarre. *H. concinna*; common, the var. *albina* is sometimes found near Minster. *H. rapicida*; dead shells at Minster. *H. pulchella*, var. *costata*; common at Margate. *H. rotundata*; Reculvers, &c. *H. caperata*; Kingsgate, &c., common. *Bulimus obscurus*; common. *Pupa umbilicata*; common at Minster, &c. *Cochlicopa lubrica*; Minster, *Achatina acicula*; Ebbsfleet and Birchington. *Clausilia rugosa*; Minster and Ebbsfleet.—*T. D. A. Cockerell.*

SALMON OVA FOR NEW ZEALAND.—Additional attempts have been made to convey salmon and trout ova to New Zealand, but have hitherto failed, owing to the extremes of heat and cold to which the eggs have been exposed. Mr. Haslem, of Derby, has, at length, overcome the difficulty, by the introduction of moisture into the chilled air in which the eggs are kept, thus securing the proper conditions necessary to preserve the ova both from heat and fungi, and yet retard their development. The New Zealand Shipping Company's steamer Ionic has taken 60,000 salmon eggs, protected by the new method, to stock the waters of the colony. These eggs have been collected by Sir James Gibson Maitland, at his fish-hatchery at Howietown, near Stirling.

PLEASURE AND PAIN.—Memory has been the teacher of the ages, so that the avoidance of pain and the pursuit of pleasure have been the business of living things since the dawn of consciousness, and the existence of memory. It is more than probable that these prime movers of the universe have directed the mechanical forces into profitable channels, and have converted them to their use. More than this, mechanical evolution means the development of the machine that directs other machines, the brain, and the mind. Hence mechanical evolution is the evolution of intelligence. Of course the lessons of experience are in part lessons of pain, and beings that cannot act in accordance with lessons sufficiently learned will experience a maximum of suffering, and may have foundation for a private stock of pessimism of their own. But a tolerance of suffering is of various duration, and sooner or later intelligence will have its beneficent way. And as "knowledge is power," it results that the evolution of the living world and of men, has been and will be very much as they have it, and enlightened intelligence, well lived up to, has always resulted in a minimum of pain.— "C." in "American Naturalist."

BOTANY.

WATSON'S "TOPOGRAPHICAL BOTANY."—Your correspondent T. H., whose note is printed at p. 44, has quite misunderstood his Watson's "Topographical Botany," in relation to the flora of Wigtonshire. There is no such statement contained in it as that such cosmopolitan plants as the nettle, dandelion, buttercup, daisy, and primrose, are not indigenous to the county. He will find the explanation of what has puzzled him at page 41 of the introductory remarks. "For two of the vice-counties (out of 112), 74 Wigton, and 78 Peebles, we have still no records for the commonest species. For some other counties the lists of ascertained common plants are very incomplete, although not wholly blanks: in example, Cardigan, Flint, West Lancashire, Stirling, &c." If he will send a catalogue with due vouchers as to the accuracy of his names, either to the Editor of the "Journal of Botany," or the Managers of the Botanical Record Club, it will gladly be placed on record.—*J. G. Baker.*

WATSON'S "TOPOGRAPHICAL BOTANY."—In SCIENCE-GOSSIP for February, T. H. states that he is surprised that "such cosmopolitan plants as the nettle, dandelion, buttercup, daisy, and primrose, recorded as not indigenous to the county of Wigton, as given under the authority of Mr. Balfour" in the above work. T. H. has misunderstood Mr. Watson's formula; he nowhere denies, or implies, that the above plants "are not indigenous" to Wigton. In the introduction to the first edition, Mr. W. writes: "There are nine counties for which no lists of the commoner plants have been obtained, although something is known about their rarer plants." Wigton is one of these, and Mr. (Professor) Balfour's contributions had reference to the rarer plants. Between the first and second editions, lists from seven of these counties were obtained—Wigton being one of the two still left, and this has since been visited by an Oxford botanist, with the view to the compilation of such a list. If T. H. will look over the list of the county (or counties) with which he is familiar, and send any additions, corrections or notes, to Mr. J. G. Baker, or Rev. W. W. Newbould at Kew, those who, like myself, are interested in the distribution of our flora, will be very thankful, as it is only by the help of the many that approximate completeness and correctness can be obtained in a work like "Topographical Botany."—*Arthur Bennett, Croydon.*

SPECIES OF POTATOES.—Mr. J. G. Baker, in a paper read before the Linnean Society, lately expressed his opinion that out of the twenty-six species of tuber-bearing Solanums which are usually enumerated, not more than six are really distinct, viz. *Solanum tuberosum*, *S. majlia*, *S. commersoni*, *S. cardiophyllum*, *S. jamesii*, *S. oxycarpum*. Of these only one, *S. tuberosum*, is cultivated, and it is a native of the high, dry regions of the Andes. Mr. Baker attributes the deterioration of the potato to its being cultivated in too humid a climate, and from tubers only—the tuber having been unduly stimulated at the expense of the other organs of the plant. The great majority of Solanums, in fact, many hundred species, produce seeds alone, and Mr. Baker urges that in order to extend the power of climate adaptation of potato species, *S. majlia*, *S. commersoni*, and *S. cardiophyllum* should be brought into cultivation, and tried both as pure specific types and as hybridized with the various forms of *S. tuberosum*. The first-mentioned is an inhabitant of the damp coasts of Chili, as far south as lat. 44 to 45. *S. commersoni* is a low-level plant of Uruguay, and *S. cardiophyllum* is a species from the Mexican highlands.

PHYTOLOGICAL RECORD FOR MARCH.—The common chickweed (*Stellaria media*, Linn.), is very variable, still there are three or four very good and constant varieties, by some writers made into species; and the best month of the year to look out for these is March. 1. *Stellaria media*, L. Stams. 5; pets. large and conspicuous. This is the common form found in cultivated ground. 2. *S. Borœana*, Jard. Flors. devoid of petals; stams. 3; leaves very small, and crowded on the stem. A small tufted species, found in sandy soil. 3. *S. neglecta*, Weihe. Stem tall, a foot or more in length; stams. 10; leaves large, not unlike *S. nemorum*; upper l. clasping the stem. Not uncommon on the borders of woods. Betwixt Nos. 2 and 3 comes another variety, *S. pallida*, Dunn; but it is not constant. 4. *S. umbrosa*, Opitz. Stems long, but weak; flrs. numerous, in dense panicles; pedicels glabrous. Seps. lanceolate, with raised tips; pets. as long as the calyx; seeds tubercled. Frequent in shady places. A peculiar form of the primrose (*Primula caulescens*, Bab.) should be searched for early in the season. It is readily distinguished by the firs. being on scapes, instead of single peduncles; it is often misnamed in Herbaria, *P. elatior*, Jacq., the Bardfield Oxlip is, however, a widely different species.—*J. F. R.*

ORCHID FLOWERS.—Some misconception might be felt, were no reply made to Mr. Robinson's note in the January number of SCIENCE-GOSSIP, on *Orchis purpurea, militaris*, and *Simia*. My own observations on these three rare and beautiful plants, as they grow in Kent, Bucks, and Oxon, are strongly confirmatory of the opinion, usually held, that they are distinct. When once known, they are distinguishable at a glance, and I have never met with specimens that have given difficulty. Although the three plants may be separated by the form and coloration of the lip alone, there are various other distinguishing characters which are almost entirely ignored by Mr. Robinson. In "English Botany," vol. iii., Dr. Boswell speaks of having seen forms intermediate between *O. militaris* and *O. Simia* in one place only, and in this place he states that the two species grow together. Dr. Boswell accordingly attributes the occurrence of these forms to probable hybridity, and when we consider the facilities for crossing afforded by the peculiar mode in which these plants are fertilised, but little doubt will be felt that Dr. Boswell's solution is the correct one. Possibly Mr. Robinson has chanced on a batch of such hybrids? I express no opinion on Mr. Robinson's plants, not having seen them; my remarks apply to the three species as they occur wild in this country.—*W. H. Beeby.*

GEOLOGY, &c.

THE METAMORPHIC ROCKS OF SOUTH DEVON.—At a recent meeting of the Geological Society a paper was read on this subject by Professor T. G. Bonney, F.R.S., Sec. G. S. He stated that the chief petrographical problem presented by this district was whether it afforded an example of a gradual transition from slaty to foliated rocks, or whether the two groups were perfectly distinct. He described the coast from Tor Cross round by the Start Point to Prawle Point, and thence for some distance up the estuary leading to Kingsbridge. Commencing again to the north of Salcombe, on the other shore of this inlet, he described the coast round by Bolt Head, and Bolt Tail to Hope Cove. These rocks, admittedly metamorphic, consist of a rather thick mass of a dark mica-schist, and of a somewhat variable chloritic schist, which also contains a good deal of epidote. In the lower part of this are some bands of a mica-schist not materially different from the upper mass. It is possible that there are two thick masses of mica-schist, one above and one below the chloritic schist; but, for reasons given, he inclined to the view that there was only one important mass, repeated by very sharp foldings. The junction between the admittedly metamorphic group and the slaty series at Hope Cove, as well as that north of Salcombe, is clearly a fault, and the rocks on either side of it differ materially. Between the Start and Tor Cross the author believes there is also a fault, running down a valley, and so concealed. On the north side of this the rocks, though greatly contorted and exhibiting such alterations as are usual in greatly compressed rocks, cannot properly be called foliated, while on the south side all are foliated. This division he places near Hollsands, about half a mile to the south of where it is laid down on the geological map. As a further proof of the distinctness of the two series, the author pointed out that there were clear indications that the foliated series had undergone great crumpling and folding after the process of foliation had been completed. Hence that it was long anterior to the great earth movements which had affected the Palæozoic rocks of South Devon. He stated that the nature of these disturbances suggested that this district of South Devon had formed the flank of a mountain-range of some elevation, which had lain to the south. Of the foundations of this we may see traces in the crystalline gneisses of the Eddystone and of the Channel Islands, besides possibly the older rocks of South Cornwall and of Brittany. He also called attention to some very remarkable structures in the slaty series near Tor Cross, which appeared to him to throw light upon some of the structures observed at times in gneisses and other foliated rocks.

THE VOLCANIC GROUP OF ST. DAVID'S.—This is one of the most fruitful sources of geological controversy at the present time. It has been still further perplexed by additional views offered by Professor Blake, in a paper just read before the Geological Society. The result of the author's examination of the rocks in the district of St. David's which have been designated Dimetian, Arvonian, and Pebidian, is that they belong to one volcanic series, whose members are those usually recognised in eruptive areas, and whose age is anterior to, and independent of, the true Cambrian epoch. The independence of this series and the Cambrian is shown by the nature of the junction at all points of the circuit that have been seen.

ANOTHER DEEP WELL BORING NEAR LONDON.—Another deep well boring, at Richmond, has thrown extra light upon the ridge of ancient rocks which previous borings in other places had brought to light. Reckoning from the Thames level, the Richmond well boring has now reached a greater depth by 150 feet than any other. The Eocene strata, Cretaceous, Neocomean, and 87½ feet of the Great Oolite were passed through, and the new red sandstone and marl reached. No strata of the Lower Oolite were known before to exist in the London Basin, nor had the Trias been previously found, although some geologists believe the beds reached at Kentish Town and Crossness belonged to the latter formation. Fragments of anthracite coal, and coal-measure sandstone have been detected in several deposits, showing that portions of the old Palæozoic ridge consisted of carboniferous strata. Thus, as Professor Judd remarks, "the predictions of geologists have been verified, and coal has been found under London, though as yet unfortunately not *in situ*." The scent, however, is gradually getting stronger.

NOTES AND QUERIES.

HOUSE-MARTIN.—Observers of the habits of birds may be interested in the following fact which has come under my notice this autumn. Numbers of house-martins had built under the eaves of our house, and after their flight early in October, we found that they had left behind many young birds to die of cold and hunger. These young ones were fully fledged, but apparently not strong enough for a long flight. They came into the house seeking warmth and shelter—one morning there were as many as ten in one room—and they were so torpid that they made no resistance when we took them in our hands. Many were found dead in the garden; none lived more than three or four days after their desertion. The weather when the parent birds left was cold and stormy, in fact the gales had set in. Our gardener is of opinion that had the warm weather lasted a few days longer, these young birds would have been strong enough for flight.—*C. E. Tritton, Sevenoaks, Kent.*

HEDGEHOGS.—In reply to R. T. V. S. W., the following may be of some use. Hedgehogs, like many other Insectivora, are by no means limited to insect food, but prey on larger animals, as reptiles, small quadrupeds, and birds; they are fond of eggs and of milk, and in confinement will readily eat soaked bread, cooked vegetables, or porridge. It readily kills snakes, and even vipers, which it eats, beginning always at the tail. It is said to be capable of resisting in an extraordinary degree, not only the venom of serpents, but other kinds of poison, however administered. In winter the hedgehog becomes torpid, retiring to some hole at the base of a tree, beneath roots, or in some such situation. It is satisfied with a small quantity of food, and provides no winter stores; nor does any other British animal hybernate so completely. It spends the greater part of the day in sleep, and forages for worms, insects, and other petty spoil, principally in the night. It prefers small thickets, hedges, and bushy ditches, for its retreat, where it makes a hole about six or eight inches deep, and lines it with moss, grass, or leaves.—*Lennox Moore.*

HEDGEHOGS.—Hedgehogs eat snails, as I know from watching them. Once, indeed, my attention was directed to a hedgehog half-concealed among brambles, by the noise it made crunching a great snail shell. But I think they are still more fond of beetles. As a rule, they surely must sleep in the day; or else what can become of all the hedgepigs that go roaming through the woods in May and June as soon as it begins to darken? I have often stood and watched them about that time when the grass on every side of me was rustling, and the air was filled with snorts of all descriptions, merely from the vigour of these little animals hunting after their prey. Oftener than I can tell they have come poking their snouts even from under the instep of my boot, in expectation, I suppose, of finding something valuable in the crevice; and I am sure they would not be in the least averse to making similar scientific explorations into rabbit-burrows, or, for that matter, badger-holes. But tackling the rabbits is a different business. I cannot answer R. T. V. S. W.'s other questions. Among themselves hedgepigs fight desperately, and when two get into a battle they grow so excited that they puff like steam-engines. They can also screech in the most piercing manner, and when one begins it has no objection to go on for hours, with short rests.—*C. B. Moffat.*

PRESERVING CARTILAGE.—Can you, or any of your readers, inform me of a way of preserving cartilage in its natural shape, without permanent immersion in fluid?—*Frederick Rutt.*

PLANT NAMES.—Is it the latest botanical fashion, or a printer's error, to call *Barbarea vulgaris* water-cress, and *Tanacetum vulgare*, aromatic pansy? What are we to understand from the following passage:—"Along the road to this house we found several mints: *Nepeta cataria, Stachys sylvatica, Torilis anthriscus* and others;" but that the writer classes *Torilis anthriscus* (order Umbelliferæ) among the mints? See SCIENCE-GOSSIP, p. 267, December number.—*G. W. Bulman.*

THE SUNDEWS.—I think your correspondent Mr. J. P. Smythe is rather too sanguine about growing drosera in confinement. His experience comprises, according to his note, about six weeks' time. No doubt he will be able to keep it alive, and growing, to some extent, for considerably longer than this, but I doubt very much whether next year he will be able to say the same. I have grown it here in a cold frame for three years, but the second year's flowers were not so large as the first, neither were there so many leaves. The brilliant red colour of the glands also deteriorated very much. The third year it had dwindled down to almost nothing, it did not flower at all, and the leaves were exceedingly small, so I am afraid I have seen the last of it.—*A. N. Urmston, Manchester.*

CUCUMBER TREES.—In "Harper's Magazine" for last August, there is an article entitled "The Heart of the Alleghanies," in which mention is made of forests of hemlock, oak, chestnut, tulip-trees, cucumber-trees, wild-cherry, and forked pine. Can any of your readers, American or otherwise, give descriptive particulars of the cucumber-trees referred to? I have seen it stated that a leather strap buckled about the height of a man around a young tree in a few years will be away beyond your reach, but if nailed at the same distance will never get any higher, as only the outer shell or bark runs up. Is there any truth in this?—*J. F. C., Leeds.*

THE SCIENCE OF OUR GUIDE-BOOKS.—Mr. G. H. Bryan at the end of his article "September at the English Lakes," in the September number of SCIENCE-GOSSIP gave a fine specimen of scientific spelling from an old Murray's "Guide-book." A few years back being at Tintagel, weather bound, and having nothing to read but a Black's "Guide to the South Coast," date 1872 which was the whole of the coffee-room literature, he came across the following choice natural history information, p. 114. "Towards Babbicombe the *botanist* will meet with *Actinia crassicornis*, *Actinia nivea* and *Actinia mesembryanthemnm*, *Tortula tortuosa* and *didymum*, *Rhodymenia palmata*, *Laminaria digitata* and *sacharina*, *Laurentia pinnatifida* and *Plocaminm coccineum*. Along the beach towards Paignton the *geologist* may look for *Echinus miliaris*, *Trochus ziziphinius*, *Asterina gibbosa*, *Doris pilosa*, *Pholus parva*, *Pholus dactylus* and *Anthea cereus*." Whatever may be the accuracy of the information, it is certainly a little mixed.—*W. T. Suffolk, F.R.M.S.*

DRIED FLOWERS WITH COLOURS.—In answer to Mr. T. McGann's communication, I beg to state that he may find worthy information on the subject he seeks in Mr. English's admirable "Manual on the Preservation of the Larger Fungi and Wild Flowers," published by A. B. Davis, High Street, Epping, at half-a-crown.—*J. W. Williams, B.Sc.*

NOTICES TO CORRESPONDENTS.

TO CORRESPONDENTS AND EXCHANGERS.— As we now publish SCIENCE-GOSSIP earlier than heretofore, we cannot possibly insert in the following number any communications which reach us later than the 8th of the previous month.
TO ANONYMOUS QUERISTS.—We receive so many queries which do not bear the writers' names that we are forced to adhere to our rule of not noticing them.
TO DEALERS AND OTHERS.—We are always glad to treat dealers in natural history objects on the same fair and general ground as amateurs, in so far as the "exchanges" offered are fair exchanges. But it is evident that, when their offers are simply disguised advertisements, for the purpose of evading the cost of advertising, an advantage is taken of our *gratuitous* insertion of "exchanges" which cannot be tolerated.
WE request that all exchanges may be signed with name (or initials) and full address at the end.

O. J. H.—We have no doubt the object you saw in the grass was not an uncommon species of phosphorescent millepide, called *Geophilus electricus*.
T. RICHARDSON.—Your specimen is the Lancashire Asphodel (*Narthecium ossifragum*) in the seeding condition.
S. A. B.—It is intended that our coloured plates shall deal with subjects calculated to please every class of readers, zoological and botanical, as we hope you will see in future numbers. Thanks for the suggestion.
W. H. G.—A lengthy and well-illustrated paper on "The Galls of Essex," giving an account of all the commoner galls and the insects which produce them, will be found in the "Transactions of the Epping Forest, &c., Field Club." It is by Mr. E. A. Fitch, the well-known entomologist, it occurs in Part 6, vol. ii., and may be obtained from the club, Buckhurst Hill, Essex.
JOHN SMITH.—The best table we know of the Animal Kingdom, showing the most recently accepted arrangement of classes, orders, families, &c. in the fullest manner, is Mr. Francis P. Pascoe's "Zoological Classification." A handy book of reference, with the tables of the sub-kingdoms, classes, orders, &c., of the animal kingdom, their characters and lists of the families and principal genera. Published by Van Voorst, at 10s. 6d.
W. E. WARD.—The "Ornithologist and Oologist" is an American journal, published by F. B. Webster, Pautucket, R.I. The subscription (including postage) is 5s. per annum. You may obtain it of Mr. A. Cliff, 35 Osborne Road, Forest Gate, Essex.
J. A. J.—Apply to James Gardner, 29, Oxford Street, London, for Doubleday's "Classification of Insects."
J. S. HUTCHINSON.—In base agricultural manures common salt is sometimes used instead of nitrate of soda. Both of them look much alike to the eye, and both deliquesce from moisture. But the best way for you to detect the difference would be to throw a handful on the fire. If it is salt it will simply burn with the well-known blue lambent flame—if nitrate of soda, it will burn like gunpowder.

R. N. M. (Adelaide.)—A full description of how to make and use carmine and picro-carmine staining will be found at pp. 35 and 79 of Dr. Marsh's book on "Section-Cutting," published by J. A. Churchill, New Burlington Street, London.
C. D.—You will find an excellent description of the Anatomy and Physiology of Birds in the collected papers of the late Professor Garrod, also in the Volume of Cassell's "Natural History" (edited by Professor Duncan), devoted to birds. Roberts' "Popular History of the Mollusca," price 10s. 6d., published by Lovell Reeve & Co., has excellent plates, but we recommend Dr. Gwyn Jeffrey's "Marine Mollusca," a really full and complete book, with uncoloured plates of all the leading types.
K. D.—Make a strong solution of chloride of lime in hot water, and leave the corals in all night. They will take no harm, unless very delicate. Move them to and fro in the water next morning, to wash off the dirt.
DR. CUNNYNGHAME.—The "Popular Science Review" has not been issued since 1880.
W. H. CHARLES.—Accept our thanks for the article.

EXCHANGES.

"POPULAR SCIENCE REVIEW," unbound, from January, 1870, to April, 1878, both inclusive. What offers? Also "The Leisure Hour," January, 1880, to December, 1883.—E. H. Robertson, Swalcliffe, Banbury, Oxon.
WANTED, Darwin's "Descent of Man" (new) in exchange for six good plants of the bristle fern (*Trichomanes radicans*). —G. Donovan. jun., Myross Wood, Leap, co. Cork.
A COLLECTION of British birds' eggs in exchange for moths or eggs.—J. C. Machay, 15 Gordon Street, Aberdeen.
OFFERED, L. C. 7th ed., 31, 395, 588, 595, 699, 809, 1012*b*, 1021, 1024, 1043, 1149, 1287, 1288.—A. Sangster, Cattle, Oldmeldrum, N.B.
WANTED, a tourmaline, for electrical experiments; will exchange or purchase.—Robert Knight, Wellington, Somerset.
WANTED, pathological, anatomical specimens, Stirling's "Histology," or any microscopical books, in exchange for treadle, fret-saw with blowers, drills, &c. (new). List to V. Latham, 15 Thorncliffe Grove, Oxford Road, Manchester.
WANTED, books on Natural History, for "Cassell's Family Magazine" from December, 1879, to January, 1883, unbound. "Popular Educator," 6 vols., and "Technical Education," 4 vols., bound.—E. A. Snell, 70 City Road, E.C.
WANTED, double nose piece, also ⅓ or ½ objective, will exchange slides, &c. ; scale of black bream, mounted, in exchange, or any interesting slide.—H. Moulton, 37 Chancery Lane, London, W.C.
I WILL give English specimens (taken by myself) of *Helix lapicida*, var. *albida*, for examples (British) of *Vertigo pusilla*, *V. substriata*, *V. alpestris*, *V. antivertigo*.—G. Sherriff Tye, 65 Villa Road, Handsworth, Birmingham.
"THE LIFE AND EPISTLES OF ST. PAUL," best edition, and new book, offered in exchange for Gwyn Jeffrey's volume of British Marine Shells, figured plates. Address, Miss F. M. Hele, Fairlight, Elmgrove Road, Cotham, Bristol.
BRITISH land and freshwater shells offered in exchange for Transvaal stamps. Address, Miss F. M. Hele, Fairlight, Elmgrove Road, Cotham, Bristol.
WANTED, back number of SCIENCE-GOSSIP in exchange for a serpent's skin (from Demerara) about fifteen feet long.—E. Peak, The Park, Hull.
Two shillings and sixpence will be paid for a clean copy of the "Entomologist" for January, 1882 (No. 224). Address, Wm. J. V. Vandenbergh, F.R.A.S., Hornsey, Middlesex.
WANTED, "Journal of Conchology," (seeds) Hardwicke's SCIENCE-GOSSIP, Vols. i.-x. inclusive, bound or unbound, or any complete volumes, "Geological Magazine," vols. i.-viii., Decade II. In exchange for 150 species American melanian (Anculosa, Gonicharis, Pleurocera, Schizostoma and Lithosia) or 100 species American Unionidæ—including many of the rarer forms from the Southern States. Correspondence solicited. Exchanges in Indian freshwater shells also wanted.—R. Ellsworth Call, F.A.A.A.S., David City, Nebraska, U.S.A.
EXCHANGE, Ewald's "Life and Times of Lord Beaconsfield," new, "Zoologist," in parts, for 1883, clean, Dr. Pye Smith's "Geology and Scripture," for works on Natural History, ornithology preferred.—Geo. Roberts, Lofthouse, near Wakefield.
MICROGRAPHIC Dictionary wanted in exchange for mahogany slide cabinet to hold two gross. Gosse's "Marine Zoology," 2 vols., also wanted.—Wm. Tyler, 20 Geach Street, Birmingham.
"THE POPULAR SCIENTIFIC POCKET CABINET SERIES," two slides, soundings Cyclops—with Page's Text Book and Geikie's Primer—for foreign or fossil shells.—A. Loydell, 10 Aulay Street, Ossery Road, S.E.
MICROSCOPIC slides of stained botanical sections, insects, spicules, marine algæ, zoophytes and histological sections, to exchange for other good slides.—Dr. Moorhead, Errigle, Ootehill, Ireland.
WANTED, primary and lower secondary fossils ; will give in exchange fossils from cretaceous and eocene formations. Apply to F. O'Farrell, Southampton Boys' College, Moira Place, Southampton.

OFFERED post-tertiary and raised beach fossils. Wanted recent British land, freshwater and marine shells. Lists exchanged.—J. Smith, Kilwinning, Ayrshire.

WANTED, works on biology, or Stainton's "Manual;" will give in exchange any of following: SCIENCE-GOSSIP, 1880 to 1883, four years, "Familiar Wild Birds," sixteen parts, Buffon's "Natural History," Part i., original edition, and many others.—Wm. P. Ellis, Enfield Chase, West Enfield.

MOTHER-O'-PEARL shells from Tahiti, encrusted with lovely forms of polyzoa, &c., also sweepings from warehouse where the shells have been in stock. Wanted mosses, zoophytes, algæ, insects, or anything of interest. Write first to B. B. Scott, 18 Chiswell Street, Needham Road, Liverpool.

I AM very anxious to obtain correspondents in all parts of the world for exchange of plants, insects, &c.—B. B. Scott, 18 Chiswell Street, Needham Road, Liverpool, England.

WILL send well-mounted transverse section of kidney for well-mounted object. Address, E. H., 18 Salisbury Road, Dalston, London.

WANTED, objectives, micro-appliances, material, slides, and books on micro-subjects, in exchange for other slides (anatomical, pathological, botanical, micro-fungi, fern sori, diatoms, odontophores, parasites, foraminifera, polycistinæ, &c.). No cards. Please state wants and send lists to F. L. Carter, College of Medicine, Newcastle-on-Tyne.

WANTED, correspondents at home and abroad to exchange micro-material, shells, &c.—F. L. Carter, College of Medicine, Newcastle-on-Tyne.

WANTED, the following micro-fungi: *Œcidium tragopogonis, Œ. leucospermum, Œ. quadrifidum, Œ. euphorbiæ, Œ. urticæ,* and other species for mounting, for cash or exchange.—George Ward, 18 Nursling Street, Leicester.

WILL give "Imperial Lexicon," 2 vols., half calf, and "Unseen Universe," for good platyscopic lens: or these supplemented with other books or cash for good ½ objective.—Daniel Mayor, 7 Chaddock Street, Preston, Lancashire.

FIRST seventy-seven numbers of "Knowledge," ninety birds' eggs, not named, and small quantity foraminiferous sponge sand ; exchange for well-mounted diatoms : offers requested.—Henderson, 45 Eglinton Street, Glasgow.

WANTED, SCIENCE-GOSSIP, vols. i. to xiv., both inclusive, also "Nature," vols. i. to viii., both inclusive. State conditions, &c.—W. E. Martin, C.E., The Bungalows, Birchington, Kent.

A QUEKETT'S dissecting microscope to be exchanged for a Rutherford, Stirling, Daucer, or Cathcart microtome, the latter preferred. The Q. D. M. is similar to that described and figured in "Dr. Carpenter on the Microscope" but with glass table top and packs all requisites in stand, three object glasses, reflector, and condenser complete.—F. Derry, Upper Hockley Street, Birmingham.

WANTED, 1¼ inch objective (Baker's combination would do). State particulars.—W. Hollebon, Newark House, Langney Road, Eastborne.

FOR exchange, eleven parts of SCIENCE-GOSSIP for 1882, eight parts of "Entomologist," 1882, for objects of natural history.—Thomas Macrae, 41 McNeil Street, Glasgow, S.S.

J. W. will be glad to forward a few Podosoma (*Lepidocyrtus curvicollis*) to microscopists wishing to study scales.—Woburn House, Woburn Hill Green Lane, Liverpool.

WANTED, specimens of recent and fossil shells, birds' eggs, freshly taken or carefully preserved specimens of crustacea. Also works (with plates) on above and British mosses. Offered, recent shells, *Venus gallina*, with varieties *alba* and *gibba, Pholas candida, Lucinopsis undata, Tellina tenuis, T. fabula, Mactia stultorum*, and variety *Solen marginatus, Nassa reticulata, Mytilus pellucida, Syndosmya alba, Utriculus obtusus*. Also Welsh Mosses (unnamed).—C. Jefferys, Langhame, Carmarthenshire.

AUSTRIAN wood opal, serpentine, obsidian, lapis lazuli, arsenical ore, mica-schist, &c., to exchange for the following fossils, *Oldhamia radiata, O. antiqua,* graptolites, *Marsupites Milleri, Temnechinus excavatus.*—Miss Linter, Coloony House, Grosvenor Road, Twickenham.

WANTED, specimens of *Macrodon hirsonensis* (Great Oolite) and *Temnechinus* (Crag.). Good exchanges in British shells or fossils.—G. L., 8 Winchester Place, Highgate, London.

WANTED, well-mounted slides of botanical preparation; also fruits, seeds, and other objects useful in teaching botany. Can offer in exchange shells, fossils, British plants, and foreign stamps.—H. L. E., Hay-Gordon, & Co., Widnes.

COINS wanted in exchange for stamps.—F. R. E., 35 Rufford Road, Liverpool.

FOR exchange: osprey, peregrine, falcon, kite, black kite, short-eared owl, mottled owl, nightingale, icterine warbler, meadow pipit, tree sparrow, red-winged starling, hooded crow, rook, magpie, golden-winged woodpecker, downy woodpecker, yellow-billed woodpecker, belted kingfisher, stock dove, rock dove, red grouse, Virginian culin, Carolina quail, night heron, glossy ibis, Egyptian goose, Canada goose, eider duck, puffin, shag, Arctic tern, sooty tern, common gull, Richardson's skua, fork-tailed petrel.—Ralph Turnbull, 8 Cemetery Road, Crewe.

WANTED, slides of Diatomaceæ, Desmidiaceæ, Polypi, Algæ, Foraminifera, Polycystina, Spicula, &c. Please send lists of duplicates. Other slides and unmounted micro-material, also Lepidoptera.—F. A. A. Skuse, 143 Stepney Green, London, E.

A FEW freshly-dug pupæ of *S. tiliæ*, imagoes of *S. populi, B. perla*, &c., for exchange. Wanted ova of *L. dispar*. Lists exchanged.—Geo. Balding, Ruby Street, Wisbech.

U. pictorum, tumidus, A. anatina, and *v. radiata, D. polymorpha, A. cygnea* v. *Zellensis, P. carinatus,* &c., offered for varieties *albo-lineata, roseo-labiata,* and *castanea* of *nemoralis* or v. *olivacea, aurenicola, lilacina* of *hortensis.*—B. Hudson, 15 Waterloo Road, Middlesbrough.

MICROSCOPE (Wolland), five objectives, 8 in. mahogany case, with various appurtenances. What offers?—L. R. Sutherland, 5, Hillsboro' Square, Hillhead, Glasgow.

Pottia asperula, Bryum Tozeri, Tortula recurvifolia, Trichostomum littorale, Fissidens serrulatus, Eurhynckium speciosum, Sphagnum acutifolium, vars. *quinquefarium* and *squarrosulum*. Offered to British or American bryologists.—Wm. Curnon, Pembroke Cottage, Newbyn Cliff, Penzance.

DUPLICATES. *Limnæa glabra, L. peregra, Planorbis spirorbis, P. carinatus, Physa hypnorum,* and very fine *L. stagnalis* and *P. corneus*. Locality, Strensall Common, near York. Desiderata : very numerous British land, freshwater and marine shells.—W. Hewett, 26 Clarence Street, York.

WANTED, slides of *Actinonium septenarium, Asterionella formosa, Mesocena octogona, Hemiaulus antarcticus, Heliopelta Leeuwenhoeckii, Climacosphenia monilligera, Lagenæ vulgaris,* var. *semistriata, Achnanthes longipes, Pinnularia nobilis, P. viridis, P. oblonga, P. radiosa, Gyrosigma quadratum* and other diatoms, in exchange for slides of Foraminifera, parts of insects, &c.—F. A. A. Skuse, 143 Stepney Green, London, E.

FORAMINIFEROUS sand sent to any address (to any amount) on receipt of stamps to cover postage.—F. A. A. Skuse, 143 Stepney Green, London, E.

FOR microfocal slides or material preferred, the beautiful seeds of *Nicotiana Tabacum,* mounted or unmounted.—H. F. Jolly, Stow Villa, Bath.

"KNOWLEDGE," vols. i. and ii., clean, perfect and bound, in exchange for a small herbarium collection of British Mosses.—G. F. Nest, Jedburgh, N.B.

"NATURE," vols. i.–xx., bound, and xii.–xx., clean and perfect but unbound, for any natural history collections suitable for a boys' museum.—G. F. Nest, Jedburgh, N.B.

WANTED, "Popular Science Review" for 1881, 1882, and 1883.—Dr. Cunynghame, 6 Walker Street, Edinburgh.

BOOKS, ETC., RECEIVED.

"Flowers and Flower-Lore," by the Rev. Hilderic Friend, F.L.S., &c. (with illustrations) 2 vols. London: W. Swan Sonnenschein & Co.—"Proceedings of the Literary and Philosophical Society of Liverpool," vols. 35, 36, and 37, for 1880–81, 1881–82, 1882–83.—"New Commercial Plants and Drugs," by Thos. Christy, F.L.S., &c. London: Christy & Co.—"Facts Around Us." G. C. Lloyd Morgan. London : Edward Stanford.—"Studies in Microscopical Science," edited by A. C. Cole.—"The Methods of Microscopical Research," edited by A. C. Cole.—"Popular Microscopical Studies." By A. C. Cole.—"Portfolio of Drawings." No. 10. By Thomas Bolton.—"Petrological Studies." By Messrs. J. E. Ady and H. Hensoldt.—"The Gentleman's Magazine."—"Belgravia." —"The Antiquary."—"Journal of Conchology."—"The Journal of Microscopy."—"Land and Water."—"The Science Monthly."—"Midland Naturalist."—"The Inventor's Record."—"The Naturalist's World."—"Ben Brierley's Journal."—"Sunlight."—"The Medical Student." (New York.) —"Natural History Notes."—"Science."—"American Naturalist."—"Medico-Legal Journal." (New York.)—"Canadian Naturalist."—"American Monthly Microscopical Journal."—"Popular Science News."—"The Botanical Gazette."—"The Arcadian Scientist."—"The Ornithologist and Oologist."—"The Electrician."—"Revue de Botanique."—"La Feuille des Jeunes Naturalistes."—"Le Monde de la Science."—"Ciel et Terre."—"Cosmos : les Mondes." &c. &c. &c.

COMMUNICATIONS RECEIVED UP TO 12TH ULT. FROM :—
R. E. C.—B. P.—J. F. R.—W. P. C.—G. W. B.—A. H. B.—C. J. H.—T. D. A.—R. W. M.—H. M., jun.—J. G.—E. J. G.—H. M.—E. A. P.—J.—S. J.—M. B. T.—Dr. J. F.—F. M. H.—G. S. T.—P. F. L.—A. H. S.—G. H. K.—J. T.—C. K.—E. C.—C. D., jun.—E. H. R.—J. W.—E. T. D.—F. K.—A. H. W.—J. C.—S. J. C. M.—A. S.—W. H. B.—A. B.—E. A. S.—J. R. M.—M. E. T.—G.—G. O.—S. A. B.—W. H. G.—V. A. L.—E. L.—W. C. O.—N.—R. K.—E. F. B.—S. T.—A. H. S.—O. P. C.—W. T.—J. C. P.—W. J. S.—S. T.—H. T.—B. B. S.—W. B. E.—B. H.—A. H. B.—H. L. E.—K. N. K.—W. T. C.—G.—W. C. F. O. F.—A. L.—J. I. H.—J. H. B.—J. W.—C. C.—J.—S.—T. H. M.—C. M. G.—B. R. C.—J. E. L.—W. Y.—L. C. M.—O. J. H.—J. D. A.—R. N. M.—M. E. P.—G. L.—H. L. E.—W. F., jun.—D. M.—W. E. W.—K. T.—H. C. B.—J. G. B.—J. A. W.—J. H. W.—E. M.—H.—W. M.—C. F. G.—W. P.—F. H.—F. W. P.—H. P.—T. M.—J. A. J.—J. W.—H. S. G.—H. B. R.—W. H. C.—B. H.—W. C.—G. D. E.—W. H. M.—J. D.—W. R. T.—R. B.—L. R. S.—T. S.—G. B.—F. A. A. S.—H. F. J.—W. H. C.—S. B.—Dr. C.—V. A. L.—G. F. N.—W. S., &c.

GRAPHIC MICROSCOPY

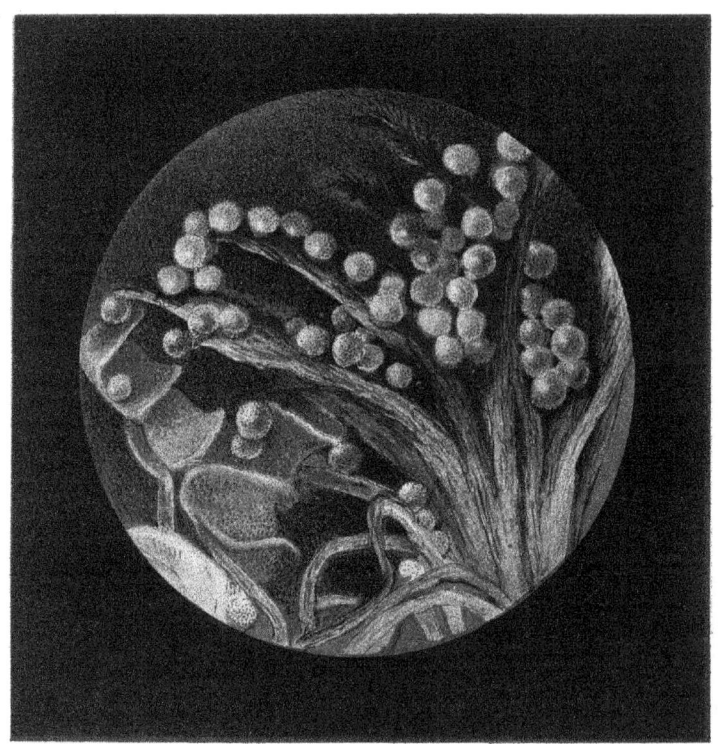

E.T.D. del. ad nat. Vincent Brooks,Day&Son,Lith.

ANTHER, POLLEN & STIGMA OF MALLOW.
× 50

GRAPHIC MICROSCOPY.

By E. T. D.

No. IV.—Pollen of Mallow.

POSSIBLY the first use the possessor of a newly-acquired microscope applies the instrument to is the examination of flowers. Easily-obtained floral envelopes of plants, arranged for observation without difficulty, offer immediate attraction, and naturally the powdery pollen-bearing anthers surmounting their filaments, arranged around the pistil and stigma, soon arrest attention. The charm of colour, form, variety, and strange contrivance, under moderate magnifying power, are easily observed, and even an experienced microscopist, with knowledge of the deeper significance of plant development, finds a pleasure in the contemplation of the general elegance disclosed in the disposition of these organs.

The flower of the common mallow (*Malva sylvestris*), picked when the anthers are ripe and just opened, discloses a combination of parts (more or less modified in other instances) which may be taken as typical. Here may be seen, on their "conspicuous" filaments or stems, the anthers (the cradle of their production), smothered with microscopic bodies, "free" pollen cells, specks filled with a fine molecular substance termed "fovilla," but each containing the element of a fructifying power in its resulting influence beyond comprehension. To meet this condition of things is the "stigma," of loose elongated cellular structure, secreting a viscid fluid, which, accumulating at the base, forms the "nectar" of bees. The mode of contact is varied and peculiar; in the mallow the process may be described in general terms. The anthers, when ripe, burst; the released grains touch the loose tissues of the stigma, and complete the fructification so far; the future process is of deeper interest, involving the development of the embryonal vesicles, and the subject then outstrips the range of "popularity," and touches on the domain of physiological botany, requiring the most careful dissection, preparation, use of chemical reagents, staining fluids, and the highest powers of the instrument.

Pollen, or "blossom-dust," varies in different flowers in shape, colour, and markings, distinguishable under the microscope—size and shape in an extraordinary degree. It is curious to contemplate that these cellular bodies, dispersed by wind, rain, the agency of insects, and other adventitious aids, keep their separate vitality, integrity, and special quality, the power of their influence being distinctly confined to the exigencies of the plants from which they emanate. Perhaps the most beautiful and perfect specimens are to be found in obscure weeds; and wild, indigenous plants, the evening primrose (*Œnothera*) for shape, the common musk-plant (*Mimulus moschatus*) for elegant marking, may be taken as typical; the wild geranium (*Pelargonium*), the hollyhock (*Althæa rosea*), and the sedums also afford peculiar forms. Colour is various: red (*Verbascum*), blue (*Epilobium*), black (Tulip).

This subject naturally leads to the strange correlation or association of necessity between insects and plants. The scattering or placing of pollen is of all importance in the production of future plants, but it cannot be, nor is it always, effected by wind or rain disturbance; but in many instances it is partially, and in other cases can only be brought about by the direct mechanical interference caused by insects (the *Hymenoptera* in particular). In the economy of bee life the collection of pollen is of vital necessity, and in every instance it aids the functions of the plant. The most ordinary observer has watched the industry of bees in collecting pollen, rifling the flower, for this "bee bread," and the no less precious nectar, thus assisting in bringing the anther and stigma into contact. This is common knowledge, but the association, although explainable on general theory (and other instances of widely diverse but necessitous dependencies exist), is strangely mysterious. Those who have watched and who understand the necessities of bees, cannot have failed to observe that their eagerness for the collection of pollen comes exactly at the right time, when storage for the nourishment of the young larvæ and the fructification of the plant are coincident. Who has not seen the bee rolling in the blossom and rifling the rich contents, leaving for the hive with golden pellets stored in receptacles on thighs (adapted for the very purpose), but considers the influence of the importance of the insects sweeping over the stigma of the plant, the few fructifying uncollected grains? It has been stated on the best authority that one community of bees will collect fifteen pounds' weight of pollen or "bee-bread" in one season.

"Brush'd from each anther's crown, the mealy gold,
With morning dew, the light-fang'd artists mould,
Fill with the foodful load, their hollow'd thigh,
And to their nurselings bear the rich supply."

To the fructification of the cryptogamous or flowerless plants it is unnecessary to refer, beyond observing that the primary process is obscured to ordinary vision, and requires for its revelation the highest powers of the microscope. The whole of the reproductive elements are involved in each individual spore, or particle of "fern dust." Under favourable circumstances these minute specks produce a growing expansion, "prothallus," in which is contained a peculiar condition of cellular contact leading to the development of the future plant.

Prothalli may be produced or propagated by strewing spores of ferns on a porous brick, kept moistened with moderate warmth under a bell-glass, and may, while growing, be arranged on glass slips for microscopic observation under high powers. Pollen grains and the adjacent parts retain their form and position when carefully dried, and may be conveniently seen [(as an opaque object) under low powers, and easily mounted and preserved for future reference and comparison.

Crouch End.

GEESE MIGRATING.—Having seen a note in SCIENCE-GOSSIP about geese migrating, it may interest the writer to know that about a fortnight ago, a flock of thirty-two geese was observed flying, very high, over Wrenbury, Cheshire. They were flying towards the north, leaving Combermere on their left, in "single file," one of their number keeping a little to the right, apparently a kind of "off-skirmisher" or leader. They were cackling as they flew.—*M. E. T.*

HOW TO KEEP SMALL MARINE AQUARIA.

THE difficulties, real and imaginary, that present themselves in connection with marine aquaria and the keeping of marine animals generally, often deter the would-be marine zoologist from any further attempt to retain life and health in a confined tank, after his first experiment has terminated, as it too often does, in decomposition and disaster.

The following observations may therefore be of some service in at any rate enabling others to do what I have recently done, namely, to keep marine animals in glass jars of only one pint or so capacity, not only in a healthy condition, but that too with little or no trouble, for a considerable time; the period in the case of my own jars being, as I write this, just four months and a half.

In the middle of October, 1883, I obtained a few marine animals which I thought I would try and keep alive for a short period; my only receptacles at the time were two glass bottles of something over one pint capacity, with mouths about half the diameter of the bottle. My stock of sea-water was about two quarts. I put a pint of water into each of the bottles, and placed a small stone in each, to which were rooted nice small tufts of *Ulva latissima*. In bottle No. I I introduced a full-grown mussel, *Mytilus edulis*, and one winkle, *Littorina littorea*. Into No. 2 I placed two beadlet anemones, *Actinia mesembryanthemum*, one *Actinoloba dianthus* (small), three very young mussels and one winkle. I placed the jars on a table close to a window which faced due east, and covered the mouths each with a small bell glass, which by resting upon the shoulders of the bottles excluded all dust. The remainder of the water I corked up in another bottle and kept it in the dark. I commenced my care of these miniature tanks by diligently syringing in order to aerate the water; it certainly did that, but it also broke off fragments of the algæ, and in a few days things began to look bad, and I had arrived at that stage when the experiment is usually given up and the decomposing mass is thrown away. However, although the water was tainted and things were not looking promising, I knew I had a reserve tank to fall back upon, so I gave the bottles a good shake up and poured off the sea-water, which was full of bits, and very dirty, replacing it with fresh sea-water. The effect was miraculous; the anemones "came out," the winkles resumed their travels, and once more matters looked favourable.

I then carefully strained the tainted water, which even after that operation was very thick, and put it into the reserve bottle, shaking it vigorously once a day; in about a week it had got quite clear again. But to return to the jars; there was no doubt that syringing, as applied to pints of sea-water, was simply courting failure, so I left jar No. I quite alone to see the result, and this is it. From that day to this, a good four months, the water in this jar has remained as clear as crystal, and its living occupants are apparently as healthy as the day they were taken from their home on the shore.

Jar No. 2, in consequence of my experiments in feeding the actinia, did from time to time become thick as regards the water, but by replacing this from the reserve bottle and carefully straining the other and keeping in the dark, I have so far succeeded that my two *A. mesembryanthemum* have, by producing young, become four, and no occupant has died, but health and vigour is seen in all.

Now I have, no doubt, been favoured so far with suitable weather for such experiments, and it may be a very different matter when the atmosphere stands at a temperature of 60° or 70° instead of 40° or 50°, but I intend then to stand each jar in a large saucer of water, and to partially cover the jars with an envelope of coarse canvas or flannel, which, by being kept constantly wet by capillary attraction, will produce coolness by evaporation, and thus preserve a suitable temperature for the contents of the jars.

Of course such an envelope will curtail the view of the live stock, but the canvas coat may be made removable.

Much interest is afforded by a small tank of this description, and the trouble connected with it is almost *nil*. I shall here give any one further details as to starting a series of pint tanks if desired, and I hope at some other time to report progress, and to make some observations on the habits of the inmates.

EDWARD LOVETT.

Addiscombe, Croydon.

MINERALOGICAL STUDIES IN THE COUNTY OF DUBLIN.

CONSIDERING the vast importance in modern times of mineral products in connection with manufactures or domestic requirements, it is surprising that the science of mineralogy is not more generally cultivated by those having a taste for natural science. There are few districts in any country so barren as not to afford some points of interest for the student who diligently seeks them. Germany, ever in the van of scientific knowledge, has done much to make this study accessible and attractive to the masses, as may be seen from the works of Groth, Naumann, Plattner, and others. America, whose great mineral resources offer an ample field for the development of information, is also doing good work, nor is France lagging behind in the matter; but in the United Kingdom few appear to think the mineral world worthy of attention, as the paucity of scientific works on the subject would seem to indicate, and for one student who devotes himself to minerals, there are a dozen who pursue botany, zoology, or other kindred subjects to the exclusion of everything else.

And yet there are few branches of natural science more fascinating to the lover of nature; but its difficulties are often magnified by those who have not taken the trouble to dip more than superficially into it. No doubt, for the proper study of mineralogy, it is necessary to use the knowledge of the mathematician, the physicist, and the chemist; but, given the inclination, there is no reason why the earnest student should find this more difficult to acquire than any other branch of learning. By the aid of a good elementary work, a piece of road metal, picked up by the wayside, may give a valuable insight into crystallography; and for the expenditure of a few shillings, a simple set of apparatus for blowpipe analysis may be placed on his study table, which will afford him a never-failing source of interest, and may possibly lead to even more substantial results in after life.

Having said thus much by way of preface, I now propose to give a few notes of a couple of excursions recently made in the neighbourhood of Dublin in search of objects for analysis. In the first of these excursions in August last, I had the advantage of being accompanied by Professor J. P. O'Reilly, C.E., of the Royal College of Science here, whose long experience, both at home and on the Continent, has qualified him to be regarded as a leading authority on mineralogy. The vicinity of Dublin is not remarkable for minerals of a workable kind, but there is still much of interest to be found, and, at Professor O'Reilly's suggestion, we selected the hill of Feltrim, a locality within easy reach of Dublin, where a rather remarkable quarry exists. A run of half an hour by the Great Northern Railway brought us to Malahide, a watering-place possessing many advantages of situation and salubrity. Feltrim is situated a short distance off, and consists of a singular-looking ridge running nearly east to west, and rising almost abruptly from the surrounding country. It is surmounted by a ruined windmill, the whole forming a most desirable locality for artistic students, from a combination of rich colour, and the splendid view of Dublin Bay to be had from the summit. The district is in the lower part of the carboniferous limestone formation which extends so generally throughout Ireland, and is highly fossiliferous, containing examples of *Fenestella antiqua, Productus aculeatus, P. mesolobus, P. semireticulatus, Spirifera lineata, S. striata, Euomphalus pentangulatus*, and many other species too numerous to mention. Large masses of the limestone here are dolomitised to a considerable extent. The undolomitised portion, that is to say, the masses lying between the dolomitic ribs, as they may be called, furnish an excellent lime, and in a large heap of material rejected by the quarrymen, we found many objects of interest. Crystals of calcite were abundant, many good-sized sections affording excellent examples of the character-

E 2

istic rhombohedral cleavage, whilst in the dolomitised portion the curved or "saddleback" crystals were equally well developed. The prevailing form was as shewn in Fig. 41, which is often most distinctly recognised in the variety of dolomite known as pearl spar, but occasional groupings were found in the form indicated by Fig. 42.

Isolated portions of the dolomite afforded some very interesting metallic characters. Emerald green patches abundantly scattered over the surface seemed to indicate the presence of carbonate of copper (afterwards proved by analysis); and smaller spots of a blood-red colour more sparsely distributed, appeared to denote the oxide of the same metal; but owing to their position on the surface, this could not be verified without damaging the specimen more or less. The prevailing colour of the specimens being light brown inclining to ochre, suggested the presence of sesqui-oxide of iron, and that this existed in large quantity was proved by the strong yellow colour of the solution, when a portion was afterwards pulverised and dissolved in hydrochloric acid; a dark coloured mass resulted from this, which, on being heated under the blowpipe, became magnetic.

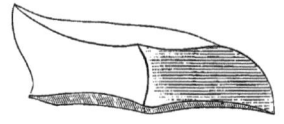

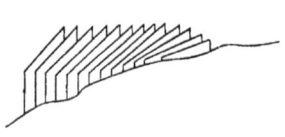

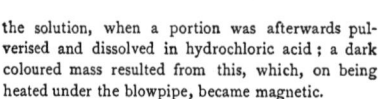

Perhaps the most interesting feature observed was a series of dendritic marks on some of the specimens obtained, a few being not inferior in elegance to the delicate tracery observed on moss agates, and some of the finer sandstones. These are now well known to be caused by infiltration of the metals iron or manganese when in a state of solution, and that the latter was probably the agent, appeared by the manganese reaction being very decidedly given when a small portion was tested in the usual way with carbonate of soda on platinum foil. Iron, however, existed in such quantity in every specimen tested, that it was impossible to say it had no part in contributing to the forms mentioned.

Many other points of interest were also noted on that occasion, but space will not allow of their being dwelt on further at present. On the same railway line, but some miles nearer to town, is another locality known as Killester, where there is a small exposure of the limestone formation laid bare in making the cutting, and now occasionally used as a quarry. The strata are immediately overlaid by a boulder clay, the stones of which have a markedly rounded form, and are scored with deep striæ, giving, it is believed, strong evidence of glacial action in the neighbourhood. The stone in general is of a shaly character, though there are some good beds. They are intersected at intervals by numerous vertical joints, all highly mineralised, the crystals of carbonate of lime being very well developed. The prevailing form was the following combination of the hexagonal prism with rhombohedrons (Fig. 43), some of the crystals reaching a length of from two to three inches. Scalenohedrons were also abundant, but not nearly so large in size. The crystalline forms of calcite being very numerous, other combinations of the prism and rhombohedron, &c., occurred, but not with the same frequency. It was curious, however, to note an occasional single crystal of quartz standing amid the calcite, often so closely resembling it in form and translucency as to render it no easy matter to distinguish the difference until the unerring test of hardness was applied by the penknife. A large heap of broken stone and shale was in a corner of the quarry, and on turning over some of the crystalline surfaces, copper pyrites were found freely distributed, evidenced by their yellow colour, sphenoidal crystals of the tetragonal system, and the comparative ease with which they yielded to the knife. Bitter spar was equally abundant in much the same positions. In fact, throughout the whole district I have mentioned, dolomitised limestone, accompanied by copper pyrites, is a prominent feature.

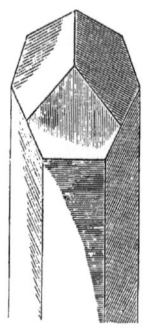

Perceiving something to glisten slightly in a bed of shale, I proceeded to dig it out with my cold chisel, which I succeeded in doing with a little trouble. On looking at the lump more closely, I saw that it had the peculiar pale brassy colour of iron pyrites, but was so imbedded in the limestone that no idea could be obtained of its crystalline form. The calcareous mass, however, being afterwards dissolved by acid, the lump of metal was obtained perfectly clean. It appeared to consist of an aggregated mass of cubical faces, the striæ on which were very strongly marked. On some of these faces was a decided bevelment parallel with the striæ, showing a distinct approach to the pentagonal dodecahedron with which crystallographers associate these markings when so placed.

Before leaving the quarry, I picked up a thin piece of limestone crystallised on both sides, on the upper surface of which was a single crystal of small size, but of different form, imbedded amongst the crystals of carbonate of lime. It was of a reddish colour, and of adamantine lustre. Its whole appearance was so different from that of the surrounding matrix that it was impossible to overlook it, though it might be termed even minute in size. From its position it was difficult to make out the crystalline form, although there were distinct facets. However, on looking over the specimen with a strong lens, traces of similar pieces could be perceived on the opposite side, but more easily accessible. Strong muriatic acid dissolved sufficient of the calcite to enable the form of the reddish-coloured matter to be observed. There were distinct indications of the faces of the rhombic dodecahedron, though somewhat distorted, and in a fragmentary piece, the cleavage in this direction was well marked. This having given a little help, I began to suspect what the substance was, but it was necessary to apply the blowpipe test. Lifting out a small portion with the forceps, I placed it on charcoal in the oxydising flame. It became white, and on applying a drop of nitrate of cobalt in the reducing flame, a beautiful green colour resulted, proving that the minute crystals were blende, or sulphide of zinc.

These few facts will go far to show how interesting is the science of determinative mineralogy. The substances met with in the localities mentioned are of common occurrence, and their distinctive characters easily mastered. The student's powers of exact observation are greatly improved, and were such branches of natural science more prominently taught in our schools, young men starting to seek a career in distant places, might have a new avenue to fortune opened up by a knowledge of the common things that lie on the surface of an apparently wild and barren country. I hope, with the Editor's permission, to contribute some further notes of mineralogical excursions in various parts of the county of Dublin.
Dublin. - W. McC. O'NEILL.

EXPERIMENTS made by Pasteur and others, inoculated with the virus of hydrophobia, have been so far successful that twenty-three dogs were rendered absolutely safe by the process.

A PRE-HISTORIC CITY.

PROFESSOR P. W. NORRIS, Assistant United States Ethnologist, has recently discovered the ruins of an ancient city, five miles in extent, in the Kanacoha valley near the city of Charleston. The Professor commenced his explorations of the ancient mounds in August last, and has just completed them. During the process of the excavation of these mounds he has collected upwards of 4000 fine specimens of various articles, which he intends to place in the National Museum at Washington. Seven of the mounds which he opened were from twenty to thirty-five feet in height, and from 300 to 540 feet in circumference at the base. He opened altogether fifty-six mounds in the Kanacoha valley. The articles he procured consisted principally of about thirty specimens of steatite and sandstone pottery and pipes, many lance and arrow heads, hatchets, fish-darts, celts, gorgets, hematite iron paint hatchets and paint cups, several hundred pieces of shell money, bone and horn punches used for dressing the flint arrow heads into proper shape, twenty-one bracelets made of copper, one copper breast plate, copper crowns, and many copper heads. All the copper was heavily coated with verdigris. In a mound thirty-five feet high and 545 feet around its base was a vault twelve feet square and ten feet high, the walls of which had been supported by black walnut timbers. In the centre of this vault, lying horizontally on its back, was a giant skeleton, seven feet six inches long, and measuring nineteen inches through the breast under the arms. On each wrist were six large copper bracelets, four of which had been enclosed in cloth or dressed skin. Under the skull was a stone lance head. There was a copper gorget upon the breast with two holes in it. This gorget, which was four inches square, is regarded as having been a badge of authority. In the right hand was a hematite iron hatchet having a four-inch blade. In the left hand were several lance heads, six inches in length, of flint manufacture. Leaning backwards in a dark coffin, which stood somewhat standing from a perpendicular position, was another skeleton, in such a position in relation to the first as to let the left hand extend over its head. On this left hand were two copper bracelets. In the right hand were a bunch of lance heads, similar to those of the giant. In each corner of this vault was a warrior, enclosed in his dark coffin, and standing nearly erect, with a stone hatchet and lances in or near his hands. Nearly one hundred various specimens of arms and ornaments were found in this one vault. In another mound the remains of a large sized warrior was found lying flat on his back, with a copper crown covering his head and neck, ornamented with sea-shell and bone heads. On one side of this warrior lay five others, with their feet all pointing towards him; while, the other side, were five women, as indicated by their size and ornaments, also having their feet pointed towards the central warrior. About these skeletons were found various weapons and ornaments. The question occurs, how came these skeletons there, five male and five female, all evidently buried at the same time with the chief? Could they have been entombed alive? There were ten cemented double cisterns at the head of each skeleton, all containing water. For thirty feet above, and more than that distance around this spot there was hard earth, dry like mortar, which had to be excavated with picks. There is another mound at about two miles distance, twenty-five feet high, and 306 feet in circumference. The top of this mound is a small flat plane, forty feet in diameter. Its top and sides, to a depth of above two feet, were covered with the natural soil. The entire remainder of the mound was composed of burned ashes, which had all been deposited in dark vessels containing about half a bushel each. About the centre of the mound were two large skeletons, in a sitting posture, facing each other, and their hands extended, palms upward, towards one another. Resting upon their hands was a curious altar, made of stone, of about two feet in diameter, the concave side up, and filled with ashes. On its top was a flat stone cover with two holes in it, and having on it the ancient totem marks. Down even with the natural surface of the ground in this mound was found an immense slightly concave altar; the centre of which was filled to about the depth of six inches with fine ashes. Around, farther from the centre, there were ashes and bits of human bones, piled up to the depth of nearly two feet. The Professor says there cannot be the least doubt but that these ashes are the cremated remains of human beings.

DIPTON BURN.

GLOUCESTERSHIRE SLUGS.

AMONG the most valuable and interesting of the consignments of living slugs which I have lately received, are some from Mr. E. J. Elliott, of Stroud, who has collected for me on both sides of the line which divides East from West Gloucestershire. As these areas are kept separate in our system of topographical conchology (which is borrowed from Watson's well-known scheme for Topographical Botany), I will keep the records for them distinct and commence with the species which Mr. Elliott obtained on the east (or rather north) side of the Thames and Severn Canal, which forms the dividing line.

The EAST GLOUCESTERSHIRE specimens, then, were of five species, viz., *Arion hortensis*, Fér., *Amalia marginata*, Müll., *Limax flavus*, L., *L. maximus*, L., and *L. agrestis*, L. Of the specimens of *Arion hortensis*, there is nothing further to remark than that they were juvenile specimens, very deep grey colour, with the foot-sole brilliant orange. The numerous specimens sent of *Amalia marginata* exhibited a considerable range of shades of brownish-yellow colour, and a number of them were referable to Millet's variety *rustica*, which has the ground colour grey without any admixture of brown or yellow, and is usually smaller in size. Of *L. flavus* were sent one full-grown and one diminutive young example. The specimens of *L. maximus* were very varied and of great interest. Several were of the var. *vulgaris* of Moquin, which is but another name for the typical form; this form is described as having the fasciæ of the back black, but it is to be observed that these examples of Mr. Elliott's had the fasciæ simply of an intensification of the cinereous ground colour and could not be called black. There were a few examples of the var. *rufescens*, one of the var. *obscura*, and a few of the var. *maculata*, but the gem of the collection was a very remarkable form which in its markings belonged to the var. *Johnstoni*, but differed from it completely in the matter of the ground colour, which, instead of being of the usual brownish or cinereous hue, was uniformly of a pure clear pellucid lilac tint. Mr. Elliott has since informed me that this form occurs in his brother's garden, the specimen not being an isolated one. I have proposed in the Journal of Conchology that for the present this form be named var. *Johnstoni*, sub-var. *lilacina*. Of the numerous examples of *Limax agrestis* or the common field-slug many pertained to the typical form as defined by Lessona and Pollonera (*cinereus immaculatus, capite tentaculisque brunneis*), others to the var. *albida*, and others to Moquin's var. *sylvatica*, while one was of the var. *lilacina* of Moquin.

WEST GLOUCESTERSHIRE SLUGS.—The specimens which Mr. Elliott collected for me on the West Gloucestershire side of Stroud included *Arion ater*, *A. hortensis*, *Amalia marginata*, *Lehmannia arborum*, *Limax maximus*, and *L. agrestis*. Of the first named species, *Arion ater*, were the nameable varieties *succinea* (the yellow form, immature as usual), and *rufa* (of a dark rich chocolate brown), and numerous other forms—some being pale unicolorous dun, and two small ones very pale greenish. Of *A. hortensis* numerous examples of the type, mostly juveniles, very dark coloured with the foot-sole orange, in one specimen very vividly so, were sent, and also a number of var. *fasciata*. In addition to these were specimens of the puzzling form of Arion whose specific location is as yet dubious. Of *Amalia marginata* there was but one specimen, a dark coloured one. Of *Lehmannia arborum* was sent one specimen, Mr. Elliott remarking that the species is plentiful on beeches in the woods after showers. Of *Limax maximus* were numerous typical examples, and one of the var. *obscura*. This latter was very dark ash-colour, with only a faint trace of banding on the body, and one solitary spot on the shield. The common field slug (*Limax agrestis*) was present in the consignment in great variety and abundance. The type as defined by Lessona, var. *sylvatica* as defined by Moquin-Tandon, and var. *tristis*, also of Moquin, were in abundance, and with them were a few lilac specimens referable to var. *lilacina* and one of Mr. Butterell's black var. *nigra*.

Altogether the consignments were of much interest, and had the great merit that the common species were sent in sufficient number to permit of an efficient study of their variation. I may conclude this note by stating that the line which parts the two divisions of Gloucestershire is traced along the Thames and Severn Canal, to the point where it joins the Severn; then up that river as far as Tewkesbury.

I shall be glad to receive similar consignments from other districts.

WM. DENISON ROEBUCK.
Sunny Bank, Leeds.

ON DRYING FLOWERS.

SEVERAL queries with reference to this subject have appeared in recent numbers of SCIENCE-GOSSIP. Hence there need be no apology for describing an original method, which is fairly successful, and is at least both simple and inexpensive.

It is easily understood that, as a great help to success, the plants should be gathered when the sun is shining upon them. If obtained at other times, they should be kept in a room some hours before putting into the press.

As a basis of operations, obtain a well-seasoned piece of board, about a foot wide, and eighteen inches long. On this place several layers of paper, not necessarily botanical drying paper, but any unglazed kind will do. Old newspapers answer admirably. Then carefully place a limited number of specimens on this, or possibly one, if large, exercising caution in properly disposing the petals and foliage. Over this place several more thicknesses of paper, on which other specimens can be arranged, as before, repeating the process, till discretion, or experience, suggest that a sufficient number have been put in. Then place on the top a stout piece of mill-board, slightly less in size than the wooden base. Next, obtain a piece of good cord, and make a running noose near one end, then bring the cord under the press; repeat the cross tie, near the other end, finally bringing it to the point of starting, and drawing it up quite slightly. Or, in other words, tie up, in the usual way for a long parcel, care being taken that the investing cord should have considerable tension upon the press, so as to keep the organs of the plant in their places during drying, but not with too severe a strain, by which the delicate parts would be crushed.

Then comes the question of rapid drying. Culinary heat does well. Excellent results have been obtained by putting the presses on the plate rack over a kitchen stove. Or a box of sand may be placed in a similar position, and the presses immersed therein, keeping the sand quite warm. If access can be had to the boiler for an engine, or the stoves of a greenhouse, these may be utilised, but the great secret of success is to dry them rapidly. I have known the hare-bell (*Campanula rotundifolia*) prepared in twenty-four hours, with its colours in good state. As a rule, for ordinary plants, two or three days are sufficient, and there need be no necessity for changing the papers, when once the specimens are put in. This is a great advantage over the old method, as the delicate organs are liable to injury during the process of transference.

Some may be tempted, possibly, to use a wooden board both above and below, but this does not answer, as the moisture cannot easily escape. The experience of several years teaches one, that, by using mill-board at the top, the moisture is enabled to pass off rapidly, which is an essential point in ensuring success. Otherwise the plants may be almost stewed in their own fluids. Another rule for guidance may be thus expressed :—" much paper, few plants." As the presses are so inexpensive, they can easily be multiplied if necessary.

That this method if carried out carefully, will ensure a fair amount of success, is attested by specimens that were gathered in 1879, still retaining their colours almost as fresh as when first collected. Amongst these may be enumerated several of the Ranunculaceæ, and as a better test, the petals of the white variety of musk mallow (*Malva moschata*), which are perfect in their preservation. Good results are also obtainable with the green winged orchis (*Orchis morio*) which shows the various shades from creamy white to deep purple, in different specimens. Some orchids are best treated by being killed in steam, not immersed in boiling water, as this saturates the tissues with moisture. These plants, being succulent, may require changing once or twice.

Your correspondent Mr. McGann asks for blue or red flowers that will retain their colours. Let him try *Salvinia splendens* and *Delphinium Ajacis*, and some of the exotic Euphorbiaceæ. Our native Ranunculaceæ, and Polygalaceæ, contain species which are very useful for the purpose desired. Other groups of plants seem to resist all efforts to preserve them in anything like their original beauty; such as the Rubiaceæ, parasites, saprophytes, also the field geranium (*G. pratense*), and water forget-me-not. The field forget-me-not is much better for drying.

The above hints, carried out with patience and perseverance, will ensure sufficient success to repay the labour expended.

J. SAUNDERS.
Luton.

THE BRITISH FRESH-WATER MITES.

DURING the past season, I have met with four males, belonging to the family Arrenurus, which are not figured in my former papers. I now send you sketches, and short descriptions.

The gatherings, with one exception, were taken by Ed. Scudamore, Esq., B.A., and are not from my immediate neighbourhood. One of the mites, so far as I know, is new to Science, and I have therefore named it; if, however, it has been before described, I should be glad to be referred to the description, and withdraw my name, which is only provisional, and I feel not so good an one as it deserves, for it is a most beautiful and distinct species.

1. *Arrenurus tricuspidator.*— A figure of that which I take to be the female of this mite was given in the February number of SCIENCE-GOSSIP for 1883, page 37.

It is of a beautiful red colour, like vermilion, only transparent—it has the peculiar process on the last joint but two, of the hind leg, very highly developed. As in all the males of Arrenurus, the tail part is very characteristic.

2. *Arrenurus truncatellus.*— This very curious creature is at once recognised by its unusually elongated form, and the two peculiar little tubercles, at the end of the tail. It is of a lovely green colour, with vermilion eyes, surrounded by a yellow-green colour. Cœca of a warm brown colour. The process on the last joint but two of the hind legs is well developed.

3. *Arrenurus integrator.*—This is rather a small mite—it is of a lovely blue colour, with lightish

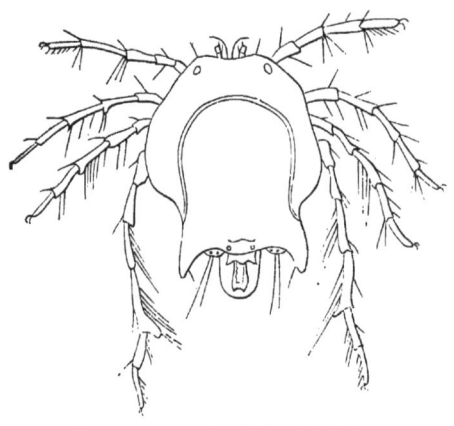

Fig. 44.—*Arrenurus tricuspidator*, ♂, ⅔ objective.

Fig. 45.—*Arrenurus truncatellus*, ♂.

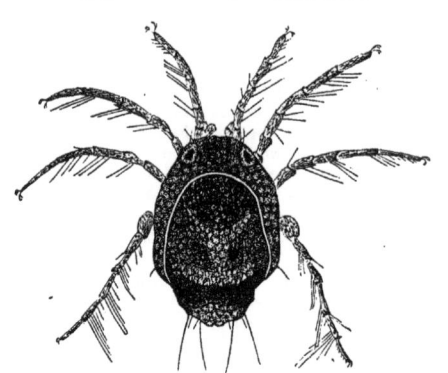

Fig. 46.—*Arrenurus integrator*, ♂, ⅔ objective.

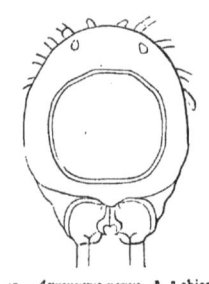

Fig. 47.—*Arrenurus novus*, ♂, ⅔ objective. The positions of the legs are only indicated.

coloured legs, and without the spur, on the last joint but two of the hind leg.

4. *Arrenurus novus.*—This mite, which I have never seen figured or described, is a very beautiful one, and retains its form (and, to some extent, its colour) when mounted in balsam, without pressure. Nothing short of a coloured figure could give any adequate idea of its beauty, under the microscope. The tail part is very characteristic, the central line being dark, but the curved process on either side is of glassy transparency, the last joint but two of the hind leg is quite destitute of the spur so highly developed in the two first-described mites.

C. F. GEORGE.

RECOLLECTIONS OF AUSTRALIAN ENTOMOLOGY.

By W. T. GREENE, M.D., F.Z.S., AUTHOR OF "PARROTS IN CAPTIVITY," &c.

[*Continued from page* 33.]

THAT Australia is a land of paradoxes—a reputation it has long enjoyed—could not, I think, be better exemplified than by a comparison between the spiders of our own country and those that are most commonly to be met with at our antipodes; where (reversing the, to us, natural order of things) the fly preys upon the spider, and not the spider on the fly.

A man I once knew on the goldfields used to be greatly annoyed by a loud, intermittent buzzing in his hut, and, for a long time, was quite unable to trace it to its source; one day, however, he discovered that it was caused by a large black fly, marked with bright yellow transverse bars on its body, that had taken up its abode underneath his table, and a queer abode it was, as I can testify, for I was by when it was discovered and broken up.

Upon examining the table, as soon as it had been found out whence the wearisome noise proceeded, on its under surface was discovered a large patch of tenacious yellow clay that must have been brought from a distance, as nothing at all resembling it in appearance was to be found in the neighbourhood. Upon breaking up this patch, it was found to be hollow, and to consist of several chambers—sixteen, as well as I remember—each filled up with a number of small spiders of various kinds, not dead, but paralysed and incapable of motion; and, in addition, each cell contained a grub, or egg, presumably of the fly-architect; the spiders, evidently, being a provision laid up for her offspring, whilst in their immature state, by their provident and industrious parent.

Australia contains prodigious numbers of ants of every hue and colour. Some huge monsters, of a bright red hue, are nearly two inches in length, and are, very appropriately, named "soldiers": there are black ants, too, of equal size, which are as appro-priately denominated "niggers;" both kinds are armed with a formidable sting, fully a quarter of an inch in length; and wage a perpetual war against each other. The wound they inflict with the caudal appendage is immediately fatal to the small creatures that constitute their natural prey, and exquisitely painful to man and the larger animals.

Other ants are so minute as to be barely perceptible to the naked eye, and, like their colossal congeners, are red and black, but, unlike them, are devoid of stings, and have an outrageous passion for sugar. Between these extremes there are ants of every size, all more or less objectionable to the human denizens of the country.

One species, which is extremely active, is about half an inch in length, black, with a strange pair of yellow jaws, in the use of which the owner is very expert, as well as with an extremely formidable sting. It is so ferocious as to have received the name of "bull-dog ant" from the colonists, which is singularly well bestowed; for once it has fastened on a foe, the death of the victim will alone cause it to release its hold; for even when crushed to atoms itself the sting and nippers remain pertinaciously fixed in the wounds. A new arrival who should chance to sit down on the nest of these creatures would have cause to remember it—it is not a comfortable seat.

There is a terrible story current among the "old hands" of the colony concerning these formidable insects, to the effect that in the early days of the penal settlement in New South Wales, some bushrangers had suspicions, whether well grounded or not, that one of their companions was playing them false, and reporting their proceedings to the police; on this unfortunate a drum-head court martial (if I may use the expression) was one day held, and he was sentenced unanimously, by his "pals," to be tied to a bull-dog nest, and there left to his fate.

That this atrocious deed was actually perpetrated there can be no manner of doubt; for some months afterwards the skeleton of a man was discovered in the bush, still bound to a stake that had been driven into the centre of the heap of sand, small stones, twigs and fragments of grass-stems that go to make up the residence of these terrible and much-dreaded insects; and, moreover, if I mistake not, one of the party subsequently made a full confession of the crime, and his own participation in it. These ants will anatomise a snake, or a small bird, that has been placed near their habitation, in the most beautiful manner, leaving nothing but the bones, and these as white as snow; but, notwithstanding their talent in this respect, the bull-dogs are decidedly objectionable neighbours.

Another variety of ant, rather smaller than that just mentioned, and of a dull brick-red colour, has a head nearly half the length of its body, which is armed with a pair of the most formidable-looking nippers possible, which, however, are perfectly harm less, for their possessor neither bites nor stings, and is the least objectionable member of the ant family with which I came in contact during my travels through the Australian bush.

Another species is soft and white, and chiefly inhabits decaying wood, from whence it is eagerly picked by many kinds of birds; the common poultry, for instance, being especially partial to it, although they will not touch the other kinds of ants. I do not know how this species contrives to exist, as it never appears to leave its home in some decaying log, either by night or day.

There are also red or reddish-purple ants, with long feeble-looking legs, with which, however, their owners contrive to run very fast. These creatures do not sting, but, in common with several other varieties, exhale a most disgusting odour when alarmed; and, as they have a great taste for sugar and sweet things in general, they are rather unpleasant visitors in a tent or hut; not even the ingenuity they display in getting at their prey compensating the poor bushman for the damage done to his stores.

In one place where I had been greatly plagued by these depredators, I suspended a bag of sugar to the ridge pole of the tent I was inhabiting at the time, and went out, thinking it was quite secure from their attack; but on my return, some hours afterwards, I found that a regular highway had been established between my sugar bag and the nearest ant-hill, along which some thousands of the long-legged inhabitants were hurrying to and fro, carrying off my property as fast as ever they could. Although much annoyed, I could not forbear watching them for a moment, as they struggled, heavily laden, up the string that suspended the bag from the pole of the tent, stopping every now and then to rest, and permit their descending companions to pass lightly over their distended bodies.

I next placed my sugar in a basin set in a saucer full of water on the table, and thought it would be safe; but it was not: for the ants brought up little bits of straw and grass, with which they formed bridges across the gulf that separated them from the object of their desires, which they then carried away in triumph.

As a *dernier ressort*, I placed the sugar basin in a large tin plate full of water, forming a moat too wide to be bridged over by the ingenious insects; but they were not to be outwitted, for they crawled up the overhanging side of the tent and let themselves drop on to the coveted sweets, where, however, they were obliged to remain until I came in, when I put an end to the thievish propensities of some hundreds of them by pitching the whole into the fire.

The peppermint ants are little black creatures, about a quarter of an inch in length, that generally inhabit old decayed trees of the same name, from the interior of which a tap or two on the bark will cause them to emerge in countless myriads, tainting the air around with the insufferable odour they exhale. These little pests are as fond of sweet things as those above described, and are also very partial to meat, particularly when cooked; I need scarcely add that whatever they have touched is utterly unfitted afterwards for human food.

All the preceding kinds of ants are diurnal in their habits, excepting the soft white variety, and retire to their abodes with the setting of the sun; but there are nocturnal ants, too, and ghastly-looking beings they are.

The day ants are gregarious, and always hunt in packs, but their nocturnal relatives are unsocial creatures, and startle the digger, singly, as he sits at his solitary tea, or reads his novel by candle-light, after the day's work is done. They have black, attenuated bodies, and long white legs; they appear mysteriously on the edge of the table, or the corner of the book, lift up their antennæ in a menacing manner, and directly scuttle away out of sight; they do not seem to be a numerous tribe, but to me they were more unwelcome even than the peppermint ants, although they never did me any harm; but I could not fancy them "natural," and would as soon have seen a tarantula at my board.

Of the termites I do not speak, as they are restricted to the far north, where I have never been, and so, taking leave of the ants, I pass on to another species of insect remarkable for the regularity of its appearance and disappearance—the March fly, as it is called, from invariably appearing on the first day of that month, and as constantly retiring from observation on or before the first day of April; after which latter date I do not remember ever having seen one, although thousands of them might have been visible in every direction, only the day before. The March fly is a sedate-looking, large eyed, black insect, with longitudinal white stripes running the whole length of its body; it is quite a ferocious creature, fastening with avidity on the face and hands of the colonist, and inflicting a sharp puncture with its proboscis; but beyond the pain of the wound no harm is done, as the creature instils no poison into the wound it has made; it is about the size of an English bluebottle.

Scorpions and centipedes are among the most unpleasant reminiscences of the traveller in the Australian bush, for they are very dangerous creatures, far more so than the tarantula, or giant spider, of which I have already made mention, and which is usually called the "triantelope" by the old bush hands: in former times the shepherds not unfrequently, for their own amusement, got up a duel between a couple of these nasty insects, which usually terminated in favour of the scorpion; the latter they often placed in the centre of a circle of live embers, when the creature, finding escape impossible, turned its tail over, and, stinging itself n the back, would presently expire.

The centipede lays its eggs at the beginning of the

Australian winter, and remains closely coiled round them till the return of Spring, when the young ones are hatched, and, some people say, the mother then devours as many of them as she can. Centipedes, scorpions, and tarantulas, are usually found concealed, during the day, beneath the loose bark of a dead tree, or an unstripped post or rail in a fence, and are the favourite food of the pied crow, or Australian magpie, which displays considerable ingenuity in extracting them from their lurking-places. Having, probably by the sense of smell, ascertained the retreat of one of the above-named disagreeable insects, Mag taps the spot sharply with her powerful beak, and upon the centipede, scorpion, or tarantula, as the case may be, popping out to ascertain the cause of the disturbance, snaps it up, and cracking it carefully from head to tail, swallows the dainty morsel, and flies off, with a self-congratulatory chuckle, to repeat the process elsewhere.

The Mantis (sacred, from the Greek) is a curious insect, of a green or blue colour, and varies in size from that of a grasshopper to that of a European wren. It derives its name from a habit it has of sitting upright on a leaf or branch, wrapped closely in its gauzy wings, its head turned skyward, in a contemplative, quasi-devotional attitude, while its fore-legs are crossed over each other, and partly raised, as if in prayer; it is a thorough hypocrite, nevertheless, and assumes this appearance of devotion simply to deceive the unwary flies and creatures or which it lives. It is common enough in the bush, but not very frequently seen, on account of its colour harmonising so thoroughly with that of the vegetation upon which it is usually found.

Another queer creature has been called the "walking-stick" by the colonists, from the great resemblance it bears to a piece of animated twig. There are several varieties of these abnormal-looking beings, at least they are found of various sizes and colours, and inhabit, some the water, others the dry land. They can all run with considerable agility; and it is very curious to watch the terrestrial walking-sticks making their way along from one to another of the sparsely distributed flowering tops of the indigenous Australian grasses. I have been told that these creatures develop, in course of time, into the different kinds of mantis, but have no personal knowledge of the metamorphosis, which, however, I do not consider unlikely, as there is a general resemblance between them.

The locusts, at least such is the name given in the colonies to the insects I am now about to describe, are a numerous and interesting family; some of them are no larger than a bee, while others are not less than a European tit. The larger sorts are of a bright green colour, with golden eyes, and are the most indefatigable songsters I have ever heard: the hotter the weather, the louder they chirp; and though not at all unpleasing, when heard for a short time, the concert becomes all but intolerable when kept up, as it is, without intermission from daylight to dark.

Grasshoppers are without number in the summer-time, and may be seen of almost every size and colour, hopping about in every direction, and contributing their shrill quota to the high-pitched concert of the locusts.

Many other curious insects there are in the Australian bush, which considerations of space will not permit of my even enumerating here: the following, however, are deserving of briefest mention. During the summer months the intelligent observer in the bush cannot fail to notice a little beetle, most destructive, by the way, to furniture and dry wood in general, which looks exactly as if some one had cut off its head in a slanting forward direction; and a grub, called the "carpenter," which makes itself a habitation with little bits of stick, neatly rounded off at either end, and fastened together with a species of silk woven for the purpose: which puts me in mind that there are several kinds of native silk-worms to be met with in Australia, which will, I am sure, be utilised some day.

One more reminiscence and I have done: years ago I recollect reading somewhere an account of an insect called the "burying-beetle," which performed the "last offices" for mice and small birds: but in Australia are *coleoptera* endowed with similar propensities, which I have seen ambitiously attempting the sepulture of an ox, and that, too, not without a fair prospect of success, had not their labours been interrupted.

NOTES FOR SCIENCE CLASSES.

No. VII.—Mosses.

THE mosses are humble plants, but they have no insignificant part to play in the economy of Nature, or in the colouring of the landscape; trees, rocks, and old ruins look grand under their covering; whilst the various species of Sphagnum, which grow in boggy places, perform an important part in the formation of turfy soil. These aquatic mosses grow very rapidly, so as in a very short time to occupy the whole of the pools which they inhabit. The genus Phascum are very minute species, found plentifully in fallow fields, but the large family of Hypnums are the most conspicuous, and often elegant plants, commonly seen on tree trunks, old walls, &c. The mosses can be gathered all the year round, although they vary in their period of flowering; for example, the Funaria is always in good condition for examination; on the other hand, the Phascum blossoms in early summer, and is ripe in the autumn, but the Hypnum, in many instances, takes twelve months to form the mature capsule, or theca.

The specimen selected for examination is the *Funaria hygrometrica*, L. First make a section of the stem (Fig. 48), and compare with any vascular cryptogam, such as the fern; it will be seen to differ perform the function of a vascular bundle, in the conduction of sap. Now note the leaves of *Funaria* (Fig. 49), by plucking off any of the upper ones, and place beneath a cover slip in a drop of water. They

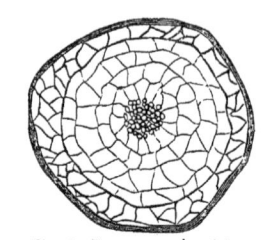

Fig. 48.—Transverse section of stem.

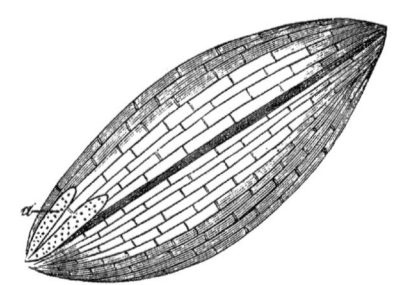

Fig. 49.—Leaf with Antheridia.

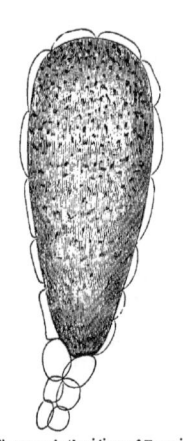

Fig. 50.—Antheridium of Funaria.

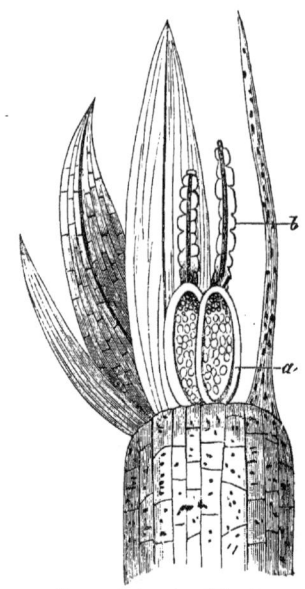

Fig. 51.—Archegonium of Funaria.

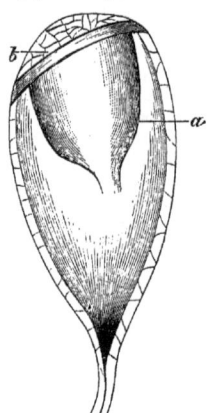

Fig. 52.—Theca of Funaria.

widely, in the absence of vascular bundles. In most mosses we find an outer layer of thick walled cells which passes into a mass of tissue in the centre. These are not sharply defined, and are said to are of a simple structure, with the exception of the midrib, and consist of a single layer of parenchyma, containing granules of chlorophyll; it originates from the bulging of a stem cell, afterwards separated by a longitudinal partition. Then carefully look out a stem bearing in the apex a quantity of differentiated leaves in a circular tuft; this is the *perigonium*, amongst which we shall find the reproductive organs. Pluck off a few of the leaves with a fine pair of forceps, near the centre, and search for the *antheridia* (Fig. 49, *a*), or a longitudinal section may be made; but I have found it far easier to point out the male organs as directed above. The student must be careful not to mistake the *paraphyses* for the

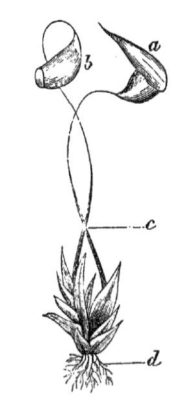

Fig. 53.—*Funaria hygrometrica*, L.

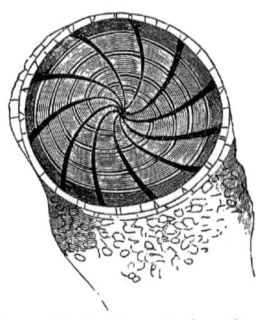

Fig. 54.—Mouth of Theca, showing peristome.

antheridia; the former are filiform structures, or abortive leaves, the antheridia are on short stems. Place the antheridium beneath a higher power (Fig. 50). It is seen to be a stalked sac, composed of a layer of chlorophyll, bearing cells when young, but they assume a reddish tint before bursting. They are filled with very minute antherozoids. On another stem, but taller than the last, will be found the archegonia (Fig. 51). Make a section by holding the stem betwixt the thumb and finger, and gently pushing the razor from you, then float out the sections in a bowl of water, select a few, carefully spreading them out with a needle on the slide, then search for the *archegonia*. It consists of two portions, the lower ovate (Fig. 51, *a*), and the upper, or neck, of archegonium (Fig. 51, *b*). The archegonium is ruptured by the fertilised oosphere, often in such a way that, while the lower part remains as a sheath, the neck is elevated as a cap now known as the *calyptra* (Fig. 53, *a*), on the top of the *theca* (Fig. 53, *b*) or capsule. On the top of the theca is a small lid, or *operculum*. When this is removed, the mouth or stoma is seen surrounded by a beautiful series of teeth called the *peristome* (Fig. 54); the stalk supporting the theca is the *seta* (Fig. 53, *c*). Now prepare a section of the theca. Fig. 52, *a* is the columella, and Fig. 52, *b* the operculum, beneath which is the peristome. When the spores germinate, it sends out a filiform body, known as the *protenema*, or proembryo, on which the young plant is developed. The root hairs, which will be found at the base of the stem (Fig. 53, *d*), and which take the place of true roots, are called *rhizoids*, play an important part in the economy of these plants. Detached leaves of the Funaria placed on moist soil, will produce the protenema.

J. F. R.

ABOUT MOSQUITOS.

I HAVE had my attention lately called to mosquitos in a practical way. On the 25th of January, this year, during a short run to the Mediterranean coast, I was bitten in quite cold weather at Alassio by one. The venom was feeble, and the swelling, with very slight irritation, lasted only two or three days. There are several points of interest about this insect pest on which light needs to be thrown. I find that Dr. Hassal, in his book on San Remo, denies that any poison is injected into the wound made by the mosquito, and supposes that the irritation is caused simply by the depth to which the mouth or jaws penetrate. The ordinary opinion among naturalists certainly has been that the insect has the power of pouring into the wound a secretion to make the blood flow more freely. Réaumur, besides, saw the watery drop frequently on occasions when his hand was bitten. And certainly the effects of the bite on many persons seem to be unaccountable, on the theory of the mere depth of the wound. A friend's maid had her arm swollen so terribly that the dress had to be cut to get it off. I have myself (at Alexandria, in January, a good many years ago) suffered pain and irritation which could not be described by any word less strong than severe, and the pain and irritation lasted seven or eight days. My two companions were untouched, or, at least, unhurt, though exposed in the same way as myself. The famous General Gordon testifies to the intolerable plague mosquitos are in the Soudan. They attacked every part of the skin which was tight, and seemed to find cane-bottomed chairs peculiarly convenient. Another point that requires clearing up is the defence against them. In "Euterpe," xcv. Herodotus says: "The Egyptians are provided with a remedy against gnats (mosquitos) of which there are a surprising number. They who live in marshy grounds use a net with which they fish by day, and which they render useful by night. They cover their beds with their nets and sleep securely beneath them. If they slept in their common habits, or under linen, the gnats would not fail to torment them, which they do not even attempt through a net."

Mosquitos do undoubtedly find their way in by rents in a mosquito curtain far smaller than the mesh of an ordinary fishing net. I have seen somewhere the testimony of a clergyman who protected himself from gnats by a 1¼ inch mesh fixed over the aperture of the window, provided the room was lighted only from one side. The gnats come through if there are back or side lights. Has this been confirmed by the experience of other observers? It would also be interesting and important to know if chestnut trees planted near houses drive away mosquitos, as they are said to do.

J. J. MUIR, F.L.S.

RED HILL.

THERE are few localities within so short a distance of London, that can include in their floras such an endless variety as the rapidly increasing town of Red Hill. Like most other spots so dear to the true botanist, it is fast being spoilt, not only by the continual enclosing of ground for building purposes, but also by that numerous class of so-called amateurs, who ruthlessly destroy and root out, solely for the sake of possession, the few rarities that are left to us in England. It is a well-known fact that no county can boast of such pretty rural landscapes as Surrey; the secret beauty of these views lies in the luxuriant vegetation. The diversity of soils in the neighbourhood of Red Hill, as well as its position, being on the southern slope of the North Downs, are sufficient to account for its extreme richness in prizes of botanical interest. It will be as well in giving the following list of a few of the rarer plants that can be found in this most delightful locality, simply to name them, and leave it for those who are really interested to find the habitats for themselves; the search will be as much enjoyed as the find, and whoever like to try can come home with a well-filled vasculum. Soon the time for active work will commence, when that pretty little drooping flower, the snowdrop, can be found in comparative abundance close to the railway; a very scarce variety of cuckoo flower with double blossoms may be met with in Gatton Park, and a little later on the well-known four-leaved herb Paris, occurs not far distant. As regards the natural order Orchidaceæ, Red Hill can boast of as long a list as any locality of the same size in the kingdom. The following were all found last year within a radius of about four miles: early purple, green-winged, spotted, man, fly, tway-blade, musk, butterfly, hellebore, pyramidal, and marsh orchis. On some marshy ground near Reigate, the bogbean, and that curious little insectivorous plant the round-leaved sundew, grow abundantly, also the delicate little ivy-leaved campanula. To come a little nearer home, on a high bank on the road to the Reigate Hills, the beautiful silvery stellate blossoms of the drooping Star of Bethlehem may be noticed by a close observer, though only two small clumps, consisting of two or three plants each, are there; yet it is a sufficiently valuable prize to encourage one to a thorough search. Along the slopes of the hills many flowers may be found, notably the deadly nightshade, toothwort, columbine, and Martagon lily; the exact spots, though, are only known to a few local botanists. Towards harvest time in some fields where the soil is very sandy, the fashionable corn marigold grows in profusion, in places quite killing the colour of the crops with its golden blossoms. In concluding these few lines, the only advice that can be offered to our friends in the botanical department about Red Hill, is to go and see for themselves, and, if they wish to preserve their bodies in a perfect condition, to beware of man traps and spring guns, which, if the notice boards speak truly, are rather plentiful in that neighbourhood.

J. R. M.

SCIENCE-GOSSIP.

THE statement as to the atmosphere of Italy containing an unusual amount of peroxide of hydrogen, made by Dr. Moffat, has been contradicted by Mr. Lennox Browne, who shows that the atmosphere of that country contains no more than that of any other country. Our readers will remember that the "$tenor$ voice" of Italians was believed to be due to this compound, and an artificial preparation of it was stated to be capable of producing artificial tenor voices!

IN a paper recently read before the Chemical Society, by Sir J. B. Lawes and Dr. Gilbert, on certain experiments conducted at Rothamsted, those gentlemen showed that the influence of the season on the composition of the ash of plants is much more marked than the influence of the manures.

INCANDESCENT lamps immersed in water are now used in the powder-houses of the Royal Factory at Waltham Abbey, where before no artificial light whatever was allowed.

MR. M. W. HARRINGTON, an American astronomer, is of opinion that the physical condition of the little planet Vesta is similar to that of our moon, being devoid of water, and also of a perceptible atmosphere.

WHITE bronze is being manufactured in Germany from the powder of the mother-of-pearl inside meleagrina.

THE vapour of bisulphide of carbon is being employed in New York instead of steam. It boils at 118°, and its volume at the same temperature at which water is converted into steam is approximately as three to two compared with water vapour. Moreover, its volume increases in the heat.

AT a meeting of the Royal Society of South Wales, Professor Liversidge produced some flints, and said white powdery limestone had been brought from New Britain so exactly resembling chalk that he thought true chalk of Cretaceous age occurred in the South Sea Islands. If this is the case, it upsets a good deal that has been said about the "permanency" of the great ocean basins.

CROPS of the freshwater sponge (*Spongilla fluviatilis*) have formed in such immense quantities in one of the Boston (U.S.) waterworks storage tanks, that the odour and taste of the water became so offensive the water could hardly be drunk.

PROFESSOR HULL and his party have returned from their survey of the Jordan Valley. One of the river-terraces of the Jordan was found at a height of 600 feet above the present level of the Dead Sea. Professor Hull will doubtless put the results fully before the scientific world before long.

AT a meeting of the Entomological Society the other day, several members expressed their opinion that the butterflies of this country are becoming scarcer. This is especially the case in the neighbourhood of large towns. Can it be from the smoke and disengaged gases thrown into the atmosphere? If the insects go, our most beautiful flowers will soon follow.

A FRENCH geologist, M. Chaper, is said to have found the matrix of the Indian diamonds at Naizam, near Bellay, in the Madras Presidency. The matrix is composed of pegmatite—a binary granite, composed of felspar and quartz. The diamonds were found chiefly where the rock was traversed by veins of felspar, and quartz containing epidote. The diamonds are weathered out of their matrix into the soil.

MR. CHARLES COLLINS (nephew of Mr. C. Collins, the well-known microscope manufacturer) is issuing a special set of excellently mounted slides illustrating the skins and scales of fishes. As the reader will see by referring to the earlier volumes of SCIENCE-GOSSIP (where the scales of many British fishes are illustrated) they are very beautiful objects.

PROFESSOR HUGHES'S lecture on 'The Theory of Magnetism,' at the Royal Institution, has created a great effect. The mechanical theory of magnetism may be deemed to be the proper style and title of that brought forward by the lecturer. The phenomena of magnetism he explained by a simple rotation of the molecules of iron, as well as of all metals, nay more, of all matter, solid, liquid, gaseous, or ether. All matter, according to his views, has inherent magnetic power, varying in degree in molecules of different nature, but not to any great extent. The lecturer demonstrated each portion of his theory by experiment, so that the effects were visible to the audience. He expressed his belief that electrical currents can be fairly classed with heat as a mode of motion. When a bar of soft iron is strongly magnetised, as in the instance of an electro-magnet, it returns, like a spring, to a neutral state upon the cessation of the inducing force. This well-known fact has long remained a mystery. All theories of magnetism have hitherto supposed that the molecules became, on the removal of the induced current, mixed or heterogeneous. Professor Hughes believes he has made a great discovery in having solved this problem, leaving no mystery any longer, as the demonstration he subsequently brought forward before the Royal Society reduced the matter within the domain of absolute fact. He proved his case before his audience at the Royal Institution in a less formal way, but quite as effectually, rendering a bar of iron sensibly neutral or polarised at will by simply turning it upside down. The mechanical inertia of the molecules was demonstrated by magnetising a bar, and then changing its polarity by the earth's influence alone. The inertia of magnetism and of electricity was illustrated by two bars of diverse hardness. Having dealt with other points of great interest, the lecturer concluded by saying that scientific men are agreed that heat is a mode of motion, and that the molecules of the most solid bar of iron can move in a certain space with comparative freedom, the oscillations being greatly increased with every rise in temperature. If, as already well known, the molecules can move in all planes, then there could be no valid objection to the idea of their rotation, in fact they were known to rotate in the act of crystallisation. Thus, according to Professor Hughes, magnetism is an endowment of every atom of matter.

THE officials for the meeting of the British Association at Montreal, on the 27th of August, have been appointed, and are as follows: President, Lord Rayleigh; Sectional President, *Mathematical and Physical Science*, Prof. Sir William Thomson; *Chemistry*, Prof. Roscoe; *Geology*, W. T. Blandford; *Biology*, Prof. Moseley; *Geography*, Col. Rhodes; *Economical Science*, Sir R. Temple; *Mechanical Science*, Sir F. J. Bramwell; *Anthropology*, Dr. E. B. Tylor.

PROFESSOR HALL is of opinion that the outer satellite of Mars, seen during the recent opposition, may always be seen by large telescopes at every opposition of Mars.

DR. ZINTGRAFF, in company with Dr. Chavanne, is about to visit the Congo and the interior of Africa. He takes with him a phonograph wherewith to fix the speech and melodies of hitherto unknown tribes, which, thus received by the instrument, will be forwarded to scientific men in Germany. The apparatus (which will be used for such a purpose for the first time) exactly corresponds with one he has in that city, so that the plates used in Africa can be sent to Berlin to be unrolled by that machine, and caused to re-emit the sounds received.

THE last number of the "Journal of Conchology" has two plates of importance and interest to naturalists, one illustrating a paper on "The Darts of the British Helicidæ," and the other giving an elaborate anatomy of *Helix aspersa*.

WE are extremely sorry to notice the death of Dr. Todhunter, F.R.S., the eminent mathematician, whose works have been before the world for so long a time that we are almost surprised to find he was only sixty-four when he died.

NOTHING could more plainly indicate the great strides made in the popularisation and wider study of natural science than that its pursuit has made it necessary for specialists to supply materials for study. Much time is saved the student by such beautiful stained preparations as those sent by Mr. C. V. Smith, of Carmarthen, showing a transverse section of the ovary of *Malva moschata*, and another presenting a vertical section of the ovary of *Digitalis purpurea*.

WE have received from Messrs. Sinel & Co., from their marine laboratory at Jersey, some mounts for microscopic examination and study of the eggs and zoea or larva of the shore-crab (*Carcinus mænas*), zoea of *Pisa tetraodon*, two days old (bred in Messrs. Sinel's tanks); young individuals of the common spider (*Epeira diadema*), and the parasite of the capercailzie. These specimens, to naturalists who have neither the time nor opportunity to collect for the purposes of verification, and who have hitherto been obliged to remain content with reading about them, are of the highest interest and value.

A PECULIAR Selachian has been taken in Japanese waters. Its head is remarkably snake-like, the eyes being placed to look sideways and downwards. It is thought, from the shape of its body, that the creature is in the habit of bending and striking forward to seize its prey, as snakes do. The teeth resemble, in some respects, those of fishes found in Devonian rocks. Professor Garman thinks it is the type of a new order, to which the name of Selachophichthyoidi might be given.

MR. C. B. MOFFAT kindly sends us a list of fifty-nine different kinds of wild flowers, all found growing within a radius of about four miles in the neighbourhood of Enniscorthy, during the first three weeks of January last. Also, another list of fifty garden flowers found in bloom during the same period.

MR. H. H. JOHNSTON has set out for Zanzibar, in order to explore Kilmanjaro, the highest mountain in Africa, and to collect examples of the fauna and flora of the mountain. The result, if successful, will be of the highest biological interest.

DR. KOCH, the distinguished German physiologist, reports from Calcutta that he has found the cholera *Bacillus* in the water of a tank. Many people affected by cholera had obtained their drinking water from this very tank.

MR. CLARENCE E. FRY has described in the "Entomologist" a Cape plant found near Table Mountain, *Physianthus albens*, which is insectivorous. Butterflies and moths are caught by the glutinous base of the corolla, and cannot withdraw themselves; whilst the petals close over them and form their shroud.

THE increasing, if not loudly expressed, feeling of dissatisfaction is manifesting itself among the science teachers of Great Britain at the way they and their services are treated by the South Kensington Bureaucracy. The directors of that establishment appear to think that science teachers exist for their benefit, whereas the establishment really exists for the benefit of science and those who teach it.

PROFESSOR MARSH has just described a new and strange fossil Saurian, from the Oolitic strata of North America. It is the type of a new family, and has been named *Diplodocus longus*. Its zoological position is intermediate between the gigantic Atlantosaurs and the Mososaurs. The size of the skull indicates an animal probably forty or fifty feet long. Its teeth suggest that it was herbivorous, and its food probably succulent aquatic vegetation. The shape of its skull is singularly like that of a horse.

G$_{UY}$OT, the French scientist, well known for his works on physical geography, which have been translated into English, has just died.

MR. WORTHINGTON G. SMITH, Mr. J. E. Greenhill, and Professor Rupert Jones, F.R.S., delivered interesting addresses before the Geologists' Association on March 7th, on the subject of "Implementiferous valley-gravels," when a large collection of implements from the gravels was exhibited.

M. J. B. SCHNETZLER has described an aerial alga which inhabits the bark of the vine in the Canton de Vaud. It occurs as a brownish-red powder, penetrating the fissures of the bark. It is named *Chroolepus umbrinum*.

AT the anniversary meeting of the Geological Society, the Wollaston Gold Medal was presented to Professor A. Gaudry, the eminent French geologist; the Murchison Gold Medal to Dr. Henry Woodward, F.R.S., for his valuable researches in fossil crustacea; and the Lyell Gold Medal to Dr. Joseph Leidy, the celebrated American palæontologist.

IN the recent number of the "Canadian Entomologist," there appears one of Mr. W. H. Edward's suggestive and exhaustive papers, on the "History of the Preparatory Stages of *Colias Eurydice*," with remarks upon the genus Meganostoma." He thinks the latter should be dropped from circulation.

MR. T. M. READE has contributed a further paper to the Geological Society on "Rock fragments from the South of Scotland imbedded in the low-level boulder-clay of Lancashire." He thinks they confirm the view that all stones in the drift of North-Western England are derived from the basins of the Irish Sea and of rivers draining into it, except some stray fragments that may have come from the Highlands of Scotland.

IN his "Science Notes" for March, contributed to "The Gentleman's Magazine," Mr. Mathieu Williams shows that after burning the stems of the Equisetum, and examining the ash, there will be found the siliceous scales, which interlock by means of teeth to form the cuticle, and are very pretty microscopic objects. He recommends the dried stem of the Equisetum for rubbing down irregularities of the teeth.

THE "Westbury House" School Ephemeris, for January, contains a list of the flowering plants of Worthing and the neighbourhood, a list of plants noticed in flower during January, 1884, and a full catalogue of species and localities of the pulmoniferous mollusca of Worthing.

THE Nottingham Naturalists' Society held their soirée and exhibition on the 6th of March. There was a capital collection of geological, mineralogical, and natural history specimens shown. A novel feature was the "Naturalists' Dinner Table," where visitors could taste rat pie, French snails, stewed squirrel, &c.

IN the recent number of "Appalachia," we have the "Flora of Mount Lafayette and Franconia Valley," by Professor Bailey.

THE editors of the "Botanical Gazette" have brought out a new edition of the Catalogue of Phænogamous and Cryptogamic Plants of Indiana, which will be useful to collectors.

IT is stated that if the *Deutzia scabra*, a very ornamental shrub, is cultivated near grape-vines, rose-bugs and other noxious insects prefer the Deutzia flowers, so that the vines are protected.

ALONG with the above we have received a check list of the Ferns of North America, intended for Herbaria.

MR. A. J. DOHERTY has kindly sent us two slides, one showing the annular rings (stained) in section of lime; and the other a cross-section of the ovary of *Rhododendron ponticum*. Both are exquisitely neat, tasteful, and beautiful objects.

THE *Trillium erectum*, a plant often found in shrubberies, is pointed out as affording a splendid opportunity for studying raphides. A portion of the petal should be placed betwixt two glass slips, and pressed until it is almost transparent; then, upon examining it under a low power, a beautiful series of needle-like crystals are revealed.

MICROSCOPY.

TO HARDEN ANIMAL TISSUES.—After the animal has been injected, I generally place it at once in equal parts of alcohol and water, and allow it to remain in it for some hours, so that the gelatine may become solid. If a carmine mass has been used, alcohol and water is the only fluid suitable for hardening, and a few drops of acetic acid should be added to prevent the carmine becoming diffused when in contact with the tissues. If Prussian blue has been injected, either alcohol, Müller's fluid, or picric acid may be used. Some recommend a $\frac{1}{4}$ per cent. solution of osmic acid.—*V. A. L.*, *Manchester*.

GREENWICH MICROSCOPICAL SOCIETY.—At the annual meeting held last month, Mr. T. W. Dannatt, of No. 21 The Circus, Greenwich, was unanimously elected hon. sec. of this Society, in place of Mr. G. D. Colsell, resigned.

CUTTING SECTIONS IN RIBBONS.—"The American Monthly Journal of Microscopy" says: The process of cutting sections in ribbons, recently introduced, is much employed in the laboratory of the John Hopkins University. The object of the process is to enable the observer to cut a series of extremely thin sections of any soft preparation, such as an embryo for example, and to mount the sections in a series in the order of succession, retaining all the parts of the specimen in their proper position. The value of the process needs no further explanation. It is carried out perfectly, and in an exceedingly simple manner. The specimen is first properly prepared, and imbedded in paraffin. The paraffin is then placed in the section cutter, which is made on the principle of the Rivet microtome, although much longer than the usual form of the latter instrument, and somewhat modified in the details of construction. Sections are then rapidly cut, by moving the knife forward and backward within proper limits, and the successive sections of paraffin, which are square, adhere together by their edges into a ribbon, which may grow to an indefinite length. It is essential that the paraffin be of the proper consistency and at the right temperature. Slides are now prepared by spreading a thin layer of shellac dissolved in creosote on one surface, to which the ribbons are now transferred, two or three being placed parallel on each slide, so that the sections may be readily examined in succession. By heating for a short time in a warm oven the sections become firmly attached to the slide, and may be mounted in balsam with very little trouble. As a result of this method of procedure we were shown a series of sections across the body of Lingula, in which the arms were shown in section precisely as in life, and in the stomach were remains of diatoms quite undisturbed by the operations of preparation.

THE DIATOMACEÆ OF NORFOLK.—It is with much pleasure we call attention to Mr. F. Kitton's important announcement that he proposes to publish, by subscription, a century of slides of the Norfolk diatoms. Mr. Kitton's world-wide reputation as a diatomist should secure a large number of subscribers to his scheme.

THE BELGIAN MICROSCOPICAL SOCIETY.—As our readers are well aware, this society is one of the most vigorous in the prosecution of microscopical research. Vol. viii. of its "Annales" has just appeared, containing papers on "The Structure of Certain Diatoms, from the cement-stone of Jutland," by MM. W. Prinz and E. Von Ermengem (beautifully and profusely illustrated, some parts up to 3000 diams.). A paper on "Terrains et Microbes," by Dr. Casse; besides abundant notes on various microscopical researches, discussions, &c.

BOLTON'S PORTFOLIO OF DRAWINGS.—Amid the mass of new and welcome materials now furnished to lovers of the microscope, we cannot forget that the name of Mr. Thomas Bolton, F.R.M.S., stands prominent as one of the earliest and most enthusiastic pioneers. No. x. of his well-known "Portfolio of Drawings" has recently appeared. As many of our readers know, this is a collection of the weekly illustrated sheets which Mr. Bolton sends out along with living specimens for microscopical examination and study. It includes *Pediastrum Boryanum* and *Encyonema prostratum*, in the vegetable kingdom, and *Raphidiophrys elegans, Hemiidinium nasutum, Chilomonas spiralis, Anthophysa vegetans, Limnocodium Sowerbii* (the freshwater jelly-fish), *Asplanchna priodonta, A. Ebbesbornii, Brachionus Bakeri, Ælosoma quaterarium, Chætogaster Limnæi, Chirocephalus diaphanus, Lynceus sphæricus, Balanus balanoides,* oyster spat, young sticklebat, &c., among the animal kingdom.

"JOURNAL OF THE ROYAL MICROSCOPICAL SOCIETY."—There are few of the periodical publications of our learned societies which come so welcomely as this "Journal," thanks to the ability with which it is edited, and the admirable abstracts of all the most important papers, in every language, which appear in the interval of its publication, in which microscopical investigation is involved. In this respect the "Journal of the Royal Microscopical Society," has undertaken and fulfilled a *rôle* which was never before attempted. The "Journal" also furnishes us with abstracts of the "Proceedings" of the Microscopical Society, and publishes in full (illustrated) all the most important papers. The part for February contains two papers of great value; the first we more particularly recommend to notice on account of its high sanitary value. "The Constituents of Sewage in the Mud of the Thames," by Lionel S. Beale, F.R.S., and "On the Mode of Vision with Objectives of Wide Aperture."

"PETROGRAPHICAL STUDIES," by Messrs. J. E. Ady and H. Hensoldt. The last two parts of this valuable work, from the exquisite neatness of the lithographed text, and the carefulness with which the plates have been coloured, indicate the intention of the authors to give the world a really good and useful production. The objects figured and described are "Paulite-Diorite," from Banff, Scotland, magnified twenty diameters; and "Pikrite," from Inchholm, Firth of Forth, magnified thirty diameters. The slides sent out with these papers are, of course, specimens of the rocks themselves, and they are cut and mounted in Mr. Hensoldt's best style of workmanship. Mr. Ady's "Popular Studies in Comparative Histology," in which he is assisted by Mr. A. J. Doherty, of the Victoria University, Manchester, commenced on March 10th.

"THE JOURNAL OF THE QUEKETT CLUB" (edited by Henry Hailes) for February, contains papers by Dr. M. C. Cooke, "On Circumnutation in Fungi," and by Mr. G. C. Karop, on a "Description of a Table for Microscopical Purposes," in addition to which there are abstracts of the Proceedings of the Society.

"STUDIES IN MICROSCOPICAL SCIENCE" (edited by A. C. Cole, F.R.M.S.).—No. 13 of vol. ii. of this now widely known work, illustrates the subject of "Fibrous Connective Tissue," by a paper on the "Tendon of Lamb," illustrated by a very beautiful coloured plate of the object, magnified seventy times. No. 14 deals with the "Primary Tissue" of plants, and is illustrated by a plate showing a transverse section through the apex of root of maize. The specimen accompanying the "Tendon" paper, and also a mounted "Prothallus of Fern" (mounted in glycerine jelly), are among the best yet sent out, the latter being especially good. It shows all the details of structure, and by the aid of a ¼ the student can perceive the undeveloped antheridia and archegonia.

A MEDIUM FOR HARD SECTIONS.—Can you tell me a good medium in which to mount hard sections, such as bone, teeth, and horn, one in which they will not change, and in which all, or nearly all, of their minute structure will not be lost; that is lacunal canaliculi and dentinal tubules, &c? Balsam, as a rule, obliterates all such detail. I have been trying glycerine jelly, but the glycerine seems to have a solvent action on the lime salts of the section, more especially of teeth; as in a short time after the mount is made there is a very fine deposit, all round the section, of what I supposed was a lime salt extracted from it by the glycerine, and again deposited.—*J. J. A.*

ZOOLOGY.

SOUTH ESSEX.—Has it ever occurred to conchologists to collect in the southern division of the county of Essex? For there do not seem to be any records whatever of its molluscan fauna, except a single casual allusion to *Helix virgata* and *H. caperata;* and it seems rather curious that one of the metropolitan counties should be so much of a terra incognita that its recorded molluscan fauna should amount to actually no more than two species.—*W. J. D. W. R. T.*

CONCHOLOGICAL NOTES.—Referring to Mr. Geo. Roberts' notes, *H. rufescens* is not mentioned in my list of local shells (SCIENCE-GOSSIP, July, 1883) as I have not myself taken it, but it is included in Mr. Ashford's "List of Shells of the Lower Tees" ("Journal of Conchology," vol. ii. p. 240), on the faith of my friend, Mr. J. W. Watson, as being found near Ginsbro, Stokesby, Ayton, &c. Mr. Roberts' remarks respecting ants and Helices made me turn up my note-book, and I find that last spring I noticed a circumstance which struck me as peculiar. It was early in the season, and I was searching for zonites under some stones at Redcar. Under several of these stones were colonies of the common sand ant, and also several specimens of *Z. cellarius, nitidulus,* and *allarius,* and less often *Pupa marginata.* In all cases the shells of *cellarius, nitidulus,* and *Pupa marginata* were tenantless and beautifully cleaned, I presumed by the ants who had eaten their guests, but out of twenty-three *S. allarius,* only one dead shell came to hand, and it occurred to me that possibly the somewhat unpleasant odour of the garlic snail had saved its life.—*B. Hudson.*

BIVALVES OUT OF THEIR ELEMENT.—During the second week of December 1883, I took half a dozen specimens of *Sphærium corneum* from a pond near Middlesbrough, and, having to attend to some particular business, on my return home, I forgot to clean my shells. They remained thus forgotten until January the 7th this year, when I stumbled upon them. They were in a tin box which also contained a handful of grass which was perfectly dry; in fact the box contained no apparent moisture. Happening to fracture one of the shells, I noticed that the animal appeared to be alive. I therefore took three of the shells and placed them in water having a temperature of 55° F., and was much gratified as well as surprised to find that two out of the three were alive and well, as, in about an hour's time, the valves were opened, and the syphons protruded. Another half-hour saw them perambulating the dish probably in search of food. This capability of sustaining life under somewhat trying circumstances for so long a period as seven weeks, must certainly serve these animals to good purpose in the struggle for existence.—*Baker Hudson.*

LAND AND FRESHWATER SHELLS IN THE MIDDLESBROUGH DISTRICT.—I am glad to be able to supplement my list of shells (SCIENCE-GOSSIP, July, 1883) taken within a radius of twelve miles from Middlesbrough by the following. *Helix nemoralis* v. *bi-marginata, H. virgata* v. *albicans* and *minor, H. caperata* v. *ornata, H. pulchella* v. *costata, H. ericetorum* v. *alba, H. sericea, Bulimus obscurus, Pupa umbilicata* v. *edentula, Vertigo edentula, Zonites nitidulus* v. *Helmii, Z. fulvus, Succinea putris, valvata, cristata, Physa hypnorum, P. fontinalis* and v. *oblonga, Planorbis nautilus,* and *Cochlicopa tridens.*—*Baker Hudson, Middlesbrough.*

CONCHOLOGICAL NOTES — BAND-MARKING. — I have noticed that many species, normally without band-markings, occasionally have traces of the bands which are present in other shells, these bands always occupying the same position on the shell as those of allied species whose bands are usually present. It also appears that one band, corresponding to No. 3 on *Helix nemoralis,* is the most often developed; it being often present in *H. nemoralis* when the other bands are absent, although I have a specimen having the band-formula 12045, but this is not a common form. This third band seems to be less constant in *H. hortensis;* a specimen having the band-formula 00300 being very uncommon, while I have a specimen whose formula is 00045; and 00000 is very common. With regard to the abnormal development of bands, I find I have specimens of *Zonites nitidulus, Bulimus montanus, Helix hispida,* and among the freshwater species, *Limnæa peregra, L. truncatula,* and *Physa hypnorum,* in which one or more bands are developed. In *Physa hypnorum* I have only noticed the band-marking on American specimens. Variety of *Helix aspersa.*—I have a very curious variety of *H. aspersa,* found at St. Mary Cray, Kent, in which the whole of the upper part of each whorl, above the lower edge of the third band, is suffused with a dark chocolate-brown colour, without any markings whatever; while the lower part of the shell, below the third band, is quite normal, giving the shell a very strange appearance.—*T. D. A. Cockerell.*

"THE BUTTERFLIES OF EUROPE," by H. C. Lang, M.D. (London: L. Reeve & Co.).—Part xvii. of the work is now out, dealing with the genera Erebia, Oeneis, and Satyrus. The plates illustrating the species of the latter genus are the most exquisite specimens of natural history art we have yet seen. To our mind this is the most attractive part yet issued.

PROVINCIAL SOCIETIES.—The Annual Report and Proceedings of the Belfast Naturalists' Field Club for 1882-83, contains a very interesting series of short descriptions of the numerous field excursions, which are a leading feature in the agenda of this society. We also find well-condensed abstracts of the following papers, in addition to the address of the Vice-President, Mr. W. H. Patterson, "Recent Examination of the Crannoges, Lough Inverness," by F. W. Lockwood; "Sensitive Plants," by T. H. Corry; "The Stone Monuments at Carrowmore," by Charles Elcock; "Rudestone Monuments in Antrim and Down," by William Gray; "Rare Plants recently found in Down and Antrim," by S. A. Stewart; "Fungi, Mushrooms, and Toadstools," by the Rev. H. W. Lett, &c. Mr. S. A. Stewart also gives a supplementary list of the mosses of the north-east of Ireland, and there is a meteorological summary for 1883. The Transactions of the Huddersfield Naturalists' Society, Part i., contains, in addition to the Annual Report, a catalogue of the Lepidoptera found in the district. The macrolepidoptera are described by Mr. S. Moseley; and the micro-lepidoptera by Mr. G. T. Porritt, F.L.S. Altogether there is a total of 666 species.

BOTANY.

AREGMA MUCRONATUM.—I observe that the late Mr. Brittain, in his "Micro-Fungi," says (p. 47), "I have never met with the rose-brand upon the cultivated or garden-rose." I send herewith a specimen, taken in 1882, from a large standard budded tree. I believe I found this fungus once before on a garden-rose, together with Lecythea; and Mr. Cooke leads us to expect we should do so.—*M. O. H.*

CHANGING THE COLOURS OF VIOLETS. — The Rev. W. Gilbert Edwards has informed me that in Paris, when they want to get their violets of a light colour, they dose them with sulphur. If you turn to Dr. M. C. Cooke's book, "A Manual of Structural Botany," it says of sulphur, "Sal, salt; pyr, fire." Now as calcareous soil is of a purely mineral nature, that is why albino flowers lose their colour on contact with calcareous soil.—*Alexander W. Ogilvy.*

A NEW FLORA OF SURREY.—Mr. W. H. Beeby has published an excellent map of Surrey, showing the districts and sub-districts into which the county has been divided for the new Flora now in preparation. Explanatory notes thereon have already appeared in the "Journal of Botany." He requests us to say that assistance may be rendered in the following ways, viz., 1. By extracts from old works, local publications, or herbaria, which are unlikely to have come under my notice. 2. New stations and recent corroboration of old stations, for the less common species. 3. Notes as to the occurrence of usually common plants in any district, as, whether common (frequent and abundant) throughout; or, frequent but not abundant; or, confined to certain parts of a district; or, apparently absent. Information concerning the usually common plants is particularly invited, as the absence of one of these may be as important a fact as the occurrence of a rare species. It is desirable that specimens of rare and critical plants should be sent in confirmation of records. Such specimens will be returned if wished, or will eventually be placed in one of our public herbaria, in order that they may be accessible to future students who may feel uncertain as to exactly what plant was intended by any name.

NEW BRITISH FUNGUS.—In September of 1882 I found on the culms of wheat stubble, near Birmingham, a Fusisporium which I could not identify. Recently while turning over the pages of Grevillea, I chanced upon *F. cereale*, Cooke (*l.c.* vi. p. 139), and recognised that my plant was the same. This was found on the sheaths of maize, in Florida, U.S.A., by Ravenel, and has not before been recorded for Britain. Description: Pallid; flocci short; spores fusiform, curved, acute, 3-5-septate, constricted, hyaline, 50–$70\,\mu \times 8\,\mu$. My specimens agreed in every respect, except that the spores were rather more variable in size.—*W. B. Grove, B.A.*

"THE SAGACITY AND MORALITY OF PLANTS; a Sketch of the Life and Conduct of the Vegetable Kingdom," by Dr. J. E. Taylor, F.L.S., F.G.S., &c. Price 7s. 6d. (London: Chatto & Windus). Our position with reference to this work prevents us doing more than saying it will be published in a few days; but the following extract from the preface will give some idea of its scope: "The reader may, if he so chooses, consider both the title of this book, and much of its contents, as a parable. But I have taken up the parable with a view of bringing the lives of plants more nearly home to us. Botany is no longer a matter of counting stamens and pistils, and expressing the classified result in a Greek-derived nomenclature; it no longer consists in merely collecting as many kinds of plants as possible, whose dried and shrivelled remains are too often only the caricatures of their once living beauty. It is now a science of *living things*, and not of mechanical automata, and

I have endeavoured to give my readers a glance at the laws of their lives. Therefore, whilst not beseeching criticism (seeing I have not written so much for learned botanists as for those who take an intelligent interest in plants), I do not deprecate it. Nobody is more conscious than the author that he has only lightly touched upon the fringe of a great subject; but if this little book is the means of rendering plants and flowers more interesting to people after they have read it than they were before, it will not have been written in vain."

ON THE CONTINUITY OF PROTOPLASM.—Will you kindly allow me to draw attention to an error which appeared in your February number of the present year? In the article entitled, "New Facts concerning the Flow of Protoplasm," it is stated, "Mr. Gardiner, however, was antedated by Mr. Thomas Hick," &c. The following is the actual state of the case. My first paper on "Open Communication between the Cells of the Pulvinus of *Mimosa pudica*" was published in the October number of the "Quarterly Journal of Microscopical Science" for 1882. Then followed a communication to the Royal Society, Nov. 11th, 1882, "On the Continuity of the Protoplasm in the Motile Organs of Leaves," in which I first make use of the term "continuity of the protoplasm." On April 10th, 1883, I communicated to the same society a paper "On the Continuity of the Protoplasm through the Walls of Vegetable Cells," which is being published in the Philosophical Transactions for that year. Finally, at Southport, Sept. 23rd, 1883, I read a paper which, since it was the first on the list, was also prior to that of Mr. Hick. To my other papers which I have published I need not refer here. As regards the subject of the continuity of the protoplasm in the Florideæ, I might mention that its occurrence has been known for some time. Besides the two papers by Professor Wright in the Transactions of the Royal Irish Academy, 1878, to which Hick refers, Thuret, in the classic work of Bornet and himself ("Etudes Phycologiques," 1878, p. 100), mentions that in many of these algæ open communication between the contents of adjacent cells exists. The most satisfactory work upon the subject was, however, carried out by Schmitz ("Sitzber. Akad. Wiss." Berlin, 1883, Feb. 22nd), who in his paper "On the Structure and Fertilisation of the Florideæ," describes the development of the thallus and the connection existing between the individual cells. He finds that usually the pits are not open, but that a delicate closing membrane is present, which is perforated in a sieve-like manner in the same way as in the cells described by myself. In certain of the corallines he is disposed to believe that open pits do occur, so that the protoplasmic thiends of adjacent cells join directly with one another, although this observation certainly requires confirmation. I believe that the mode of connection of the cells of the Florideæ is also dealt with by Professor Agardh in his "Morphology of the Florideæ ("Sv. Vetenskaps akad. Handl." xv. 1879), but at the time of writing I have not the book. With regard to Mr. Hick's results as recorded in the "Journal of Botany" for February and March, there can be little doubt that the continuity of the protoplasm in the Florideæ is not maintained by means of open pits as he has stated. By appropriate treatment it is easy to see that except in the very youngest cells a distinct pit-closing membrane is present, and I can only make the same remark concerning his research as I did in a communication to the Royal Society of December 13th, 1883, with special reference to Mr. Hillhouse's results ("Bot. Centralblatt," Nos. 16, 17, 1883), that "I am unable to agree with observers whose statements necessitate the existence of open pits."—*Walter Gardiner.*

GEOLOGY, &c.

DISCOVERIES OF VERTEBRATE REMAINS IN THE TRIASSIC STRATA OF DEVONSHIRE.—Mr. A. T. Metcalfe, F.G.S., has described before the Geological Society, some vertebrate remains consisting chiefly of portions of jaw-bones with teeth in line, probably of Labyrinthodonts, found in the Upper Sandstones (Ussher's classification), at High Peake Hill, near Sidmouth, by H. J. Carter, Esq., F.R.S. At numerous places between Budleigh Salterton and Sidmouth, Mr. Carter and Mr. Metcalfe had found a large number of isolated bone fragments. Such fragments had been submitted to a microscopical examination by Mr. Carter. In some specimens the bone structure was visible throughout; in some the bony portion had been partially removed and replaced by an infiltration of mineral matter; in others the removal of the bony portion was complete. From these facts Mr. Metcalfe drew the conclusion that a comparative abundance of vertebrate life was maintained during the Triassic period; and that the rareness of Triassic fossils was due not so much to the paucity of animal life during that period as to the fact that Triassic strata afforded no suitable conditions for the preservation of organic remains. The President said that the author, in this interesting communication, had proved that there was an abundant vertebrate fauna in the Triassic strata of Devonshire, and we could only regret that the specimens found up to the present time were all so fragmentary and imperfect.

POST-GLACIAL RAVINES IN THE CHALK-WOLDS OF LINCOLNSHIRE.—Some time ago, Mr. Jukes-Browne, F.G.S., stated that of the valleys intersecting the Chalk-Wolds some were older and some were newer than the formation of the boulder-clays (Hessle and purple clays). He now describes some cases where the modern watercourse, after flowing for some distance along the line of an ancient (pre-boulder-clay) valley, suddenly deserts that valley and passes through a ravine excavated entirely out of the chalk. These ravines are very different from the other parts of the valley traversed by the same stream, being deep and narrow cuts or trenches with steep wooded sides, and exhibiting more the scenery of Derbyshire vales than that of ordinary chalk valleys. In accounting for the origin of these ravines, Mr. Browne points out that the whole district in which they occur must once have been completely covered by the boulder-clays; and he supposes that at certain points where the ancient valleys were blocked with high mounds of drift, the streams found it easier to cut new channels through the flanking ridge of chalk, than through the obstacles in front of them.

SHAM FOSSILS.—At a recent meeting of the Geological Society, Professor Hughes described a branched structure found in the Red and White Chalk of Hunstanton, which has generally been known as *Spongia* or *Siphonia paradoxica*. The beds in which this supposed sponge occurs contain fragments of various organisms, including sponge-spicules, but no trace of structure can be found in sections of the *Spongia paradoxica*. The fragmentary state of the undoubted organic remains would indicate that they were drifted into their present position, and therefore a state of things quite unfitted for the growth of a slender branching sponge; the so-called sponge commonly occurs in layers along the bedding-planes, but frequently rises through the whole thickness of one bed and extends up into the overlying layers. It does not seem likely that it was the root of a *Siphonia* or some similar organism. Another body which has been also called *Spongia paradoxica* consists of masses of more crystalline texture, exhibiting upon weathered surfaces a net-work of small ridges enclosing cup-like depressions. These appearances were compared by the author to the weathered surfaces often seen in certain beds of the mountain limestone and in gypsum; the masses show no traces of internal structure. Sections of these bodies show exactly the same characters as the containing rock, except that the material is more compactly crystalline; it contains the same fragments of shell, &c., and the same sand and pebbles. He regarded them as of concretionary origin, and explained their symmetry of form and regularity of arrangement by their being formed at the intersections of joints with the bedding-planes or with one another. Phosphatic nodules occurred in the lower parts of the white chalk, and had these bodies been sponges they would probably have been phosphatised; but analyses have shown no marked difference in this respect between their substance and that of the surrounding rock.

NOTES AND QUERIES.

CATERPILLARS FEEDING.—In SCIENCE-GOSSIP, for last October, p. 222, the sentence occurs "Since caterpillars only feed in bright light," &c. On reading this, it occurred to me that if this statement were intended to include English caterpillars, as it doubtless is, it was incorrect, as I remember having taken caterpillars (what kinds I cannot now call to mind) in the act of feeding by night, when collecting with the aid of a bull's-eye lantern. If some of the readers of SCIENCE-GOSSIP, will state their observations on this point I shall feel greatly obliged. I should also like to ask if any one has noticed that while butterfly larvæ show a preference for feeding by day, the reverse is the case with moth larvæ, as a general rule; each thus exhibiting the same preference for light or darkness which it maintains in after life?—*J. H. B.*

ODD NAME OF PLANT.—The parish clerk here tells me of a plant which he calls "Cain and Abel," and which grows in the neighbourhood. The root, he says, consists of two parts, and if thrown into water the one will sink, while the other will swim. Can any of your readers tell me anything about this plant?—*John Hawell, Ingleby Vicarage, North-allerton.*

EFFECT OF POISON ON HEDGEHOG.—It is commonly supposed these animals are difficult to poison. I once kept one in the house as a pet, at the time we were infested with mice. I laid poison for the mice, composed of strychnine mixed with butter on bread; by some unlucky chance the hedgehog obtained one of the pieces, and licked off the whole of the butter, with the result that death ensued in an hour or two afterwards.—*James F. Robinson.*

SEASONABLE NOTES FROM CUSHENDEN.—The song-thrush commenced to sing on December 1st, on Christmas Eve, and Christmas morning; they were singing in concert. Maximum thermometer from 53° to 55° during most of the month. Daffodils and snowdrops well above ground, the latter showing colour. Hive bees flying about on December 28th. Black swan (*Cygnus atratus*) was shot on a lake in Rathlin Island, Nov. 24. Where could it have come from?—*S. A. B.*

DARWIN ON INSTINCT.—It is now some years since I first saw the announcement in the "English Mechanic," on the authority of Mr. Darwin, that insects do not "feign death," that the disposition of their limbs is not the same in danger as in death, but merely the same as when remaining motionless. This was so contrary to what I expected at the time that I took every opportunity of observing the attitude of beetles under fear. But for seeing this statement repeated in "Nature" for December 6th, my observations might have been for ever unpublished; but I am now able to say that as far as certain genera of beetles extend, the disposition of the limbs of insects is exactly the same in danger as in death. It is well known that beetles belonging to the genera *Anchomenus*, *Bembidium* and *Pterostichus*, will generally "take to their heels" when in danger, but sometimes these will remain motionless, at least for a time, as if considering what to do. In neither case does Mr. Darwin say what insects he experimented upon, but it is well known that such genera as *Cryptorhynchus*, *Cœloides*, *Centhorhynchus*, and

Coccinella, adopt a method to deceive known as "feigning death," from the way their legs and antennæ are disposed under fear. With your permission, I will give two instances, out of many, showing the nature of my observation, in which I feel sure the attitudes were the same in life as. in death. I received some specimens of *Cryptorhynchus lapathi* as dead, and during the operation of setting no difference could be distinguished. The insect turned out to be alive, and had to be thoroughly killed and reset. In this case there was ample opportunity of noticing the disposition of rostrum and legs. In an hibernating specimen of *Byrrhus pilula* the under side had to be clearly examined with a lens to distinguish legs and antennæ, and I had almost made up my mind that it was worthless, when the surrounding warmth caused it to unfold its limbs. Thus it is with many of the species in the above-named genera; the legs so exactly fit the segments of the body, and the antennæ and rostrum fit into grooves like the blade of a pocket-knife, that they never fold them in any other direction, either when killed, or when they fold them in fear. It seems most natural, I think, that on any contraction of the muscles the limbs should be brought into these natural receptacles. Those insects having grooves and indentations to fold away their limbs, can best simulate death. I am not prepared to say how far the insects are conscious of their action, but certainly, seeing the structure of beetles, there is nothing unreasonable in supposing the same disposition of limbs should take place in fear as in death. Is a beetle conscious of becoming inconspicuous? if so, may it not also be conscious of putting on an appearance found by experience to make them most inconspicuous, namely, that of death? Some species of Cœloides look just like the buds of nettle-flowers, and *Cryptorhynchus lapathi* looks exactly like a bit of grey lichen when rolled up on being disturbed, or when dead.—*G. Robson.*

ANIMALS AND BIRDS IN JERSEY.—As supplement to the note of your contributor J. J. B. in January number, I beg to say that the stoat, which is unfortunately too common here, is invariably white in the winter. I obtained several specimens last winter, and one now before me—awaiting the scalpel —is in that state. The hoopoe is of far more frequent occurrence than your contributor's note would imply. I have known two specimens killed within the last two or three months—one by the station-master at Les Marais, near the railway, the other by Mr. J. Romeril, near his house at Longueville. I also had the pleasure of watching a specimen on two occasions during the summer, near the latter locality; this last could easily have been added to the list of specimens obtained, but the possibility of its being breeding made it of course sacred. I have known altogether nine specimens of the bird killed here.—*J. Sinel, Jersey.*

EARLY BIRDS, &c.—On January 4th I heard the missel thrush (*Turdus viscivorus*) and the song thrush (*Turdus musicus*) singing beautifully. At the same time I noticed the blackbirds (*Turdus merula*) and some starlings (*Sturnus vulgaris*) pairing. Two pairs of the latter were busy searching the eaves of my house for nesting places. On the same day I found the common toad (*Bufo communis*) which had left his place of hibernation to look for food among a small plantation of gooseberry trees in the garden. The weather at the time was mild and calm, with glimpses of sunshine between slight showers of rain. The thermometer on the ground registered as high as 51°.—*J. H. M.*

A MIDSUMMER RAMBLE OVER THE SURREY DOWNS.—My critic calls my attention to the circumstance that I accidentally wrote "I see a purple emperor on an oak spray," for "I am accustomed to see a purple emperor on an oak spray." The churring of the fern-owl in the day-time I believe to be correct; my note book informs me it was at the commencement of June that my attention was drawn to it. The bird pairs at the commencement of July. The ramble in question was taken in 1882, at that period of the year.—*The Author of Insect Variety.*

NOTICES TO CORRESPONDENTS.

TO CORRESPONDENTS AND EXCHANGERS.—As we now publish SCIENCE-GOSSIP earlier than heretofore, we cannot possibly insert in the following number any communications which reach us later than the 8th of the previous month.
TO ANONYMOUS QUERISTS.—We receive so many queries which do not bear the writers' names that we are forced to adhere to our rule of not noticing them.
TO DEALERS AND OTHERS.—We are always glad to treat dealers in natural history objects on the same fair and general ground as amateurs, in so far as the "exchanges" offered are fair exchanges. But it is evident that, when their offers are simply disguised advertisements, for the purpose of evading the cost of advertising, an advantage is taken of our *gratuitous* insertion of "exchanges" which cannot be tolerated.
WE request that all exchanges may be signed with name (or initials) and full address at the end.

H. C. C.—Apply to Mr. H. G. Fordham, Odsey Grange, near Royston, for Report of the Local Scientific Societies Committee.
C. H. O. C.—Your paper will appear shortly.
W. J. NORTON.—A capital book on British diptera is Curtis's 'Monograph,' containing 103 plates. The price, however, is high—about £5. It may be had of L. Reeve & Co.
W. H. R.—It is rather difficult to answer off-hand, "What is an orchid?" Of course there are strong peculiar distinctions which constitute an orchid, which you may see in any botanical manual giving the characteristics of the order. Perhaps the most striking feature in an orchid flower are the two pollinia, or pollen-masses, and the flat stigma.
A. H. T.—Dr. M. C. Cooke is now bringing out a work on British Desmidiaceæ in parts. See Davis's "Practical Microscopy" for projecting microscopic objects, &c. A visitor is admitted to the R. M. Society's meetings by a Fellow.
E. T. SCOTT.—Please send us the peculiar wing-cases you allude to in your letter of the 1st ult. for examination.
H. C. BROOKE.—You would be able to get most, if not all, of the live animals you require of Mr. E. Wade Wilton, Northfield Villas, Leeds. Stoats and weasels you would get from any gamekeeper. St. John, the author of "Wild Sports of the Highlands," has been dead some years.
W. G. H. TAYLOR.—We cannot lay our hands upon the vols. of the "Leisure Hour" containing the articles on "Primitive Man" alluded to in the paper of last month. The reference was given as related to the author. You had best write to the editor of "The Leisure Hour" for further information.
W. CROSS.—Admission to the societies you mention is obtained, first on the recommendation of three Fellows, from personal knowledge, and then on paying an entrance fee of six guineas. The annual subscription varies, but it is generally two guineas.
MISS L. CLARE.—Tenby is a capital hunting-ground for a naturalist. Get Gosse's "Tenby."
W. A. S.—The black fungus you sent us is a dead and dried specimen of a species of Polyporus.
W. POTTER.—Your specimen is a branch of the acacia, with male or staminate flowers.
F. TREBOR.—The names of the Pre-Cambrian systems of rocks you refer to, were given by Dr. Hicks in a paper read before the Geological Society, NOV. 22, 1876. The Dimetian he named after the ancient local name (Dimetia) of that part of Wales. Pebidian is called after the name of the hundred (Pebidiac) in which these rocks are chiefly exposed. Menevian rocks are named from the ancient Menevia (St. David's).
C. G. HALL.—Your plant is *Salicornia herbacea*, var. *pusilla*.
W. L. BALMBRA.—Your specimens are: No. 1, "Cockscomb" (*Pennatula phosphorea*), one of the alcyonarian zoophytes; No. 2, "Sea-fir" (*Sertularia abietina*), one of the hydroid zoophytes.
J. W. WILLIAMS.—We have no doubt you would be able to get specimens of sand-lizards from Messrs. J. Sinel & Co., Naturalists, St. Helier's, Jersey.

EXCHANGES.

OFFERED, small air-pump, Valentine's knife, Lawson's dissecting microscope without instruments, for ¼ inch objective, Gosse's works, or good micro mounts. Write first.—H. A. Francis, 12 Aberdeen Terrace, Clifton, Bristol.

FOR skin of chameleon and ditto of blue shark, send address to J. Sinel, Davitt Place, Jersey.

WANTED, some good micro-slides in exchange for SCIENCE-GOSSIP (unbound) for 1880, 1881, and 1882: January to June, 1883; May to December, 1879; and April to October, 1878; or any portion of them.—H. Thomson, 22 Brooke Road, Stoke Newington, N.

WANTED, well-mounted micro slides of mosses, hepaticæ, ferns, and grasses. State lowest price, or exchange required. —J. R. Murdoch, 24 Blenheim Place, Leeds.

GOOD and well-dried specimens of British plants and grasses. Mosses and grasses wanted. Will give in exchange scientific books or cash.—J. R. Murdoch, 24 Blenheim Place, Leeds.

STOW'S microscope lamp with pine case in first-rate condition. What offers in micro apparatus?—Geo. Ward, Wallwood Nursery, Leytonstone, E.

Helix pisana and var. alba, *Helix ericetorum*, &c. For other land and freshwater or marine shells, &c.—F. W. Adamsdown, Cardiff.

OFFERED the following hairs: ourang-outang (*Simia satyrus*, Borneo); chimpanzee (Sierra Leone); platypus (New South Wales); echidna (under side); lemur (*Propithecus holomelas*, Madagascar). Desiderata, mounted or unmounted; animal objects.—H. Thrupp, Murbury, Axminster, Devon.

CRUSTACEA, echinoderms, mollusca, &c. Also microscopic preparations of the embryo forms of the same—parasites and spiders. Desiderata numerous.—Edward Lovett, 43 Clyde Road, Croydon.

WANTED, one or two tree frogs.—A. Pittis, Carisbrooke Road, Newport, Isle of Wight.

FORAMINIFERA (selected) from Scilly Isles, Pothcurno Cove, and Penzance; quartz sand and very thin sections of potassic ferricyanide, in exchange for well-mounted slides of other foraminifera, soundings, &c., or parasites (the latter preferred). —F. E. Hillman, 1 Harcourt Road, Wallington, Surrey.

COLLECTION of plants (about 150), also a few eggs for microscopical slides.—C. H. Goodman, Lesnes Heath, Kent.

WANTED *Neritina viridis*; will give *N. zebra*, *N. ustillata*, *Nerita tesselata*, *Oliva pura*, *pygmæa* or *reticulata*. Also wanted species of Cypræa and Oliva.—J. Harvey Bloom, Westbury House School, Worthing.

WANTED tropical land shells; also fossils from the eocene, miocene, and pliocene from France and Italy. Offered, British marine shells and British land and freshwater shells, also fossils. —Miss F. Hele, Fairlight, Elmgrove Road, Cotham, Bristol.

DUPLICATES—*Sphærium corneum*, and var. *nucleus*, *Sp. rivicola*, *Pisidium amnicum*, *Neritina fluviatilis*, *Bythinia tentaculata*, *B. Leachii*, *Physa hypnorum*, *P. fontinalis*, and *Limnæa glabra*, Desiderata: British land and freshwater shells, especially varieties, also British marine shells.—Robert B. Cook, 44 St. John Street, Lord Mayor's Walk, York.

SIX well-mounted slides for good stage micrometer, or exchange for other slides. Lists to T. Turnell, 19 Park Terrace, Regent's Park, N.W.

RARE British birds' eggs offered for a well-mounted specimen of stormy petrel; also for various British land and freshwater shells, and fossils—ammonites, belemnites, fruits, and shark's teeth, specially wanted.—R. Standen, Goosnargh, Preston, Lancashire.

WANTED, back volumes of "Zoologist," also of "Ibis."— Rev. H. A. Macpherson, 3 St. James's Road, Carlisle.

VERY superior anatomical and pathological microscopic slides in exchange for good magic lantern slides or offers.—Henry Vial, Crediton, Devon.

EXOTIC lepidoptera. Numerous fresh duplicates, particularly in Papilio, Catagramma, Cybdelis, and Romalæosoma; great atlas moths, bred; wings of all the most brilliant species for microscopic purposes.—J. C. Hudson, Railway Terrace, Cross Lane, near Manchester.

FOR exchange, 13 Nos. of "Zoologist," 21 Nos. of "Journal of the Linnean Society—Botany;" 24 Nos. of "Journal of the Linnean Society—Zoology." Wanted, botanical slides and Davis's "Practical Microscopy."—W. Rose, Abergavenny.

"THE Day after Death," "The World before the Deluge," "Primitive Man," by Louis Figuier, "Punch" (unbound), Nos. 13 to 22 inclusive; odd numbers of "Popular Science Review," containing excellent plates. Wanted, foreign marine, land and freshwater shells, British fossils, mammalian skulls.— Miss Liuter, Arragon Close, Twickenham.

OFFERED, during the forthcoming season, fresh or dried, Lond. Cat., 7th edit.—2, 45, 115*b*, 121, 130, 147, 164, 218, 275, 366, 367, 406, 453, 457, 495, 519, 521, 534, 538, 539, 631, 772, 829, 859, 906, 924, 1040, 1124, 1275, 1330, 1349, 1361, 1401, 1412, 1422, 1438, 1447, 1458, 1473, 1501, 1506, 1659, 1665, with many others. Exchange by special arrangement.—H. Ibbotson 2 Grape Lane, York.

A LARGE number of bi-pendulograph writings, ratios of the musical system, and various others, in exchange for other drawing or objects of interest chiefly microscopic.—J. J. Andrew, 2 Belgravia, Belfast, Ireland.

WANTED, photographic apparatus in exchange for microscopic slides.—S. Wells, Gladstone Terrace, Goole.

WILL send four specimens of living budding *Hydra viridis* on the receipt of a good mounted object (physiological or botanical preferred, or for nine stamps. Wanted, Balfour's, and Balfour & Foster's "Embryology," and Schafer's "Practical Histology."—T. W. Lockwood, Lobley Street, Heckmondwike, Yorkshire.

A FEW slides of *Pleurosigma attenuatum* in exchange for slides of other diatoms or micro-material. Also a number of slides of spicules, diatoms, and sections and a few tubes of cleaned diatoms. Send for list, sending list at same time.— J. J. Andrew, 2 Belgravia, Belfast, Ireland.

WILL send well-mounted slide of (dry or in balsam) diatoms, *Gomphonema geminatum*, in exchange for any interesting or instructive slide.—H. F. Jollys, Stow Villa, Bath.

SHELLS for exchange: *L. glutinosa*, *A. acicula*, *H. Cartusiana*, *Unio tumidus*; *Assiminea Erayana*, *Tellina fabula*, *T. Balthiça*, &c. Wanted, British or foreign land and freshwater species or British marine.—S. C. Cockerell, Glen Druid, Chislehurst, Kent.

SHELLS for exchange: land and freshwater, *Helix nemoralis*, var. Hybrida, &c., *H. arbustorum*, *H. virgata*, *H. caprata*, &c.: marine—*Pecten opercularis*, *Cypræus Europæus*, &c. Wants very numerous.—Charles Moxley, Beaumont Park Museum, Huddersfield.

DUPLICATES: 100 species of West American marine shells (named), and about the same number of North American land and freshwater shells. Desiderata: British land, freshwater, and marine shells (recent or fossil), or back numbers of the British natural history journals. Send lists to—Z., Box 209, Post Office, Victoria, British Columbia.

DUPLICATES: *Limnæa stagnalis* and *Planorbis corneus* (very fine); *Limnæa glabra* and *L. Peregra*, and var. *Froglis* of *L. stagnalis*; *Physa hypnorum*, *Planorbis spirorbis*, from Strensall Common, near York, which is about to be drained for Government purposes. Desiderata: very numerous.—W. Hewett, 26 Clarence Street, York.

EOCENE and a few upper cretaceous fossils. Will exchange for lower secondary and primary fossils.—F. C. Phillips, 34 Carlton Road, Fitzhughs, Southampton.

WANTED to exchange a collection of marbles and of fossils (all formations) for scientific book, micro-slides, or offers.— Dr. Thresh, Buxton, Derbyshire.

"My Schools and Schoolmasters," by Hugh Miller; "The Beauties of Swift," published 1782 (clean copies); Culpeper's "British Herbal," 129 plates, in exchange for a standard illustrated work on the British land, marine and freshwater shells. —Frederic Corkett, High Street, Winslow, Bucks.

BOOKS, ETC., RECEIVED.

"The Seven Sagas of Pre-historic Man," by J. H. Stoddart. London: Chatto & Windus.—"Flowers and Flower Lore," by the Rev. Hilderic Friend. London: W. S. Sonnenschein & Co. —"The Watch and Clockmaker's Handbook," by F. J. Britten. London: W. Kent & Co.—"Universal Attraction: Its Relation to the Chemical Elements," by W. H. Sharp. Edinburgh: E. & S. Livingstone.—"Nursery Hints," by N. E. Davies. London: Chatto & Windus.—"Atlas of the Tertiary History of the Great Cañon District," by Capt. Dutton.—"Second Annual Report of the United States Geological Survey, 1881."—"Geological and Geographical Survey of the Territories of Wyoming and Idaho," Parts I. and II., by Dr. Hayden.—"Tertiary History of the Great Cañon District," by Capt. Dutton; Washington: Government Printing Office—"Studies in Microscopical Science," edited by A. C. Cole.—"The Methods of Microscopical Research," edited by A. C. Cole.—"Popular Microscopical Studies." By A. C. Cole.—"Petrological Studies." By Messrs. J. E. Ady and H. Hensoldt.—"The Gentleman's Magazine."—"Belgravia."—"Journal of Conchology."—"The Journal of Microscopy."—"The Science Monthly."—"Midland Naturalist."—"The Inventor's Record."—"The Naturalist's World."—"Ben Brierley's Journal."—"Sunlight."—"The Medical Student." (New York.)—"Natural History Notes."—"Science."—"American Naturalist."—"Medico-Legal Journal." (New York.)—"Canadian Naturalist."—"American Monthly Microscopical Journal."—"Popular Science News."—"The Botanical Gazette."—"The Ornithologist and Oologist." —"The Electrician."—"Horological Journal."—"Revue de Botanique."—"La Feuille des Jeunes Naturalistes."—"Le Monde de la Science."—"Ciel et Terre."—"Cosmos: les Mondes." &c. &c.

COMMUNICATIONS RECEIVED UP TO 11TH ULT. FROM:— F. K.—C. A. B.—T. N. H. S.—B. B. S.—W. M. E.—H. R.— J. H. B.—H. C. J.—C. R. H. T.—E. J. E.—D. B.—W. D. R.— J. J. M.—C. J.—W. H. R.—L. L. B.—H. B. R.—F. J. F.— D. A. C.—T. S.—A. C. G.—F. E. H.—W. H. B.—R. H. T.— A. P.—E. L.—A. W. O.—H. A. F.—F. W. A.—G. W.— C. F. G.—W. J. N.—F. R. M.—F. C. C.—B. M.—H. T.— J. S.—J. J. A.—J. F. H.—W. H. P.—H. I.—J. E. L.—T. W. L. —W. K.—W. C.—J. C. H.—R. S.—H. V.—H. A. M.—W. R. —E. H. R.—H. A. S.—J. H. B.—A. J. D.—H. F. J.—E. T.— J. F. R.—L. M.—A. D.—S. C. C.—S. W.—W. H.—W. B. G. —G. W. T.—B. S. D.—W. G. H. T.—R. S. H.—W. S. —A. S. E.—R. B. C.—M. H.—T. T.—W. G., &c.

GRAPHIC MICROSCOPY.

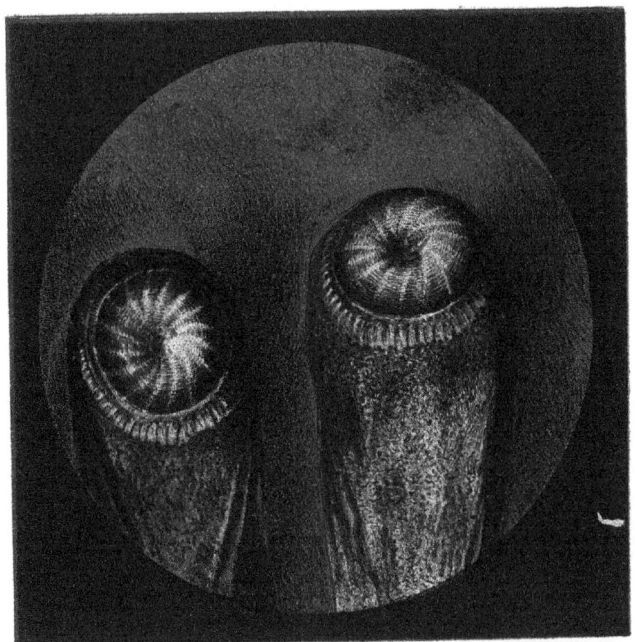

E.T.D. del. ad nat.   Vincent Brooks, Day & Son, Lith.

PERISTOME OF FUNARIA HYGROMETRICA.
× 25

GRAPHIC MICROSCOPY.

By E. T. D.

No. V.—Peristome of Funaria Hygrometrica.

THE urn-shaped capsule, or pocket, erect on the summit of its footstalk, so often seen by an observer of growing mosses, for structure and colour, is to the microscopist, and without reference to its deeper morphological interest, an object of great beauty. It is impossible to approach the subject in its amplitude, but a concise explanation of the plate is necessary.

This well-known "moss cup" is the sporangium, and in it are developed and contained the dust-like spores, protected by an elegantly perforated or reticulated tissue known as the peristome, which, in some cases, is capable of unfolding, revealing and releasing the reproductive atoms within. The whole is often covered with a calyptra, or hood—a loose extinguisher-shaped body which soon falls off; the peristome is then seen in all its beauty, when unfurled, as in Funaria, or as modified in other species, the spores escape and fulfil their function. The sporangium and its contents are the only product of the true and normal fructification of the plant. In some cases the spores are released by the decay of the outer integuments.

The sporangium under microscopic powers displays internally the most complicated structures, but externally it is crowned by the peristome, which is variously formed, either by fringes, well-fitting lappets convoluting to a centre, or in some cases of slits arranged round the margin, the structure ramifies into various alternations, their disposition being factors in the distinction of genera.

In the genus Bryum, a double peristome is found, of great interest, in *Neckera pumila*, common on trunks of trees; the twin peristome is very singular, the over and outer teeth curl back and show the inner layer like a perforated dome, united by cross bars—a microscopic exhibit of rare beauty.

Much has yet to be discovered in bryology, but the direct mode of germination from the spore, elaborated in the sporangium, is safe ground. When released the spore falls, and produces a confervoid structure (the protonema). Much might be said of the microscopical interest involved in this confervoid filament. From the protonema a differentiation of cells are evolved, from which proceed minute buds; each of these can hardly be said to grow, but they are the nidus of a leafy stem, forming at last, a small tufted group of vegetation, the foundation of "antheridia" and "archegonia." A magnification of not less than thirty diameters is required for examination. The reader may now be referred, for a further description of the ultimate development of the new plants, to page 83 in the April number, where it is well described by J. F. R., in "Notes for Science Classes."

The strange and mysterious process there detailed may be seen on the stage of the microscope, without much difficulty. A good specimen for this purpose is *Polytrichum commune*, abundant on every heath, and, at this particular season, in exactly the condition required. In the interstices of a tuft of older plants may be detected the minute objects sought for; they can be picked out with forceps and placed in a drop of water, or, better, a saturated solution of chloride of calcium, on a slip with a covering glass. The latter medium is sufficiently deliquescent to preserve an impromptu botanical preparation, with no other care, for several days or weeks, without deterioration.

In addition to the normal mode of fructification, as described by J. F. R., there are subsidiary modes, peculiarly abundant in mosses. The stems often send out branches, which root, and produce buds.

Confervoid filaments (something of the character of the mycelium of the fungi) are found, with tuberous thickenings (gemmæ) which eventually become detached, and produce new plants. The gemmæ again are not always confined to the confervoid filament, they are discovered, as the most ordinary observer of mosses may have detected, on the axils, the surface, or the tips of leaves. In time they fall off, and curiously illustrate the fact (known generally in the development of desmids) that a single released cell of a tissue is capable of erecting a new plant; the leaves of moss often throw out rootlets, but the primary function of germination emanates from the spore, a mere speck dropped from the sporangium, by the expansion of the peristome.

It may seem strange to a mind undirected to biological research, how so apparently insignificant a plant as moss can involve such mysteries. But simple as the plant appears, there is no branch of morphological botany which presents greater paradoxes, and it is an acknowledged fact that, as affecting many particulars, the entire subject under closer observation and finer instruments, requires revision and re-investigation. Taking the leaf alone, seemingly a matter of the least importance, as compared to embryology, it shows in various specialities, striking contrasts, reticular thickenings in cell walls, differences in the superimposition of tissues, diversities in the epidermis, in the arrangement of lamellæ, some inserted in the midrib, others on the whole surface of the leaf, and the more inexplicable fact that in the same leaf some cells contain chlorophyll, others do not.

Possibly no order of plants play a more important part in the economy of nature than Mosses. They form the nidus for vegetation of a higher order, being for the most part aerial, with no true root, but simply attached by filaments to their habitat, deriving nutriment and substance from the surrounding air; their decay and accumulation, collected through the agency of organic chemistry, forms a substratum of material for higher growth. The peat deposits consist of little else than the accumulated débris of centuries of the decadence of the aquatic species—Sphagnum.

The drawing for the plate was made from a mounted specimen under reflected light. If a well-ripened, dry capsule be subjected to moderate steam moisture, or breathing upon it, the teeth of the peristome (a highly hygroscopic tissue) will unfurl, separate, and stand erect, round the periphery of the sporangium, revealing the contents within.

Crouch End.

QUERY AS TO STRANGE BIRDS.—SCIENCE-GOSSIP, No. 229, p. 23. The birds seen by your correspondent, W. H. Summers, were probably lesser redpoles (*Linota linaria*).—*H. Miller, jun.*

TEN DAYS IN SCILLY.

I WAS never more surprised in my life than I was on the 8th of last May, when I stepped off the steamer that had conveyed me from Penzance to St. Mary's and landed me on the semi-tropical shores of Scilly.

The weather that had been genial in the morning on the mainland did not appear the least concerned about the reputation of the islands, for the temperature actually fell as we approached them.

But it was not the increasing coldness alone that excited my suspicions, for, in conversation with one of my fellow passengers about the tropical productions of the place, he remarked, that I had arrived too early, the mats would not be off the palms; this made me very unhappy.

Many and various have been the opinions advanced with regard to the climate, but I must here record my own—it seemed to me to be windy.

By June the 13th, I had spent five days in one of the most bracing breezes I ever had the chance of invigorating my constitution with—how it did blow! It blew on to the islands, and off the islands, and up them and down them, and then by way of change round them; it was decidedly unfavourable for entomologising, whatever it might be for witches.

From time to time I expressed my surprise to persons I met, but the promptness of their apologies and the glibness with which they were uttered made me suspect they were in the habit of it.

In very windy weather even at home, it is sometimes necessary to secure one's hat to prevent its being blown off, but here in Scilly the roofs of the houses are firmly lashed with ropes and secured with belaying pins to the walls. So violent are the gales that sweep over these Isles of the Blest.

The chief object of my visit was a hunt for beetles, but how to begin puzzled me. The moment I raised a tuft of grass, sand and beetles blew away together, with the exception of what went into my eyes; of trees there were scarcely any, and hedge-rows were rather the exception than the rule, but I did not despair. There were large slabs of stone in all directions. These suggested themselves as the only possible hiding-places for beetles, and it was under these the bulk of my captures were made, and amongst them were the following : On Annet Island—*Otiorhynchus ligneus, Hylastes palliatus.* On Scilly—*Heliopathes gibbus, Otiorhynchus sulcatus,* and *Otiorhynchus atroapterus, Calathus mollis, Broscus cephalotes Creophilus maxillosus* (var. *ciliaris,* Stephens), *Philonthus fuscus,* and *Calathus micropterus.*

One of the pleasantest excursions I made was to the Island of Annet—totally uninhabited, and the head-quarters of the sea birds.

Except in calm weather the landing is not easy, but, that difficulty over, the scenery is most wild and wonderful; it literally swarms with birds from one end to the other. The soil is burrowed; no, undermined with penguins, and the air overhead is one whirl of sea gulls.

But the most striking feature of the islands is the extensive sweep of sea-thrift, rising in pure rose-coloured hillocks against the sky, and here and there perpendicular peaks of grey stone rising out of them, all hung with lichen and crested with sea birds.

If they had only been beetles instead, what a harvest for the collector! But it was no good regretting this, so I strolled along the beach and began to hunt.

Before long my attention was attracted to the remains of wrecks scattered along the shore : here and there broken boards, bleaching in the sun, and then larger portions of vessels hurled far inland by the fury of the gales. Well! it was under the smaller débris of these wrecks that I made my principal captures.

As regards the plants I have not much to say. In one place and another I took the following: *Chara fragifera*, hitherto, I believe, only recorded in Cornwall. *Euphorbia paralias* and *Portlandica*, *Ranunculus hederaceus*, *Lenormandi*, and *tripartitus*, *Fumaria pallida*, *Lepidium campestre*, *Erodium maritimum*, and the moss *Ulota phyllantha*. Though this record of my captures is somewhat meagre, I must add that I think that later in the season, and with more propitious weather, there is much to be done in these beautiful, and, in my opinion, demi-semi-tropical islands.

By the way, I would just remark, in conclusion, that on approaching the islands, the traveller will observe a buoy with H. A. T. S. painted conspicuously with large white letters. The caution is far from unnecessary. Owing to the wind I know of no place where they require such constant attention ; future travellers will do well to take the hint.

B. P.

THE ORIGIN OF DOUBLE FLOWERS.

IN SCIENCE-GOSSIP for March, there is an interesting article on this subject by Mr. John Gibbs. I wish to suggest another view of the question which is commonly overlooked.

Every one recognizes the fact, in a general way, that organic forms do not last for ever, but have a period of infancy, a period of maturity, and a period of decay. In the case of individuals this organic law is self-evident : but investigation shows it is also the law for species, for genera, for families, for orders, probably for the organic world as a whole, and perhaps for the inorganic also. It is undoubtedly one of the fundamental laws of the universe; one of the primary modes in which force operates. Its action is illustrated under the simplest conditions in the transmission of sound or motion by a succession of waves, in which the material molecules are alternately drawn together and dispersed. It may be called the law of the wave-form.

Bearing in mind that every organic individual, species, genus, &c., is subject to this law, and represents a force-wave which has a definite initial intensity, and will, under any given conditions, run a definite course of gradual concentration, followed by dispersion, we see that it is the natural destiny of each species to have a starting-point from which it rises to its maximum of development, and then decays and disappears.

It is well known that in the inorganic world there is another great law—the law of interference, by which simple waves are compounded and modified to a large extent. Musical notes and water-waves are produced by a number of simple waves which interfere in a cumulative manner, while noises and "chopping" seas are the results of interferences which are less regularly coincident.

This law of interference applies equally to the organic world.

A species is a compound organic wave resulting from a succession of individuals, the original specific impulse being carried forward through the reproductive system to its natural climax, which is determined by the conditions of the initial impulse.

Supposing that some species of plant from its very origin, were entirely isolated, it would go on increasing in numbers, and developing in character up to a certain point, and would then decay. Each individual of the series would in like manner attain to its own maximum, and at its extreme limit would leave its energy in its seed and then decay. But as the climacteric of the species approached, the individuals attaining more complete development, those organics at the terminal growing-points which formed the reproductive elements, would now unfold into perfect petals and reproduction would gradually cease. Thus the double flower should indicate the climax of a species when reproduction is no longer necessary, and when the hitherto absorbed petals become normally developed.

But if a species is not isolated, which of course is the usual condition, the law of interference must very frequently come into operation.

The flowers become cross-fertilised. The original specific impulse is compounded more and more at each generation of individuals.

Some of the interfering waves coincide with the original and increase its intensity, others are antagonistic and neutralise it, while those which partly coincide and partly neutralise produce all sorts of intermediate results.

The final issue would support Mr. Gibbs's argument that a long course of self-fertilisation tends to produce double flowers, because such a course means non-interference, and therefore more rapid attainment of the climacteric of the original wave, while cross-fertilisation tends to produce increased vitality, more

F 2

frequent variation, and a lengthening of the life of the species.

And here I may point out, that we arrive at a probable cause of variability. Plants which are perpetually self-fertilised should vary very little, while cross-fertilisation, by compounding and modifying the original specific wave, will occasionally produce oscillations so extreme as to become the starting-points of new specific waves.

As a further corollary from this law, we should expect the most frequent intensifying of the specific wave when the crossing takes place between near relatives, yet not so near as to be practically equal to self-fertilisation; and the most frequent neutralising or violently disturbing of it when the crossing is between forms too distantly related, as between two species of the same genus. This is borne out by the facts, that the crossing of two flowers from one plant is little better in its effect than self-fertilisation; that the crossing of distinct plants or of two varieties has the greatest vitalising influence; while the crossing of species produces sterile mules, the male being neutralised and disintegrated.

F. T. MOTT.

ROOT ACTION OF THE LILY OF THE VALLEY.

By WILLIAM ROBERTS.

FEW plants are more interesting to the keen observer of nature than the lily of the valley. Although one of the most popular of British plants, there are many curious phenomena connected, under various conditions, with its life-history, which render it a subject worthy of careful and exhaustive inquiry. In its gradual development from a mere bud to the full maturation of the three year old crown, we have a most beautiful instance of Nature's wise economy.

It is, however, under the unnatural conditions of forcing that a singular and probably unique characteristic is displayed. Before proceeding farther, it may be well to mention that only crowns of three year old growth produce flowers; the rudiments of these are fully perfected during the summer and autumn periods to their ultimate expansion during the ensuing spring. If well-ripened crowns are taken up out of the open ground some time in October, placed in a compost of almost any nature, and immediately transferred to a moist bottom heat of from 95° to 100° Fahr., flowers will in all probability be produced at or about Christmas; but strangely enough the roots will not have made an iota of growth, remaining apparently dormant. If the fibrous roots are severed to within two or three inches of the base of the crown, it makes no difference whatever.

These facts seem strangely contradictory to orthodox views, which teach that absorption takes place almost exclusively by the spongioles, or young extremities of roots. It follows that in cutting the fibrous roots of the lily of the valley back several inches, the spongioles are also severed. Notwithstanding, both flowers and leaves are produced—in conditions neither abnormal nor monstrous, but perfect. It also follows as an obvious consequence, that spongioles are needless in the case of the lily of the valley when forced. Under

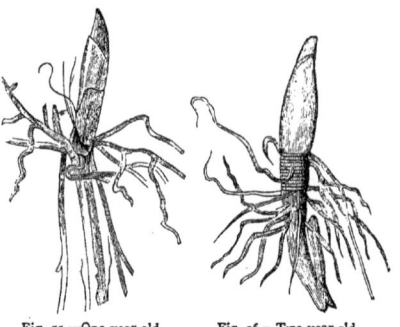

Fig. 55.—One-year-old English-grown Crown.

Fig. 56.—Two-year-old German-grown Crown.

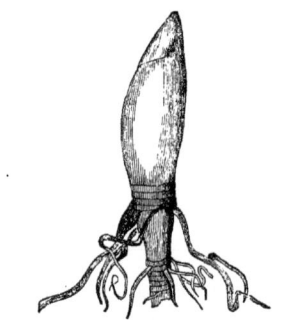

Fig. 57.—Three-year-old Crown.

no conditions have I ever found a root growth made at any other period than spring; and it is more than passing strange that flowers should be produced—under extraordinary circumstances, of course—with the roots to all appearances at rest; more especially so when these two usually make a simultaneous start into action. The subject is an interesting one, both scientifically and popularly, and well deserves close attention.

THE Rev. W. H. Dallinger, F.R.S., has been elected President of the Royal Microscopical Society.

THE ENTOMOLOGY OF HIGHGATE, MIDDLESEX.

By William J. V. Vandenbergh, F.R.A.S., F.M.S., &c.

THERE is not, I think, another locality within half-a-dozen miles of London where so many species of our commoner insects may be obtained as at Highgate.

This neighbourhood is very well known to collectors in the north of London, many of whom make it their usual hunting-ground, and apparently with the best results.

I have myself taken many hundreds of specimens here, chiefly Noctuæ at sugar, and have been in the habit of entomologising in this neighbourhood very frequently, almost always obtaining specimens which were wanted by me to fill up blanks in my collections. There are so many ways by which the London entomologists may get to Highgate that I cannot feel surprised that a large number of collectors visit the neighbourhood in quest of specimens.

It is very unusual to walk through the woods at Highgate on a genial summer evening without meeting some votary of the net and pin, but there is ample space for all, and there need be no fear on the part of a new-comer as to his finding plenty of ground to work.

Highgate has not yet entirely succumbed to the trowel of the speculative builder, as have most of the surrounding districts, but it would be too much to expect that it will be as productive, entomologically, in a few years as it is now.

Many good species of Lepidoptera are often taken in the neighbourhood, and I have no hesitation in saying that it is by far the most productive locality in the north of London.

The list of butterflies to be taken here is not a very long one, and they are all common species, viz., *Vanessa atalanta, V. Urticæ, V. Io, V. Cardui, Satyrus megæra, Gonepteryx Rhamni,* and *Anthocharis Cardamines,* &c. The common species of "skippers" are also generally abundant.

On the gas lamps at the sides of the Muswell Hill Road, leading between the woods, may be found, at the proper season of the year, an occasional specimen of *Lithosia aureola;* and its near relative, *L. quadra,* is reputed to have been taken in a like position; but I have not personally seen the latter species taken in the neighbourhood, and consequently cannot vouch for the correctness of the statement. I have taken specimens of the large emerald (*Geometra papilionaria*) off these lamps, as also specimens of *Calligenia miniata* and *Metrocampa margaritaria,* the two latter species in considerable numbers. *Cossus ligniperda* occurs in the willow-trees in the neighbourhood, and used occasionally come to regale itself upon the "sugar" spread on the palings or tree trunks for the allurement of the Noctuæ.

Nearly all the fields near the woods produce *Hepialus hectus, H. lupulinus,* and *H. humuli;* and *H. sylvinus* is reported to have been taken in the lower wood near Highgate Station.

The beautiful larvæ of *Odonestis potatoria* may often be seen in the spring feeding upon the long grass at the sides of the roads, and the perfect insect often startles the collector by flying madly at the lamps, like *Arctia Caja,* at midsummer. *Liparis auriflua, L. salicis, Hemithea thymiaria, Acidalia aversata, Anisopteryx æscularia, Eupithecia centaureata, E. castigata, Ypsipetes elutata, Cidaria fulvata, C. dotata,* and all the English species of the genus Hybernia are common at the gas lamps; whilst *Arctia mendica, Orgyia pudibunda, Eurymene dolobraria, Pericallia syringaria, Odontopera bidentata, Crocallis elinguaria, Ennomos tiliaria, Himera pennaria, Phorodesma bajularia, Lomaspilis marginata,* and *Anticlea badiata,* are occasional visitors.

Selenia illunaria, S. lunaria, S. illustraria, Acidalia imitaria, Corycia temerata, and *Melanippe montanata,* are also often taken at the lamps, but more frequently by "dusking."

Phigalia pilosaria occurs very commonly at the lamps in the early spring, in fact I have never seen so many taken in any other locality. "Dusking" along the hedgerows also produces *Boarmia repandata, Hemithea thymiaria, Melanippe montanata, Cidaria fulvata,* and many other common geometers, in great abundance.

Amphydasis prodromaria, A. betularia, and *Dicranura vinula* are occasionally taken by searching the tree trunks at the proper seasons, the latter species often coming to the lights.

Cilix spinula, Ptilodontis palpina, and *Notodonta camelina* are sparingly taken at the lamps, as are also nearly all the common species of "pugs."

Perhaps the most beautiful species of the Noctuidæ to be commonly taken at Highgate are the buff-arches (*Gonophora derasa*) and the peach blossom (*Thyatira batis*) both of which come freely to "sugar" in the months of June and July, the former species commencing to make its appearance somewhat later than *T. batis.* Both these species are very active on the wing, and it is as well to have the net always ready at hand, in order to secure them should they forsake the sugar and take to flight.

The woods in the neighbourhood are private property, and the owners, or their employés, appear to be of opinion that large trees are natural provisions for the proper exhibition of notice boards and bills warning trespassers of the dire consequences that may ensue on any attempt at a ramble without due authority. They are, however, very good in this respect, and a very slight deviation from the beaten tracks is as a rule allowed by courtesy to respectable persons. It is perhaps, however, unnecessary and unadvisable to say anything farther on the subject, except to add that good work may be done without in any way becoming liable to the operation of the law of trespass.

A small pathway by the side of the branch railway to the Alexandra Palace (to the left of the bridge over which the Muswell Hill Road passes) is a very productive locality, and "sugar" is, as a rule, very successful there. The broad rails and posts dividing the path from the wood could not have been rendered more suitable for the proper exhibition of the entomologists' sweets had they been specially constructed for that purpose.

The following species of Noctuæ may be taken in and on the borders of all the woods in the neighbourhood :—*Gonophora derasa, Thyatira batis, Cymatophora duplaris, Bryophila perla, Xylophasia lithoxylea, X. hepatica, X. scolopacina, Dipterygia pinastri, Heliophobus popularis, Miana strigilis, M. fasciuncula, Grammesia trilinea, Rusina tenebrosa, Agrotis suffusa, Tryphæna innthina, T. fimbria, T. Orbona, Noctua plecta, triangulum, brunnea, festiva, Rubi* and *baja; Tæniocampa Gothica,* and *T. cruda; Cerastis spadicea, Cosmia diffinis, Miselia oxyacanthæ, Phlogophora meticulosa, Euplexia lucipara, Aplecta nebulosa, Hadena proteus, H. oleracea, Plusia chrysitis, Gonoptera libatrix, Amphipyra pyramidea, A. Tragopogonis, Nænia typica, Mania maura, Catocala nupta,* and almost all the common English Noctuæ, the larvæ of which may, as a rule, also be found by carefully searching their respective food plants at the proper seasons.

Students of the Hymenoptera and Diptera will find ample scope for work in and near the woods at Highgate and in the lanes which abound in the locality.

The lower wood is especially productive of Tenthredinidæ (saw-flies) and it is best to collect them while in the larval state both by "beating" and "sweeping." I have noticed that when using the sweeping net on a warm evening during the spring, the larvæ of the saw-flies are very abundant, although they can be obtained at all times throughout the spring and summer months, and in considerable numbers.

The gall-flies (Cynipsidæ) are also abundant in this neighbourhood, and the perfect insects may, as a rule, be easily bred, by simply placing the galls in gallipots covered with a piece of glass. The tops of the gallipots should be ground on a piece of flat stone in order to make them perfectly level, as by this means the glass fits more accurately and the damp galls are not so liable to be dried by evaporation.

The galls of *Cynips quercus baccarum* and *Cynips quercus folii* will be found by the collector without any difficulty on the oaks, as also will the galls of *Cynips terminalis* on the oak twigs. These should be slightly damped from time to time, if the collector desires to breed the perfect insects. The common spangle galls (*Neuroterus lenticularis*) are of course also common on the oak leaves, and the perfect insects may be bred without difficulty by the method before suggested.

The coleopterist will also find a ramble in this locality productive of many interesting species, and among other common beetles to be taken here may be mentioned, *Carabus violaceus, Ocypus olens, Necrophorus vespillo, Cetonia aurata, Lucanus cervus* and *Geotrupes stercorarius*; the common cockchafer (*Melolontha vulgaris*), often becoming a perfect pest. By beating along the hedgerows and in the woods the collector may easily obtain large numbers of weevils, and "sweeping" on the low lying portions of the woods will be found to well repay the collector for his labour.

The Neuroptera are not very well represented at Highgate. Most of the species of this order require a very considerable amount of moisture, and Highgate being located on high ground is not particularly favoured in this respect.

The neuropterist will however meet with some old friends, especially among the dragon-flies, which are common near the few ponds in the neighbourhood.

Libellula depressa is perhaps the commonest species here, and specimens of this beautiful insect may occasionally be found drying their wings on the bushes near the ditches, at the borders of the woods.

The Dipterous fauna of Highgate appears to be large and comprehensive, perhaps even more so than is the case with the Lepidoptera, so that the collector of this extensive, but generally neglected, order, need have no fear that he will have to return with empty boxes.

There are, of course, many insects commonly taken at Highgate and in the vicinity which have been omitted from the foregoing lists, as anything like a complete catalogue would require an organised investigation, involving great labour and extending over a number of years. It will also be obvious to the reader that a large number of species of Lepidoptera which I have referred to as coming to the lamps, &c., may be taken in a variety of ways, such as "dusking" and "sugaring."

Coldford Wood is probably the best collecting ground at Highgate, but it is rigorously preserved, and unfortunately entomologists are not tolerated within its boundaries.

Commencing from Highgate Station on the Great Northern Railway, the gas lamps referred to in these notes extend along Muswell Hill Road, Tatterdown Lane, Fortis-green Road, St. James's Lane, Muswell Hill, Spaniards' Lane (leading to Hampstead Heath), and in fact all the lanes and roads in the immediate neighbourhood.

Highgate is easily and rapidly reached from all parts of London by means of the Great Northern Railway, and the tramway to the Archway Tavern, near Highgate Archway; and this short paper may, for that reason, be useful to entomologists to whom such a locality may be a desideratum.

NOTES ON NEW BOOKS.

WITH a generosity which stands forth in bold contrast to the niggardly stinginess with which similar British productions are dealt out to scientific journals, we have received from the United States Government the following handsome volumes: "Atlas to accompany the Tertiary History of the Grand Cañon District" (crowded with beautiful maps and large coloured plates of the chief scenic features); "Tertiary History of the Grand Cañon District," by Capt. Dutton; "Second Annual Report of the United States Geological Survey", (J. W. Powell, Director); "Geological and Geographical Survey of the Territories of Wyoming and Idaho" (for 1878 and 1879, two large vols.); "Maps and Panoramas, Twelfth Annual Report of the Survey of the Territories." These volumes are always acceptable to English geologists; and Capt. Dutton's handsome volume would be so to any one, apart from its graphic descriptions, who values first-class illustrations. The reading of this volume has produced a strange effect on our mind—the most prominent idea being the strong wish that every one of our readers could share the delight we experienced in reading the author's clear and bold geological explanation of this wonderful and unique Cañon country, which rises sometimes to an impassioned strain of reverent enthusiasm, as he describes the most striking scenic features. We have not read anything more graphic, as regards scenery, outside Ruskin. The illustrations comprise photographs, chromoliths, and woodcuts. The artist, Mr. W. H. Holmes, has evidently been as much impressed with the country to be described as Capt. Dutton himself. One remarkable feature in the illustrations of these volumes, is the clever and artistic manner with which there is presented a panoramic view of a country, and at the same time, the outcrop of the different strata, so that the eye takes in at once the relation of the scenic features to the underlying rock-masses.

Whilst everybody complains of the high price of our own Geological Survey Memoirs, and the sublime indifference shown to public opinion in the matter, as to whether they are known or not, it is pleasing to find the Trustees of the British Museum recognising the position in which that Institution stands as the great national scientific guide. We have just received the latest of the numerous handsome and elaborate volumes on different scientific subjects which are issued by the Trustees from time to time under the unpretending title of "Catalogues." The volume just mentioned is entitled "Catalogue of the Fossil Sponges in the Geological Department of the British Museum (Natural History)," by George Jennings Hinde, Ph.D., F.G.S. It is illustrated by thirty-eight large lithographic plates, showing both the form and structure of the fossil sponges from the Silurian to the Cretaceous formations. Sponges are now classed according to their microscopic structures, and this is followed in the present volume. No English palæontologist is better fitted for the work than Dr. Hinde, and he has produced a book which all workers will be grateful to him for.

Proceedings of the Literary and Philosophical Society of Liverpool, vols. 35, 36, and 37 for 1880-81, 1881-82, and 1882-83. This society is one of the oldest of its kind in the country, and the yearly volumes always contain a large variety of able articles, mostly from the pens of the many well-known and eminent men who reside in Liverpool and its neighbourhood. Science, art, and literature crowd each other in juxtaposition in these pages, so that we find a paper on "The Peculiar Development of an Egg of the common Fowl," side by side with another "On English Caricature Art." But all are well written, and many are original and valuable contributions both to science and literature. In vol. 35, the paper by Mr. R. C. Johnson on "Recent Researches into the Movements and Dimensions of the Stellar Universe," and that by Mr. A. J. Mott's "Notes on Easter Island," are both especially noteworthy. The Address of the President (Mr. Edward Davies), in vol. 37, is on "The Unity of Life;" and in this part there is a very important paper by Mr. Alfred H. Mason, on "Odours, Perfumes, and Flavours," giving their physiology, philosophy, history, sources, and preparation. Vol. 36 contains the President's address on "Chemical Force," and a paper "On the Velocities of Gases," by A. J. Mott; an incisive and discriminative paper on "Some Popular Misconceptions of Darwinism," by the Rev. S. F. Williams; and a powerfully-written article "On the Justifiability of Scientific Experiments on Living Animals," by Dr. F. Pollard, in which experimenters are defended.

Flowers and Flower Lore, 2 vols., by the Rev. Hilderic Friend (London: W. S. Sonnenschein & Co.). The title suggests to the reader that this ought to be a very readable book; but it is more, it is very attractive in every way, beautifully illustrated (indeed, we have rarely seen better woodcuts of plants), well printed, and artistically bound. The author is not only well informed in his subject—he is possessed with it. Such a writer, therefore, makes himself listened to. In every way the work before us is a delightful one, which all lovers of flowers will hasten to read.

Nursery Hints, by N. E. Davies, L.R.C.P. (London: Chatto & Windus). This valuable little guide is written especially for mothers. It is crowded with excellent recipes and instructions, and in the hands of an intelligent mother one cannot help thinking such a cheap little book would reduce the doctors' bills. There is hardly anything relating to the health of children which cannot be found here.

The Watch and Clockmakers' Handbook, by F. J. Britten (London: W. Kent & Co.). In every respect this is a unique book. It is at once a dictionary of every term used in watch and clock-making (we had no idea before there was a tenth part of them), and also a guide to young workmen, from the author's rich fund of information. Moreover, it is a brief biography of great watchmakers and horologists. The illustrations are both good and numerous. Every intelligent young watchmaker ought to have this book within his reach.

Universal Attraction: Its Relation to the Chemical Elements, by W. H. Sharp (Edinburgh: E. & S. Livingstone). A small and puzzling book, occupying only 53 pages, of which 11 are devoted to the preface. The author thinks that gravitation is propagated by wave motion, and that not only does the earth move, but that its atoms also move, and he seems to imply that these atomic movements are the most significant.

The Seven Sagas of Pre-historic Man, by James H. Stoddart (London: Chatto & Windus). At length Milton has found a competitor! He dealt with the "First Pair," and all their paradisaical arrangements in language which will never die. But science has not proved such a strict defender of the "letter" of sacred myths as Milton was. Consequently the "first pair" of human beings has been pushed farther back in time, and we fear strict anthropology does not allow paradisaical conditions to the environment of the latter! The semi-simian Palæolithic man was undoubtedly ancestor to us of the "Steel Age." What a strange line of descent! It includes the cave-dwellers of Kent's Cavern, the Neolithic men, the Lake-dwellers, and the skin-clad hordes who nearly repelled Cæsar. Mr. Stoddart has undertaken the rôle of bard to all these types of humanity. And his "Seven Sagas" is a scientifically accurate and pithy statement of the progress of mankind, from its dim dawn to its modern day. Will any one say this does not afford poetic scope? Mr. Stoddart's rhymes are easy, and thoroughly suited to his subject. He shows the power of a true poet in thoroughly identifying himself with his work, and accordingly we have a terse poetical history of the progress of Humanity, including "The Earliest Man," "The Cave Man," "The Neolithic Farmer," "The Early Man of America," "The Aryan Migration," "The Burning of the Crannog," and "The Last Sacrifice." We anticipate a general all-round consensus of opinion in favour of the high poetical (and we may also add scientific) merit of this attractively got up little book.

London Birds and London Insects, by T. Digby Piggott (London: Harrison & Sons, Pall Mall). This is a two shilling, parchment-clad brochure, artistically got up in every way, both as regards type and illustrations, and, in addition to the very pleasant, but somewhat irregularly arranged chapters, it gives a list of London birds, &c. The author tells us this little book has been strung together for holiday amusements. It is the holiday characteristic which gives it its chief charm.

Facts around us, by C. Lloyd Morgan, F.G.S. (London: Edward Stanford). A nice little book, giving simple readings in inorganic science, with description of experiments to corroborate them. It contains 40 short chapters, but they relate mostly to chemistry. The information conveyed is as clear as it is full and terse, and is altogether such as we should have expected from the author of "Water and its Teachings."

Mineralogy, by J. H. Collins, F.G.S. (London: W. Collins & Co.). This is one of the extraordinarily cheap volumes of Collins' "Advanced Series." The author's name is known as one of the best mineralogists of the day, and only lately he was hon. secretary to the Mineralogical Society of Great Britain. It is written for the use of practical working miners, quarrymen, and field geologists (and will be especially useful to the latter class), and at the same time it has been compiled with a view to being used as a text-book by those who select mineralogy as a subject for examination under the department of Science and Art. Perhaps its scope would be better indicated by calling it a "Dictionary of Minerals;" and its usefulness would at the same time be pointed out to geological readers, who want a cheap, handy, and accurate book of reference. To all such, we cordially recommend this little volume. It is abundantly illustrated.

The Student's Handbook to Physical Geology, by A. J. Jukes-Browne, F.G.S., &c. (London: George Bell & Sons). Such a manual as the present has long been needed, for there was a great scarcity of good and trustworthy books intermediate between merely elementary treatises and the advanced but expensive works fit chiefly for specialists. Of Mr. Jukes-Browne's ability to undertake this task there can be no doubt. His elementary work on geology, written many years ago, has well held its place among a throng of competitors; and we feel that this more elaborate handbook will gain a position at once in geological literature. The book is divided into three parts; the first part deals with "Dynamical Geology" (including volcanic, meteorological, &c., action, and the mode in which various deposits are formed); the second part takes up Lithology and Petrology (a most valuable and ably worked out helper to the student); and the third part deals with Physiographical Geology, such as the origin of mountain chains and continents, "Earth-Sculpture," &c.

MR. MUIRHEAD, of Port Kil, Kilcreggan, recently sent to Sir William Thomson some meteoric dust he noticed to be discolouring the snow on the 1st of March. It was noticed over an area of 810 square miles, and Mr. Muirhead found the percentage to be about four grains to the square foot. This would give a shower of meteoric dust over the above area weighing no less than 5,760 tons.

AN INTERESTING BIT OF LINCOLNSHIRE GEOLOGY.

THE county of Lincoln has, until very recent years, lain in a sadly neglected condition, as far as its geology has been concerned. It is true that Mr. Judd paid considerable attention to it about twenty years ago; but his observations constituted almost the only valuable information extant, until the shire was taken in hand by the staff of the geological survey, and even now there is no authorised map of the county, geologically coloured, to be had. It was about the last of the counties to receive attention from the survey, and their results are not yet published. Mr. Jukes-Browne, has, however, communicated two or three papers to the Geological Society during the past twelve months, and it is to his observations that I am indebted for the explanation of the very curious phenomena about to be mentioned.

On the eastern side of Chalk Wolds of Lincolnshire, there are several very striking valleys cut so deeply down through the chalk, and with such precipitous sides, that Mr. Jukes-Browne has spoken of them as ravines. They are most romantic in appearance, usually clothed with growths of pine, and suggest an alpine character far more than the rounded scenery of chalk districts. In some cases the depth is 100 feet, and often the sides slope at an angle of $35°$-$38°$, and occasionally are steeper than that. In nearly every case a brook flows along the bottom of the valley, and a most remarkable feature about the matter is that, standing near the entrance to any one of these valleys, the spectator is puzzled to see why the brooks have cut through the chalk at all. One of these places is in the neighbourhood of Louth, and is called Hubbard's Hills. It is the prettiest of several pretty walks outside the town, and as less than three miles walking takes one through it and home again, it is a very popular resort. I spent my boyhood close to it, and many times I have sat upon the top of a grassy slope giving a view down the gorge, and have often wondered why the babbling brook at my feet cut through the high hill of chalk, when it might so easily have kept its course along low ground which is continuous from its own bed at the point where it enters the hill, to the course of another stream which it might have joined. Here at my feet was the brook; there on my right was a continuation of its broad valley; and there on my left was the high chalk hill through which the stream had made its way. It is true that if it had followed the broad valley, it would have had some twenty feet of clay to excavate; but it has actually removed 100 feet in depth of chalk. How did this come about? Well, this is Mr. Jukes-Browne's explanation: Along the eastern flank of the Lincolnshire Wolds lies a band of boulder-clay—an extension southward into the county of the Purple and Hessle Boulder Clays. This band is about three miles wide, and flanks the wolds between their base and the marsh lands farther east. Wherever the wolds have been intersected by rivers, the boulder-clay pushes itself far up their valleys, indicating at the present day the positions of the ancient river channels. The clays, sands, and gravels of which this drift is composed seem to have been left where they are by coast ice—at least that is the supposition—and to belong to the very latest portion of the glacial period. Originally they were so massive as to cover even the higher hills of chalk on that side of the ridge, but they appear to have so "draped" the undulations of chalk that most of the subsequent streams followed the old depressions. In some places, however, the old river

Fig. 58.

channels seem to have been choked by the glacial detritus in such a way that the brooks found an easier exit in new directions. Here, for instance, at the southern end of Hubbard's Hills, the boulder-clay occupies the whole of the broad valley along which the river Lud flows for some distance. It is continuous down the valley beyond the point where the latter is forsaken by the river, and with the channel of another stream which occupies the valley lower down. The bed of this lower stream was no doubt once also the bed of the Lud, which now runs at the bottom of Hubbard's Hills. Between the two streams the old valley is filled up with boulder-clay, and the Louth and Lincoln Railway at that point roughly represents the course of the stream when the waters of Tathwell and Withcall (see maps of county) ran down where Bishop's Bridge is now. Of course very much of the boulder-clay has been removed since it was left by the retreating ice; but when the present streams flowed over the original summit of it, the

Lud seems to have found it easier to flow northward over what is now Hubbard's Hills, than to keep to the direction of the older valley, and it has since cut its way deep down into the hill as already shown. After passing through the hills, it joins another stream which has behaved in a precisely identical manner. This other stream, rising at Welton village, pursues an ancient valley in the old chalk, now occupied by the drift for about a mile, and then suddenly turns southward and runs through a ravine known as Welton Vale, a place of very romantic beauty. Where it turns into the vale it leaves what was, clearly enough, its ancient bed, for the old depression is very palpable round by the front of Elkington Hall, and down to a lower portion of the present stream near to where it joins the brook from Hubbard's Hills. That old depression is still filled with boulder-clay, and along the greater part of its distance is occupied by another and smaller stream. Near Swaby is another example of these post-glacial ravines in the chalk, and at Hatcliffe there are four of them, one of which has since been deserted by the brook which excavated it.

W. MAWER.

THE OUTER SKELETON OF THE COCKROACH.*

By Professor L. C. MIALL and ALFRED DENNY.

WHEN the skin of an insect is boiled successively in acids, alkalies, alcohol, and ether, an insoluble residue known as Chitin ($C_9 H_{15} NO_6$) is obtained. It may be recognized and sufficiently separated by its resistance to boiling liquor potassæ. Chitin forms less than one-half by weight of the integument, but it is so coherent and uniformly distributed, that when isolated by chemical reagents, and even when cautiously calcined, it retains its original organized form. The colour which it frequently exhibits is not due to any essential ingredient; it may be diminished or even destroyed by various bleaching processes. The colouring-matter of the chitin of the cockroach, which is amber-yellow in thin sheets and blackish-brown in dense masses, is particularly stable and difficult of removal. Its composition does not appear to have been ascertained; it is white when first secreted, but darkens on exposure to air. Fresh-moulted cockroaches are white, but gradually darken in three or four hours. Lowne[†] observes that in the blow-fly the pigment is "first to be met with in the fat-bodies of the larvæ. These are perfectly white, but when cut from the larva, and exposed to the air, they rapidly assume an inky blackness. . . . When the perfect insect emerges from the pupa, and respiration again commences, the integument is nearly white, or a faint ashy colour prevails. This soon gives place to the characteristic blue or violet tint, first immediately around those portions most largely supplied with air vessels." Professor Moseley* tells us that, thinking it just within the limits of possibility that the brown coloration of the cockroach might be due to the presence of silver, he analysed one pound weight of Blatta. He found no silver, but plenty of iron, and a remarkable quantity of manganese. That light has some action upon the colouring matter seems to be indicated by the fact that in a newly-moulted cockroach the dorsal surface darkens first.

The chitinous exoskeleton is rather an exudation than a true tissue. It is not made up of cells, but of many superposed laminæ, secreted by an underlying epithelium, or "chitinogenous layer." This consists of a single layer of rounded, nucleated cells, resting upon a basement membrane. A cross-section of the chitinous layer, or "cuticle," examined with a high power, shows extremely close and fine lines perpendicular to the laminæ. Here and there an unusually long, flask-shaped, epithelial cell projects through the cuticle, and forms for itself an elongate chitinous sheath, commonly articulated at the base; such hollow sheaths form the hairs or setæ of insects—structures quite different histologically from the hairs of vertebrates.

Like other Arthropoda, insects shed their chitinous cuticle from time to time. A new cuticle, at first soft and colourless, is previously secreted, and from it the old one gradually becomes detached. The setæ probably serve the same purpose as the "casting-hairs" described by Braun in the cray-fish, and by Cartier in certain reptiles,[†] that is, they mechanically loosen the old skin by pushing beneath it. The integument about to be shed splits along the back of the cockroach, from the head to the end of the thorax, and the animal draws its limbs out of their discarded sheaths with much effort. Nearly at the same time the chitinous lining, which occupies a great part of the alimentary canal, and of the tracheal tubes, is cast.

The head of the cockroach, as seen from the front, is pear-shaped, having a semi-circular outline above, and narrowing downwards. A side-view shows that the front and back are flattish, while the top and sides are regularly rounded. In the living animal the face is usually almost vertical, but it can be tilted till the lower end projects considerably forward. The mouth, surrounded by gnathites or foot-jaws, opens below. On the hinder surface is the occipital foramen, by which the head communicates with the

---

* For "The Natural History of the Cockroaches," see this Journal, March, 1884.
† "Anatomy of the Blow-fly," p. 11.

* "Q. J. Micr. Sci.," 1871, p. 394.
† A condensed and popular account of these researches will be found in Semper's "Animal Life," p. 20.

thorax. A rather long neck allows the head to be retracted beneath the pronotum (first dorsal shield of the thorax) or protruded beneath it.

On the front of the head we observe the clypeus, which occupies a large central tract, extending almost completely across the widest part of the face. It is divided above by a sharply bent suture from the two epicranial plates, which form the top of the head as well as a great part of its back and sides. The labrum hangs like a flap from its lower edge. A little above the articulation of the labrum the width of the clypeus is suddenly reduced, as if a squarish piece had been cut out of each lower corner. In the re-entrant angle so formed the ginglymus, or anterior articulation of the mandible, is situated.

The labrum is narrower than the clypeus, and of squarish shape, the lower angles being rounded. It hangs downwards, with a slight inclination backwards towards the mouth, whose front wall it forms. On each side, about halfway between the lateral margin and the middle line, the posterior surface of the labrum is strengthened by a vertical chitinous slip set with large setæ. Each of these plates passes above into a ring, from the upper and outer part of which a short lever passes upwards, and gives attachment to a muscle (*levator menti*).

The top and back of the head are defended by the two epicranial plates, which meet along the middle line, but diverge widely as they descend upon the posterior surface, thus enclosing a large opening, the occipital foramen. Beyond the foramen, they pass still further downwards, their inner edges receding in a sharp curve from the vertical line, and end below in cavities for the articulation of the mandibular condyles.*

The sides of the head are completed by the eyes and the genæ. The large compound eye is bounded above by the epicranium; in front by a narrow band which connects the epicranium with the clypeus; behind, by the gena. The gena passes downwards between the eye and the epicranial plate, then curves forwards beneath the eye, and just appears upon the front of the face, being loosely connected at this point with the clypeus. Its lower edge overlaps the base of the mandible, and encloses the extensor mandibulæ.

The occipital foramen has the form of a heraldic shield. Its lateral margin is strengthened by a rim continuous with the tentorium, or internal skeleton of the head. Below, the foramen is completed by the upper edge of the tentorial plate, which nearly coincides with the upper edge of the submentum (basal piece of the second pair of maxillæ); a cleft, however, divides the two, through which nerve-commissures pass from the sub-œsophageal to the first thoracic ganglion. Through the occipital foramen pass the œsophagus, the aorta, and the tracheal tubes, for the supply of air to the head.

The internal skeleton of the head consists of a nearly transparent chitinous septum, named *tentorium* by Burmeister, which extends downwards and forwards from the lower border of the occipital foramen. In front it gives off two long crura, or props, which pass to the ginglymus, and are reflected thence upon the inner surface of the clypeus, ascending as high as the antennary socket, round which they form a kind of rim. Each crus is twisted, so that the front surface becomes first internal and then posterior, as it passes towards the clypeus. The form of the tentorium is in other respects readily understood from the figure (fig. 59). Its lower surface is strengthened by a median keel which gives attachment to muscles. The œsophagus passes upwards between its anterior crura, the long flexor of the mandible lies on each side of the central plate; the supra-œsophageal ganglion rests on the plate above, and the sub-œsophageal ganglion lies below it, the nerve-cords which unite the two, passing through the circular aperture. A similar internal chitinous skeleton occurs in the heads of other Orthoptera, as well as in Neuroptera, and Lepidoptera. Palmén ("Morphologie des Tracheensystems," p. 103) thinks that it represents a pair of stigmata or spiracles, which have thus become modified for muscular attachment, their respiratory function being wholly lost. In Ephemera he finds that the tentorium breaks across the middle when the skin is changed, and each half is drawn out from the head, like the chitinous lining of a tracheal tube.

A pair of antennæ spring from the front of the head. In the male of the common cockroach they are a little longer than the body; in the female rather shorter. From seventy-five to ninety joints are usually found, and the three basal joints are larger than the rest. Up to about the thirtieth, the joints are about twice as wide as long; from this point they become more elongate. The joints are connected by flexible membranes, and provided with stiff, forward-directed bristles. The ordinary position of the antennæ is forwards and outwards.

Three pairs of appendages are specially modified as gnathites, or jaws. These are, in order from before backwards, the mandibles, the maxillæ, and the labium, or second pair of maxillæ.

The mandibles are powerful, single-jointed jaws, each of which is articulated by a convex "condyle" to the lower end of the epicranial plate, and again by a concave "ginglymus" to the clypeus. The opposable inner edges are armed with strong tooth-like processes of dense chitin, which interlock when the mandibles close; those towards the tip of the mandible are sharp, while others are blunt, as if for crushing. Each mandible can be moved through an angle of about 30°. The powerful flexor of the

---

* One of the few points in which we have to differ from the admirable description of the cockroach given in Huxley's "Comparative Anatomy of Invertebrated Animals" relates to the articulation of the mandible, which is there said to be carried by the gena.

mandible arises within the epicranial vault; its fibres converge to a chitinous tendon, which passes outside the central plate of the tentorium, and at a lower level through a fold on the lower border of the gena, and is inserted close to the outer side of the condyle of the mandible.

The anterior maxillæ lie behind the mandibles, and like them are unconnected with each other.

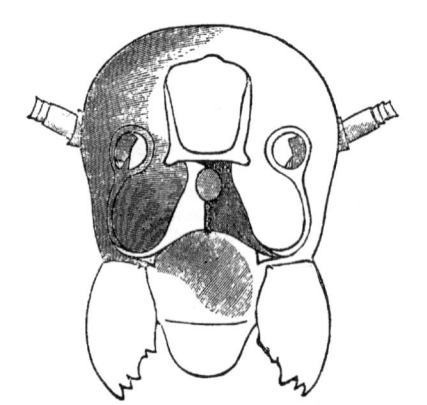

Fig. 59.—Fore-half of head, with tentorium, seen from behind. × 12.

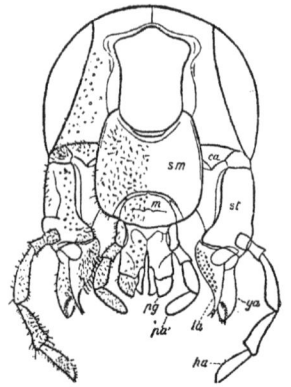

Fig. 62.—Head of *P. orientalis*, back view. *ca*, cardo; *st*, stipes; *ga*, galea; *la*, lacinia; *pa*, palp; *sm*, submentum; *m*, mentum; *pg*, paraglossa. × 10.

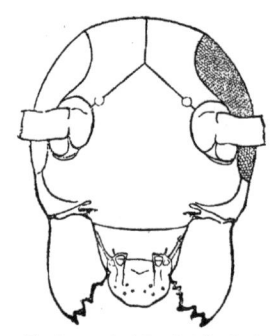

Fig. 60.—Head of *P. orientalis*; front view. × 10.

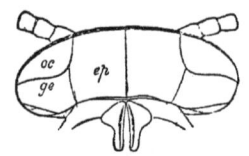

Fig. 61.—Head of *P. orientalis*; top view. *ep*, epicranial plate; *oc*, eye; *ge*, gena. × 10.

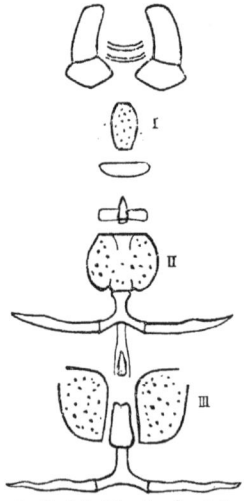

Fig. 63.—Ventral plates of neck and thorax of male cockroach. I., prosternum; II., mesosternum; III., metasternum. × 6.

clypeus, being finally inserted near the ginglymus. A short flexor arises from the crus of the tentorium. The extensor muscle arises from the side of the head, passes through the fold formed by the lower end of They retain much more of the primitive structure of a gnathite than the mandibles, in which parts quite distinct in the maxillæ are condensed or suppressed. The constituent pieces are seen in fig. 62. Of these, the five-jointed palp forms a flexible limb used in the prehension of food and in exploration; the lacinia carries a double-pointed tooth of dense chitin, while the galea is soft, clothed with fine setæ, and probably sensory in function.

On the hinder surface of the head, below the occipital foramen, is the labium, which represents the second pair of maxillæ, fused together in their basal half, but retaining elsewhere sufficient resemblance to the less-modified anterior pair to permit of the identification of its component parts. The palp is three-jointed, and carried upon a separate basal joint, or palpiger, and the lacinia and galea (known as paraglossæ) are readily made out.

We cannot here discuss the difficult question as to the theoretical composition of the head. It bears three pairs of appendages which unquestionably denote as many typical segments; the antennæ, though not developed precisely in the same way or from similar parts, may be taken to represent a peculiar fourth segment, but there is no reason to suppose that other segments are represented in the head of any insect.

The neck is a narrow cylindrical tube, with a flexible wall, strengthened by eight plates, viz., two dorsal (slightly united together), two lateral on each side, and two ventral. The peculiar shape of the lateral and ventral plates may be seen in fig. 63.

The elements of the thoracic exoskeleton are simpler in the cockroach than in insects of powerful flight, where adaptive changes greatly obscure the primitive arrangement. There are three segments, each defended by a dorsal plate (*tergum*) and a ventral plate (*sternum*). The sterna are often divided into lateral halves. Of the three terga the first (*pronotum*) is the largest; it has a wide free edge on each side, projects forwards over the neck, and when the head is retracted, covers this also, its semi-circular fore-edge then forming the apparent head-end of the animal. All the terga are dense and opaque in the female; in the male the middle one (*mesonotum*) and the hindmost (*metanotum*) are thin and semi-transparent, being ordinarily overlaid by the wing-cases. While the thoracic terga diminish backwards, the sterna increase in extent and firmness, proportionally to the size of the attached legs. The prosternum is small and coffin-shaped; the mesosternum partly divided into lateral halves in the male, and completely so in the female. The metasternum is completely divided in both sexes, while a median piece intervenes between its lateral halves in the male. Behind the sterna, especially in the case of the second and

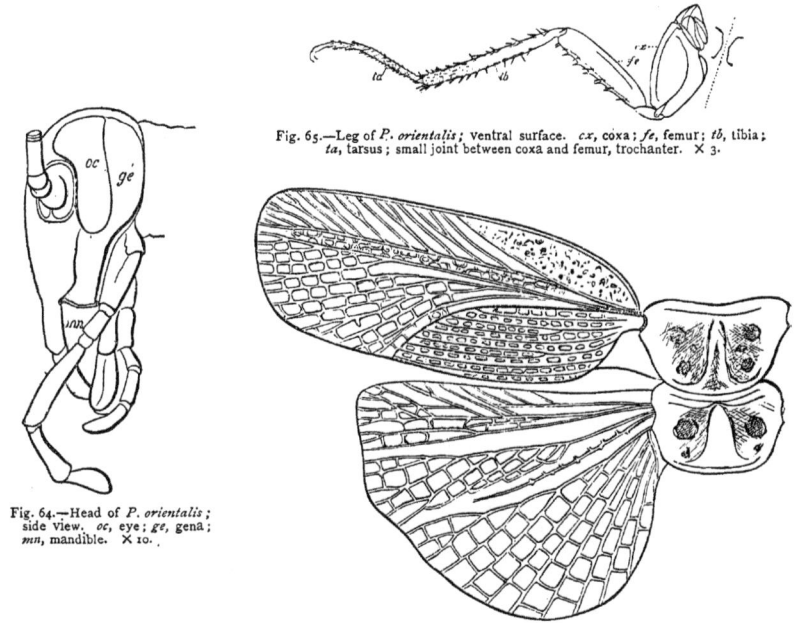

Fig. 65.—Leg of *P. orientalis*; ventral surface. *cx*, coxa; *fe*, femur; *tb*, tibia; *ta*, tarsus; small joint between coxa and femur, trochanter. × 3.

Fig. 64.—Head of *P. orientalis*; side view. *oc*, eye; *ge*, gena; *mn*, mandible. × 10.

Fig. 66.—Wings and wing-cases of male Cockroach. × 4.

third, the flexible under-surface of the thorax is inclined, so as to form a nearly vertical step. In the two hinder of these steps a chitinous prop is fixed; each is Y-shaped, with long, curved arms for muscular attachment, and a central notch, which supports the nerve-cord. A third piece of similar nature (the antefurca), which is well developed in some insects, *e.g.*, in ants, is apparently wanting in the cockroach.

Three pairs of legs are attached to the thoracic segments; they regularly increase in size from the first to the third, but hardly differ except in size; the peculiar modifications which affect the fore pair in predatory and burrowing Orthoptera (*Mantis, Gryllotalpa*), and the third pair in leaping Orthoptera (*Acridium*), being absent in the cursorial Blattina. Each leg is divided into the five segments usual in insects (see fig. 65), and there are also basal pieces, which apparently represent a sixth joint.

The legs of the cockroach, when bearing the weight of the body, are bent variously to such an extent that the abdomen rests lightly on the ground. In locomotion three of the six legs move nearly simultaneously; these are a fore and hind limb of one side, and the opposite middle leg. The legs differ in their action; the fore-leg may be compared to a grappling-iron; it is extended, seizes the ground with its claws, and is then flexed so as to draw the body forwards; the hind leg is used for shoving; and the middle leg serves to support the body while the fore and hind limb of the opposite side are pulling and pushing.

Cockroaches of both sexes are provided with wings, which, however, are only functional in the male. The wing-cases (or anterior pair of wings) of the male are carried by the second thoracic segment. As in most Orthoptera genuina, they are denser than the hind wings, and protect them when at rest. They reach to the fifth segment of the abdomen, and the left wing-case overlaps the right. Branching veins or nervures form a characteristic pattern upon the surface (see fig. 66), and it is largely by means of this pattern that many of the fossil species are identified and distinguished. The true or posterior wings are attached to the metathorax. They are membranous and flexible, but the fore-edge is stiffened, like that of the wing-cases, by additional chitinous deposit. When extended, each wing forms an irregular quadrant of a circle; when at rest, the radiating furrows of the hinder part close up fan-wise, and the inner half is folded beneath the outer. The wing reaches back as far as the hinder end of the fourth abdominal segment. The wing-cases of the female are small, and though movable, seem never to be voluntarily extended; each covers about one-third of the width of the mesonotum, and extends backwards to the middle of the metanotum. A reticulated pattern on the outer fourth of the metanotum plainly represents the hind wing; it is probably rather a

degeneration or survival than an anticipation of an organ tending towards useful completeness.

The rudimentary wing of the female cockroach illustrates the homology of the wings of insects with the free edges of thoracic terga, and the same conclusion is enforced by the study of the development of the more complete wings and wing-cases of the male. The hinder edges of the terga become produced at the later moults preceding the completely winged stage, and may even assume something of the shape and pattern of true wings; it is not, however, true, though more than once stated, that winged pupæ are common. Adults with imperfectly developed wings have been mistaken for such, though the sexual characters of the hinder part of the abdomen, which are only present in the last stage of growth, distinguish such forms from true pupæ.

The structure of the wing testifies to its origin as a fold of the chitinous integument. It is a double lamina, which often encloses a visible space at its base. The nervures, with their vessels and tracheal tubes, lie between the two layers, which, except at the base, are in close contact. Oken termed the wings of an insect "aerial gills," and this rather fanciful designation is in some degree justified by their resemblance to the tracheal gills of such aquatic larvæ as are found in Ephemeridæ, Perlidæ, Phryanidæ, &c. In the larva of *Chloeon* (*Ephemera*) *dimidiatum*, for example, the second thoracic segment carries a pair of large expansions, which ultimately are replaced by organs of aerial flight. The third thoracic segment is provided with small appendages, resembling wings in position and texture, but of no importance for locomotion; to these succeed similarly placed respiratory leaflets, the tracheal gills of the abdominal segments, which by their vigorous flapping movements bring a rush of water against their membranous and tracheated surfaces. It would be going too far to say that the primitive insect-wing was specially adapted to respiration or locomotion in water, but the aerial wing may certainly closely resemble, both in structure and development, organs converted to these uses.*

The wings of the cockroach have little functional activity. The male will sometimes fly to fresh quarters, as from one house to another. At times of sexual excitement it raises the wings, but does not rise in the air. Even in extremity it will not fly to escape from danger. Readers will recollect the Amblyrhynchus, a maritime lizard of the Galapagos, described by Darwin.† Though an excellent swimmer and diver, it will not enter the water when frightened. Darwin suggests that the reptile has no enemy on

---

* Wing-like expansions, serviceable either for locomotion, aquatic respiration, or aerial respiration, can apparently be developed or suppressed as required, not only in insects, but in many other animals. It appears from Palmén's researches that all tracheal gills are not homodynamous or equivalent. Some are dorsal in position, others ventral; some are folds of integument, others modified appendages.
† "Naturalist's Voyage," chap. xvii.

shore, whereas at sea it must often fall a prey to the numerous sharks. Hence, a settled instinct drives it to take refuge on shore in any emergency. It may be that long experience has established in the male cockroach a habit no longer reasonable, of invariably seeking its hole when menaced.

In the abdomen of the female cockroach eight terga (1–7 ; 10) are externally visible. Two more (8, 9) are readily displayed by extending the abdomen ;* they are ordinarily concealed beneath the seventh tergum. The tenth tergum is notched in the middle of its posterior margin. A pair of sixteen-jointed setose *cerci* project laterally from beneath it. Seven abdominal sterna (1–7) are externally visible. The first is quite rudimentary, and consists of a transversely oval plate ; the second is irregular and imperfectly chitinised in front ; the seventh is large, and its hinder part, which is boat-shaped, is divided into lateral halves, for facilitating the discharge of the large egg-capsule. The eighth and ninth sterna have become accessory to the female organs of reproduction. A pair of triangular " podical plates," which lie on either side of the anus, and towards the dorsal surface, may represent, as Professor Huxley conjectures, the tergum of an eleventh segment, or perhaps with equal probability, the tenth sternum.

In the male cockroach ten abdominal terga are visible without dissection, (fig. 36, p. 60), though the eighth and ninth are greatly overlapped by the seventh. The tenth tergum is hardly notched. Nine abdominal sterna are readily made out, the first being rudimentary, as in the female. The eighth is narrower than the seventh, the ninth still narrower, and largely concealed by the eighth ; its covered anterior part is thin and transparent, the exposed part denser. This forms the extreme end of the body, except that the small subanal styles project beyond it. The podical plates resemble those of the female.

The differences between the male and female abdomen may be tabulated thus :—

Female.	Male.
Abdomen broader.	Abdomen narrower.
Terga 8, 9 not externally visible.	Terga 8, 9 externally visible.
The 10th tergum notched.	The 10th tergum hardly notched.
The 7th sternum divided behind.	The 7th sternum undivided.
The external outlet of the rectum and vulva between the 10th tergum and the 7th sternum.	The outlet between the 10th tergum and the 9th sternum.
No subanal styles.	Subanal styles.

[ERRATUM.—Fig. 40, p. 61, has been inaccurately copied, and should be cancelled.]

* These are partly visible in fig. 37, p. 60.

SCIENCE-GOSSIP.

AT the annual general meeting of the Hackney Microscopical and Natural History Society, held on the 19th of March, at the Morley Hall, Hackney, a valuable microscope was presented to the hon. secretary by the members. The president, Dr. M. C. Cooke, in presenting the testimonial, made some highly eulogistic remarks upon the energy and unremitting attention given by the hon. secretary during the seven years of the existence of the society, to which he ascribed its present flourishing condition. A silver plate bearing the following inscription was, attached to the instrument : "Presented to Collis Willmott, Esq., by members of the Hackney Microscopical and Natural History Society, in appreciation of his services as hon. secretary. 19th March, 1884."

THE readers of SCIENCE-GOSSIP will sympathise in the loss of our old contributor, Mr. H. W. Kidd, of Godalming, carried off by scarlet fever on the 23rd of March. Born a paralytic, he mastered his infirmity so as to become a good scholar and most indefatigable collector of the plants, rocks, and shells, that lend their beauty to the seclusion of his native town. When I became acquainted with him in 1876 he had found fame as a student of gall makers, and we have worked together at local entomology, geology, and botany. Latterly we were corresponding on antiquarian subjects, and the last of his letters on the pilgrim marks in the rural churches, lies before me unanswered. The peat balls cast up on the coast also engaged his attention, and induced him last autumn to take a journey to Littlehampton.—*A. H. Swinton.*

MR. W. J. KNOWLES exhibited a chipped flint implement at the last meeting of the Anthropological Institute, which he had found in undisturbed boulder clay in Ireland. He thought this carried the age of man into the glacial period.

WE are glad to see that our old friend and contributor, Mr. W. Saville-Kent, F.L.S., author of "A Manual of the Infusoria," &c., has been appointed Inspector of Fisheries to the Governmeat of Tasmania. No better man for the post could have been found in the British Islands.

A VERY interesting communication was lately made before the Paris Academy of Sciences, on Observations made at the Observatory of Nice, by M. Perrotin, of Saturn and Uranus. The outer ring of Saturn appears to consist of three distinct rings, slightly diminishing in breadth outwardly, and each apparently made up of numerous subdivisions. Uranus, as seen on the 18th of March, presented in some respects the general aspect of Mars, with dark spots towards the centre, and a white speck like the pole of that planet at the angle of position 380° on the edge of the disk.

It was stated at a recent meeting of the Linnean Society of South Wales, that the total number of species of Australian fishes described up to the present time is 1291.

M. REGNARD has been making some very interesting experiments on the influence of high pressure on living organisms. They are particularly striking, as bearing on the pressure which the sea-water must exercise at great depths of the ocean. It was found that soluble ferments were unaffected by extreme pressure. Starch, at 1000 atmospheres, was converted into sugar; algæ at 600 ditto were decomposed and the carbonic acid liberated; infusoria, leeches, and mollusca were rendered insensible at a pressure of 600 atmospheres, but recovered when the pressure was removed; fishes possessed of swimming bladders resisted 100 atmospheres, but became insensible at 200, and died at 300 ditto.

NUMBERS of our readers will be grieved to hear of the death of Mr. James Fullagar, of Canterbury, hon. sec. of the East Kent Natural History Society. Our Journal has been enriched for many years by contributions both from his pen and his pencil, all of them the result of original investigation and observation. Few men were ever more respected by all who knew him (and this included many men of scientific distinction), and, although he died at the ripe age of 77, he will be much missed and regretted for a long time to come.

ANOTHER of our scientific celebrities has "joined the majority"—Professor Allen Thomson, the distinguished naturalist. He was one of the brilliant circle which included the names of Edward Forbes, Goodsir, Carpenter, &c. In 1879 he was President of the British Association Meeting at Sheffield, and his inaugural address bore on Protoplasm, and will not soon be forgotten by those who heard it.

OUR short obituary notices are far too numerous this month. We have the sad duty to chronicle the death of a man well known in Manchester and Lancashire scientific circles—Mr. James Parker, formerly assistant curator at the Museum of the Manchester Natural History Society. A kinder, gentler, or more modest man never lived. Fossils (especially those of the "Mountain Limestone," as he always loved to call that formation) were more to him than meat or drink, gold or raiment. We received our first lessons in Fossils at his hands, twenty-five years ago; and therefore his memory will always have a sweet savour. Mr. Parker discovered many species new to science. He was well known to Lyell, Murchison, John Phillips, Edward Forbes, and others of the older school. His knowledge of Carboniferous crinoids and brachiopods was remarkable. Part of his collection, we believe, went to Dublin more than twenty years ago. He died as he had lived, peacefully, and without a pang.

ON Friday April 6, Mr. J. J. H. Teall, F.G.S., read a paper before the Geologists' Association on "Some Modern Petrological Methods." On Easter Monday and Tuesday (April 14th and 15th) the members made their annual excursion to Lincoln, under the directorship of Mr. W. H. Dalton, F.G.S., Mr. A. Strahan, F.G.S., and Mr. W. D. Carr, and thoroughly explored all the known sections of the Lias, Oolitic, and Neocomian strata in the district.

SIR RICHARD OWEN has just described a new labyrinthodont amphibian, by the name of *Rhytidosteus Capensis*. The remarkable feature about these South African reptilia is that they exhibit mammalian characters.

EVERY Naturalist will of course sympathise with the motives of Professor Bryce's Bill in Parliament for securing "access to mountain and moorlands in Scotland," to natural history pursuits; but—is it necessary? We contend that these places are accessible, and everyone acquainted with the Law of Trespass knows that neither landowner nor gamekeeper can hinder a man from wandering over them. All the former could do would be to sue him for the "damage done," but it would be difficult to appraise the "damage done" in collecting alpine flowers or insects, or specimens of quartz and mica-schist. There is no law in our litigated land so little understood as that of "Trespass," and none so utterly harmless. Hence, Professor Bryce's Bill seems to us like erecting a steam-hammer to crack nuts. The gamekeeper cannot even force a man off the premises without being liable to an action for assault.

EVERYBODY knows the story of Charles the Second and the Royal Society about the gold-fish, and why the latter didn't make the vase of water run over when they were introduced into it. The savants suggested a score of reasons, but—none of them tried the experiment! All took it for granted. It seems to us some of our scientists are doing the same thing with regard to the newspaper report that an isolated mountain in Algeria, 800 feet high, is sinking, and a cavity is forming round its base. Has any trustworthy geologist authenticated this statement? And yet we see strong inferences being drawn about over-weighting parts (such as London) of the earth's crust. The earth's crust is laden with hundreds of hills and mountains, much higher than the unfortunate hill in Algeria, but they don't sink, or form depressions around their bases. Indeed, our hills and mountains are getting lighter every year by reason of denudation. So that their sinking ought to be less probable as time rolls on.

A VERY succinct and suggestive pamphlet has been written by Mr. F. P. Pascoe, F.L.S., called, "Notes on Natural Selection and the Origin of Species." It is a very handy little guide to the Darwinian theory.

A Russian chemist has succeeded in solidifying alchohol at a temperature of 130° Centigrade. It was transformed into a solid white body.

Professor Landolt exhibited a solid cylinder of carbonic acid at the Berlin Physical Society, which he had made an hour before. The flakes of solid carbonic acid were hammered in a cylindrical vessel into a solid cylinder. The solid gas could be touched by the hand. It resembled common chalk.

Mr. C. B. Plowright expresses his opinion that "canker" on apple-trees is due to a fungus, probably to a species of Nectria.

Malgutti's "Elementary Chemistry," Fresenius's "Chemical Analysis," have just been translated into Chinese, for the use of Chinese students at the Imperial Colleges. Western ideas are extending. The first minister has himself written a preface to the first mentioned work.

The French viticulturists are said to be importing Californian vines to replace those destroyed by phylloxera.

Mr. Harvie-Brown has shown that both common trout and brown trout were transformed into *Salmo ferox* when transferred to a barren loch, where there was an abundance of food.

The Australian representatives assembled at Sydney have been discussing various schemes for getting rid of the rabbit pest. These rodents have been spreading with inconceivable rapidity, at the rate of two and three hundred miles in three or four years. They consume so much of the herbage as to have educed the wool in one "run" from 800 to 300 bales.

Mr. N. Trübner, the well-known publisher and introducer of American scientific works into Great Britain, has just died at the age of sixty-seven.

No. II. of "The Acadian Scientist," has appeared. It is addressed chiefly to Canadian naturalists, but has some good practical papers on subjects of a general character.

Captain Lupton has just delivered an excellent lecture before the Stratford-on-Avon Literary Society on "The Natural History of the District," which was reported at full length in the "Stratford Chronicle" of March 14th.

In the American journal, "Psyche," Dr. C. V. Riley has written a very graceful "Tribute to the Memory of John Lawrence Leconte."

No. IV. of the "Rochester Naturalist" has appeared. This is the quarterly record of the Rochester Naturalists' Club, and contains two capital papers on the "Chalk in the Medway Valley," by A. W. Hood, and on the "Destruction of Birds, and Consequent Increase of Insects." There are a good many valuable notes besides.

Under the title of "Notes from my Aquarium," I. to VI., Mr. George Brook, F.L.S., has published an attractive pamphlet relating how he successfully managed his salt-water aquarium, and giving a well-written account of the various animals and their parasites which lived in the tanks. There are also some valuable observations on the development from the eggs of various podura.

No. I. of a new American serial has appeared under the title of "The Microscopical Bulletin and Opticians' Circular."

Two swallows were seen at Kelso, in Roxburghshire, on the 27th of March.

Mr. F. Enock has issued a beautiful slide of the apterous female of the horned aphis (Ceratophis), as figured in Science-Gossip in October last.

We are glad to welcome back an old friend under a new name. The "Canadian Naturalist," discontinued last year, has re-appeared under the title of "The Canadian Record of Natural History and Geology," and is published by the Natural History Society of Montreal. The first part is out, and contains articles by Principal Dawson, J. T. Donald, and others.

The Postal Microscopical Society.—This well-known society now publishes "The Journal of Microscopy and Natural Science," edited by Mr. Alfred Allen. Part 10 of vol. iii. has appeared, containing the following papers : "On *Psychoptera paludosa*," by Mr. A. Hammond, F.L.S. (illustrated) ; "The Foraminifera of Galway," by F. P. Balkwill and F. W. Millett ; "The Palpi of Freshwater Mites as Aids to distinguishing Sub-Families," by C. F. George (illustrated) ; "Diamonds and their History," by James A. Forster ; "A Bit of Groundsel," by the Rev. H. W. Lett, M.A. ; "An Inexpensive Turntable," by E. J. E. Creese ; "Stylops," by Mr. V. R. Perkins. In addition to these original papers we have the chapter headed "Half-an-Hour at the Microscope with Mr. Tuffen West," reports of microscopical societies, selected notes from the Society's note-books, reviews, current notices, queries, &c., making up a thoroughly good and readable number to all interested in natural history pursuits.

Sir John Lubbock persists in teaching his dog to read ! He writes to say that when the dog is hungry it always brings him the card with "Food" printed upon it. When it wants to go out it brings the card marked "Out," and when it wants a bone it brings another card on which Sir John has had that word printed. "Water" is asked for in a similar way ; and the dog goes along a whole row of printed cards, and eventually selects one containing the word which expresses what it wants. So well has the dog advanced in its literary studies that its master now intends to teach it simple arithmetic !

THE Report of the South London Entomological Society, for 1883, has appeared. It contains the address of the President (Mr. W. West), and a record of the captures made during the year by the members.

AT a meeting of the College of Physicians recently held in Philadelphia, Professor Harrison Allen spoke of an interesting discovery, by which spoken words can be represented by a series of curved lines or receiving surface composed of white paper covered with soot. The experiments were suggested from watching the movements of the soft palate, when conducting experiments connected with the human throat. By means of this device, Professor Allen is enabled to register upon the surface of the sooted paper the lines and curves which represent the various phonetic sounds of the human voice. In diagnosing cases of disease of the palate, this Invention may prove of great value. Singularly enough, by its aid, Professor Allen has already discovered that many of the sounds which have long been considered by elocutionists and others to be formed by the direct action of the lips, the teeth, or the tongue, are in reality formed primarily by the direct action of the palate. The instrument has been named the Palate-Myograph.

AN important gathering of English scientists, chiefly biologists, has taken place at the rooms of the Royal Society, under the Presidency of Professor Huxley, when resolutions were passed to the effect that biological laboratories, similar to those existing in France, Austria, Italy and America, were urgently needed on the British coast, where accurate researches may be carried on leading to the improvement of zoological and botanical science, and to the increase of our knowledge regarding the food, life, habits, and condition of our British food-fishes, and mollusca especially, and the animal and vegetable resources of the sea in general. The meeting eventually resolved itself into a society, to be henceforth known as "The Society for the Biological Investigation of the Coasts of the United Kingdom."

MICROSCOPY.

CARBOLIC ACID AND CEMENT.—Will some reader of SCIENCE-GOSSIP oblige by saying what cement will not be dissolved by carbolic acid when forming the medium for mounting insects? Carbolic acid acts upon chitine in an extraordinary manner, showing a kind of imbrication on one side (of some insects, notably parasites), and reticulation on the other. The insect when treated with oil of cloves or aniseed, after the carbolic acid, loses its interesting appearance, and the imbricated character is almost destroyed. Again, if treated with glycerine, after carbolic acid, the change is greater still. Will some one help?—*W. H.*

THE BACILLUS OF CHOLERA.—Professor Koch, whose investigations into the nature and history of the lower fungi have already gained him a world-wide renown, has recently discovered the Bacillus of Cholera in tank-water, in one of the suburbs of Calcutta. Accompanied by two other savants he came to India, at the instance of the German Government, for the purpose of studying cholera, in what may be its original home. For some time after his arrival he made little progress. The bacillus detected by him in the lower intestines of cholera patients in Egypt was readily found in several bodies which he examined at one of the hospitals in Calcutta; but, beyond ascertaining that it did not exist in the intestines of persons who suffered from either diarrhœa or dysentery, he was unable to trace the cholera bacillus further. The disease, shortly after this, broke out in the neighbourhood of a tank at Balliaghatte, in the suburbs of Calcutta. Professor Koch and his colleagues submitted the water of the tank in question to microscopic examination, and found that it teemed with the bacillus of cholera; and further that as the bacilli in the water decreased in number, the disease decreased in the vicinity of the tank. Professor Koch has not been able to communicate the complaint by inoculation to cats, dogs, &c. I have never heard of dogs being attacked with cholera, but it has recently been announced that an epidemic resembling that disease carried off hundreds of cats in Guzerat. The question suggests itself naturally—is the disease communicable by inoculation? The German Commissioners will, however, return to India next winter, and continue their investigations, and extend them to the study of malarial fevers. In the meantime I have no doubt some of Professor Koch's stained and mounted preparations will find their way into the cabinets of European microscopists; and that we in Calcutta will first learn from drawings in English scientific journals what is the peculiar form and fashion of the deadly saprophyte which has been discovered in our midst.—*W. J. S., Calcutta.*

MOUNTING CHITINE.—Can any reader inform me by what process I can render thin sections of chitine transparent, and what is the best medium to mount them in?—*B. H.*

"PETROLOGICAL STUDIES."—The last two slides sent out from Messrs. Ady & Hensoldt's morphological laboratory fully sustain the high character which these petrological studies have already earned among geologists and mineralogists. The first is a specimen of Pitchstone, from the Isle of Arran, and is accompanied by a gracefully-written and very interesting essay, with a coloured plate showing the composition of the mineral as it appears when × 150, and a detailed explanation of each substance. The second slide is a beautifully-prepared section of Anamesite, from Craiglockhart, Scotland. The physiological preparations have been commenced, one showing a section of the common lichen (*Physcia stellaris*) through a mature apothecium, plainly reveals both brown spores and green gonidia. It is a very striking object, and of great value to the botanical student.

MR. COLE'S "STUDIES."—No. 15 of the "Studies in Microscopical Science" deals with Adipose tissue, its nature and method of preparation. The beautiful coloured illustration shows a portion of such tissue magnified 250. No. 7 of the "Popular Microscopical Studies" is a well-written and condensed essay on "The Common Bulrush," accompanied by an exquisitely-coloured plate showing the transverse section of the stem of this plant × 75, double stained. Both the above papers were accompanied by slides of the objects treated upon, mounted in Mr. Cole's characteristically neat and artistic manner.

ZOOLOGY.

HYALINA DRAPALNALDI, *Drap.*—In one of the back numbers of SCIENCE-GOSSIP a paper was written about a Zonites (?), found near Clifton. It was stated that beyond a question the shell is really *Hyalina Drapalnaldi*, Drap., of the European Continent. During the last twelve months I have been keeping some of these shells in captivity. During the summer I found it easy to feed them and they rapidly grew and increased in number, eating what my other family of snails rejected, or feeding voraciously on any dead snail in their house. The winter treatment at first perplexed me, as I found they would not hybernate like my Helicidæ, but remained wide awake and hungry also. I had read that all Zonites are fond of old meat bones. I tried, but my Hyalinas would have nothing to do with bones. One of my sisters then suggested carefully chopped raw meat—beef and mutton—which proved most successful. As soon as the meat is put into their house, a little crowd crawls forth from the moss, &c., and, in the course of a few hours, demolishes the supply. My Hyalinas are now fat and flourishing, and my colony promises to be greatly enlarged. I find them already laying large clusters of eggs.—*F. M. Hele, Bristol.*

SHRIMPS IN AQUARIA.—Shrimps and gobies, and such creatures as spend their days for the most part at the bottom of the water, require special arrangements to keep them successfully for any length of time. They succeed exceedingly well in a broad shallow tank, or even a large baking dish with the bottom covered about two inches deep with sea sand. In these I have kept them in good health for a year or more, whereas I have found they very soon perish if kept in a deep tank or vase. I had one shrimp in particular which lived fourteen months in a large baking dish half-filled with sand, and covered by scarcely an inch of water. In this it grew from little more than half an inch in length (its size when I first had it) to something over three inches. It was very tame and familiar, and would readily take tiny morsels of meat off a quill with its claws, and this with a very pretty deliberateness. It would then scoop itself out a hollow in the sand and settle down into it, and there it would remain motionless, but for an occasional sweeping round of its long antennæ, for hours together while it ate and digested its food. It and its brother and sister shrimps were always restless however at night, and a candle taken to the aquarium after dark would disclose them silently moving about. Some persons may however object to keeping shrimps and gobies in a tank so shallow that it can only be looked at from above. I usually have a dark chamber in my tank, recommended by Mr. Lloyd in his pamphlet on Aquaria and Aquaria Management. In an aquarium made on this principle I have had the water unchanged for years, and it is still as good as ever and full of living creatures. It is however fair to say that I have also a vase with no dark chamber containing anemones, which, with the water they are living in, I procured at Brighton in the summer of 1876. So it does not seem to be absolutely essential, and the arrangement shown above economises the water—a great object to those that live inland.—*Albert Waters, B.A., Cambridge.*

CONCHOLOGICAL NOTES.—A few days ago I visited Hammersmith, and found *Clausilia biplicata* fairly common on a bank near the Thames, living at the roots of *Urtica dioica*, *Lamium album*, and *Nepeta glechoma*. I found six or seven full-grown living specimens, and young shells appear to be common. In company with *C. biplicata* I found *Helix hortensis*, and its variety *lutea*, *Helix rufescens* and var. *alba* (the var. *alba* being about as common as the type), *Limax agrestis*, &c. On March 30 I visited Kew Gardens, and found *Planorbis carinatus*, *P. corneus*, *P. vortex*, *Limnæa palustris*, *L. peregra* (and a monstrosity having a wide umbilicus), *L. stagnalis*, and *Bythinia tentaculata* in one of the tanks. I also found *Zonites crystallinus*, *Carychium minimum*, *Achatina acicula*, *Helix pulchella*, *H. rufescens*, *Neritina fluviatilis*, &c., in the rejectamenta of the Thames. I have on two occasions (once at the Reculvers in Kent, and once at Hendon in Middlesex) found *Helix rotundata* living in an ants' nest. Although this species has no garlic odour, the ants do not appear to molest them. Mr. Hudson's remarks on *Sphærium corneum* are very interesting. There is a ditch at Bickley, in Kent, which contains *Pisidium pusillum*. This ditch is often dried up for weeks during the hotter months of the year, yet the little bivalves always appear in large numbers when the ditch begins to be filled with water. In my description of a variety of *Helix aspersa*, on page 91, I forgot to mention that in speaking of "the lower edge of the 3rd band," I meant the situation of the lower edge of the 3rd band of a specimen having five bands, as in *H. nemoralis*, as in my variety the 3rd band was obsolete, being taken up in the chocolate colour of the upper part of the whorl.—*T. D. A. Cockerell, 51 Woodstock Road, Bedford Park, Chiswick.*

RARE SPECIES OF TROCHUS.—I shall be glad if any of your readers can tell me where I could procure a specimen of *Trochus Duminyi*, figured and thus described in Jeffrey's "Marine Conchology:"— "Shell orbicular, rather solid, but semi-transparent and somewhat glossy, colour white—spire scarcely raised, but apex well defined—named in honour of Professor Duminy of Ajaccio, who discovered it at Bundoran, co. Donegal, Ireland, the only locality at which it has been observed. Could any conchologist assist me to find this shell, by sending me specimen for comparison?—*H. Allingham, Ballyshannon.*

BOTANY.

DOUBLE FLOWERS.—Being interested as to the origin of double flowers, I have read John Gibbs's articles on that subject with great pleasure, and quite agree with him as to self-fertilised flowers being sometimes the cause of doubles. In the spring, 1881, I selected one plant of sweet william (*Dianthus barbatus*), with very dark green foliage, from a number of seedlings, and planted it, throwing away all the rest, my object being to ascertain if the flowers from the offspring would be of the same colour as the parent. This plant was of robust habit and carried a fine broad head of dark red flowers; the seed being saved, were sown in 1882, the offspring flowered in 1883, all the flowers being a dark red colour of various shades, but I was gladly surprised to find one a very fine double. I think the experience of other observers would be of universal interest.—*W. Sim, Fyvie.*

THE BEHAVIOUR OF PLANTS.—As you are publishing a volume on the behaviour of plants, I am tempted to take the liberty of telling you about my arbutus. As it grows near my study window, and is, according to Miller, very similar as regards fertilisation to Erica, which he describes, the flowers this season got more of my attention than ever before. I need not tell you what Müller says, but this in addition. I noticed in sunny weather that many of the flowers were tenanted by small flies of various kinds, and the question arose, with what object does the flower entice and harbour these otherwise unwelcome guests? I could not at the time see any reason for such a course, but not long after, there happened to follow several days of such cold wet weather as necessarily led to all fertilising insects staying at home, and the query arose in my mind, how is fertilisation going on now? On examining the flowers, I found that every one of the tenants had flown, but an interesting sight presented itself: evidence that they had not been there in vain. They had shaken out a great deal of pollen, and this was retained for future use by the fine silvery hairs that abound in the interior. Still the question arose as to how the stigma was to avail itself of them, so as in the absence of insects to effect cross-fertilisation, to fertilise itself. But on closer inspection I found one style bending back near to one side of the flower, then another doing the same, and in addition curving forward so as to shorten the distance from base to apex, and, as a consequence, bring the stigma into contact with the captive pollen. On further examination, I found various degrees of curvature until some were bent even to a right angle, and all had obtained a plentiful supply of what they were apparently seeking after. I ought to have said that the little flies seemed to have an irksome time of it, for, with scarcely room to move, they had to push hard to get about, thanks to the broad hairy filaments and confined space; so much the better, however, for the flower, which, by this means, secured a good shaking for the anthers. But another surprise awaited me. At the beginning of January, when all my flowers had dropped off, I happened to visit our cemetery, where I found an arbutus still in flower. On examining some of these, I found that the style had saved itself the trouble (and evident waste of time and labour) of growing beyond the corolla, and then having to bend itself back again, by the very simple expedient of ceasing to grow in proper time. In breaking off one or two of the horn-like appendages of the anthers, I was delighted to find (on touching them with a needle) that they were dry, hard, and elastic, just proper for being shaken by the bees; but my delight was increased to wonder, when, under the microscope, they each exhibited hundreds of small projections, thus, by a roughened surface, still further securing the evident object of their existence.—*J. Wallis.*

PEZIZA SUMNERIA.—A few days ago I gathered at Roehampton some specimens of a fungus of the genus Peziza, which Dr. M. C. Cooke, to whom I showed one, identified as the *Peziza Sumneria*. This fungus is, I believe, of very rare occurrence in England the first specimen being found, I think, in 1870, in Warwick, and I have heard of its being seen then at Chiswick. I should be glad to know whether it has occurred in other English localities. Its form is that common to the genus Peziza, viz. cup-shaped; the interior being smooth and white, and the exterior of a dark colour. It is subterranean in habit, the fungus being completely buried under ground, except the rim of the cup, which is on a level with the surface of the ground; it thus gives it the appearance of a small hollow in the soil, of one or two inches diameter; this hollow sometimes gets filled up with soil or stones, so that it is almost impossible without close observation to detect the fungus at all. The specimens I found were growing near an elm-tree, among a great quantity of the *Arum maculatum*, and in close proximity to a yew, and once before when this fungus was found, it was growing under a cedar. Perhaps they have some liking for the neighbourhood of conifers. Some of the specimens I gathered, and which I left to dry, have shrivelled up to a great extent, so that it might be difficult to recognise them, but some which were left in a tin collecting case, have lost their cup-shape, and become almost flat, thus making it very easy to preserve them in a collection.—*J. H. Wright.*

THE SUNDEWS.—I do not think there is any difficulty in cultivating the Droseras in confinement or captivity, at any rate not such as your contributor, the writer of the curious article on "The Sundews" (one column of which alone refers to the subject, and that only to one species) seems to meet with. All three British species grow abundantly on Goole Moors, and I have repeatedly kept them for observation for some time without the least trouble or extra care. The other day I saw in the house of a friend very fine specimens growing in Bohemian dishes (placed in a window facing south-east), which had been taken from the moor when very young and kept for many weeks under daily observation; these were flowering in fine condition, and as perfectly healthy as any individual living on its native heath. If E. T. D. or any of your readers wish to study these curious plants, they may do so at their leisure by taking up the plant with a ball of peat attached and placed in a dish, afterwards keeping it constantly moist. *Drosera rotundifolia* may be small and local, but it is surely not "minute;" this term must be a misnomer when applied to species the size of any of the sundews.—*Thomas Birks, jun., Goole.*

ODD NAME OF PLANT.—Might I be allowed to suggest that persons interested in curious plant-names will find Messrs. Holland & Britten's "Dictionary of English Plant-Names," a perfect storehouse of out-of-the-way information? That work contains (pt. I., p. 81), the following entry: "CAIN AND ABEL. The tubers of *Orchis latifolia*, L., 'Cain being the heavy one.' *E. Bord.* Bot. E. Bord. See ADAM AND EVE." Turning to the place indicated we find a similar entry. Several plants bear names of like nature either on account of their having bi-coloured flowers or for some other striking peculiarity. A chapter on "Rustic Plant Names" will be found in "Flowers and Flower Lore" (Sonnenschein & Co. Paternoster Square).—*Hilderic Friend, F.L.S., Brackley, Northants.*

GEOLOGY, &c.

A FOSSIL ANTELOPE IN THE CRAG BEDS.—At the last meeting of the Geological Society, Mr. E. T. Newton, F.G.S., described the occurrence of part of the skull and horn-core of a smal lcavicorn ruminant, which had been obtained by Mr. H. B. Woodward from the Norwich Crag of Thorpe. The presence of a frontal fossa with a foramen passing directly into the orbit, was held to indicate an affinity with the antelopes; and after comparison with the available recent specimens in the British Museum and Royal College of Surgeons, it was regarded as most near to the gazelles—*Gazella dorcas, G. subgutturosa, G. picticauda,* and *G. Bennettii,* being most like the fossil, and agreeing with it in having the skulls more or less compressed in the frontal region, nearly upright horns, and a well-marked frontal fossa and foramen, but differing in the form of the fossa and in the position of the pit on the pedicle. On the whole, *G. Bennettii* was regarded as nearest to the fossil. Among the known fossil forms only a few were thought sufficiently near to render a comparison with them necessary; the following, however, were mentioned, and attention called to the points in which they differed from the Norwich specimen, namely, *Antilope deperdita, A. brevicornis, A. porrecticornis, Tragoceros Valenciennesi,* and *Palæoryx parvidens.* Seeing that all the important characters of this fossil are found among the recent gazelles, it is referred to that genus; but as it differs in certain points from each of them, it is necessary to give it a new specific name; the author therefore called it *Gazella Anglica.* Fortunately, this interesting discovery is corroborated by two other similar examples of horn-cores with frontals from the same locality and horizon. One of them is in the British Museum, and the other in the possession of Dr. Arthur King, of Norwich. A short appendix, by Mr. H. B. Woodward, on the horizon from which these fossil gazelles were obtained, was also read. In the discussion which followed, Mr. Lydekker agreed with the author that the species was a gazelle. He remarked that the hyæna occurring in the crag was an African type, and that further comparison of the present species with African gazelles was desirable. Mr. Blanford remarked that the present paper was the outcome of an excellent piece of palæontological work. After noticing the present distribution of the genus gazella, he pointed out that nearly all the species were inhabitants of plains, and most of deserts, and that the occurrence of this species in the crag might perhaps indicate the condition of England in Pliocene times. Professor Prestwich said that this species was particularly valuable, because the specimen was not derived. He noticed that a species of antelope, belonging to the genus Saïga already found in France and Belgium, occurred in Britain in Pleistocene times. Mr. H. B. Woodward remarked on the valuable corroboration furnished by the specimen subsequently obtained by Dr. A. King from the same locality.

THE ORIGIN OF DIAMONDS.—In SCIENCE-GOSSIP for April, it is recorded that M. Chaper has traced diamonds to their matrix, having found them in "pegmatite." I am not sure what his claim in connection with this discovery exactly is. If he thinks that he is absolutely the first who has anywhere traced the diamond to any matrix, I need hardly remind you that he is quite mistaken. Westrop's Manual names three Brazilian rocks in which diamonds have been found, viz., "Grès Psammite," Stacolumite and Hornblende. But perhaps M. Chaper claims only the distinction of having been (apparently) the first to find the diamond *in situ* in the old world, and in a different rock from those above mentioned. Possibly you may think the matter worthy to be a little cleared up. If so, an explanation in S. G. will be much valued by many readers.—*H. J. Moule.*

THE GEOLOGISTS' ASSOCIATION.—No. 4 of vol. viii. of the Proceedings of this society (edited by Professor G. S. Boulger), contains some unusually good papers as follows :—" On the Geology of the district in North Wales to be visited during the Lay excursion," by Dr. Hicks ; "The Chalk, its distribution and subdivisions," by Professor Morris ; "On a Section of the Lower London Tertiaries at Park Hill, Croydon," by Mr. H. M. Klaassen. "Note on Coryphodon Remains" (illustrated) by E. T. Newton, besides accounts of the geological excursions made to the Medway Valley, Bangor, Snowdon, Holyhead, &c.

NOTES AND QUERIES.

THE RECENT SUNSETS.—The abnormal displays of colour in the sunsets of the autumn of 1883 have been described as seen from many places. I do not think many of them have come from Ireland ; at all events I would briefly tell the incidents of the sunset, as seen at Stoneyford, in the county of Kilkenny, and, afterwards, on Sunday, December 30th. By the almanacks the sun set on that day at 3.47 P.M. The sky was a clear grey-blue, wind N.N.W., and calm at sunset. There was nothing remarkable in the actual setting of the sun, no redness or glow preceded or accompanied the sinking of the sun's orb, nothing but a yellowish diffused tint over the S.S.W. sky, but shortly after glows appeared, and by 4.20 a band of ruddy orange lay along the horizon, and above it a space of yellowish opal-tinted green, higher still an arc of rose colour, the upper limit of which extended about half-way to the zenith, fading upwards into purple pearl-tinted grey. By 4.30 the arc of rose grew rose red, and although the sun had three-quarters of an hour a burst of red lit up the landscape. The leafless trees shone red, and the fields and water were lit up by the ruddy glow like sunlight. By 4.40 the rose red arc had taken the place of the opal green and touched the horizon ; this rose-red glow also appeared on the opposite (N.E.) side of the sky, but more faint than that over the place of sunset. By 5 o'clock the rose-red glow had sunk lower to the western horizon, and the sky was a purple blue, above it. The sunset seemed over. One hour and fourteen minutes after sunset, but still bright twilight. However, by 5.10 the rose-red glow flashed again up western horizon, spreading upwards gradually to the zenith. By 5.30 this glow was almost flame coloured— a deep rose-red. The stars were now shining in the zenith. At 5.40 the glow was like the reflection of a fire-flame colour, and extended pretty high towards the zenith, stars shone green through it, and the new moon (a day old) showed its thin sickle, a wan green, through the glow a little above the S.S.W. point of the horizon ; bright twilight still, nearly two hours after sunset. At 5.50 the glow still remained like the reflection of a great fire, but rapidly faded, and by 6 o'clock all was gone, but there was still twilight.—*James Graves.*

THE STORM OF JANUARY 26TH.—It may be of interest to some of your readers to know that after the severe storm of wind, rain, and snow, which occurred on the night of the 26th inst., the windows assumed a very dirty appearance ; on closer examination, they were found to be covered with a deposit of chloride of sodium or common salt. On evaporating some of the rain which fell on that night upon a microscopic slide, and placing it under the microscope, crystals of chloride of sodium were found in large numbers. The same thing was noticed upon melting and evaporating some of the snow which also fell that night. The wind blew from a westerly direction, and it is supposed that the salt observed in the rain and snow must have come from the sea, which is distant thirty miles from here due west, the spray having been carried inland by the force of the wind. The windows when the sun shone upon them sparkled as if they had been covered with frost, the light catching the angles of the crystals.—*J. C. Smith, Edenhall, near Penrith, Cumberland.*

ARE WATER-VOLES ENTIRELY VEGETABLE FEEDERS ?—Whilst walking on the banks of a canal, I noticed in the runs of the water-vole (*Cervicola amphibius*) numerous small heaps of the empty shells of water-snails, all of which had their lips very much notched, as if they had been broken by the teeth of some small quadruped. It seems to indicate that water-voles are not entirely vegetable feeders, as they are generally believed to be. I should be glad to hear if any of the readers of SCIENCE-GOSSIP have observed the same thing.—*F. H. Parrott, Aylesbury.*

STRIPED HAWK MOTH.—Early last June, whilst a gentleman and myself were walking on the cliffs here, my friend captured a fine specimen of the striped hawk-moth (*Deilephila Livornica*). Judging from the geographical position of the place, the appearance of the insect, and the fact of the ladies' bedstraw (on which the caterpillar is said to feed) growing abundantly here, there can be no doubt of this being a genuine British example of this extremely rare moth.—*C. Jefferys, Langharne, Carmarthenshire.*

SWELLING ON HORSE'S LEG.—Will any reader kindly explain the following :—For some time I have noticed that several of the horses belonging to the Nottingham Tram Company have at times large swellings behind the elbow, which appear as if filled with some fluid. These swellings are sometimes of great size, but do not seem to cause the animal any inconvenience. I have only noticed it in the tram horses, and have never seen it in any others. I should very much like to know the cause of these swellings, and if they cause the animal any pain.—*W. Finch, jun.*

ANOTHER BAT.—Mr. Barlow's interesting account of his pet "Tommy the Bat," in a recent GOSSIP, reminds me of a similar capture made in a bedroom when I was living at Twickenham. I placed the "strange visitor" in a disused aquarium covered with perforated zinc. Like Mr. Barlow's, he treated flies rather contemptuously, but ate moths of all kinds very voraciously, fluffy wings and all. I did not try the meat and milk diet, or possibly I should have succeeded in keeping him longer. As it was, he died about a fortnight after the commencement of his captivity.—*W. Matthieu Williams, Stonebridge Park, N.W.*

WHAT IS A WHITE ELEPHANT?—Mr. C. P. Sanderson, "Superintendent of the Government Elephant Catching Operations in Bengal," says in a letter to the "Times," respecting the Barnum animal, "Neither in the general colour of his body, in the flesh-coloured blotchings on his face, ears, and chest, nor in the smallest particular does he differ one whit from the hundreds of elephants of the Commissariat and Forest Department, which may be seen any day in India and Burmah, carrying the baggage of troops, or dragging timber down to the banks of rivers." "The value of such an elephant in Burmah or India is from £150 to £200." "The testimony of all trustworthy observers who have seen the sacred, so-called white elephants of the Kings of Burmah or Siam, proves that they are but ordinary elephants possessed of certain whimsical 'lucky' marks." "My own experience, as well as that of many of the older native hunters attached to the elephant-catching establishment at Dacca, satisfies me personally, that there is not, nor ever was, such a creature as a white elephant, in the ordinary acceptation of the term." He then speaks of "two very young elephants (newly captured) of a dirty cream-colour; one died, the other turned as black as his fellows in a few years." The "Newcastle Weekly Chronicle" gives a quotation from Mr. Crawfurd's "Embassy to the Court of Ava" (apparently referring to about the year 1820): "Our attention was chiefly attracted to the celebrated white elephant, which was immediately in front of the palace. It is the only one in possession of the King of Ava, notwithstanding his titles (Lord of the White Elephants), whereas his Majesty of Siam had six when I was in that country. The Burman white elephant was of a cream colour, and by no means so complete an Albino as any of those shown to us in Siam." He then mentions their rarity, and says, "Several of a light tint, but not deserving the name of white, have been taken within the last twenty years" (in Burmah). These statements effectually dispose of Mr. Barnum's prodigy, but the fact of Mr. Sanderson having satisfied himself as to the non-existence of a white elephant is not quite satisfactory to people in general, and any information on the subject would just now be very acceptable.—*W. Gain, Tuxford.*

ANATOMICAL PREPARATION.—Can you, or any of your readers, inform me where I can obtain fresh animal tissues, (either human or from lower animals) for Microscopic preparation?—*W. H. Pratt, Nottingham.*

THE ROOK (CORVUS FRUGILEGUS).—Is the name rook, applied to this bird, universal in the British Islands? The carrion crow (*Corvus corone*) is, I believe, often confounded with the rook, in poetry and novels, from the writers often not knowing they are distinct species. I have heard persons remark they thought the names rook and crow referred to the same bird. Of course the term "scare-crow" proves that "crow" is another name for the rook, as it is the latter that frequents the corn-fields, the crow being a more solitary bird, and not generally associating with the rook. Which name is most commonly used for *Corvus frugilegus*, crow or rook?—*Henry Lamb, Maidstone.*

NOTICES TO CORRESPONDENTS.

TO CORRESPONDENTS AND EXCHANGERS.—As we now publish SCIENCE-GOSSIP earlier than heretofore, we cannot possibly insert in the following number any communications which reach us later than the 8th of the previous month.

To ANONYMOUS QUERISTS.—We receive so many queries which do not bear the writers' names that we are forced to adhere to our rule of not noticing them.

To DEALERS AND OTHERS.—We are always glad to treat dealers in natural history objects on the same fair and general ground as amateurs, in so far as the "exchanges" offered are fair exchanges. But it is evident that, when their offers are simply disguised advertisements, for the purpose of evading the cost of advertising, an advantage is taken of our *gratuitous* insertion of "exchanges" which cannot be tolerated.

WE request that all exchanges may be signed with name (or initials) and full address at the end.

J. H. C. R.—Thanks for specimens of *Lathræa squamaria* and primrose.

Dr. T. S. R.—The swellings at the buds of the stem of the apple-tree are galls, probably caused by a species of Andricus. Wash, or brush the bark of the trees with a mixture of petroleum and water, stirring well before using.

W. P. Q.—Write to Dr. M. C. Cooke, 146 Junction Road, Upper Holloway, N., for particulars.

H. W. PARRITT.—The usual way to obtain skeletons of small mammals, &c , is to skin the bodies and place them near an ants' nest. Tadpoles are also capital nibblers, and will pick the bones quite clean. See chapter on "Bones" in "Collecting and Preserving Natural History Objects," price 3s. 6d., published by W. H. Allen & Co.

F. T.—Tripp's " Mosses" is a good book, although we prefer Hobkirk's "British Mosses" (without plates, however). The " British Moss Flora," now being issued in parts by Dr. Braithwaite (beautifully illustrated), promises to be the best work of its kind when completed. Rossiter's "Scientific Dictionary" is very good as far as it goes. We want a really good one, and the new dictionary being issued by Cassell and Co. promises to be very complete as regards the portion devoted to scientific terms.

J. BLACK.—The insect of which you sent us a sketch, brought from Syria, is evidently a species of Notonecta, one of the water bugs.

A. W. LYONS.—All the exchanges in our columns are inserted gratis, unless more than three lines in length.

S. G.—Thanks for the specimen of *Matricaria parthenium*. All the plants seem to have forgotten the proper time of flowering, owing to the last three mild winters. Look at the list of plants in flower in our February number.

I. T. M.—Get Watt's "Manual of Chemistry." You would have to go through a course of practical instruction at some good laboratory before you were fit.

S. S. desires to know the publisher of Blewer's "Flora of Surrey." Perhaps some reader will oblige by sending it.

H. A. F.—The parasite has been sent on to be named. The answer will be given shortly.

F. RAYNER.—Sprinkle the eggs with benzine to destroy the maggots.

R. H. BATTERBEE.—There is no reason why the wallflower should not be the wild kind under the circumstances you mention. It is a plant which rapidly spreads when the conditions are favourable.

J. B. BESSELL.—Sir Joseph Hooker's "Student's Flora of the British Islands" would be of great help to you. Hobkirk's "British Mosses," and Miss Ridley's little book on Ferns, would be useful for those plants.

EXCHANGES.

OFFERED, thirty-six well-mounted human pathological slides, beautiful slides of marine algæ in fruit (50 varieties), many other good slides. Wanted, cabinets or boxes for slides.—T. H. Buffham, Connaught Road, Walthamstow.

FOR exchange, large solar microscope in mahogany case. Geological collection and microscopic slides preferred.—A. E. Palmer, Tettenhall Wood, Wolverhampton.

MICROSCOPIC slides of wood sections, animal sections of hairs for exchange for other slides.—J. E. Nowers, 69 Barnstone Road, Burton-on-Trent.

"NORTHERN MICROSCOPIST," first 24 numbers in exchange for Bolton's 10 parts of Portfolio of drawings. — Robert McLaughlin, Rookery, Hathern, Leicestershire.

VALENTINE'S section knife in perfect condition, also pocket microscope, for exchange; desideratum polariscope.—John R. Marten, 2 Lower Seymour Street, Portman Square, W.

I WILL give genuine Chinese coins in exchange for plants from the Holy Land, Europe, or the East, or for rare English fossils.—Rev. Hilderic Friend, F.L.S., Brackley.

WANTED, Mr. Cameron's papers on "Gaelic Names of Plants," published in the "Scottish Naturalist" for 1881. Will give copy of "Devonshire Plant Names," Chinese coins, or cash.—Rev. H. Friend, F.L.S., Brackley.

PORTIONS of Roman urns, Samian ware and early English pottery, also palæolithic and neolithic flakes (all from London and suburban district) in exchange for ancient stone implements and flakes from other district or palæozoic or earlier mesozoic fossils.—G. F. L., 49 Beech Street, E.C.

OFFERED, old numbers of SCIENCE-GOSSIP, years 1880 and 1881. Also for exchange, shells (post-tertiary), from the raised beach Isle of Portland, and fossils from the chalk marl, Portland stone, Kimeridge clay, coral ray, Oxford clay, cornbrash, great oolite, fuller's earth, inferior oolite, lias and rhætic formations. — F. Sumner, 136 Walter Street, Oxford.

J. W. will be glad to forward a few *Lepidocyrtus curvicollis* to microscopists wishing to study scales. — Woburn House, Woburn Hill, Green Lane, Liverpool.

MARINE diatoms from Pensacola, Gulf of Mexico, well mounted, for crustaceans or minute marine forms in fluid.—A. W. Griffin, Saville Row, Bath.

FOR slide of *Rhizosolenia Shrubzolii* send one of *Aulacollis* var. *Petersii*, or any well-cleaned slide of diatoms from guanoes. —A. W. Griffin, Saville Row, Bath.

SCIENCE-GOSSIP for 1872, bound, 1873 to 1879, inclusive, and 1883, unbound, for exchange. Wanted microscopic slides or material.—Frank Adams, High Street, Stoke-upon-Trent.

ANSTED'S "Physical Geography," Milner's "Gallery of Geography," "Home Naturalist," "Field Naturalists' Handbook," and well-mounted slides. Wanted, Davis's "Practical Microscopy, Rimmer's "Land and Freshwater Shells," polariscope and micro apparatus.—J. C. Blackshaw, 4 Ranelagh Road, Wolverhampton.

FORAMINIFERA from the coast of Galway, Ireland, and gold sand with gem pebbles from the Ovens Diggings, Australia, mounted; for other good mounts, insect preparations preferred; objects of animal physiology or diatoms not acceptable.—C. Croydon, Pato Point, Torpoint, Cornwall.

A FEW portions of the backbone of a very large whale, which I will exchange for British birds' eggs, side blown, or other natural history objects.—A. Foster, Rodger Street, Anstruther.

WANTED, Rimmer's "Land and Freshwater Shells," Rye's "Beetles," and Rossiter's "Scientific Dictionary;" or exchange. —Frank Tufnall, Amity Street, Reading.

WANTED, rare British marine shells, also land and freshwater—especially varieties of *H. aspersa*. Offered good exchange in other shells.—C. Jefferys, Hill House, Langharne, Carmarthenshire.

PALATE of Limpet (*Patella vulgata*) unmounted, any quantity. What offers? Unmounted objects not required.—C. Jefferys, Hill House, Langharne, Carmarthenshire.

BEAUTIFUL specimen of common scoter (*Oidemia nigra*) in case, by professional stuffer. What offers?—C. Jefferys, Hill House, Langharne, Carmarthenshire.

"SKETCHES OF BIRD LIFE," by Harting, in exchange for birds' eggs, coins, or some good birds' eggs plates.—B. Mason, 8 Warwick Road, Stratford-on-Avon.

WANTED, micro-photographs in exchange for various slides. —B Berry, Dudfleet, Horbury, Wakefield.

Volvox globator—wanted, a good gathering in exchange for first class botanical slides, by C. V. Smith, Carmarthen.

TWENTY-EIGHT sorts of micro-seeds, 24 kinds of plant hairs, and 18 species of fern soil to exchange for mounted objects.— Rev. H. W. Lett, Lurgan, Ireland.

A GOOD ¼ in. object glass, with adjustment; also a 2 in., by Andrew Ross, both as perfect as possible; what offers in exchange?—T. B. Forty, Buckingham.

A FEW slides of diseased human lung (pneumonia) in exchange for other well-mounted slides of interest.—S. C. L., 276 Middleton Road, Oldham.

WILL send six living budding specimens of *Hydra viridis* on receipt of a good mounted object.—J. W. Lockwood, Lobley Street, Heckmondwike, Yorkshire.

FORAMINIFERA.—Will send well-mounted slide of *Tineoporus baculatus*, two varieties, Torres Straits, for any well-mounted slide of diatoms.—W. Aldridge, Westow Street, Upper Norwood, London, S.E.

WANTED, stamps or silver coin, reign James I. England, in exchange for emu's egg or books.—J. T. Millar, Inverkeithing, Fifeshire.

SHELLS for exchange: *Unio tumidus, L. glutinosa, paludina, vivipara, A. acicula, C. biplicata*, &c. Send lists of duplicates and desiderata to—S. C. Cockerell, 51 Woodstock Road, Bedford Park, Chiswick, London.

SCIENCE-GOSSIP, vols. 1881, 1882, and first four numbers of 1883, in exchange either for older vo's., books on natural history, specimens, or apparatus.—S. B. Axford, 114 Ebury Street, London, S W.

FORAMINIFERA material (good) wanted in exchange for well-mounted slides.—A. C. Tipple, 35 Alexander Road, Upper Holloway, N.

SWIFT'S sea-side and clinical microscope, with 1 in. objective, spot lens, and tripod stand, in morocco case, 6½ in. × 3, equal to new. What offers?—J. C. P. Brown, 31 Derby Street, Moss Side, Manchester.

A FEW slides of *Bacillus tuberculosus*, the Bacilli stained red. and surrounding tissue blue; also slides of ostracoda and foraminifera. Lists in exchange.—Samuel M. Malcolmson, M.D., Union Hospital, Belfast.

DUPLICATES: *Pisidium pusillum, Planorbis nautileus*, and var. *cristata, Pl. nitidus, Pl. vortex, Pl. carinatus, Pl. complanatus, Pl. contortus*, and *Ancylus lacustris*. Desiderata: Varieties of British land and freshwater shells; also many British marine shells.—Robert B. Cook, 44 St. John Street, Lord Mayor's Walk, York.

WANTED, "British Graphideæ" and "British Umbilicariæ." by Rev. W. A. Leighton.—F. Bower, 6 Dryden Street, C.-on-M., Manchester.

CARBONIFEROUS fossils, including fishes' teeth, scales, &c., from the coal, in exchange for fossils from other formations.— Wm. Hallam, Worsboro' Dale, Barnsley.

WANTED, sea-weeds from the south coast of England.— W. L. Balmain, Castle Street, Warkworth, Northumberland.

OFFERED, *Testacella maugei*, preserved in spirit for microscopical purposes, in exchange for good tropical land shells, or marine.—Jessie Hele, Fairlight, Elmgrove Road, Catham, Bristol.

WANTED, a few specimens of shells and other objects from earth mounds, tumuli, &c. A liberal and varied exchange offered.—Ed. Lovett, 43 Clyde Road, Croydon.

BOOKS, ETC., RECEIVED.

"The Student's Handbook to Physical Geology," by A. Jukes-Brown, F.G.S. London: G. Bell & Sons.—"Mineralogy," by J. H. Collins. London: W. Collins & Co.—"Bulletin of the United States Geological Survey," No. 1, "Reports of Observations and Experiments under the Direction of the Entomologist," by Dr. C. V. Riley.—"Thoughts on the Interdependence of Water and Electricity." London: W. Ridgway.— "Report of Botanical Exchange Club for 1882."—"Transactions of the Ottawa Field Naturalists' Club for 1882–83."—"Annual Report of the Wellington College Natural Science Society, 1883." —"Annual Report of the Nottingham Naturalists' Society, 1883."—"Proceedings of the Geologists' Association," No. 4, Vol. 8. "The Journal of Microscopy," Vol. 3, Part 10.— "Studies in Microscopical Science," edited by A. C. Cole.— "The Methods of Microscopical Research," edited by A. C. Cole.—"Popular Microscopical Studies." By A. C. Cole. —"Petrological Studies." By Messrs. J. E. Ady & H. Hensoldt. —"The Gentleman's Magazine."—"Belgravia."—"Journal of Conchology."—"The Journal of Microscopy."—"The Science Monthly."—"Midland Naturalist."—"The Inventor's Record." "Ben Brierley's Journal."—"Science."—"American Naturalist."—"Canadian Naturalist."—"The Canadian Record of Natural History and Geology," Vol. 1, No. 1.—"Science Record."—"American Monthly Microscopical Journal."— "The Microscopical Bulletin."—"Popular Science News."— "The Botanical Gazette."—"The Ornithologist and Oologist." —"The Electrician."—"Revue de Botanique."—"La Feuille des Jeunes Naturalistes."—"Le Monde de la Science."— "Journal of the Royal Microscopical Society," Vol. 4, Part 2. —"Bulletin of the California Academy of Sciences," No. 1, &c.

COMMUNICATIONS RECEIVED UP TO 10TH ULT. FROM:— J. F.—H. C. R.—Dr. T. S. R.—R. H.—G. T. F. A. —C. F. H.—F. S.—J. W.—F. T. M.—J. W.—T. R. R.— A. W. G.—W. H.—J. P.—E. H. W.—M. W.—W.— W. J. S.—R. L.—J. G.—J. H. W.—J. R. M.—T. H. B.— F. M. H.—H. F.—J. B.—A. H. S.—C. P.—G. F. L.—B. T.— G. C. D.—J. W.—E. & G. F.—T. B. R.—J. M. B. T.— H. C. B.—A. H. W.—W. M. C. O'N.—J. W. W.—J. B. B.— A. E. P.—R. M. L.—J. E. N.—G. T. G.—J. J. O. G. D.— C. C.—W. M.—T. D. A. C.—A. A.—E. T. D.—J. C. B.— H. J. M.—J. H.—A. F.—J. M.—B. H.—F. T.—W. P. Q. H. W. P.—J. Y.—R. B.—H. D. G.—H. W. L.—W. H. M.— B. M.—A. A. C. S.—C. J.—C. R.—A. H. W.—H. A.— W. B. & S.—C. V. S.—B. B.—W. J.—J. M.—S. C. L.— A. W. G.—T. B. F.—W. A.—R. L.—T. W. L.—S. B.— F. C.—A. C. T.—A. W. O.—A. S.—W. H.—W. A. L.—F. B. —R. B. C.—Dr. S. M. M.—T. C. P. B.—G. M.—S. C. C.— J. M. T.—C. B.—J. G. B.—J. W. P.—T. H.—H. L.—F. R.— W. B. G.—R. H. B.—T. H. M.—J. H. E.—L. W. L. B.— F. C., &c.

GRAPHIC MICROSCOPY.

E.T.D. del. ad nat.   Vincent Brooks, Day & Son, Lith.

PUPA OF LOCUST, ONE DAY OLD.
× 20

GRAPHIC MICROSCOPY.

By E. T. D.

No. VI.—Pupa of Locust, One Day Old.

THE "Complete metamorphosis" of an insect conveys the idea of a sequence, starting from an egg, producing an apodal caterpillar, or maggot (the larval feeding state, when the chief business of life is performed), leading to the chrysalid, a period of repose, requiring no food, eventually breaking or bursting into the imago, the perfect insect —a serial transition from a simple and elementary condition, to one complex and compound.

In the Orthoptera (the order comprising the locusts) and in the Hemiptera (bugs, plant-lice), the true apodal larval state is masked, the vermiform condition being developed in the egg—thus, the young Orthopteran and Hemipteran issues at once into actual life, with perfectly developed jointed legs, eyes, antennæ, and maxillary organs. The metamorphosis the Locust undergoes from the potential germ to the procreative imago, may be as varied in order as that of the butterfly, but the initiative, and more important changes occur in the egg, which, when hatched, assumes not a larval appearance, as generally understood, but the exact likeness and habits of its parents; this active pupa or nymph, by successive moults, attains a stage when the wings expand to their full size, the circulation of the blood through the nervures is arrested, and the development perfected.

It is clear that the pupa stage, which, in the butterfly is passive and embryonic, is, in the locust, active and voracious, whilst the condition of the larval state is reversed; the egg, when hatched, reveals a creature not only in the semblance of the perfect insect, with similar instincts, desiring and capable of obtaining the same food; but so far superior in development to what is popularly understood as a larva, or even a pupa, as to be able to perform *every* function of the mature condition, except flight, the wings and their cases only bursting through the thoracic rings after the second moult. It seems, therefore, a matter of mere nomenclature whether it should be designated larva (a ghost), or pupa (a doll), the plate representing a locust one day emerged from the egg, at once suggests an intermingling of both characters. It has been named "Pupa."

Burmeister, ignoring the term larva, designates this condition of the winged genera of the orders Orthoptera and Hemiptera "sub-incomplete pupæ," and Lamarck, in describing this tenacity of similarity of the insect in its progress from the egg to perfection, adopts the term "nymphæ." The familiar cockroach (Blatta) of the same order may be seen from the moment of emerging from the egg, in all stages of growth, with all its parents' well-known social domestic qualities, powers of appetite, and activity.

The Orthoptera although by no means an extensive Order, either as to genus or species, is of very general distribution; representatives of the genera Gryllus (grasshoppers) and Acheta (crickets) are found in most countries. *Locusta cruciata*, genus Œdipoda, mostly exotic, is the familiar locust of the East, and of tradition; its powers of increase and devastation are matters of history and common knowledge. The eggs are deposited in cases or tubes of earth about an inch and a half deep, previously made by the female, softening the material with a watery glutinous secretion; each tube holds fifty to sixty eggs, they are always placed in hilly country, and invariably in uncultivated soil. It is said, by an actual observer, that when depositing the eggs, the female receives a certain homage or attention, being,

No. 234.—June 1884.

during the performance, surrounded and protected by others. No further care is then taken. These cylindrical tubes of earth are found in erect position in great numbers; the cases or cysts are hard and tenacious, effectually protecting the contents. The eggs do not, however, escape the attacks of enemies; a fly of the order Bombylites, which, both in its larval and pupal state inhabits the earth, destroys immense numbers.

In the south of Europe rewards are given for the capture of eggs and perfect insects. When hatched, the young at once begin the business of life, with all the destructive instincts and ferocious powers of their progenitors. In Kirby and Spence are found interesting and reliable descriptions of their habits, and of the devastation they cause in every state of their career. Like other Orthopterans, they associate together, assembling and moving onwards over a country in one direction without deviation. If any impediment causes dispersion they collect again, re-organise their ranks, and follow the same route; in this way they advance without halting. In May they retire into plains or hill-sides and deposit eggs, which are hatched in June. The broods then collect and march forward: nothing in the form of vegetation escapes their ravages; leaves or succulent substances failing, they attack solid wood, and when swarms congregate " the sound of their jaws may be palpably heard." In a month they complete full growth, and power of flight.

Dr. Moffatt, in "Missionary Labour in South Africa," refers to the locust as the plague of the country, and after graphically describing their peculiarities, unappeasable appetites, and the dismay their presence creates, says, "they are on the whole not bad food, when well fed; almost as good as shrimps." Native tribes "fatten on them."

The drawing is from a very perfect specimen *hatched in England*, and mounted in balsam, without pressure, by Mr. Enock; the eggs in their cases, or "pods," were presented to him by the late Sir Sidney Smith Saunders, who procured them direct from Troad.

Crouch End.

A GENUINE BRITISH EARTHQUAKE.

By The Editor.

THE morning of April 22nd will not soon be forgotten by the dwellers in southern Suffolk and northern Essex. At eighteen minutes past nine, the first shock of an earthquake was felt, succeeded by two others. Buildings rocked to and fro, and, in Essex, some partly toppled down, and others were moved on their foundations.

I was in the upper part of my house at Ipswich when the first shock came. It was preceded by a noise like that of a waggon rumbling through the streets. Then followed a new sensation; and, in the space of a few seconds, a new experience was gained for life. All my old instinctive notions of the solidity and strength of a well-built English house vanished in a moment. The walls and floors were converted into india-rubber of the most elastic kind, and a kind of nausea accompanied the change of experience.

It was a terrible morning that, for news came in all round, but especially from the neighbourhood of Colchester, of fearful damage done by the earthquake in the district where it had reached its "seismic vertical," or climax. We, in England, have regarded earthquakes as something exotic, and as if they could not occur in our densely-populated country; and now, all on a sudden, we felt that a couple more oscillations like those of the morning would have shaken down every house.

People in distant parts of England may think the newspapers made the most of the occasion. But I can bear personal testimony to the fact, that, on the whole, they understated the event. I have been all over the district most affected, for it was not every day one had the opportunity of making personal acquaintance with an earthquake, and I for one do not wish the acquaintance to become any closer.

The London Clay of Suffolk and Essex is full of faults, and these dislocations even occur in our drift beds, so that British earthquakes are not novel or original phenomena, although they have seldom come with any violence within historic times; nor is there any geological reasons to suppose they will become any more frequent in the future—in spite of the alarmist sermons preached in Colchester, to "improve" the event. Indeed, there is reason to suppose the earthquake has some connection with the ridge of Palæozoic rocks known to extend beneath the eastern counties. It is the bellied down chain of hills connecting the Mendips in Somersetshire with the Ardennes in Belgium, and has been repeatedly reached by deep well borings at Crossness, Kentish Town, in Tottenham Court Road, Ware, Harwich, and recently at Richmond, where it lies at depths of from about 900 to 1,100 feet, covered chiefly by Cretaceous and Tertiary strata.

The following communications I sent to "Nature," giving my personal investigations and experience of the earthquake.

On Wednesday morning, the day after the earthquake, I determined to start upon its track. In Ipswich, little or no visible harm has been done; but no sooner had I arrived in Colchester, and commenced to walk through the town, from the chief station to the Hythe, than abundant evidence of the ruin wrought by it was visible. Chimneys were totally thrown down, and the brickwork had crashed through the frail roofs. Others were standing, but they looked as if they had been struck by lightning. Their upper parts were splintered, and laterally expanded. I could not help noticing that nearly all the houses whose chimneys were wrecked were the oldest—hardly any of the modern, cheaply-built, cottages being affected, contrary to my expectation. At Wivenhoe I found the appearance of the town best expressed by the remark that, "It looked as if it had been bombarded." That was the first idea which rose in my mind.

Hardly a house was untouched, inside or out. The newest houses seemed to be externally least affected, but they made up for this inside. They looked as if they had been given a few half turns, and then shaken up. The plaster had been detached from all the walls, the roofs were rent and loosened all along the cornices, and the framework of the windows was everywhere splintered. The battlements of the grand old church had been thrown down, and about fifteen tons of rubbish lay among the crushed headstones and the delicate and abundant grave flowers. Here there was evidence of a semi-rotatory motion on the part of the earthquake. The beautiful Independent chapel is so utterly wrecked within and without, that it will all have to come down. The streets were full of bricks, mortar, and tiles, although with characteristic English tidiness and diligence, the terror-stricken inhabitants were already clearing away the debris. I noticed several houses with rents at the bases of their walls, and in such of the chimneys as remained standing, they were frequent. One thing struck me—the rents sprang at an angle of about 30° at the bases of the buildings, whilst in the chimneys, this was increased to from 40° to 45°. The old ferryman related his experience to me, after the manner of an old salt. He was just bringing his boat to the shore, when the shock occurred. "It seemed just like three seas," he said ; a capital and vivid idea of the wave motion.

Crossing the river, I made my way through Fingrinhoe village, and on to Langenhoe. I did not see a single house on the road, large or small, for a distance of about four miles, that had escaped untouched. The fine old Jacobean hall at Fingrinhoe has lost the upper part of the front elevation. Here I found some of the chimneys that had been left standing twisted on their pediments. I carefully noted this on the way, and on examining those of the massive chimneys of the rectory at Langenhoe, the torsion was very plainly visible. The twist had come from the south, for the faces of the chimneys which had previously looked in that direction were now turned almost south-easterly. I did not set out a minute too soon to note these circumstances, for all the builders of the country-side were already abroad, and in a few days all the evidences of earthquake-action of the greatest value to seismologists were completely obliterated. Thus, I found a very intelligent builder from Colchester on the lawn of the Langenhoe Rectory, giving orders for having the twisted chimneys removed, and I have no doubt they were all taken down within twenty-four hours. He had been driving all over the disturbed country-side, and told me that wherever the big chimneys had been left they were twisted from the south-south-west to the north-north-east, especially in the contiguous villages of Peldon and Abberton. This, I think, settles the original direction of the earthquake wave, and also establishes its rotatory character.

Langenhoe church is an utter ruin, and all that yet stands will have to come down. It is a sad sight to see this picturesque, ivy-clad old church—standing so prettily overlooking the creeks where the ancient Danish Vikings landed in the dawn of our modern history ; but a comparatively few years before the church was built, now so utterly ruined. The porch on the north side is of brick, and a modern structure. Two large rents run up, one on each side of the doorway, at an angle of about 32°. They run from opposite directions, and meet just above the keystone of the arch. Here another large rent parallel with the ground traverses the masonry. It seemed to me that the first earthquake shock which rent the brickwork sprang from the western corner, and was reflected so as to form the opposite rent after striking and lifting up and forming the parallel crack above mentioned. The battlements of Langenhoe Church unlike those of Wivenhoe, have been shaken down, but while those of Wivenhoe were thrown upon the ground on the west side, those of Langenhoe Church were thrown on the nave, that is, in an opposite or easterly direction. They crashed through the roof, and carried a gallery with them ; the concussion, meantime, bursting out the upper part of the chancel end. Am I right in thinking that this pitching forward of the loosened rubbish in opposite directions, as exemplified in these two churches, taken in connection with the overwhelming proof of rotatory motion, indicates that the movement of the earthquake had swerved right round between Wivenhoe and Langenhoe? In that case it also suggests the local character of the earthquake : Langenhoe and the adjacent villages, with the Isle of Mersea close by and in full view, appear to form the chief area of disturbance. So far as I have been able to learn, the clocks stopped by the shocks were those facing the north. The newspapers referred to various cracks and fissures in the ground at Langenhoe, Abberton, Mersea, and elsewhere, as having been caused by the earthquake. I saw numbers of them, but in every instance they were the ordinary cracks which always appear in the London clay during a drought, or after a spell of dry weather like that of the preceding few weeks. In none of the instances I saw had the fissures anything to do with the earthquake. The local character of the area of chief disturbance is not only indicated by the different directions in which the rubbish was thrown from the battlements of Wivenhoe and Langenhoe churches relatively, but also by the fact that whilst the western side of Mersea Island suffered severely, the eastern side was only slightly affected in comparison.

G 2

From more recent observations I concluded that the seismic vertical was at or near Dr. Green's house, close to the strood or causeway which connects the mainland of Essex with Mersea Island. The house was built in 1860, and is therefore new. I may here observe that (as I hinted before) the modern, cheaply-built cottages were not so much affected as the more ancient ones. The chimneys, walls, &c., of the latter were invariably destroyed, damaged, or cracked—those of the former seldom so. I was much surprised at this. The first thought naturally was that these "jerry built" houses would be shaken down like a pack of cards. Is it that their very looseness of structure is in their favour, as compared with the stronger built cottages of two and three hundred years ago? I have somewhere seen that in earthquake-visited centres, the houses most secured from destruction are the loosely-built, low edifices. One can speak plainly on this matter, as no premium is required to encourage the development of "jerry building." Dr. Green's house is literally split and cracked in all directions, and the splits and cracks are the most vertical of any to be seen. The entire building was twisted on its foundations. At the south-west corner this is visible to the amount of about one inch and a half. Dr. Green informed me he was lifted up, as if from behind, and shot violently forward. A friend of mine remarks (and I noticed the same fact in my note book, but omitted enclosing it in my first communication) that the railway cutting at Wivenhoe appears to have broken the continuity of the undulations, for the houses contiguous to it are comparatively uninjured. A noteworthy fact in connection with the recent earthquake, to which I can personally testify, and which appears to be the general experience of all the most trustworthy observers I have come across, is that the sounds or noises preceded the oscillations for an appreciable period of time. Mallet's experiments showed that the shock of an explosion travels through wet sand at the rate of 951 feet per second. In Ipswich we are situated chiefly on drift sands and London clay, and allowing that the earthquake shocks travelled through these strata at a more rapid rate, it is not likely to have been much more rapid. As sound travels at the rate of 1118 feet per second, it is very probable that the noise accompanying the earth-movements preceded the oscillations.

Mr. Wilkins, the well-known yacht-builder at Wivenhoe, tells me he was standing at the time the earthquake occurred in the yard, and his first impression was that a new yacht he was looking at was heeling over, and he called out so to his workmen in the shop close by. Then followed the crash of the tall chimney and the rending of the walls. The workshop has an upper floor with windows on each side, and, as he stood in the yard, Mr. Wilkins says the oscillatory waves were such that he was enabled to look right through these windows, so as to see the falling chimneys of the buildings on the other side. He calculates there must have been a rise and fall of the ground of two feet nine inches to have enabled him to do this.

In view of Mr. Topley's suggestion that the earthquake probably has some connection with the ridge of palæozoic rocks [which underlies the eastern counties, it would be interesting to know if the shocks were felt in the Boulonnais and the Ardennes, as they were in the Mendip Hills—the other end of the chain which is let down beneath London, and covered up with chalk and Tertiary strata.

J. E. TAYLOR.

FREE-SWIMMING ROTIFERS.

IN the course of my study of the rotifera, I often come across forms, especially in the more minute free-swimming ones, which I am unable satisfactorily to make out. The Micro. Dict. does little beyond giving generic differences, while "Pritchard," an invaluable book, is yet far from being all that a student of these animals could wish. A short

Fig. 67.—Side view.

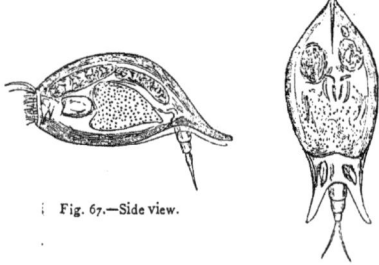

Fig. 68.—Dorsal view.

time ago an announcement was made, I believe at a meeting of the Manchester Microscopical Society, that some authority was engaged in preparing a manual on this subject. If this could receive an authoritative confirmation, it would gladden the hearts of many students of these animals. From Saville Kent's preface to "A Manual of the Infusoria," we gather that this work was originally intended to be based upon the same lines as "Pritchard's Infusoria," but there was an accumulation of such a quantity of material, as to render a more limited scope desirable. If from the materials which he had undoubtedly gathered together, he would supplement his magnificent work by one on the rotifera, especially if it could be issued in shilling monthly parts, he would confer a boon on hundreds if not thousands, of enthusiastic workers on this subject.

I send sketches from my "Note Book," of a rotifer

which I am unable to name, and shall be glad if any of your readers can assist me. Fig. 67, side view; Fig. 68, dorsal view. I have never seen more than a very limited number of them, and these always from one locality, a shaded well, having a north-west aspect. Above, and along the surface of the water, the stones are covered with mosses and diatoms, and *Batrachiospermum moniliforme* is very plentiful. I get there several rotifers, and among them, very sparingly, my unknown specimen. Its lorica is cylindrical, slightly compressed, open on the ventral side, and with faint indications of a dorsal ridge anteriorly. On the dorsal view, it is pointed anteriorly, rapidly widening to about the middle, then narrowing posteriorly, which is terminated by two spines, curving upwards and outwards. I have only been able to detect one eye. The gizzard or masticatory organ is very difficult to make out, from the rotund conformation of the animal. Rotatory organ with two superior hooks, and two inferior large cilia, or, perhaps, the latter may be designated setæ. Tail-foot, forked; toes about as long as the foot. If new, which perhaps is hardly probable, I propose to call it *Colurus navicularis*, a name expressive of its general resemblance, on its dorsal aspect, to a ship or boat. On a subsequent occasion, I may invite the assistance of brother naturalists in naming others of the free-swimming rotiferæ.

J. E. LORD.
Rawtenstall, near Manchester.

GOSSIP ON CURRENT TOPICS.

BY W. MATTIEU WILLIAMS, F.R.A.S.

THE Society of Arts is gradually developing into a practical science parliament where all kinds of scientific novelties that have reached the stage of useful application may be brought forward by their inventors and publicly discussed, criticised with any degree of courteous severity by those who either object to them on scientific grounds, or who have opposing interests. This is very desirable, and hitherto the debates have been characterised by an amount of straightforward and concise adherence to the subject on hand that contrasts very favourably with some recent debates in "another place."

One of the longest and most animated of these followed (with adjournments) the paper read by Dr. Percy Frankland (son of Professor Frankland) on March 13th. The question raised is one of much moment, viz. whether the water of a river, such as the Thames, when once polluted by sewage can be rendered fit for drinking purposes, either by the oxidation incident to its own flow, or by artificial filtration. Dr. Frankland contended that it cannot, and therefore that the Thames' supplies to London should be abandoned; while many eminent engineers and a few chemists very positively contradicted both his data and his conclusions. It should be understood that most of these are concerned in the construction of filter beds and other engineering appliances and processes for river-water purification, or in schemes for chemical precipitation. It is quite right, of course, where such large interests are concerned, that full opportunities for defence should be afforded against an attack upon those interests. Dr. Frankland's paper certainly is such an attack, and a very serious one, as we cannot submit to be poisoned merely for the sake of maintaining the value of water companies' shares. The advocates of river supply make out a fair case so long as the drainage areas of the upper waters are free from epidemics, but whether or not the specific microzoa producing such diseases as cholera and typhoid fever can be oxidized to death, or kept back by filter beds, or exterminated by A. B. C. or other processes, still seems doubtful. Those of our readers who are interested in the subject should not fail to read the paper and the discussion which are fully reported in the society's journal. Some very caustic comments on the paper are published in the Journal of Science for May, under the title of "The Ghost of the Season." The writer pleads for the A. B. C. process.

The interesting pictorial and mechanical display made by the water companies at the Health Exhibition is another form of reply, though not directly put forward as such. The pretty pictures of the "shining river" do not, however, display the underground tributaries. But to be very critical on the water companies just now, is like hitting a man when he is down.

Apropos of typhoid and cholera microzoa, we see that Dr. B. W. Richardson is sounding a note of warning, not the usual note of warning against invasion by these creatures, but against the tendency to ignore "all the preceding clinical history," and treat it "as nothing in presence of bacillus." The subject is treated in his usual quaint and picturesque style in the May number of his "Asclepiad." He asks: "Can this greedily absorbed hypothesis come to any good? Upon the evidence of how many or how few men does it rest? On what reasoning does it rest? Who has separated, in relation to it, coincidence from causation?" Whether "we have either before us a revolution in discovery, momentous in character, or one of the most dangerous of speculations that was ever revived out of the mad past—Dwight's animalcular hypothesis with less excuse for it," it is certainly desirable that some protest or scepticism should be agitated against the risk we encounter from the natural tendency of medical aspirants to perpetrate fashionable follies in their desire to be well up to the level of latest discoveries, and the most advanced practice; to follow, in short, the medical fashion of the day.

The observations of Dr. Henry F. Walker, of New York, concerning the absence of earth-worms in regions where man has not settled are very curious.

He states that it is well known to settlers on the virgin soil of the United States that no earth-worms are found on the first tillage of the ground, not even in natural meadows; that they are first found in the vicinity of the stable yard, then in soil enriched by stable manure, from which they spread out to all the soils around whether cultivated or native. It appears that Dr. Walker pays little regard to the red men, or other earlier aboriginal specimens of humanity when he says that until a place has been inhabited for five years by man it is useless to look for the earth-worm.

More evidence is demanded to render this broad generalisation worthy of acceptance. If the facts are confirmed as regards America, they rather indicate that the earth-worm is an importation from the Old World, than the companion of man as described by Dr. Walker. Many generations of man have lived and died on most of those parts of the American continent that are now covered with so-called "virgin soil," and some of these people have kept horses.

Possibly some of our readers may be able to supply facts of their own observation in support or refutation of the generality of Dr. Walker's observations;—Are these earth-worms in the virgin soils of Australia?

Mr. Griffin W. Vyse, in his paper on "Routes through Afghanistan," read before the Indian section of the Society of Arts on 28th March, supplies a curious illustration of newspaper geography. As he says, "it seems hardly credible that one of our leading journals should have stated, only a few weeks ago, that there are only two passes from Afghanistan into India," the fact being that we now know of 289 routes. Between the Khyber and Bolan passes, "lie all those celebrated routes distinguished as being about the oldest highways in the world, traversed for thousands of years by the countless generations of early traders," especially when Babylon was at the height of its glory, and the riches of India were transmitted to the great Emporium of the west. Mr. Vyse describes these routes in detail, showing what they have been, and may be again, and refuting the newspaper descriptions of Afghanistan as a land of rocks and stones. Instead of this, he describes it as a land that "still remains full of hidden riches, and of mineral wealth untold." He affirms that "the natural elements of its ancient beauty and life still exist in the marvellous fertility of its soil, and in the manliness of character of some of the people, and expresses his opinion that "the day cannot be far distant when this province will again become one of the most prosperous in the East." Kandahar was once the capital of Central Asia, is 3000 years old, was named after Alexander the Great who visited it, Ishkandahar, Alickjalandar, Kandahar.

Our little earthquake of April 22nd is evidently doing something towards the extension of popular scientific education, as well as demonstrating the demand for its extension. The favourite popular theory seems to be that the earth was visited by an electric shock; this is quite in accordance with the general practice of ascribing everything that is mysterious to "electrical influence." We heard a learned shopkeeper in Holborn describe the exact path of the shock from the corner of Fetter Lane, across the road, striking some houses and omitting others, affecting only the second floor in one case without at all shaking the upper and lower parts of the house. He had investigated the subject by careful inquiry, and his conclusions were based on the fact that it was felt in some places and not in others. That the possibility of feeling such a tremor depended upon the quiescence at the moment of the person questioned did not occur to this investigator. Mr. C. E. de Rance's observations on the effect of the earthquake on the supply of underground waters in the regions affected are of a very different character, and, like all that he has to tell us on this subject of underground waters, are very interesting. They are communicated in a letter to "Nature" of May 8th, from which we may conclude that he is making further investigations in this direction. So far, the general effect has been to increase the supplies, by opening or widening of fissures in the rocks that obstruct the rising or free movement of the water in question.

The red glows are not so brilliant as they were, but they persist to a degree that must be very puzzling to those who were satisfied with the Krakatoa theory. The subsidence of the dust must be slow indeed if they are to continue not only through all the winter, with its snows and rain, but all the spring, and into the coming summer. The English sunset display on Easter Sunday was nearly equal to the most brilliant of its predecessors of the previous autumn.

A CHAPTER ON MARINE DENUDATION.

BY C. H. OCTAVIUS CURTIS.

MY object in writing this paper is not to propound any new theory of denudation, nor is it my intention to enter into a discussion on the relative extent of subaerial or marine denudation, but simply to call attention to a large field of research in Dynamical Geology, open to all geological students, and which does not require any expensive apparatus for its solution.

All readers of Lyell's great work must have felt a little astonishment when, for the first time, they perused his chapters on the Denuding Action of the Sea (ch. xx. and xxi.), for although, having no doubt noticed in their seaside excursions, the effect produced by the tremendous force with which the waves break on the shore, still the fact that our little island is year by year being encroached upon to the extent of hundreds of acres, is at first rather amazing; but that such must be the case is evident, when we consider that in many parts of England the sea is encroaching on the land at the rate of a yard, or even more, a year. If further evidence was wanted, a few moments' conversation with a coastguardsman or old seaside villager would convince the most sceptical of the correctness of my statement, for I have often had pointed out to me areas where in place of harvests of corn, only seaweed can now be gathered, and where instead of village communities, molluscan colonies now exist.

The object of this article is to enlist all those who are scientifically inclined, and who either live at, or visit seaside districts, in the work of collecting records of the past work of this powerful denuding agent, and watching the extent to which it is acting at the present time. We cannot all be Darwins or Lyells, but we can all assist in that equally important work of collecting materials without which no theory worthy of the name can be propounded. The collecting of evidence, to be of any scientific value, must be undertaken in a very careful and patient manner; and I think it will not be out of place if we briefly consider a few cases where attention has been paid to the subject, in order that those who are inclined to assist in the work may more clearly understand the nature of the evidence required. Yorkshire furnishes an excellent example, for, from the Mouth of the Tees to the Humber, the whole coast is undergoing rapid denudation. Professor Phillips wrote in 1853: "For many years the rate at which the cliffs recede from Bridlington to Spurn, a distance of thirty-six miles, has been found by measurement to equal, on an average, two and a quarter yards annually, which, upon thirty-six miles of coast, would amount to about thirty acres a year. At this rate, the coast has lost one mile in breadth since the Norman Conquest." But this is by no means an excessive rate of demolition, for Pennant, by reference to old maps of the same country, finds a number of villages marked, which are now only represented by sand banks in the sea. The whole coast of Norfolk and Suffolk furnishes similar examples. Mention need only be made of Dunwich, once a large city with twelve churches, and the most important seaport on the east coast, which is now only represented by an unimportant village; but there is no need to give further examples. The Reculver church of Kent, the Hordwell Cliffs of Hampshire, and many other examples might be easily found, but we have already, in our geological text-books, become acquainted with them, and it would be foreign to my object to burden this paper with details, as I think we have already sufficient examples to show that the sea is yearly bringing nearer home the truth that "Britannia rules the waves."

A very natural question for my reader now to ask, would be: Why do you plead for observers, when you own that we already have such good records? My answer is, that these only refer to limited districts of our extensive islandic coast, whereas a full record is required of its whole past extent; and again, we must not be satisfied with what we have, but should bear in mind that, as every year masses of strata are becoming denuded, the records of their rich flora and fauna are being lost irrevocably by the triturating action of the waves. When we bear in mind certain formations, for example, the crags of the east coast and the Purbeck beds, it becomes evident that unless proper accounts are kept, and continual visits made to such places of destruction, whole series of fossils (many of unknown species and genera) are becoming lost to science, and thus leaving missing links in the chain of palæontological development, which might have been found.

I think I have now shown sufficiently that there is plenty of room for such work, and I can confidently promise all those who join in it, that they will be fully repaid for their trouble by the new facts which will be daily brought to view, and by the knowledge that while investigating the records of the past, they are each adding their little to the advance of one of the greatest of sciences.

In closing my remarks, a few hints will not be amiss as to the most important points to be attended to in such research. These I will class under three heads.

1. *The extent of Denudation, past and present.*—The former extent of the sea coast may be determined with some degree of accuracy, by hunting up old maps, consulting church registers and tithe rolls, by questioning old residents (great care should be taken on this point, as the human species is very subject to the exaggerative). Another point of importance is to settle the question as to whether the denudation has been regular, or whether there have been any excessive years, and if so to determine the dates of the same. The present extent of denudation is a matter for personal observation.

2. *The Nature of the Denudation.*—Under this head should be considered the questions of how far the marine work has been aided by sub-aerial action, nature of rocks, &c.

3. *The Palæontological Problem*—Which involves the collection of specimens, and determining and reaching the same.

Should all these points be well attended to, the value of the work will be great to science, while the investigator will feel that all has not been in vain.

TWO CURIOUS PHENOMENA.—The following facts are sufficiently remarkable to be chronicled. A lady, breaking her egg at breakfast, found inside it and attached to the shell, a small piece of printed paper, on which the printing was still visible. In a garden at Acton there is an apple-tree which has at the present time half-a-dozen apples remaining upon it. The blossom of this year is not out yet.—*R. H. N. B*

THE GENTIANS OF THE ALPS.
GENTIANA (L.).

CALYX tubular or campanulate; 4–10 divisions, but usually corresponding with lobes of corolla. Corolla entire; funnel or salver-shaped; 4–5-cleft; mouth smooth or ciliated.

nearly to the base, lobes somewhat lanceolate; flowers stalked, in whorls and terminal cluster, star-like; anthers free; leaves opposite, elliptical, strongly ribbed, and the upper ones sessile. Plant 2–4 ft. high, common in Alpine pastures; root valuable as a tonic, for which peasants dig it up and sell to the chemists.

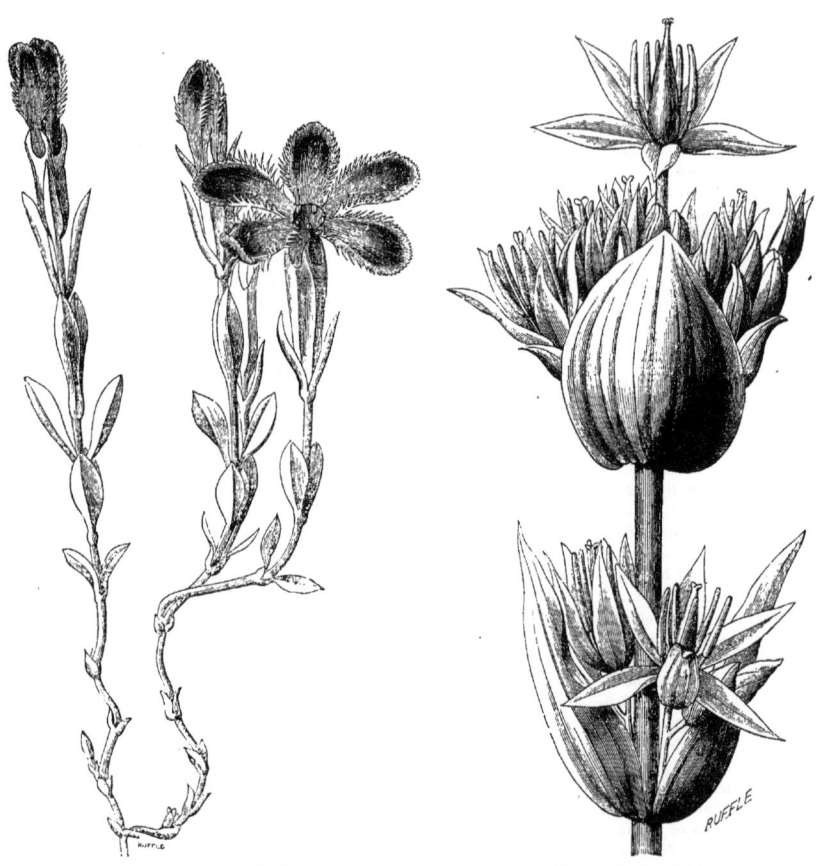

Fig. 69.—*Gentiana ciliata*.

Fig. 70.—*Gentiana lutea*.

Stamens 4–5; stigmas sessile, 2-cleft; capsule 1 cell, 2 valves.

α 1. Flowers yellow or purple; clustered either at the top of the stem, or in axils of the leaves.
β 2. Flowers blue, fringed at the mouth.
γ 3. Flowers blue, smooth.

α.

1. *G. lutea*, L. (yellow gentian). Corolla divided

A variety occurs, *G. luteo-punctata*, having yellow petals spotted with violet, and lobes of corolla obtuse.

2. *G. purpurea*, L. (purple gentian). Reddish-purple corolla, not deeply divided, having petals spotted; flowers generally twice verticillated, the terminal cluster being most profuse; leaves opposite, strongly ribbed, upper ones sessile.

3. *G. Burseri*, Lap. (Burser's gentian). Flowers yellow, in terminal cluster and funnel-shaped; co-

rolla cut one-third of its length; leaves egg-shaped. Plant 9–14 inches high.

4. *G. Thomasii* (Thomasi's gentian). Terminal cluster of bright orange-red flowers funnel-shaped; petals spreading; calyx slightly inflated. Stated by Weber to be a hybrid between *G. purpurea* and *G. lutea*.

5. *G. Pannonica*, L. (Hungarian gentian). Divisions of calyx remarkably narrow and pointed, equal in number to lobes of corolla, which are usually 6; leaves opposite, oblong and narrow. Violet colour with many darker spots.

6. *G. punctata*, L. (spotted gentian). Divisions of calyx unequal, much shorter than the last species; flowers yellow, much spotted; leaves ovate.

7. *G. Charpentieri* (Charpentier's gentian). Probably a hybrid between *G. lutea* and *G. punctata*; terminal cluster of closely set orange-yellow flowers; calyx toothed; leaves ribbed, opposite, and somewhat narrow.

8. *G. Gaudiniana* (Gaudin's gentian). Terminal cluster of rich purple flowers; leaves linear. Probably a hybrid between *G. purpurea* and *G. punctata*.

Fig. 71.—*Gentiana Charpentieri*.

Fig. 72.—*Gentiana gracilis*.

For excellent coloured drawings of *G. Burseri*, *G. Pannonica*, *G. Charpentieri*, *G. Gaudiniana*, vide "Weber's Alpen Flora," vol. iii.

β.

9. *G. nana*, All. (dwarf gentian). Corolla 5-cleft; plant very small; single blue flower, usually terminal; stalked leaves in minute pairs o n the stem, 3 or 4 at the root (*G. tenella*, Rottb.).

10. *G. Germanica*, Willd. (small autumn gentian). Differing from *G. amarella* chiefly in size; calyx 5-cleft, lobes of corolla lanceolate, capsule slightly stalked; leaves oval, lanceolate, pale purple (*G. angustifolia*, Will.).

11. *G. obtusifolia*, Willd. (obtuse-leaved gentian)

Flowers pale purple, terminal. Altogether of looser growth than last species, well marked by obtuse-shaped opposite leaves (another less distinct variety is known as *G. uliginosa*).

12. *G. campestris*, L. (field gentian). Differs from *G. amarella*, in having corolla 4-cleft, flowers more blue.

13. *G. glacialis* (small alpine gentian). Corolla 4-cleft, intensely blue, terminal flower on each branchlet, several times larger than *G. nana;* long stem with few pairs of elliptical leaves ; 4 or 5 root leaves.

γ.

14. *G. ciliata*, L. (hairy gentian). Well marked by fringed corolla ; deep purple-blue flowers single, or several upright branchlets ; leaves lanceolate, lower ones broader.

15. *G. cruciata*, L. (cross-leaved gentian). Corolla 4-cleft ; anthers free ; flowers sessile, in whorls and terminal clusters ; leaves connate, deep blue flowers.

16. *G. asclepiadea*, L. (swallow-wort gentian). A handsome plant with upright stem, 12-18 inches high ; corolla 5-cleft, with minute points between each division, blue with white stripes ; sometimes lilac leaves ovate, lanceolate, pale in colour, not connate ; grows in moist situations.

17. *G. pneumonanthe*, L. (marsh gentian). Upright plant, with deep coloured leaves, green or yellow outside, stalked ; leaves linear-lanceolate.

18. *G. acaulis*, L. (gentianella or stemless gentian). The beautiful deep coloured plant, with bell-shaped corolla, so well known to all frequenters of Alpine meadows ; also cultivated extensively in English gardens. *G. angustifolia* is a narrow-leaved variety.

19. *G. Froehlichii* (Froehlich's gentian). A small single-flowered plant ; corolla 6-cleft, blue ; radical leaves extremely narrow and numerous, leaves on stem resembling the segments of the calyx ; upper part of the root somewhat imbricated ; extremely rare (figured in Weber).

20. *G. frigida*, Weber (frigid gentian). Flower white, spotted with blue, upright ; corolla 5–6-cleft, having small points between segments as in *G. asclepiadea*; leaves lanceolate, partly clasping stem. Very rare, found chiefly in Styria (figured in Weber).

21. *G. excisa*, Presl (notched gentian). Differs from *G. acaulis* in having a decided stem, and the root-leaves broadly ovate.

22. *G. verna*, L. (spring gentian). Corolla 5-cleft, deep blue, or occasionally white ; calyx somewhat swollen ; leaves ovate, lanceolate, root-leaves forming a rosette. The earliest of Swiss spring flowers, carpeting the green mountain-sides with brilliant blue ; plant about 2 inches high, single flowers, or growing in tufts.

23. *G. pumila*, Vill. (small gentian). Deep blue corolla 5-cleft ; calyx smaller than *G. verna;* leaves very narrow, pairs on stem ; small rosette at the base.

24. *G. imbricata*, Schl. (imbricated gentian). Differs from the preceding species in having small blunt leaves thickly imbricated.

25. *G. brachyphylla*, Vill. (short-leaved gentian). Segments of corolla almost dissevered, root leaves only ovate ; small pair leaflets below calyx, stem very short.

26. *G. Bavarica*, L. (Bavarian gentian). Longer stem than last species, leaves in pairs on the stem, obovate, inferior leaves serrated, and partly imbricated ; purple-blue or light deep blue.

27. *G. utriculosa*, L. (inflated gentian). A well-marked species with stellate, intensely blue, denticulated corolla ; calyx inflated in a remarkable mannner ; leaves ovate, obtuse ; single flower on each plant.

28. *G. nivalis*, L. (snowy gentian). Calyx not inflated but slightly angular ; leaves small, egg-shaped ; flowers very small, blue, resembling *G. nana*, but larger, having several flowers and large leaves.

29. *G. prostrata*, Weber (prostrate gentian). Plant branched, adhering to the ground ; stem covered with numerous pairs of leaflets ; exterior of blue corolla of a curiously pale colour, grayish blue.

G. prostrata, *nivalis*, *Bavarica*, *brachyphylla*, *imbricata* and *pumila*, are all well figured in Weber's "Alpen Planten," vol. iii.

Attention may also be called to excellent plates given in "Wild Flowers in Switzerland," by H. C. W., published last year by Sampson Low & Co., the page devoted to species of gentiana being especially good. It is to be hoped the writer will continue her labours, and give us a second volume of the Alpine plants.

C. PARKINSON, F.G.S.

THE PEDIGREE OF THE ELEPHANT.

BY R. LYDEKKER, M.A., F.G.S.

A TIME like the present, when the public mind has been strongly directed to the subject of elephants, by the all-absorbing question whether Mr. Barnum's elephant be white or black, or even whether there be such creatures as white elephants, is one when we are all likely to be anxious to know as much as possible of the history of these interesting and sagacious animals ; and is, therefore, peculiarly well suited for the consideration of the subject of the present sketch. In such a subject it is of course absolutely necessary to introduce a certain amount of anatomical details, which are apt to be somewhat wearisome to the non-zoological reader ; these details have, however, been made as simple as possible, while the use of dry technical terms has been as far as possible avoided.

The reader must, in imagination, transport himself to the far north of India, to the foot of the mighty

Himalaya, and carry back his thoughts to a time long ere the first Aryan invader rushed down from the arid highlands of Central Asia on to the fertile plains of Hindustan; long also ere the earliest of the pre-Aryan aborigines had settled in the latter region, and even to a time when the very mountains that now form the foot of the Himalaya had no existence, and when the Himalaya proper was but a faint shadow of its present mighty self. At that chronologically remote, although geologically recent epoch —considerably before man, as far as we yet know, had made his first appearance on the globe—the regions at the foot of the Himalaya, from the Ganges on the east to the Indus on the west, instead of consisting as now of arid ridges and valleys formed of uptilted strata of clay and sandstone, frequently rising to a height of several thousand feet above the sea-level, were low swampy tracts, covered doubtless in places with thick forest and jungle, and in others consisting of wide, grassy plains, through which the mighty rivers of India flowed in their seaward course. Over that vast expanse of country, which it is quite probable was not then parched by a heat so fierce and intense as that of the Punjab of to-day, there roamed at will a vast assemblage of huge animals; the like of which the world has never since beheld. As they roamed and wandered, untrammelled by any fear of man and his lethal weapons, some of them from time to time met their death from various natural causes, or from the attacks of one another. Frequently their entire bodies became suddenly engulfed alive in the treacherous quicksands which (as we know from the evidence of the strata of soft silvery sandstone) then, as now, formed the river beds; or their already bleached skeletons and bones became gradually entombed in the mud and clay of the swamps and morasses. There they remained for countless ages, during which the soil of these old river-beds and swamps has been gradually upraised to form the present mountains at the foot of the Himalaya: and now in every valley and every gorge of these regions there may be found numbers of the petrified bones of the former denizens of plain and forest, washed out from the solid rock by its slow decay, and waiting but the magic hand of the comparative anatomist to make them tell their marvellous story of the glorious profusion to which the animal life of that long past epoch attained.

In that old Siwalik* epoch, as it is generally called, there roamed over those regions and the present barren plains and deserts of the Punjab and Sind, countless herds of stately giraffes, close akin to those which to-day people the mimosa-groves of Africa, accompanied by droves of extinct species of horses, antelopes, and oxen; while rhinoceroses of several species lurked in the thicker jungle, and ponderous hippopotami wallowed in the lagoons and rivers, as they do now in those of Africa. There, too, might be seen, had there but been human eye to see, countless herds of swine; some of the species reaching to the size of a hippopotamus, while others were scarcely larger than the existing pigmy-hog of Northern India, lately made known to us by the specimens in the Zoological Gardens. Strangest among the stranger forms of quadrupeds, was the ponderous sivathere, rivalling the elephant in bulk, and bearing on his forehead two pairs of horns or antlers—truly a formidable beast. Side by side with these, and a host of other herbivores, which it would be tedious to enumerate, there roamed a variety of large carnivores; some like the tigers, hyænas, and wolves of the present day, and others, like the sabre-toothed tiger of Europe, belonging to types which have completely passed away.

Leaving, however, all the other animals, our attention may be directed more particularly to one group which was especially strongly represented in this old Indian fauna. This group is that of the elephants, of which there were no less than twelve distinct species existing at the time in Northern India; a number far larger than is known in the fossil state from a single region of any other part of the globe, and presenting a marvellous contrast to the group at the present day; when, as we all know it is represented by but two species, confined respectively to the Indian and African regions.

These twelve fossil Indian elephants, from the characters of their grinding-teeth (which, as the parts most frequently preserved, are of the most importance in the determination of extinct animals), may be readily divided into three main sections, which it will be convenient to term respectively true elephants, intermediate elephants, and primitive elephants. The first section includes only two species, to one of which has been assigned the name of the Sutledj elephant, and to the other that of the flat-headed elephant; these names having been taken from the district where the remains were first found, and from a peculiarity in the structure of one of the species. The Sutledj elephant in its main characters was not unlike the living Indian elephant, having tusks in the upper jaw, a short chin to the lower jaw, and the grinding-teeth of great size and depth. Of these teeth, as in living elephants, only one complete tooth was in use at any one time, the back teeth coming into use only as the early ones wore away and fell out. As it is necessary, in order to comprehend the points alluded to in the sequel, to have a clear idea of the structure of these grinding-teeth, a few lines must be devoted to that purpose. These teeth consist of a number of closely packed plates of three different bony substances, and the best idea that the non-anatomical reader can form of their structure is to take a strip of brown cardboard, some three feet in length by four inches in width, and a similar strip of white cardboard; he should then lay the former beneath the latter,

* So called from the Siwalik Hills, in the neighbourhood of the Ganges valley.

and form the two into seven folds, about five inches in height. The folded cards should then be placed on a table with their folds in a vertical direction, with a space of about half an inch between the summits of each ridge, and the intervening troughs or valleys should then be filled to the level of the ridges with fine sand or sawdust. If it be then assumed that the hollows beneath the folds of the brown cardboard are filled up with the same substance, the reader will have a very fair model of a part of the grinding-tooth of a true elephant; a perfect tooth consisting only of a greater number of similar folds. The folds of brown cardboard which form the base of the tooth,

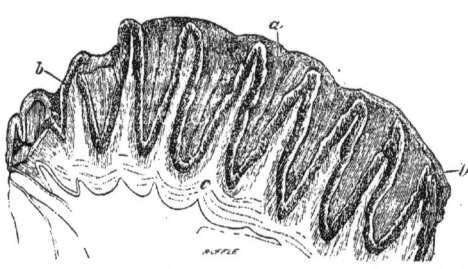

Fig. 73.—Vertical section of a large upper grinding-tooth of the flat-headed elephant, to show the structure of the teeth in the true elephants: *a*, cement; *b*, enamel; *c*, ivory. In the living Indian elephants the folds are taller and more closely packed together. (After Falconer and Cantley.)

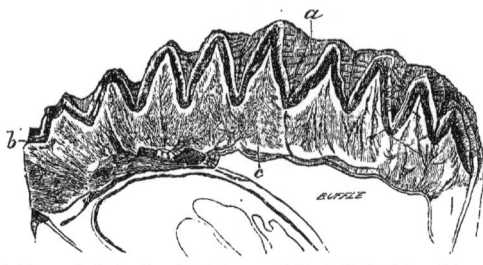

Fig. 74.—Vertical section of the last upper grinding-tooth of an intermediate elephant: *a*, cement; *b*, enamel; *c*, ivory. (After Falconer and Cantley.)

correspond to the ivory of the elephant's grinder; the overlying white cardboard to the enamel, and the sand or sawdust, to an external substance termed cement. It will be readily seen that if a horizontal section be made of such a structure, the exposed surface will consist of layers or plates of the three different substances arranged in the following sequence, viz. brown cardboard, white cardboard, sand, white cardboard, brown cardboard; this surface corresponding to the worn surface of an elephant's grinder, where the order of arrangement of the different substances is the same, viz. ivory, enamel, cement, enamel, ivory. The different degrees of hardness of these three substances produce on

the grinding surface a number of fine ridges, admirably adapted for trituration. In the Indian elephant as many as twenty-four of these folds, or plates, are contained in a single last grinder, while the middle teeth contain from twelve to eighteen of such plates. In the Sutledj elephant the number of these plates is somewhat less, and their vertical height is not so great; from which it may be inferred that in this respect the animal was somewhat lower in the scale than its living Indian congener. In the flat-headed fossil elephant the plates of the grinding-teeth were even yet lower and thicker, as if the folds of the cardboard had been shorter, and not folded quite so close; this type of tooth being more like that of the existing African elephant, which differs widely in this respect from his Indian cousin.

We now come to the consideration of the intermediate elephants, of which four species are known, and have been respectively named Ganesa's elephant, the round-headed elephant, the remarkable elephant, and Clift's elephant. These animals were to all intents and purposes elephants; having but two tusks, and jaws of the same form and structure as those of the existing species; their grinding-teeth are, however, of a much simpler structure than those even of the flat-headed elephant, the ridges being much fewer in number, and lower in height; their elevation being indeed not more than two inches, and the interval between their summits as much as an inch and a half. The cement in the hollows between the ridges is, moreover, present in much smaller quantity, so that the hollows or valleys themselves are more or less completely open. In most of the intermediate elephants the number of ridges in the middle grinders is from about eight to eleven, and twelve or thirteen in the last tooth; but in Clift's elephant the number is reduced to six in the middle teeth, and to seven or eight in the last. From this circumstance the reader will not fail to see that, as regards their teeth, these elephants are on much lower grade than the true elephants, although there is an almost complete transition from Clift's elephant, which is the lowest form, through the round-headed and Ganesa's elephant to the simplest type of true elephant, like the flat-headed species.

Of the primitive elephants, or mastodons as they are more generally called, there are six Indian species, three of which have more complex grinders than the others, and thus effect a regular transition from the intermediate elephants to the latter. The three species with the more complex teeth have been respectively named the broad-toothed elephant, the Siwalik elephant, and the Perim elephant. In the first the ridges are still lower and wider than in Clift's elephant, and the cement is almost entirely absent from the valleys; the number of ridges being four or five in the middle teeth, and five or six in the last. To represent one of these teeth with cardboard, we should simply require the brown and white strips (without the sand), which we should have to form into four folds of about an inch and a half in height, placing the summit of one ridge at a distance of about two inches from the next one : the absence of the sand would leave the intervening valleys quite

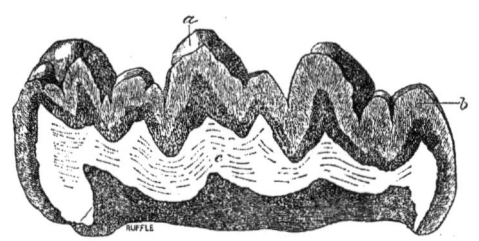

Fig. 75.—Vertical section of a lower grinding-tooth of the narrow-toothed elephant: *a*, cement ; *b*, enamel ; *c*, ivory. (After Gaudry.)

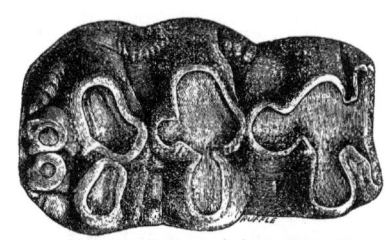

Fig. 76.—Ground plan of a lower grinding-tooth of the narrow-toothed elephant, showing the three low ridges (with lateral expansion) into which the crown is divided.

wide and open. In spite of the very simple structure of its teeth, which the reader will see are just one step farther away from the true elephants than those of Clift's elephant, the animal was otherwise much the same as a modern elephant, having the same number of tusks, and the same short chin, although it is not impossible that its trunk may have been somewhat shorter than that of our living elephants. In the Siwalik elephant the number of ridges in the teeth was nearly the same, although there are certain differences in their plan of structure, into which it is unnecessary for our present purpose to inquire. In the last of these three elephants, or the Perim elephant, the teeth are very similar to those of the Siwalik elephant, but in the form of its chin this animal is markedly different from any of those noticed above. Thus the extremity of the lower jaw, instead of being rounded off, is produced into a spout-like process about six inches in length, which in some individuals (probably males) was furnished with a pair of small cylindrical tusks. This animal differed then from the modern elephants, not only in the much simpler structure of its grinding-teeth, but also in having four instead of two tusks, and in its long and pointed chin. From the circumstance that the middle grinding-teeth in the three above-mentioned species of elephants have only four ridges, they are frequently spoken of as four-ridged elephants.

(*To be continued.*)

A LIVE HERBARIUM.

IT is generally admitted that dried plants are very unsatisfactory things. The leaves often retain a certain amount of their elegant form, and the loss of colour is not great ; but only the most experienced and skilful hands can keep the flowers in a fairly presentable condition. As a rule, it is deplorable to behold the livid masses which were once golden marigolds and spotless lilies, or the pale shrivelled issues which appear as the ghosts of glowing gentians and sky-blue harebells. It is therefore rather astonishing that botanists have not given more attention to growing our native plants. Such a small bit of garden, and such a little attention will do much in this way, and many of our plants will flourish in town gardens. This method of forming a herbarium has also this great advantage. The whole growth of the plant can be watched : the first opening leaf, the formation of the flower bud, the expansion of the blossom, the method of fertilisation, and the perfecting of the seed. Such a garden moreover is a source of endless pleasure, from the pleasant memories it wakes in the mind. A garden which has no associations but seed catalogues and market-gardeners has not half the attractions of a garden in which every plant reminds its owner of some enjoyable ramble, some romantic spot where it was gathered ; while the fact of the flowers appearing year by year at the same time gives double strength to the spell of association. Walking at this time in my very tiny plot of town garden, the rising flower stems of *Primula farinosa* and *Geranium sanguineum* bring to my memory the grand limestone mountains of Wharfedale. *Genista Anglica*, just opening its pretty yellow blooms, tells of a certain familiar wind-swept common ; a bit of white Cochlearia is eloquent of the romantic cliffs of Staithes, and the delicate fronds of *Lastrea thelypteris* just uncurling, recall a picture of the dense rich bogland whence I brought them. It is a mistake to suppose that nothing but "florists' flowers" look well in a garden. Few things look gayer than our familiar blue forget-me-nots, our yellow iris, our scarlet poppies. But there is no need to confine ourselves to British plants, and it is particularly instructive to grow exotics by the side of their native allies. In my garden a plant of *Paris quadrifolia* from the Yorkshire dales is opening its curious flower by the side of the lovely *Trillium grandiflorum*—its Canadian cousin. The blooming season appears identical. Similarly *Primula farinosa* is growing by its beautiful Himalayan sister, *Primula rosea*, and our wood anemone finds a congenial comparison with blue blossoms of *A. apennina*.

I do not of course suggest the substitution of a live herbarium for a dead one, but that it is a most desirable adjunct to it, more beautiful, more instructive, and far more enjoyable, is, I think, a fact not nearly sufficiently recognised. Science is every day bringing horticulture into closer union with botany.

WM. C. HEY.

St. Olave's Vicarage, York.

SCIENCE-GOSSIP.

AN important lecture has just been delivered before the Royal Society of Dublin, by Dr. O. J. Lodge, on "Dust-free Spaces," in which he gave a summary of certain researches he has lately been engaged in, which may ultimately prove of practical importance. It is well known that we owe the blue of the sky to the distribution of ultra-microscopical dust particles—that is, to diffusion in the lower strata of the atmosphere of foreign bodies. What we call *smoke* is only the dust given off during combustion. A cloud or mist is only so much water-dust. But a fog is something more; it is due to water vapour having been condensed around each dust-particle as a nucleus. Dr. Lodge and his colleague, Mr. Clark, have found out how to disperse fogs. All our readers know what they are in London, Manchester, Liverpool, and elsewhere, and of late years they have been increasing both in volume and density, especially in the metropolis, until the phrase, "as thick as a London fog," has become a proverb. Evidently Dr. Lodge is not without hope that it can be artificially dispersed, and moreover he means to try. On a small scale he shows that the air can be cleared of dust (and therefore of fog) by discharging electricity into it! Experiments made with turpentine, smoke, vapour, magnesium ribbon, steam, and artificially composed "London fog," carried out before the audience, demonstrated that a charge of electricity clarified the air containing them. Dr. Lodge thinks it may be possible to clear the air of our railway tunnels, by simply discharging electricity into it, and that some impression may also be made on a London fog by sufficiently powerful electrical discharges. He referred to the old saying that "Thunder clears the air," in proof of the influence which the electricity developed during a thunderstorm has upon the previously murky atmosphere. He intends to experiment in a genuine London fog with large machines, and is of opinion that in this respect the underground railway offers him a tempting field for experiment.

A COPY has reached us of Mr. J. W. Moll's paper (in French) on the Potétomètre, an apparatus devised for the purpose of measuring the aspiration of water by plants.

THE last number of the Proceedings of the Geologists' Association, contains the following papers: The President's address "On the Succession in the Archæan Rocks of America, compared with that in the Pre-Cambrian Rocks of Europe;" "On the Methods which have been devised for the rapid determination of the Specific Gravity of Minerals and Rocks," by Professor Judd; and "Description of the Origin and Distribution of the Water-worn Chalk-gravel of the Yorkshire Chalk Hills, &c."

"THE NATURALIST'S WORLD" is now permanently enlarged to twenty pages monthly, and this month's number contains, among other interesting matters, an article on "Cycling for Naturalists."

MANY years ago, a discussion took place in the columns of SCIENCE-GOSSIP as to who was the author of the once famous "Vestiges of Creation," and it was then conclusively shown to be the work of the late Robert Chambers. In a new edition of this work, Mr. Alexander Ireland, through whose hands the MS. passed, publicly declares Robert Chambers to have been the author.

THE Rev. C. W. Markham writes to say, that whilst some labourers were excavating clay on some level land in the Valley of Aucholme, near Brigg, in Lincolnshire, they discovered, seven feet below the surface, an ancient wooden way, composed of beams of oak laid transversely. Six feet of solid clay lay over it, and Mr. Markham thinks it is of neolithic age.

WE have received the last issues, respectively, of the catalogues of scientific books issued by Mr. W. Wesley, Essex Street, and Mr. W. P. Collins, 157 Great Portland Street. They include both new and second-hand volumes, besides notable papers, journals, &c.

WE have received the catalogue of scientific apparatus and chemicals issued by Mr. Thomas Laurie, 31 Paternoster Row, valuable as showing what a vast number of appliances are now placed at the disposition of students.

A NEW locality for emeralds has been found in North Carolina. The crystals are pale green, and occur in decomposed black mica, associated with quartz, rutile, &c.

UNDER the title of "Tricycles of the year 1884," Mr. H. H. Griffin, of the London Athletic Club, has written the Fifth chronicle of the new inventions and improvements introduced last season and a record of the progress in the manufacture of tricycles. This handy little book is intended to assist purchasers in the choice of a machine. It is published by L. Upcott Gill.

MOST of our readers have doubtless heard of the fact that clover does not seed in New Zealand because there are no humble bees there to fertilise the flowers, and that various futile attempts have been made to introduce these insects. At length Mr. J. C. Firth has succeeded in receiving a consignment of humble bees, which were brought over in a torpid state, in a chilled room. Mr. Firth thinks that if he is successful with their acclimatisation it will save him a thousand pounds a year in clover-seed, which at present has all to be imported.

WE utilise cats to get rid of mice, and the settlers in Dakotah, according to the Rev. Dr. McCook, turn the thatching ant to similar account. These ants are very insectivorous, and the settlers avail themselves of this habit on their part to rid their clothing of vermin. Garments so infected, if left in the vicinity of the formicaries, are quickly and perfectly cleaned of both parasites and eggs.

WE have received from Mr. William Vick, of Ipswich, seven capital photographs of the chief buildings near Mersea Island which were wrecked by the late earthquake. They are as follows: Langenhoe church (various views of the interior and exterior); Peldon church; the Rose Inn, Peldon; the miller's house, Peldon; and the mill showing the fractured shaft of the chimney partly twisted. By this means we get a much more accurate idea of the havoc caused by the earthquake than mere sketching could supply us with.

"THE Butterflies of Europe," by Dr. H. C. Lang, has now reached its eighteenth part, which continues specific descriptions of the genus Satyrus and commences with those of Pararge. The coloured illustrations keep up their high artistic character.

ON May 2nd, Mr. W. Topley, F.G.S., read a valuable paper before the Geologists' Association on "The Agricultural Geology of England and Wales, with reference to the Drift Maps of the Geological Survey."

A TRANSLATION of Dr. Margo's thoughtful paper on "The Classification of the Animal Kingdom, with reference to the Newer Zoological Systems," appears in the last number of "The Annals and Magazine of Natural History."

PROFESSOR HAECKEL, Dr. Kowalevsky, and Dr. Schwendener, have been elected foreign members of the Linnean Society.

UNDER the title of "Notes from my Aquarium," Mr. George Brook, F.L.S., has reprinted a series of twenty-four short articles. They are all eminently practical, for Mr. Brook built an aquarium for himself, thirty feet by eighteen feet, containing four tanks. This is connected with two underground reservoirs of 2000 gallons each, so that his aquarium is on an unusually large scale for a private individual. These "notes" deal with practical observations, some on fish parasites, others on rare fishes, mollusca, &c. We regard the pamphlet as a valuable contribution to practical zoology.

A MUSEUM of archæological and ethnological antiquities has been established and was formally opened at Cambridge, on the 6th ult. Professor E. B. Tylor, Dr. John Evans, and other celebrated ethnologists were present at the ceremony.

MR. THOMAS BOLTON has exhibited before the Royal Microscopical Society a rhizopod (*Clathrulina elegans*) from Epping Forest, which employs a fourth method of reproduction, in the formation of flagellate monads.

THE white spots on the planet Venus have been the special object of observation by M. Trouvelot, and he recently stated before the Paris Academy of Sciences that he had taken no fewer than 249 observations of them. The northern spot alone was visible on April 5th. These spots do not appear to be affected by the diurnal rotation of the planet, and M. Trouvelot thinks it must be because its axis passes through them or very close to them. He expresses his opinion that they are the summits of high mountains projecting above the cloudy envelope which covers Venus.

THOSE who have the opportunity should not lose it, of visiting the establishment of Mr. William Bull, F.L.S., the celebrated introducer of exotic plants, at Chelsea, as his annual exhibition of new, rare, and beautiful orchids is now open, and will continue so until the end of July.

AMONG the wonders of cheap scientific literature are the penny handbooks on collecting objects, published by W. Swan Sonnenschein & Co., all of them written by men of high eminence in their several departments of research. Thus we have one on Flowering Plants, by Mr. James Britten, F.L.S.; on Butterflies, Beetles, and Insects, by Mr. W. F. Kirby; on Shells, by Mr. B. Woodward, F.G.S.; on British Birds, by R. B. Sharp, F.L.S.; on Postage Stamps, by Mr. W. T. Ogilvy; and on Greek and Roman Coins, by B. Y. Head, M.R.A.S., &c.

WE are much pleased to welcome a second edition of Mr. Edward Step's "Plant-Life" (London: J. Fisher Unwin). Few books on popular botany have been so successful in including so wide a range of the subject with such complete accuracy. It well deserves its success.

THE third Report of the United States Entomological Commission, relating to the Rocky Mountain locust, the western cricket, the army worm, canker worms, the Hessian fly, together with descriptions of larvæ of injurious forest insects, studies of the embryological development of the locust and other insects, &c., has just been issued from the United States printing office. The chief contributors are Messrs. Riley, Packard, and Thomas, who have turned out a useful and handsome volume.

THE International Food and Health Exhibition was opened by the Duke of Cambridge on May 8th, in the building occupied last year by the Fisheries Exhibition, the aquarium of which is still in working order and on view.

THE Linnean Society of New South Wales have offered a prize of £100 for the best essay on "The Life-History of the Bacillus of Typhoid Fever."

IN the Transactions of the New Zealand Institute, Mr. J. A. Pond describes the occurrence of a vein of Platinum, in octahedral crystals, associated with quartz in the Thames gold district.

A NATURALISTS' society has been formed at Clevedon, Somersetshire, for the study of the fauna and flora of the neighbourhood, and the formation of a local museum. President, Sir E. H. Elton, Bart., Clevedon Court; hon. secretary and curator, R. J. Morgan Esq., Wellington House, Clevedon.

IT will be remembered that the Dutch Government appointed a commission to investigate the nature and results of the eruption at Krakatoa, in August last. The Report has just been presented by Dr. Verbeek, and it is a remarkable illustration of scientific courage and adventure. At the same time it shows that the eruption was on a scale which might almost be called catastrophic. Krakatoa lies on a rent or fissure in the crust of the earth which runs across the Straits of Sunda. Dr. Verbeek thinks that sea water may have been admitted thus to the molten matter beneath, so as to form steam at high pressure. The sound of the volcanic explosion of August last was heard over a space equal to one-sixth of the earth's circumference. So violent were the air-waves caused by the explosion that walls were rent by them (not by earthquakes) at the distance of 830 kilometres away. One air-wave was propelled from Krakatoa which travelled no less than three and a quarter times round the circumference of the earth. The large tidal wave appears to have been caused by the northern part of the mountain giving way. There only remains the southern part, which has been cut in two from the very top, and forms on the north side a magnificent precipitous cliff more than 2,500 feet high. In the place where the fallen part once stood there is now everywhere deep sea, in some places as much as 1,000 feet deep. The ashes thrown out by the eruption must have been enormous, especially if we are correct in assuming the brilliant sunsets are due to the finer parts suspended in the atmosphere. Within a circle of fifteen kilometres' radius from the mountain, the layers of volcanic ashes thus ejected cover the ground from 60 to 80 metres thick. The known surface over which the ashes were projected, as calculated by Dr. Verbeek, is 750,000 square kilometres—apart from other unknown areas where they also descended. He thinks that the finer particles, propelled by the wind, have made a journey round the world. The vapour was condensed to water, and froze in the cold currents. The refraction through these innumerable ice crystals, Dr. Verbeek thinks, caused the beautiful and red glows of our phenomenal sunsets. He calculates that the quantity of solid substances ejected by the volcano was 18 cubic kilometres. Dr. Verbeek and his staff thoroughly explored the island last October, when the heat was so great as almost to stifle them.

MICROSCOPY.

IMPROVEMENTS IN MICROSCOPIC SLIDES.—Mr. B. Piffard has obtained a patent specification under the above title, and has sent us a slide made in accordance with it as a sample. A round recess is made in the glass slip of the necessary depth. In this the object to be mounted is placed, and covered with the usual thin disk of microscopic glass. In this way the upper surface is perfectly smooth, as the surface of the disk is even with that of the slip. No danger occurs, therefore, of the cover-glass and object being knocked off; and the hollowing of the recess, which is done by means of a diamond, causes a very beautiful diffusion of light, such as we have not noticed in ordinary mounts. For beauty of finish and neatness, there can be no doubt that Mr. Piffard's plan is superior to any yet brought out.

JOURNAL OF THE ROYAL MICROSCOPICAL SOCIETY.—The April part of this welcome and excellently edited serial contains besides the current researches relating to zoology, botany, and other departments of microscopical research, the following papers: "Observations on the Life-History of *Stephanoceros Eichornii*" (illustrated), by Mr. T. B. Rossetter; the President's Address, by Professor Duncan; "On the Mineral Cyprusite," by Julien Deby; "List of Desmidieæ found in Gatherings made in the Neighbourhood of Lake Windermere during 1883," by Mr. J. P. Bisset; and on "The Formation and Growth of Cells in the Genus *Polysiphonia*," by Mr. George Massee.

THE MARKINGS ON TEST DIATOMS.—These have always been a bone of contention amongst microscopists, some considering them depressions, whilst others affirm that they are elevations. Twenty years ago, the former idea was the most prevalent, and I remember the late Richard Beck teaching me the right way to show a Pleurosigma as a pearly shell dotted over with dark pits or depressions ; but now it is the fashion to exhibit them as transparent silicious valves with bead-like elevations. For many years I have been of opinion that both the exterior surfaces of the valves are perfectly smooth, and devoid of markings of any kind ; but that they are composed of two layers or laminæ, on the interior surface of one of which are attached a more or less net-like arrangement of ribs, like those on the under side of a *Victoria regia* lily-leaf ; and that these, resting on the inner surface of the other laminæ, form, as it were, so many interspaces, through which the light being thrown at different angles, produces varying effects. I do not remember who suggested this theory ; but I was led to the conclusion by observing that, after having focussed the surface of the valve as exactly as possible, I had invariably to focus downwards before the markings appeared. I need scarcely remark that in the valves of arachnoidiscus, and of many of the Coscinodisci, there are evidently two distinct laminæ, and, from analogy, we may presume it may be the case also in other genera, though they are not so easily separable. I have brought forward the subject again, as the other night I was showing to an intelligent relative, but a novice at the microscope, the strong crossed markings on the very coarse navicula rhomboides of the cherry-field deposit. He saw them distinctly, but playing with the fine adjustment to improve them, he exclaimed, "There they are, much better, more like beads ! but irregularly scattered over the shell !" On looking through the instrument, I found he had focussed up to the surface of the valve, which happened to be an abnormally abraded one, and he had taken these abrasions for the beads. He had afterwards to focus considerably downwards, so much so, that the superficial abrasions were lost sight of, before he could recover these markings. Q.E.D. as we used to say in Euclid.—*Fred. H. Lang.*

A NEW MICROTOME.—Mr. A. B. Chapman has just registered a very ingeniously constructed Microtome, which has for its cutting surface two parallel glass plates cemented to a block of mahogany, through which is inserted a brass cylinder at right angles to the glass plates ; in this cylinder (which forms the "well" of the Microtome) an accurately fitted brass plug works, carrying on its top a flat-headed table-like piece which entirely prevents the imbedding agent from rising or turning round while the sections are being cut. The plug is moved up and down by a brass disc, which revolves between the block of mahogany mentioned above, and a similar block underneath. The brass disc is graduated on the edge of its upper surface, each graduation representing a movement of ·0005 in. of the plug. The Microtome has a base-board which can be firmly clamped to a table, and the whole is so conveniently arranged that every operation or adjustment can be made at once, the whole being in view on the table. The instrument is beautifully finished, and fitted with a clamp into a neat mahogany box.

CARBOLIC ACID AND CEMENT.—While working in the Strassburg microscopical laboratories three years ago, I mounted several fresh-water Algæ in a weak carbolic-acid solution, using asphaltum for the cement ; their condition is still perfectly good. I would suggest that after treating the insect in a strong solution of carbolic acid, to produce the effects on the chitine narrated by W. H., he should steep in a 2 per cent. solution, then mount in the same solution and cement with asphaltum ; the cell, if necessary, can also be made of asphalt cement.—*B. Sç., Plymouth.*

THE MICROSCOPICAL SOCIETY OF LIVERPOOL.— At the last meeting of this society, a paper on the "Larval Forms of the Echinodermata" was read by Professor Herdman. The lecturer referred to the somewhat isolated systematic position, and the wide distribution in space and time of the Echinodermata, and to the doubts which still exist as to their relationship with other groups of the Invertebrata. He then, after briefly describing some of the more important anatomical peculiarities of the Echinoderm, proceeded to discuss the embryology of certain members of the five classes living in the seas of the present day. Commencing with the Holothuroidea (or sea cucumbers), the development of *Holothuria tubulosa* and of *Cucumaria doliolum* was traced, from the fertilisation of the ovum to the Auricularia stage, according to the observations of Selenka. The characters of the Auricularia were then discussed, and its metamorphosis through the pupa stage into the young Holothurian described. The development of the Asteroidea or star-fishes was exemplified by Prof. Alexander Agassiz's investigation of various species of Asteracanthion. The evolution of the Bipinnaria and Brachiolaria stages was traced, and their structure compared with that of the Auricularia. Then the remarkable transformation of the free swimming larva into the young star-fish was briefly described, and the ultimate fate of the ciliated processes discussed. The lecturer then passed on to the Ophiuroidea (Brittle stars), and in Echinodea (sea urchins), in which the larva has the form of a Pluteus characterised by the possession of a large post-anal lobe, and a provisional calcareous skeleton. The recent investigations of Apostolides and others were discussed, and the course of development of the various larval organs was compared with that which had been previously described in the Holothurids and Asterids. In describing the metamorphosis of the Pluteus into the adult Echinoderm, the importance of Metschoukoff's recent discovery of the action of Amœboid mesoderm cells in absorbing temporary larval structures such as the calcareous spicules was pointed out. The life-history of Comatula, one of the Crinoïdea, was illustrated from the older observations of Sir Wyville Thomson, and the more recent work of Götto, and was traced from the pseudembryo to the pentacrinoid stage, and from that to the adult condition. In conclusion, the various larval forms were compared, and their possible bearing upon the classification of the group and its relationship to the other invertebrata was briefly discussed.

MR. COLES' STUDIES.—Numbers 16 and 17 of the second vol. of the "Studies in Microscopical Science," are devoted severally to "Epidermal Tissue" and "Development of Bone," both exquisitely illustrated. That in the former part is a highly artistic coloured drawing of a transverse section of the aerial root of Dendrobium, × 130, from the well-known pencil of E. T. D., the designer of our own coloured plates. The slides sent out to illustrate these parts are among Mr. Coles's best. Part 9 of "The Methods of Microscopical Research" continues the subject of "Mounting," and gives explanations of the kinds of materials to be used. Number 8 of "Popular Microscopical Studies" is taken up with the subject of "The Intestine," and is illustrated by a coloured plate of the transverse section of the ileum of cat,(injected) × 50.

ZOOLOGY.

MOLLUSCA OF BRENTFORD DISTRICT.—The following list of shells taken yesterday in a walk from Brentford to Hanwell, may be of some use to London conchologists, as the district appears to be a good one, and would, I think, well repay a careful search. 1. Freshwater species, taken for the most part in the river Brent.—*Sphærium rivicola* (dead shells only, but from the abundance of these, this species must be very abundant in some part of the river), *S. corneum*, and also its var. *flavescens*, *Unio tumidus*, *U. pictorum*, *Anodonta anatina*, *Dreissena polymorpha* (single valves very abundant), *Paludina vivipara*, *Bythinia tentaculata*, *B. Leachii*, *Planorbis corneus*, *P. vortex*, *Physa fontinalis*, *Limnæa peregra*, *L. palustris*, and *Ancylus lacustris*. 2. Terrestrial species; these I did not search long for, and so my list is scanty.—*Arion ater*, var. *succinea* (a small specimen, why is this var. always small?), *Arion hortensis*, *Limax agrestis* (also var. mottled with black), *Succinea elegans*, *Zonites nitidulus*, *Helix aspersa*, *H. hortensis*, (also vars. *incarnata* and *lutea*), *H. cantiana* (I found two young specimens having the last whorl strongly tinged with rufous, except a band corresponding to number three in *H. nemoralis*, which was strongly marked out in white), *H. rufescens*, and *H. rotundata*. On my way home, I passed through Ealing, where there are two large ponds, one on each side of the road. The pond on the south side contains *Planorbis corneus* (some of the young specimens were covered with very fine longitudinal striæ, such as are seen in *P. albus*, but not so well marked), *P. vortex*, and *L. peregra*. The pond on the north side contains *P. corneus* (large specimens), *P. albus*, *L. peregra*, and *Physa fontinalis*.—*T. D. A. Cockerell.*

THE CRESTED NEWT.—I have to-day (April 5th) taken a specimen of the crested newt (*Triton cristatus*) having a clearly divided tail. The tail is of the ordinary fully-developed size and shape common at this time of the year, but within three-quarters of an inch of the tip there is on the right side a small tail clearly margined, and having a white band running throughout its length. This tail is about a quarter of an inch long and nearly the same in breadth, and is rather abruptly truncated. It is not in any way a division or splitting of the main tail, as it grows out quite distinct from it, and the large tail is entire. I enclose a rough sketch, natural size. Is this a common occurrence in this kind of newt?—*Alfred Sutton.*

THE SCARCITY OF BUTTERFLIES.—I notice that at a meeting of the Entomological Society, several members expressed their opinion that butterflies are becoming scarcer. I can fully indorse their opinion that it is so true; and my opinion is that it is attributable to the increasing numbers of collectors; but more so to the wholesale slaughter that some of them make. I know of two or three collectors who took last year over two hundred specimens of the orange tip, as well as scores of red admiral, painted lady, small heath, small copper, small tortoise-shell, &c. If this goes on all over the country, there is no wonder at the scarcity of these lovely creatures. I hope all naturalists' societies will urge their members to take only a few specimens of a sort, and put down this wholesale slaughter of our summer visitors.—*J. T. Jepson.*

SCALES (?) OF THE TURBOT.—Can any reader of SCIENCE-GOSSIP confirm, from his own observation, the statement made in some standard books on fishes, that the turbot has small, smooth, cycloid scales, besides the bony tubercles with which every one is acquainted? Yarrell writes that the skin of this fish is "studded with hard, roundish tubercles, the surface otherwise being covered with small, smooth scales." Dr. Günther writes of it: "Scales none, or small and cycloid." Jonathan Couch distinctly states that "the upper surface is without scales," but mentions the tubercles. Donovan writes, "The skin is covered with scales, but those are extremely small." I have never been able to discover the small cycloid scales mentioned, either by scraping away the epidermis, or by examining the skin, properly prepared, under the microscope by different illumina tions, including that by polarised light, which exhibits so clearly and beautifully the subcutaneous scales of the eel. I venture to caution intending observers that many scales may generally be found adhering to the turbot's skin; but they are those of other fishes which have been in contact with it in the trawl, or the trunk, or the fishmonger's ice-box. These should therefore be carefully washed and wiped away, before search be commenced for those which are said to belong to the turbot itself, but the existence of which I doubt. Donovan also states that the turbot has tubercles both on the upper and under side, those on the upper surface being larger and more numerous than those on the under, or white side. This is certainly incorrect. The turbot has no tubercles on the under side, except in the rare cases of malformation occasionally seen, when the fish has both sides alike dark-coloured.—*Henry Lee, Savage Club.*

PROVINCIAL SOCIETIES.—There is no more significant sign of the progress of science than the fact that local scientific societies are springing up and multiplying in every town and city of the empire. In this way local talent is made the most of, and papers are often read at the meetings of these societies equal to many of those which appear in the Transactions of the learned societies of London. Among the publications of these provincial societies we may mention the following as having lately appeared:—The Proceedings of the Norwich Geological Society, vol. I. part viii. containing papers on local geology by Messrs. F. W. Harmer, F.G.S., John Gunn, F.G.S., H. B. Woodward, F.G.S., W. Whitaker, F.G.S. (President), F. J. Bennett, F.G.S., T. V. Holmes, F.G.S., S. G. Sothern, &c. The 31st annual report of the Nottingham Naturalists' Society contains the following papers:—"Disease Germs" (illustrated), by Dr. Seaton (President); and "Bank Tokens and their Forgeries," by Mr. G. Mundon (illustrated). The 14th Annual Report of the Wellington College National Science Society contains notes and report on the local entomology, flora, and phenological observations, together with admirably condensed abstracts of the numerous papers read at the meetings. The Transactions of the Ottawa Field-Naturalists' Club show that our kinsfolk across the water are manifesting their love of Nature in the same manner that we are. The last number contains a most able address by the President (Mr. James Fletcher); papers on "The Laurentian System," by Mr. F. D. Adams; "Fishes of the Ottawa District," by Mr. H. B. Small; "Fossils from the Trenton Limestone," by Mr. W. R. Billings; "The Ducks of the Locality," by Mr. W. P. Lett; together with accurately drawn-up reports on the geology, mineralogy, palæontology, botany, conchology, entomology, ornithology, &c., all showing good and valuable work in the cause of practical science.

BOTANY.

ON THE CONTINUITY OF PROTOPLASM.—Though in no way and in no sense responsible for the paragraph to which Mr. Gardiner takes objection, I cannot allow his remarks thereon to pass altogether unnoticed. Ostensibly written on a question of priority, nearly two-thirds of his letter is an attempt to discount the value of my paper on "Protoplasmic Continuity in the Florideæ," published in the February and March numbers of "The Journal of Botany." To me this appears a somewhat irregular mode of procedure, but I do not care to dwell upon it. On the matter of priority, Mr. Gardiner writes as if his work and mine were identical. This is not the case. So far as I have yet learnt, his investigations have had reference to certain special tissues of phanerogams, while mine have dealt with the whole thallus of several leading genera of the large group of Florideæ. This being so, it seems to me that each may fairly claim priority over the other in his own branch of the inquiry, but nothing more. So much I, for my part, am very willing to allow to Mr. Gardiner, and so much I must demand for myself. Having dealt with the question of priority, Mr. Gardiner turns off at a tangent, and endeavours to discredit my results on two grounds: (i.) that they are not new, but were known before, and (ii.) that they are inaccurate. Now if he knew of these results before my work was published, and if, besides, he knew them to be inaccurate, why did he wait for the publication of my paper before taking upon himself to controvert them? Is it to be inferred that he was only led to look up the literature of the subject when he became acquainted with what I was doing? But, despite the authoritative tone assumed by Mr. Gardiner, I am not disposed to accept his view of the extent to which my results were anticipated, and I am supported by botanists, no less competent than himself, in the opinion, that none of the authorities he mentions definitely realised in their entirety the various points I have sought to establish. Chief among these are the following: (i.) that protoplasmic continuity is very widely, if not universally distributed in the Florideæ; (ii.) that where found it obtains practically over the whole thallus; (iii.) that the connecting threads are permanent and not merely temporary structures; (iv.) that the threads retain their vital activity, growing in thickness with the growth of the cells; and (v.) that they give rise to differentiated structures. Be this, however, as it may, I am content that my statements can now be examined and tested by all who care to do so, and to their judgment I leave them. That Mr. Gardiner has not given them attentive perusal I can hardly believe, but I had difficulty in repelling a suspicion to that effect when I read in his letter that "there can be little doubt that the continuity of the protoplasm in the

Florideæ is not maintained by open pits as he (Mr. Hick) has stated. By appropriate treatment it is easy to see that, except in the very youngest cells, a distinct pit-closing membrane is present." Will the readers of SCIENCE-GOSSIP believe that I have myself described in several species the existence of an extremely delicate membrane, crossing the connecting threads, which I presume is what is here termed the "pit-closing membrane"? But, at the same time, I have advanced reasons for believing that this is developed by and within the cord, that it is not an ordinary cellulose partition, and that it offers no interruption to the continuity of the protoplasm, so long as the cells are alive. Where no such membrane was detected, as occurred in *Chondrus crispus*, *Gigartina mamillosa*, and *Petrocelis cruenta*, and other forms, none was described as a matter of course. If Mr. Gardiner has been more successful, both he and botanical science are to be congratulated. In that case, I am sure he will not object to inform the readers of SCIENCE-GOSSIP by what "appropriate methods" "it is easy to see a distinct pit-closing membrane" in these species.—*Thomas Hick*.

THE BOTANICAL EXCHANGE CLUB.—The report of this club, drawn up for the observations and collections of 1882, has just been published. The total number of plants received for distribution was 3440, from thirty-five contributors. The "Notes on the Plants gathered in 1882" are most interesting to all botanists, every new variety of a species and every new habitat being authoritatively set down.

PECULIAR NARCISSUS.—I enclose a specimen of *Narcissus pseudo-narcissus*, found in a field near Ashopton, Derbyshire. Instead of six stamens it has only four normal, and another, the filament of which is united to the corona through all its length, and the anther lobe has a winged appendage on either side; besides which there arises from its base a leafy appendage. The parts of the perianth are only four. It is evidently approaching the double form, and was found growing among a great many others, a large bunch of which was gathered, but no abnormality could be detected in any of them.—*G. A. Grierson*.

EARLY PUCCINIA.—During an excursion to Blackwell, Worcestershire, on the 22nd of March last, I found the leaves of a large patch of *Adoxa moschatellina* plentifully besprinkled with the pustules of *Puccinia adoxæ*. Is this unusually early appearance to be put down to the mildness of the season?—*W. B. Grove, B.A.*

THE SUNDEWS.—There need not, as Mr. Birks points out, be any difficulty in the cultivation of these curious and interesting insectivorous plants. They should, however, be taken up with plenty of soil attached. To my mind, the common *Drosera* *rotundifolia*, and particularly *D. longifolia*, with its long-petioled, radical, spatulate, and very obtuse leaves, are very well worth growing, if only for curiosity's sake. I have rarely seen the flowers of the common species fully expanded, even amongst thousands of specimens; it would be interesting to know if this circumstance is frequent.—*W. Roberts*, 157, *Camden Grove North, Peckham.*

THE ORIGIN OF [DOUBLE FLOWERS.—In Mr. Mott's paper last month, at page 99, col. 2, line 24 from bottom, for *organics* read *organs*. Page 99, col. 2, line 19 from bottom, for *absorbed* read *aborted*. Page 100, col. 1, line 22, for *male* read *wave*.

VEGETABLE TERATOLOGY.—You will perhaps consider the following lusus naturæ worthy of a corner in your pages. It consisted of three leaves of the ordinary domestic cabbage, each of which bore upon their midrib from two to six secondary stalks. The smaller of these secondary stalks supported a miniature leaf resembling the parent leaf, but with more undulated and divided margins; the larger again subdivided into similar tertiary growths; and these again, in one or two instances, into further quaternary (if that is the correct word) ones.—*Augustin Ley*.

CONVALLARIA RUBRA.—I shall be glad to learn, from any reader of SCIENCE-GOSSIP, or through your Notes to Correspondents, what has become of *Convallaria rubra*? It is mentioned in Paxton's "Botanical Dictionary" as a native of Britain; and a gentleman told me a few days ago that it grew at Skipton in 1845. But there is no mention of such a plant in Hooker's "Student's Flora," nor in the London Catalogue. Has it become extinct?—*F. J. George.*

ŒCIDIUM ARI.—I am pleased to say that I have just found this fungus in some quantity at Limpley Stoke, near here. This is the first season I have been fortunate enough to obtain specimens, for though yearly on the look-out for this Œcidium, I have, till the present time, failed to find it. As *Œcidium Ari* is by no means common, I shall be pleased to send specimens to any one sending me a stamped and addressed envelope.—*Charles F. W. T. Williams, B.A., Bath.*

PECULIAR APPEARANCE OF THE MOON.—My son and I saw the following remarkable appearance of the moon in the west, at Burnham, on Saturday February 2nd, 1884. The best description I can give of rough figure sent is that it was the shape of a boomerang, the north and largest end with a rather narrowed but deep notch, cut as if with a fret-saw. My wife saw it at the same time with my son and myself. It retained this shape as long as we looked at it. Two hours later intense frost set in.—*A. H. B.*

GEOLOGY, &c.

GALENITE IN SANDSTONE.—At a cutting to the south of Paisley, made by the company of the Paisley and Johnstone Canal Railway, which is in course of making, galenite is found in the sandstone there. Galenite, which is the sulphuret of lead (otherwise called galena), is found in rocks of different formation, but its occurrence in the sandstone is not common. It is mentioned in the third volume of Mineralogy by Robert Jamieson, second edition 1816, as occurring "in small veins or disseminated in the grey sandstone of the coal formation in the Lothians and Fifeshire." The same is mentioned by Gregg and Lettsom (1858) in Manual of Mineralogy of Great Britain and Ireland, but by them another locality is given, viz. Cumberhead, Lanarkshire, and there also in sandstone of the coal formation. After this other writers in mineralogy have copied the same localities. In the locality near Paisley, where the galenite has been discovered, the sandstone is also of the coal formation, in which there are many fossil plants of the carboniferous period. At the point where the mineral is found most abundantly there is a fault, and one portion of the strata has a dip of nineteen degrees to the south. In the fault slickenside is well shown, and in cavities in the fault there are clay and septaria; on breaking some of these I also found galenite disseminated through the structure in addition to calcite or carbonate of lime. In the sandstone the galenite is in small veins, or disseminated, more generally the latter. This being so, is it not reasonable to suppose that lead in some form exists near by? Considering the composition of the earth's crust, lead is to be regarded as scarce. There are sixteen of the known elements that make up ninety-nine parts of the earth's crust, and lead is among the number that compose part of the one-hundredth part of the earth's crust, iron being excepted. Iron, being thus common in the earth's crust, is often found in fossils in the form of pyrites, the iron having united with the sulphur from the animal decomposition and formed the pyrites. Galenite is simply a chemical union of lead and sulphur, yet it is not common in fossils, but does occur in localities where lead exists. Sir H. De La Beche, in "Geological Observer" (1853), mentions galenite occurring in cavities left by fossil molluscs in lias near Merthyr Mawr, Glamorganshire, also in cracks in fossil wood at Dunraven Castle, but points out that lead exists in some form near by. Galenite is formed artificially by heating the oxide or silicate of lead with vapour of sulphur, or by suspending sulphate of lead in a bag in water charged with carbonic anhydride in which putrid fermentation is kept up (as by keeping an oyster in it), the galenite incrusting upon its shell). The galenite found near Paisley, from analyses I have made, shows in addition to the lead and sulphur, some iron and silver.—*Taylor, Sub-Curator, Museum, Paisley.*

A NEW BRITISH MINERAL.—The last number of the "Mineralogical Magazine" contains a note by Mr. Arthur Smith Woodward, of the British Museum, on the occurrence of Evansite at a locality near Macclesfield, in east Cheshire. Its chief interest lies in the fact that this mineral has never before been recorded as occurring in the British Isles, and does not appear to have been recognised hitherto in any part of the world but the province of Gömör Comität, Hungary. It was first described twenty years ago by Mr. David Forbes from the latter locality, and the original specimens are now in the Museum of the Owens College, Manchester. Evansite is a highly hydrated phosphate of alumina, containing as much as 40 per cent. of water, and usually very free from impurities. It is glassy in appearance, very similar to the hydrated form of silica known as Hyalite, and likewise covers the sides of fissures with a nearly transparent film of varying thickness, rising in innumerable little lumps, which impart to it characteristics that are unmistakable; it can be readily distinguished from Hyalite in the field, in consequence of its exceedingly soft and brittle nature. The mineral near Macclesfield is associated with oxides of iron and manganese, calcite, pearl spar, and zinc-blende, and the conditions are very similar to those under which it occurs in Hungary. As such conditions are not uncommon, particularly in the Carboniferous formation, it seems strange that Evansite has not been observed in this country before, and those who are interested in mineralogical matters will do well to search carefully for it among the crystalline incrustations in the fissures of rocks exhibiting the peculiarities referred to.

NOTES AND QUERIES.

"CAIN AND ABEL."—The plant known by this "odd name:" the "Dictionary of English Plant-Names," by Messrs. Britten and Holland, under the head of "Cain and Abel," says "the tubers of *Orchis latifolia*, Lin. Cain being the heavy one," with a reference indicating that the name is in use in the "Eastern Borders," which "comprehend the whole of Berwickshire, the liberties of Berwick, North Durham, and the immediately adjacent parts of Northumberland and Roxburghshire." The plant also goes by the name of "Adam and Eve," the tuber which sinks being Adam and that which swims being Eve. But this name is applied also to *Orchis maculata, Orchis mascula, Arum maculatum, Pulmonaria officinalis*, and *Aconita Napellus*, principally apparently, because these plants bear flowers of two different shades, dark and light—the dark, of course, being "Adam," and the light being "Eve." And the latter (*Aconita N.*) because "when the hood of the flower is lifted up there is an appearance of two little figures."—*Chas. Browne.*

A REVERSED HELIX ASPERSA AT BRISTOL.—On the 12th of March I found a very good, although a dead specimen of *Helix aspersa* reversed, in a lane near Bristol. It may be interesting to some conchologists who read this paper, to hear that about four years ago I found a dead reversed *Helix aspersa* not five minutes' walk from this spot, in the same lane, on the left-hand side to the nearest village; that almost opposite the hedge six years ago, I found a fine reversed yellow *Helix hortensis;* that a quarter of an hour's walk further on, my sister, Miss Jessie Hele, found a few years ago a young reversed *Helix aspersa*, which she reared to maturity. And I, where she found this shell, took my second yellow *Helix hortensis* reversed. So that in one district of Bristol, within half an hour's walk of one part of the lane to to the other, we have taken three reversed *Helix aspersa*, and two plain yellow *Helix hortensis* reversed. A little further off to the right of this lane I found about the same time a beautiful reversed banded *Helix hortensis*; and at Keynsham, near Bristol, my sister found a *Helix hortensis*, plain yellow, reversed. Shortly after, my servant at Yates found a fine reversed *Helix aspersa*. So, taking a circle of about twenty to thirty miles round Bristol, in six years we have added from the neighbourhood to our collections four reversed *Helix aspersa* and five reversed *Helix hortensis*. Some of the most curious monstrosities in the British Museum of *Helix aspersa*, I believe, were found near Bristol; and were it not for the snail-eaters would help this city, I believe we should add many more rare discoveries, to the interest of the Bristol neighbourhood in the eyes of conchologists.—*Fanny M. Hele, Bristol.*

THE ROOK.—As far as my experience goes, "Rook" is an exclusively English word. I never heard it used in Scotland or Ireland by country people. "Rooks" are always "Craws" there, causing no little confusion. I think, however, but am not sure, that many Scotchmen have a distinction of their own, calling rooks, crows; and crows, corbies. I have also often heard them call crows, hoody craws. These were the carrion crows, not the royston kind.—*H. J. Moule.*

CURIOUS HABIT OF FISH.—It has struck me that perhaps some of the readers of SCIENCE-GOSSIP could explain the following curious freak :—Several gold fish which I have kept in my aquarium, after a time have acquired a very ridiculous—and, it appears, fatal—habit. They swim up to the top of the water with great velocity, thrust their heads out of the water, and take in a mouthful of air. Then, swimming towards the bottom of the tank (which requires great exertion, as the air in the mouth buoys the head upwards), they discharge the air from their mouths, in a succession of bubbles. This performance evidently causes great exhaustion, as the fish is hardly able to swim for some time after. This habit after a time ends with the death of the fish. Now what can be the motive for such a fatal habit? I have now one of my fish which is constantly at it. I expect he will die ere long. I may as well say it is not for want of fresh water, as the other fish are all right and they take no notice of his freaks; and he will do the same, on being put into a tank of fresh, clean water.—*W. Finch, jun.*

ABUNDANCE OF STAR-FISHES.—Visiting the shore between here and Pendine (Carmarthen Bay) on February 2nd, after several days of rough weather, I was surprised at the immense numbers of starfish left by the receding tide. For a distance of fully a mile and a half by one hundred to two hundred yards in width, the shore was heaped with scores of thousands of *Uraster rubens* (five finger stars) and *Ophiocoma rosula* (brittle star), and for miles the sands were covered with shells, such as *Ceratisolen legumen, Solen ensis, S. siliqua, B. undatum, Tellina tenuis,* *Venus gallina, Mactra stultorum,* occurring by millions, and many other species in less numbers. —*C. Jefferys, Langharne, Carmarthenshire.*

EIDER DUCK.—On visiting the Ferne Islands on the 24th of May last year (1883), for the purpose of collecting a few specimens of birds' eggs, having previously obtained the permission of one of the Association for the Preservation of the Birds and Eggs of these islands to do so, I had my attention drawn by the "keeper" who accompanied me to an eider duck (*Somateria mollissima*) which had dropped an egg into the nest of a lesser black-backed gull (*Larus fuscus*), and taken possession of the latter's nest and three eggs, and was sitting there; she had occupied the same for three or four days to the knowledge of my guide. I secured the eggs. In my experience of egg collecting, extending over thirty years, with the exception of a pheasant taking possession of a partridge's nest after laying two eggs among the other's thirteen, I have not found similar instances.—*R. Turnbull.*

DOMED NESTS.—I was much interested with Mr. George Roberts's remarks on domed nests. Nevertheless, I do not agree with them in every particular. In the first place, I do not think that the rook has a greater number of enemies than the grouse. The latter, in its moorland home, has foes to contend with that never molest the rook. Secondly, the advantage of the domed nest is said to be apparent, from the number of individuals of the species which build in this manner. With us, the linnet, greenfinch, hedge sparrow, and numbers of others, are quite as numerous as any of the dome builders. I also fail to see the disadvantage of these birds being unable to see the approach of enemies; that is to say, I do not think this disadvantage exists. Take the willow wren as an example. The nest is usually placed against a stump, or bank side, and hence the enemies of the bird must approach from the front to obtain admittance to the nest. As the bird sits with its breast towards the opening, it easily sees anything approaching from this side. Several birds that have very conspicuous eggs, lay them in very exposed nests. The magpie's eggs are quite as dark as those of the rook; and the eggs of the greenfinch as likely to catch the eye, as those of the willow wren or house sparrow. I may add, that with us the long tailed titmouse always has two openings into its nest; and I can confirm the fact of the robin frequently making an arch over one half of its nest, in the absence of overhanging foliage. There is also very often a heap of dead leaves in front of the nest.—*J. A. Wheldon.*

GOLD-FISH.—The experience of G. M. in keeping gold fish is very similar to my own, but as I have lately established an aquarium differing from his in some interesting particulars, I venture to say in as few words as possible in what this difference consists. When I started the aquarium—about three months ago—I set it in order and have never since touched it, except occasionally adding a little water to make up for the evaporation. Finding it difficult to obtain an aquarium made exactly in the manner I desired, I made one myself. It measures twenty-four inches long, twelve inches wide, and thirteen inches deep, nearly a double cube. The bell glass aquarium is neither so picturesque nor so useful as a square shape. There is certainly more comfort in the latter. The framework is of zinc, the bottom of slate, and the glass is $\frac{3}{16}$ sheet glass, not plate, although the latter is better when procurable. The rockwork compound of pumice-stone built up with Portland cement occupies the two back corners extending partly across the back and two sides. It is always necessary, when using cements, to take care that the poisonous matter contained in them is extracted by soaking the cemented things for several weeks. After I had built up the rock-work I allowed it to lie in water exposed to the full glare of the sun for probably ten or twelve weeks. The last two or three weeks I did not change the water, and by the time I was ready to use it, a green vegetable growth covered almost the whole of it—just the thing required. I then placed it in position in the aquarium, with river sand and pebbles about an inch thick and sloping from the back. I then filled gently with water supplied by the water company, and after standing thus in front of the window with an eastern aspect for perhaps two or three weeks, I put in two small gold fish about four inches long and a small stickleback, and there they have remained ever since, and are as lively and apparently as healthy as when at large in the river. On the inside of the back glass, a deep green vegetation has grown, and it now begins to hang in some places like curtains of green down, imparting a very pleasing appearance and serving to make it a more natural home. There need be no fear whatever in not changing the water, if only the green vegetation be allowed to grow upon the rock-work, the two sides and the back, keeping the front glass clean for viewing the interior. It will shortly appear dulling the glass and becoming thicker week after week. The fish are fed twice a week with bread, very sparingly, and once a week I treat them to a few worms obtained from a heap of chicken's dung which we keep for garden use. I have no snails to serve as scavengers, for, although serviceable in some respects, they would entirely defeat the end for which I established my aquarium, by eating off the glass and the rock-work vegetation which is here used instead of water-plants. Besides, they are great destroyers of plant life, and as we can do without the snail and the plants, we have thus more room for the fish to move about in. Mr. Easton would do well to consult the little works on the aquarium by Mr. Taylor, the editor of SCIENCE-GOSSIP, or by Shirley Hibberd, both of them being very simple and accurate in their instructions.—*Walter T. Cooper.*

STRANGE HABIT OF A BULLFINCH.—I have a bullfinch which is kept in an aviary, with several other birds. I have noticed that he has a habit of filling his large lower mandible with millet-seed; and then, retiring to a secluded corner of the cage, he cracks and eats it at his leisure. Has this monkey-like habit been observed before?—*H. B. R.*

NOTICES TO CORRESPONDENTS.

TO CORRESPONDENTS AND EXCHANGERS.—As we now publish SCIENCE-GOSSIP earlier than heretofore, we cannot possibly insert in the following number any communications which reach us later than the 8th of the previous month.

TO ANONYMOUS QUERISTS.—We receive so many queries which do not bear the writers' names that we are forced to adhere to our rule of not noticing them.

TO DEALERS AND OTHERS.—We are always glad to treat dealers in natural history objects on the same fair and general ground as amateurs, in so far as the "exchanges" offered are fair exchanges. But it is evident that, when their offers are simply disguised advertisements, for the purpose of evading the cost of advertising, an advantage is taken of our *gratuitous* insertion of "exchanges" which cannot be tolerated.

WE request that all exchanges may be signed with name (or nitials) and full address at the end.

H. DAVY.—Baird's beautifully-illustrated work on "British Entomostraca" is the chief book on the subject. You may get a copy through some second-hand scientific bookseller, such as Wesley or Collins.

B. M. WATKINS.—Thanks for the teratological specimen of the common daisy. It is due to what Dr. Mastin calls "fasciation," or fusion in growth of the stems.

J. J. A.—The common leech is carnivorous, and preys on aquatic larvæ, fish, and frogs.

J. HODGE.—Your specimen is the crumb-of-bread sponge (*Halichondria panicea*).

C. D.—Rye's "British Beetles" is by far the best of all our less expensive books.

S. S.—Brewer's (not Blewer's) "Guide to the Flora of Surrey" is published by Mr. J. Van Voorst, 1 Paternoster Row.

E. F. B.—The following are good books for your purpose:— Staveley's "British Insects," Wood's "Insects at Home," and Curtis' "Entomology" for general entomology, and the Catalogues of the British Museum on the orders of insects other than Lepidoptera and Coleoptera.

W. W. W.—You will obtain all particulars concerning the zoological station at Naples by applying to the director, Dr. Dohrn. You would meet several English students there.

H. ALLINGHAM.—Chenu's "Conchology" (in French, 2 vols.) is crowded with plates, both coloured and plain. You may get a copy through some second-hand bookseller for about 30s. Wood's "Index Testaceologicus" contains about 2800 figures, and may be had for about £3. These are the cheapest we know, apart from the smaller manuals.

D. F.—One of the British Museum publications is a "Catalogue of Myriapoda." Mr. C. F. George's valuable papers on "British Freshwater Mites" have appeared in SCIENCE-GOSSIP as follows: September and December, 1882; January, February, April, and August, 1883; besides those of the current year.

H. C. CRAVEN.—Rimmer's work on "British Land and Freshwater Shells" is the best we know. Tate's "Land and Freshwater Molluscs" may be had for about 5s.

S. B.—You may get Mr. Brooks' pamphlet, "Notes on my Aquarium," by addressing the author at Fernbrook, Huddersfield. Consult Taylor's "Aquarium: its Principles, Structure, and Management," published by Messrs. W. H. Allen & Co., price 4s. 6d.

W. MOULTON.—Accept our best thanks for the two capital photographs taken with the microscope direct from natural objects.

W. T. HAYDON.—Accept our best thanks for the specimen of *Lathræa squamaria*, which fully bears out your opinion.

F. C. (Barnett).—It is a moss, *Hypnum abietinum*.

J. N. H. S.—You will find all the information you need regarding the "ginger beer" plant, in Berkeley's "Cryptogamia;" we should imagine it to be the Torula.

C. K. F. (Preston).—The specimen you forward contains a large amount of xylem, so it can scarcely be what you imagine; it must be a part of the trunk in a state of decay. If you make a thin section with a razor, and mount in balsam, you will find it makes a pretty microscopic object.

H. D. G. (Croydon).—Many of the Agarics have a close resemblance, hence it is difficult to tell the species from a drawing only. Could you send the actual specimen, then we should indeed be happy to aid you in its identification.

W. G. W. (Brighton).—The one on the anemone is the *Puccinia anemones*, which was many years ago, in Ray's time, mistaken for the sori of a fern. The other on the pilewort leaf is a rust, *Uromyces ficariæ*.

J. Y. (South Park Hill, Croydon).—The nurseryman must have made some mistake, for the one with small leaves is the *Kalmia latifolia*, an Ericaceous plant, the other is certainly spurge laurel (*Daphne laureola*, Linn.). It is very curious and strange, and is worth watching as a freak of nature. Many thanks for the information and the trouble you have taken.

EXCHANGES.

SPLENDIDLY-preserved and correctly-named Swiss Alpine plants. Price 6d. each.—Address, Dr. B., care of Editor of SCIENCE-GOSSIP, 214 Piccadilly.

WANTED, botanical press, to take ordinary sized sheets; must be in perfectly good order. State size, price or exchange, to W. G. Woollcombe, The College, Brighton.

ELEVEN vols. of SCIENCE-GOSSIP, 1873-83; fourteen vols. of "English Mechanic," 14-27, unbound. Wanted, micro objective or other accessories.—W. E. Harper, Norfolk Road, Maidenhead.

DAWSON'S "Chain of Life," Bentley's "Botany," Attfield's "Chemistry," "Knowledge," 1884. What offers?—E. Kerup Bowden Lane, Market Harborough.

LAND and freshwater shells offered in exchange for foreign postage stamps.—E. Collier, 74 Yarburgh Street, Moss Side, Manchester.

Limnaea stagnalis and *Planorbis corneus* (very fine), *L. glabra, L. peregra, Planorbis spirorbis,* and *Physa hypnorum,* also eggs of dabchick, plover, teal, coot, wild duck, and waterhen (side blown). Desiderata, other shells and eggs (the latter side blown).—W. Hewett, 26 Clarence Street, York.

WANTED, Cooke's "Handbook to British Fungi," 2 vols.—W. Mills, 1 Ranheillor Place, Edinburgh.

FOR pure gatherings of Volvox (*Protococcus viridis*) and Hydrodictyon, I will give cash or first-class microscopical slides.—Rev. J. E. Vize, Forden Vicarage, Welshpool.

FOR exchange, nests and clutches of eggs of common bunting, meadow pipit, skylark, willow wren, willow warbler, hedge accentor, mistletoe thrush, song thrush, blackbird, chaffinch, greenfinch, yellowhammer, robin. Would be glad for offer of British birds' eggs.—Robert Barker, 11 Fowend Street, Groves, York.

MICROSCOPIC slides in great variety, well mounted, in exchange for other well-mounted slides. Lists exchanged.—C. Croydon, Pato Point, Torpoint, Cornwall.

LARGE variety of good micro slides to exchange for other good ones. Send list.—J. E. Fawcett, Rawdon, near Leeds.

"KNOWLEDGE," Nos. 33 to 87, minus No. 63, clean, unbound. Will exchange for good modern work on botany.—A. Wright, 9 High Street, King's Lynn.

WANTED, common land and freshwater shells from all parts of the country; can offer a large selection in exchange. Send lists of duplicates and desiderata to Sydney C. Cockerell, 51 Woodstock Road, Bedford Park, Chiswick, W.

WANTED a good microscope. Send particulars to S. C. C., 51 Woodstock Road, Bedford Park, W.

WANTED, "Student's Elements of Geology," by Lyell. Particulars and price to F. C., care of Mr. Girling, High Street, Harlow, Essex.

LEIGHTON'S "Lichen Flora of Great Britain," 3rd edition; Berkeley's "British Mosses," 1863; and Weinhold's "Physics;" all nearly new; what offers?—Flos, Westbury House, Liverpool Gardens, Worthing.

THE "English Entomologist," exhibiting all the coleopterous insects found in England by T. Martyn, with 546 coloured figures (1792); wanted in exchange, Newman's "British Butterflies and Moths" (new) and Rye's "British Beetles" (new).—C. Donovan, jun., Westview, Glandare, Leap, co. Cork.

WHAT offers (in micro slides preferred) for a bottle containing five snakes, three lizards, and a white frog, in spirit from Africa, also a fine specimen of the weaver bird's nest?—Killgour, 63 Dalfield Walk, Dundee, N.B.

FOR exchange, "Nature," vol. viii. (bound), vols. ix.-xiii., unbound. Wanted, works on botany or micro apparatus.—A. Wise, Seaton Villas, Leytonstone, E.

BRITISH birds' eggs and freshwater shells for newly-collected specimens of stag beetle unset and mole cricket. For list of desiderata, apply—Robert Barker, 11 Fowend Street, Groves, York.

EXOTIC Lepidoptera, many rare species in duplicate; collectors please send lists in exchange; also for microscopic purposes wings of *Papilios Buddha* and *polyctor, Apatura laurentia, Agraulis moneta, Morpho menelaus,* and *Urania rhypheus.*—J. C. Hudson, Railway Terrace, Cross Lane, Manchester.

DIATOMACEÆ: Heliopelta, Arachnoidiscus, Triceratum, and other forms, selected and arranged in groups. Best finished slides in exchange for good gatherings of *Campylodiscus costatus, C. clypeus, Pleurosigma formosum,* or rich deposits.—W. White, 17 York Street, Nottingham.

FOR hairs of *Ornithorhynchus paradoxus,* or Pinna shell (for sections showing prisms), send stamped addressed envelope to W. White, 17 York Street, Nottingham.

WANTED, diatoms on sea-weeds and in muds from all the tropic seas. Offered, a large quantity in fine selected diatoms and other slides, or cash.—J. C. Rinnböck, 14 Simmering, Wien, Austria.

"HEALTH," vols. i. and ii., clean and complete, unbound; wanted, mineralogical and zoological specimens.—Send lists to J. H. W. Laing, Tay Port, N.B.

DIATOMACEÆ: wanted, a small quantity of Glenshira sand, containing the forms described by Professor Gregory; would give in exchange mounts of good recent collections from Skye and Orkney, the latter containing many of the new forms found by him in the Firth of Clyde and Loch Fine.—Isaac Robinson, Hertford.

TUBES of *Melicerta ringens* (Rotifer), *Volvox globator,* green hydra, and living diatoms; also micro slides in exchange for other forms, slides, and micro materials.—John J. Andrew, 2 Belgravia, Belfast.

WANTED, Foraminifera, Polycistina, Spicules, and other good micro material, for good mounted slides or cash.—B. Wells, Dalman Road, Forest Hill.

WANTED, blue and orange or purple and yellow stage selenites; would give in exchange four well-mounted slides diatoms for either.—Mathie, 42 McKinlay Street, Glasgow.

CONTINENTAL plants, dried and named, to exchange for British species. Correspondence invited.—E. Straker, Kenley, Surrey.

FOR well-cut and mounted sections of cats' or dogs' gray whiskers (good polar objects), pollen of *Nicotiana tabacum,* send any interesting or instructive slide; list of duplicates exchanged.—H. F. Jolly, Stow Villa, Bath.

WANTED, micro apparatus and material, books or jewellery, in exchange for well-mounted parasites, crystals, and wooden slides of botanical, marine, and forams, recommended for dark objects by Dr. Carpenter.—S. Harrison, 12 Dalmain Road, Forest Hill.

OFFERED, *Helix virgata, H. caperata, H. rupestris, H. pisana, H. sericea, Bulimus acutus, Mytilus pellucida, Cardium echinatum, Venus gallina, Tapes pullastra, Lucinopsis undata, Tellina tenuis, T. fabula, Mactra stultorum, M. subtruncata, Scrobicularia piperata, Donax villatus, Solen vagina, Ceratisolen legumen, Natica catena, Cerithium reticulatum, Utriculus obtusus, Actæon tornatilis.* Wanted, many species of marine shells and well-mounted micro slides.—C. Jefferys, Hill House, Langharne, Carmarthenshire.

OFFERED, palate of limpet, spines of echinus, and twelve named species of zoophytes, unmounted, in exchange for well-mounted micro slides or offers.—C. Jefferys, Hill House, Langharne, Carmarthenshire.

WANTED, Alder & Hancock's "Nudibranchiate Mollusca," parts 6 and 7, Bate & Westwood's "Sessile-eyed Crustacea," Moggeridge's "Ants and Spiders," and other good natural history works in exchange for Stark's "British Mosses," Berkeley's "Cryptogamic Botany," Laishley's "Birds' Eggs," &c.—C. A. Grimes, Dover.

WANTED, a few ounces of roughly-cleaned (for frames) chalk from Gravesend and other parts of Kent, also a good gathering of Volvox, for good microscopic slides of desmids and other algæ.—W. West, 15 Horton Lane, Bradford.

FIRST-CLASS anatomical and pathological microscopic slides offered in exchange for good magic lantern slides.—Henry Vial, Crediton, Devon.

COLLECTION of dried specimens of foreign fruits, barks, &c.; also "Human Race," by Louis Figuier. Wanted, Cambrian, Lower Silurians, and Paris basin fossils, or offers.—Miss Linter, Artagon Close, Richmond Road, Twickenham.

A MAHOGANY cabinet to hold two gross of slides, in exchange for six dozen good slides.—William Tylar, 20 Geach Street, Birmingham.

SEVERAL species of British freshwater shells, either living or dead (for aquarium or collection), in exchange for British land, freshwater, and marine shells.—Robert B. Cook, 44 St. John Street, Lord Mayor's Walk, York.

OFFERED for exchange, eggs of the pochard, capercaillie, long-eared owl, red-breasted merganser, kestrel, sparrow hawk, carrion crow, &c.—R. Brodie, Robin Hood's Bay, Whitby.

PHOTOGRAPHS of microscopic objects in exchange for slides; several duplicate slides also in hand; lists exchanged.—W. Moulton, 37 Chancery Lane, London, W.C.

BOOKS, ETC., RECEIVED.

"Amongst the Wild Flowers," by the Rev. H. Wood. London: W. S. Sonnenschein & Co.—United States Commission of Fisheries, "Commissioners' Report for 1880," Washington.—"Transactions of the New York Academy of Science for 1882-83."—"Third Annual Report of the United States Geological Survey, 1881-82," by J. W. Powell, Director, Washington.—"Studies in Microscopical Science," edited by A. C. Cole.—"The Methods of Microscopical Research," edited by A. C. Cole.—"Popular Microscopical Studies." By A. C. Cole.—"The Asclepiad," No. 1, edited by Dr. Richardson.—"The Gentleman's Magazine."—"Belgravia."—"The Journal of Microscopy."—"The Science Monthly."—"Midland Naturalist."—"Ben Brierley's Journal."—"Sunlight."—"Bulletin of the California Academy of Science," No. 1, Feb. 1884.—"Natural History Notes."—"Science."—"American Naturalist."—"Medico-Legal Journal" (New York).—"Canadian Entomologist."—"American Monthly Microscopical Journal."—"Popular Science News."—"The Botanical Gazette."—"Revue de Botanique."—"La Feuille des Jeunes Naturalistes."—"Le Monde de la Science."—"Cosmos: les Mondes." &c. &c. &c.

COMMUNICATIONS RECEIVED UP TO 10TH ULT. FROM:—R. B.—S. B. A.—R. B.—T. H.—H. L.—J. G. B. A.—W. C. R.—A. H. S.—W. W. W.—J. T. J.—C. C. A.—Miss J. Y.—W. H.—G. A. G.—W. J.—G. M.—W. M.—P. L.—W. H.—S. C. L.—R. W. R.—J. H.—W. B. G.—H. D. G.—W. G. W.—C. K. F.—C. B.—W. B. B.—J. M. T.—P. A. E. H.—H. E. C.—J. E. V.—E. C.—E. F. B.—R. H. N. B.—A. L.—C. R.—B. M. W.—E. C.—F. T. M.—M. H. R.—F. H. L.—H. D.—C. B.—H. D. J. M.—W. C. H.—W. G. W.—W. E. H.—D. F.—E. K.—R. B.—I. C. T.—A. P. W.—C. D. jun.—R. A. R. B.—J. C. H.—J. A. W.—W. W.—P. K.—F. C.—W. R.—F. A. A. S.—A. W.—S. C. C.—C. B.—W.—J. E. F. J. E.—I. E.—J. H. W. L.—J. G. C.—E. B. B.—J. P.—F. J. G.—S. H.—W. W.—J. C. R.—Mrs. E. L. O. M.—J. C. M.—G.—W.—R. T. M.—C. J.—R. H. B.—A. A.—H. F. J.—VI E. R.—R. B. C.—H. U.—G. S. S.—J. E. L.—J. F. G.—C. A. G.—E. S.—R. B.—A. J. O.—G. O.—C. F. W. T. W.—S. B.—H. B.—R. H. B.—&c.

GRAPHIC MICROSCOPY.

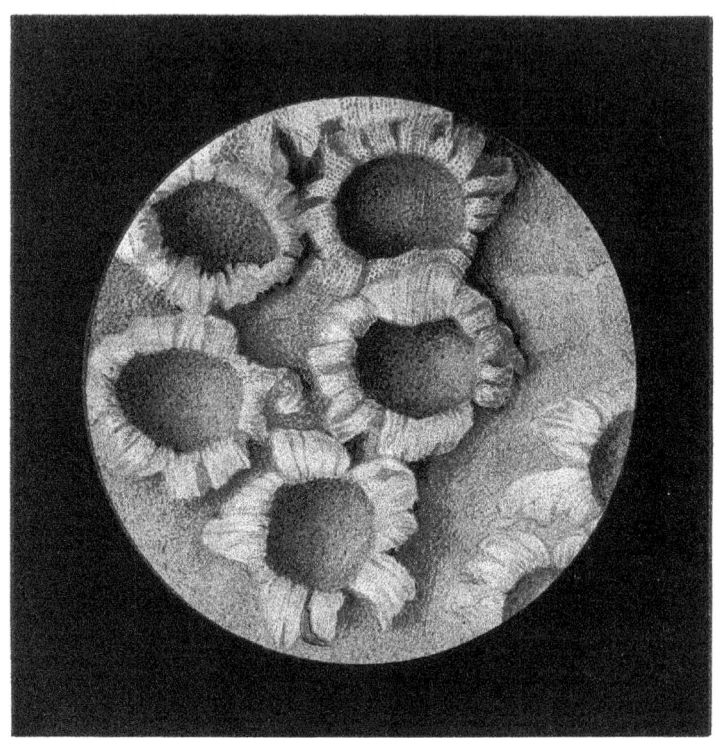

E.T.D. del. ad. nat.
Vincent Brooks Day & Son, lith.

CLUSTER CUPS.
× 50.

GRAPHIC MICROSCOPY.

By E. T. D.

No. VII.—Cluster Cups: Œcidium quadrifidum.[1]

WHEN the microscope opens a new world for investigation, an intelligent observer soon discovers that his interest does not pause at mere admiration; a curiosity is aroused, a wish created to enter deeper into the physiological wonders involved. It would be difficult to touch any particular branch of microscopy where this desire is not excited, but in the contemplation of obscure and minute forms of vegetation it exists in a high degree.

Although the Fungi seem to show unusual simplicity and homogeneity of structure, the great interest they create depends on their most intricate complexity, in connection with their reproduction; and when touching upon the life history of such a subject as the present, a short and "popular" description becomes insufficient; most interesting technical details may however be found in many papers published in this journal during the last few years.

Upon the leaves and stems of rose, gooseberry, buckthorn, nettle, common spurge, and of many herbaceous plants may be seen minute spots of a bright orange colour; the Peridia, or spore-bearing receptacles of Œcidium, a genus of Uredinei (Coniomycetous fungi). These growths are parasitic within the tissues of leaves, and the subject of the plate is the fructification, the peridia, which have burst through the epidermis of the leaf and appear as "cups" split at the margins, a feature in classification) and containing the germinating spores; found within as a yellow, and in other cases, red or brown powder.

These fungi may be observed in an early stage before the emergence of the cup, and are discovered by changes and deformity produced on the surface of the leaves of the infected plant; appearing as minute elevations, or blisters, they increase in size and tension, until the epidermis gives way and the developments beneath burst through. The "cup," or peridium, is then revealed, as an irregular sac, broken on the edge in radiating fissures; when ripe and curled back, it forms an elegant microscopic object.

If a perpendicular section of the cup and the parts beneath be made, or the cuticle of the leaf carefully peeled off, it will be found that the tissues of the host are distorted by the growth of a filamentous structure, the mycelium of the fungus which ramifies among the intercellular passages, and in certain points crowds into a compact mass, immediately beneath the epidermis; by the tension of these felted accumulations the leaf splits, and giving way, a few filaments push through the opening, forming the cupped spot, the granular masses below escape, and eventually develop the fructification. There is obviously neither root nor leaves; nothing exists as absorbents of nourishment, but the interlacing fibres, penetrating between the cells of the leaf (unlike the lichens, which derive support solely from the air), these fungi are entirely sustained by the organic substances diverted from the necessities of the plant; this vegetative felty mass, mycelium or "spawn," is branched and colourless when growing in the earth or in vegetable structures or substances, but when appearing in decomposing fluids it forms cloudy flocks; the reproductive fruit emanating from its various conditions of habitat, is extremely diverse in appearance, as may be seen in the mushroom, the puffball, mildews, and smuts.

In the Œcidiacei it takes the form of these minute cups; the spores within are in enormous numbers; seen in a perpendicular section, they are found arranged in regular order, as most delicate cellules. Each spore is capable of infection, and becoming the starting-point of a new existence, and considering that the nidus of its future development is the tissue of a living plant, it may be difficult to conceive how it finds admittance. It is supposed that the spore reaches the host through the stomata, or respiratory pores found on leaves, but a difficulty here presents itself—it is possible the stomata may not permit entrance, and it is ascertained that the spore itself has first to undergo a change, and become "vesicular." In Dr. Taylor's Book on "The Sagacity and Morality of Plants," is represented (Fig. 90) the cuticle of a wheat leaf with two spores of *Puccinia graminis* (a mildew) throwing out filaments (the vascular condition) penetrating the stomata of the cuticle; but beyond this actual access to the tissues, it has been proved that "seeds" of wheat may be inoculated with spores of "bunt," the result only appearing in its fungoid character, when the corn has attained considerable growth. It would therefore appear that the germ, or some after condition of it, must be absorbed into the seed, follow its evolution, and necessarily traverse, or be carried forward by, the growing tissues. Its progress and development is obscured, no bright "cup" bursts through the cuticle of the wheat to disclose its presence, and, although the mischief may be suspected, by a weakness and discoloration in the plant, it is frequently never discovered until the grain has ripened, when instead of sweet starch, the fetid black bunt dust is found.

There are many species of British Œcidiacei easily procured, affording beautiful objects for the microscope; freshly gathered they are brilliant in colour, curious in arrangement, and interesting in connection with the disruptions they cause in the tissues of their host. An extensive collection of the microscopic forms may be "cabineted" without difficulty, slips of thin wood of the well-known size perforated in the centre are the most convenient means of preserving opaque botanical subjects not requiring very high powers, such as fungi, lichens, sori of ferns, pollens, &c. The method is simple; a piece of gummed paper beneath the perforation, and a thin glass cover fastened on the surface, is enough. They are light, pack closely; and the cell not being absolutely airtight, the inner side of the covering glass does not become dimmed by moisture. The specimen should be dry before enclosure, such minute objects are more secure and accessible for examination than when preserved in an herbarium. As a proof, the subject of the plate has been drawn (under reflected light) from a fragment, mounted in this way, fifteen years ago, procured from the collection of three hundred specimens published or issued, by Dr. M. C. Cooke, under the title, "Fungi Britannici Exsiccati," and it appears by his note to have been collected in 1866; it has lost, for drawing purposes, nothing in colour or character, but what may be supplied from remembrance of living examples. Dr. Cooke's manual, "Rust, Smut, Mildew, and Mould," containing many coloured and well-described plates, is a most valuable book of reference for the identification of Microscopic Fungi.

Living minute fungoid growths may be preserved, and developments watched, by keeping the leaves and stems of the plants on which they are found, in a box half full of damp peat earth, and decayed wood, covered with a sheet of glass.

Crouch End.

THE PEDIGREE OF THE ELEPHANT.

By R. LYDEKER, M.A., F.G.S.

[*Continued from page* 133.]

THE last group of elephants with which we have to deal is another section of primitive elephants, in which the number of ridges is reduced to three in the middle teeth, and to four in the last. This section also comprises three Indian species to which the names of Falconer's elephant, Pandion's elephant, and the narrow-toothed elephant, have been applied. In the first species it is believed that there were tusks in the upper jaw only, and the form of the chin was probably not very unlike that of the modern elephants; but in the second and third species the chin, or extremity of the lower jaw, was produced into an enormous spout-like process, which in Pandion's elephant was as long as the whole of the rest of the jaw, and in the males carried a very large pair of flat-sided tusks. Owing to the smaller size of the grinding-teeth in these elephants, two, or even parts of three of these teeth, were in use at the same time. These and other primitive elephants also differed from the modern true elephants, in that the hinder milk-grinders were succeeded vertically by teeth of the second series, after the manner of most other quadrupeds. These second teeth were, however, much smaller than the milk-teeth they replaced, which is quite contrary to the condition prevailing in other quadrupeds.

In the narrow-toothed and Pandion's elephant we have, so to speak, elephants reduced to their lowest denomination; and it is worth while to contrast them a little more fully with the true elephants, when we shall find that they differ by their middle grinding-teeth, having only three broad, low ridges, separated by wide, open valleys, instead of consisting of some twelve or thirteen closely packed, tall, thin plates, with the narrow intervening valleys completely filled up. Their enormously elongated lower jaw, with its large tusks, indicates a corresponding elongation of the upper jaw, whence it is pretty evident that they were furnished with a very long, snout-like mouth, not unlike that of a huge pig. That these animals were provided with a trunk, or proboscis, is also tolerably evident, since with long projecting tusks in both jaws they would be unable to obtain their food without the assistance of such an organ. The long snout-like mouth must, however, have detracted posteriorly from the length of the trunk, and judging from the analogy of the simpler grinders of the primitive elephants, it is highly probable that their trunk was an altogether shorter and less perfect organ than the trunk of the true elephants, and was perhaps more like the proboscis of a tapir somewhat exaggerated. It is quite evident that the free use of the proboscis of Pandion's elephant must have been considerably impeded by the presence of the large tusks in the lower jaw, which would greatly prevent its lateral motion. The skeleton of the narrow-toothed elephant, although in most respects very like that of an existing elephant, has been found to indicate an animal with a rather longer body.

We have now to mention another important circumstance in connection with these fossil elephants, namely, that all the three-ridged primitive elephants are found in beds below those in which the intermediate and true elephants occur, and are therefore older, or in other words, existed before they were born or thought of. Some of the four-ridged primitive species are found both with the old three-ridged forms, and with the newer intermediate and true elephants. At the present day it is only the true elephants that remain.

We are now in a position to apply the facts which we have acquired to the explanation of the mutual relations of all these elephants; and since we have seen that a gradual and almost imperceptible transition may be traced from the existing true elephants to their fossil congeners, through these again to the intermediate elephants, and thence again to the four-ridged, and finally to the three-ridged primitive forms, and that the transition from the complex to the simple coincides with the order of the appearance and disappearance of the different groups, the conclusion is forced upon us that the true elephants are the highly modified direct descendants of the three-ridged primitive elephants, through the intervention of the four-ridged primitive elephants and the intermediate elephants. Although it is certain that many of the forms with which we are acquainted are not on the direct line of descent, and it is highly probable that many of the links in the chain of that descent have been irretrievably lost, yet in respect of the characters of the grinding-teeth, there is such a graduated transition from those of the narrow-toothed elephant, through the broad-toothed species, to Clift's, and then to the round-headed and Ganesa's elephant, whence the series leads readily on to the flat-headed and Sutledj species, and thence to the existing Indian elephant, that there is not the slightest difficulty in seeing how the evolution took place. In the gradual shortening of the mouth and chin, and the disappearance of the lower tusks there is a similar transition from Pandion's elephant through the Perim elephants to the intermediate and true elephants.

We may accordingly take it as proved that the pedigree of the existing true elephants, which are animals strangely isolated from all others of the present day, may be traced back as far as the primitive elephants, and it remains to inquire whether the latter show any more signs of relationship to other groups of animals than is exhibited by our elephants of to-day. This question may be most decidedly answered in the affirmative; and any one who will take the trouble to visit the galleries of the Natural History Museum at South Kensington may readily satisfy himself that there is much less difference between the grinding-teeth of a three-ridged primitive elephant and a pig or hippopotamus, than there is between those of the primitive and true elephants. The presence in the primitive elephants of a second set of teeth of smaller size than the milk-teeth which they replace, and practically useless, is an evident proof of the descent of the elephants from a group of animals in which all the teeth were of the comparatively small size of those of the pigs. The smaller absolute size of the teeth of the primitive, when compared with those of the true elephants is another circumstance leading to the same conclusion; while the fact that in the primitive elephants more than one grinding-tooth is in use at the same time, also indicates a transition from the true elephants towards the pigs. The same inference is indicated by the longer body of the narrow-toothed elephant. The long muzzle, and the two pairs of tusks of the simpler primitive elephants are also characters in which these animals exhibit relationship to the pig-like animals. None of the latter are, however, yet known in which the front teeth had assumed the characters of tusks,[*] although there is an approach to this in the hippopotamus.

In the structure of their limbs, all the elephants, as far as is known, are indeed very different from the pig-like animals; and in this respect resemble certain primitive fossil animals from North America, all of which present the same simple limb-structure. From the latter fact, coupled with other circumstances which it is unnecessary to mention here, it has been concluded that all the earlier hoofed quadrupeds had a similar simple limb-structure, and the most probable reason why this has been retained in the elephants is that the huge bulk of these animals has obviated the necessity of their seeking safety in flight from their animal foes, so that there has not arisen the necessity for the development of a more elaborate limb like that of the horse or deer.

It seems, therefore, probable that the elephants and the pigs (with which the hippopotamus is in-

---

[*] The tusks of the elephant do not correspond to the so-called tusks of the pig, but to the front teeth.

cluded) are diverging branches of a common stock intimately connected with the above-mentioned extinct animals of North America, although it unfortunately happens that the connection between the primitive elephants and the pigs is still a missing link. That the remains of such an animal will, however, turn up some day or other, the writer is fully convinced, and he ventures to predict that, when it does, it will be found to possess a long muzzle, with front teeth still more projecting than in the hippopotamus, and a proboscis somewhat like that of the tapir. It will have, moreover, grinding-teeth not unlike those of the pig, but of rather more complex structure, and with a tendency for the earlier ones to wear out before the hindmost have come into use. That in this respect there is but a step from the pigs to this hypothetical animal is indicated by the circumstance that there is a nearer approach to this condition in the teeth of certain African pigs, than in those of any other existing animal.

Curiously enough, there are some very remarkable indications of an affinity between the elephants and the gnawing animals (porcupines, squirrels, &c.), and it thus seems probable that the primitive hoofed animals and the primitive gnawing animals were related; but the closeness of this relation it is at present impossible to indicate.

Finally, it may be asked, in what respects have the existing elephants gained an advantage over their primitive ancestors, so that the former remain while the latter have perished. Firstly and foremostly, the elephant of to-day has gained enormously by the elaborate structure and great size of his grinding-teeth, which take an immense time to wear out, and consequently permit of his attaining the great age—sometimes as much as 150 years—to which he reaches. The simple and low-crowned grinders of the primitive elephants must, on the other hand, have worn out at a much earlier period; and the term of existence of their owners must perforce have been limited at the outside to something like forty or fifty years, or about twenty years more than the life of the hippopotamus, which has still simpler teeth. The modern elephants have probably also gained an advantage in the more perfect development of their proboscis, and in the loss of their lower tusks and the shortening of their jaws, which has enabled them to make more effective use of that marvellous organ. The elaborate structure of their grinding-teeth has, moreover, doubtless produced more perfect mastication of their food, and has thereby conduced to that remarkable longevity in which they exceed all other animals.

Whether the Indian elephant has attained the extreme of perfection to which it was possible for such a type of organization to reach, or whether, had he been unimpeded by the advent of man, he would have advanced still higher in the scale, is a question which, however interesting, can never be solved. That the long course of evolution which we have traced above is already at an end, is certain; and that the day is not far distant when the last member in the wondrous chain shall have disappeared for ever, is tolerably evident to those who have watched how fast our larger quadrupeds are vanishing before the advancing strides of civilization.

OPHIOGLOSSUM VULGATUM, VAR. AMBIGUUM.

By WILLIAM ROBERTS.

THIS exceedingly interesting little adder's-tongue was discovered at Scilly about twenty years ago by Mr. F. Townsend, who described it in the "Journal of Botany" as being the same as a form noticed in "Flore des Environs de Paris," 2nd ed. p. 577. Mr. Townsend first published it as occurring

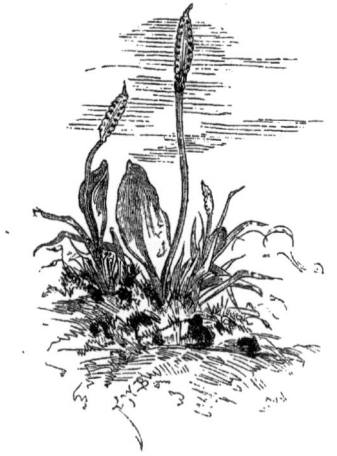

Fig. 77.—*Ophioglossum vulgatum*, var. *ambiguum*.

in St. Agnes, but Mr. Ralph informs me that both himself and Messrs. Curnow and Tellam searched for it there but in vain. To Mr. Curnow belongs the honour of discovering it in plenty, and extending over a considerable space of ground at intervals at the neighbouring isle of St. Martin's. This pretty addition to the British Cryptogamic flora rarely grows more than an inch and a half high, and has ovate or ovate-lanceolate barren fronds which are channelled, recurved, and attenuated below; whilst the fertile frond produces from twelve to twenty capsules or spore cases.

As arranged by Baker, the genus Ophioglossum, inclusive of the Ophioderma of Endl. and Rhizoglossum of Presl, contains ten species; but many authorities consider, and very properly so, that most of these so-called species are but mere forms of the common *O. vulgatum*. The genus itself is easily defined as having the sessile capsules arranged in two rows so as to form a narrow close spike. But the identity of several species is not so clear. The Scillonian form is a case in point. Mr. Baker and most other writers expressly state that the flowering period of the typical *O. Lusitanicum* is during winter. M. J. A. Henriques, director of the botanical gardens at Coimbra, Portugal, kindly sent me some living specimens in fruit, gathered in that district during the early part of summer, three or four years since. The Cornish plants fruit about the same time. This fact, when it is considered how extremely alike in general appearance Lusitanicum and Ambiguum are, seems to point to the conclusion that those found in Scilly are the true British representatives of the typical Lusitanicum, and the Guernsey one the real varieties. Both these are quite consistent as regards their fruiting season, however closely they may be allied in other respects; Scilly and Guernsey being so much alike in their comparative mildness, that the reason of these two plants fruiting at various periods cannot be ascribed to climatical effect. The geographical range of what is usually considered as *O. Lusitanicum* is very large; it may be found on the sandy coasts of Europe and Africa, and was first discovered to be a British plant so late as 1854. Mr. F. W. Burbidge has written me stating that the author of the "Cybele Hibernica," considers a specimen lately found in the west of Ireland as precisely like those growing at Scilly. It is extremely possible that this most interesting object will be found at various other British habitats if carefully searched after.

PLANT NOTES.

A PALATABLE wine is still made by some of our Cheshire dairy-maids from the flowering tops of the borage, which is called courage-cup. I am reminded that the word borage, now applied to this species, is merely a corruption of the word corage or courage (from cor and ago). Pliny says, "If the leaves and flowers of borage be put into wine, and that wine drunken, it driveth away all heavy sadness and dull melancholy." Burton, in "Anatomy of Melancholy" writes :

> Borage and hellebore fill two scenes;
> Sovereign plants to purge the veins
> Of melancholy, and cheer the heart
> Of those black fumes which make it smart.

We hear a proverb in the North of England :

> An elder stake, and a hazel-heather,
> Will make a hedge, to last for ever;

which means that the elder, being a rapid growing shrub, soon fills up the hedge, if planted with more slowly growing shrubs. But old ladies tell us, or rather give us quite another version of the reason why we so often find it near farm-houses; they say it is one of the plants which ward off the evil-eye, and demons, and no witch can cross the field where the elder, mountain-ash, or laurel grows. Jamieson states, "This shrub was supposed to possess great virtue, in warding off the force of charms, and witchcraft. Hence it was a custom to plant it round country houses, and barn-yards."

Again, why is the elder called Bourtree? Skinner, I believe, makes the name Boretre, because the young shoots are hollow: however, we incline to the explanation given in the statistical account of Stirlingshire, where it states, "It is no stranger in many parts of the parish. The branches cause an agreeable shade, hence the propriety of the name bowertree (bourtree)."

Can any of our readers tell us what species Drayton refers to under the name of Clote, in the following lines :

> This is the *clote*, bearing a yellow flower;
> And this black-horehound, both are very good
> For sheep, or shepherd, bitten by a wood [mad]
> Dog's envenomed tooth.

It is said to refer only to the cross-wort (*Galiu cruciatum*), others incline rather to the weasel's-snout (*Galeobdolon luteum*).

The blue-bottle (*Centaurea cyanus*) is turned into blewart in the northern counties (Danish *bla-wort*; Dutch, *korn bloemster*).

> When the blewart bears a pearl,
> And the daisy turns a pea,
> And the bonny lucken gowan
> Has folded up her e'e;
> Then the laverock frae the blue lift,
> Drops down, and thinks nae shame
> To woo his bonny lassie,
> When the kye comes hame.

Bilberry, our Manchester whimberry, is from the Danish *böllbær*, really bilberry; Swedish *blabær*; Scot. *blaeberry*, but the English form is without doubt taken from the Dan. *böllbær*, the first syllable of which is pronounced BIL. Some time since I was out for a day's botanising with an American professor, he declared our plant was called blueberry all throughout the States. The syllable blue is attached to many of our well-known plants, often in a very confusing manner. One or two examples may be cited; for instance, there cannot be a doubt that our English bluebell is the *Hyacinthus non-scriptus*, whilst the Scotch is *Campanula rotundifolia*. A little confusion has been created by the lines taken from the "Lady of the Lake":

> E'en the slight *Hairbell* raised its head,
> Elastic from her airy tread,

here the ancient Scottish bluebell is given a new name. It must be a very airy or fairy tread indeed if it did not snap the stalk of the hyacinthus, whilst the campanula can submit and recover from any amount of the "airy footfall."

One of our common toadstools is named Blewit,

according to Berkeley, but in Cheshire the name given to the *Agaricus personatus* is strangely like blue-hats, a not inapt title, for it is supposed to be either a fairy's hat or parasol :

> The lovely flowers of Scotland
> All others that excel;
> The thistle's purple bonnet,
> And the bonny heather-bell.
> Oh, they're the flowers of Scotland
> All others that excel;
> For the thistle in her bonnet blue
> Still nods out o'er the fell,
> And dares the proudest foeman
> To tread the heather-bell."

Another confusion exists in the use of the word Tetter, as applied to the white bryony; it is tetter berries, used in the skin disease known as tetters; whilst the tetter wort is the *Chelidonium majus*. In Ray's time, the latter term had become obsolete, but, according to Halliwell, it is now commonly used for the greater celandine.

When St. Bernard founded his abbey, near Clairvaux, he and his thirteen companions lived on barley, or cockle-bread, with boiled beech leaves as vegetables, while they were employed grubbing up the forest, and in building huts for their habitation. Was the cockle, so distinguished, the corn-cockle now so much dreaded as a farmer's pest? We have heard the word cockle applied to the garden Nigella; this could not evidently be the plant intended, it must be the *Lychnis githago*, whose seeds have somewhat a resemblance to cockles.

JAMES F. ROBINSON.
Frodsham.

THE ALIMENTARY CANAL OF THE COCKROACH.*

By PROFESSOR L. C. MIALL AND ALFRED DENNY.

THE alimentary canal of the cockroach measures about 2¾ inches in length, and is therefore about 2¾ times the length of the body. In herbivorous insects the relative length of the alimentary canal may be much greater than this; it is five times the length of the body in Hydrophilus. Parts of the canal are specialised for different digestive offices, and their order and relative size are given in the following table :—

Œsophagus and crop	.95 in.
Gizzard	.1
Chylific stomach	.5
Small intestine	.1
Colon	.875
Rectum	.25
	2.775

The principal appendages of the alimentary canal

---

* For "The Natural History of the Cockroach," and the "Outer Skeleton of the Cockroach," see this journal, March and May, 1884.

are the salivary glands, the cæcal diverticula of the stomach, and the malpighian tubules.

Considered with respect to its mode of formation, the alimentary canal of all but the very simplest animals falls into three sections, viz., (1) the mesenteron, or primitive digestive cavity, lined by hypoblast; (2) the stomodæum, or mouth-section, lined by epiblast, continuous with that of the external surface; and (3) the proctodæum, or anal section, also lined by epiblast folded inwards from the anus, just as the epiblast of the stomodæum is folded in from the mouth. The mesenteron of the cockroach is very short, as in other Arthropoda, and includes only the chylific stomach with its diverticula. The whole region in front forms the stomodæum, and is lined by a chitinous layer continuous with the outer integument; the proctodæum includes the malpighian tubules, and extends thence to the anus, being lined throughout with a chitinous layer. Wilde (Wiegmann's Archiv, 1877) states that the moult of the chitinous lining of the crop follows that of the outer integument.

The mouth of the cockroach is enclosed between the labrum in front and the labium behind, while it is bounded laterally by the mandibles and first pair of maxillæ. The chitinous lining is thrown into many folds, some of which are loose and obliterated by distention, while others are permanent and filled with solid tissues. The lingua is such a permanent fold, lying like a tongue upon the posterior wall of the cavity and reaching as far as the external opening. The thin chitinous surface of the lingua is hairy, like other parts of the mouth, and stiffened by special chitinous rods or bands. The salivary ducts open by a common orifice on its hinder surface. Above, the mouth leads into a narrow gullet or œsophagus, with longitudinally folded walls, which traverses the nervous ring, and then passes through the occipital foramen to the neck and thorax. Here it gradually dilates into the long and capacious crop, whose large, rounded end occupies the fore-part of the abdomen. When empty or half-empty, the wall of the crop contracts, and is thrown into longitudinal folds, which disappear on distention. Numerous tracheal tubes ramify upon its outer surface, and appear as fine white threads upon a greenish-grey ground.

Four layers can be distinguished in the wall of the crop, viz., (1) the muscular, (2) the connective, (3) the epithelial, and (4) the chitinous layer. The external or muscular layer consists of a single series of annular fibres, usually distant more than their own width from each other, and enclosing a still looser stratum of longitudinal fibres. Here and there the bundles divide. In most animals the muscles of organic life, subservient to nutrition and reproduction, are very largely composed of plain or unstriped fibres. In Arthropoda (with the exception of the anomalous Peripatus) this is not generally the case, and the muscular fibres of the alimentary canal belong to the striped variety. The connective tissue layer forms a thin, structureless basement membrane, firmly united in the œsophagus and crop to the muscular layer and the epithelium. The epithelium consists of scattered nucleated cells, rounded or oval. Where they are deficient the chitinous lining comes in contact, or nearly so, with the basement membrane. These epithelial cells, homologues of the chitinogenous cells of the integument, secrete the transparent and structureless chitinous lining. Hairs (setæ) of elongate, conical form, and often articulated at the base, like the large setæ of the outer skin, are abundant. In the œsophagus they are very long and grouped in bundles along sinuous transverse lines. In the crop, the hairs become shorter and the sinuous lines run into a polygonal network.

The gizzard has externally the form of a blunt cone, attached by its base to the hinder end of the crop, and produced at the other end into a narrow tube ($\frac{1}{4}$ to $\frac{1}{3}$ in. long), which projects into the chylific stomach. Its muscular wall is thick, and consists of many layers of annular fibres, while the internal cavity is nearly closed by radiating folds of the chitinous lining. Six of the principal folds, the so-called "teeth," are much stronger than the rest, and project so far inwards that they nearly meet. They vary in form, but are generally triangular in cross-section and irregularly quadrilateral in side-view. Between each pair are three much less prominent folds, and between these again are slight risings of the chitinous lining. A ridge runs along each side of the base of each principal tooth, and the minor folds as well as part of the principal teeth are covered with fine hairs. The central one of each set of three secondary folds is produced behind into a spoon-shaped process, which extends considerably beyond the rest, and gradually subsides till it hardly projects from the internal surface of the gizzard. Behind each large tooth (i.e. towards the chylific stomach) is a rounded cushion set closely with hairs, and again beyond this a second and smaller cushion. There are therefore twelve cushions, in two rows. Still further back the cushions are succeeded by six longitudinal folds, between which are smaller intermediate folds. The whole forms an elaborate machine for crushing and straining the food, and recalls the gastric mill and pyloric strainer of the crayfish. The powerful annular muscles approximate the teeth and folds, closing the passage, while small longitudinal muscles, which can be traced from the chitinous teeth to the cushions, possibly retract these last, and open a passage for the food. The long, tapering extremity of the gizzard is enclosed within the anterior extremity of the chylific stomach, and forms a kind of circular valve.

The chylific stomach is a simple cylindrical tube, provided at its anterior end with eight (sometimes fewer) cæcal tubes, and opening behind into the intestine. Its muscular coat consists of a loose layer of longitudinal fibres, enclosing annular fibres. Internal to these is a basement membrane, which supports an epithelium consisting of elongate cells clustered into regular eminences, and separated by deep cavities. Large simple glands open into the bottom of the cavities. The epithelium forms no chitinous lining in the chylific stomach or cæcal tubes, and this peculiarity no doubt promotes absorption of soluble food in this part of the alimentary canal. Short processes resembling thick, stiff cilia, are given off from the free ends of the epithelial cells.

At the hinder end of the chylific stomach is a very short tube about half the diameter of the stomach, and known as the small intestine. To it are attached 60 or 70 long and fine tubules, the malpighian tubules. The small intestine has the same general structure as the œsophagus and crop; its chitinous lining is hairy, and thrown into longitudinal folds. There are usually six primary folds, which have a radiate arrangement and nearly meet in the centre; they are often subdivided into minor folds. At the junction of the small intestine with the colon is a circular valve, which nearly closes the passage. A few long setæ project from the free edge of the valve.

From the circular valve the colon extends for nearly an inch. Its diameter is somewhat greater than that of the chylific stomach, and uniform throughout, except for a lateral diverticulum or cæcum, which is occasionally but not constantly present towards its rectal end. The fore-part of the colon is thrown into a loose spiral coil. A constriction divides the colon from the next division of the alimentary canal, the rectum.

The rectum is about $\frac{1}{4}$ inch long, and is dilated in the middle when distended. Six conspicuous longitudinal folds project into the lumen of the tube. These folds are characterised by an unusual development of the epithelium, the cells lengthening towards the centre of each, while they are altogether wanting in the intermediate spaces, where the chitinous lining blends with the basement membrane, and both are thrown into sharp longitudinal corrugations. Between the six epithelial bands and the muscular layer are as many triangular spaces, in which ramify tracheal tubes and fine nerves for the supply of the epithelium. The chitinous layer is finely setose. The muscular layer consists of annular fibres, strengthened externally by longitudinal fibres along the interspaces between the six primary folds.

The corrugated and non-epitheliated interspaces may be supposed to favour distention of the rectal chamber, while the great size of the cells of the bands of epithelium is perhaps due to their limited extent. Leydig (Lehrbuch der Histologie) attributed to these rectal bands a respiratory function, and compared them to the epithelial folds of the rectum of libellulid larvæ, which, as is well known, respire by admitting fresh supplies of water into this cavity. It is an obvious objection that cockroaches and other insects in which the rectal bands are well developed do not take water into the intestine at all. Gegenbaur has therefore modified Leydig's hypothesis. He suggests, (Grundzüge d. Vergl. Anat.) that the functional rectal folds of dragon-flies and the non-functional folds of terrestrial insects are both survivals of tracheal gills, which were the only primitive organs of respiration of insects. The late appearance of the rectal folds and the much earlier appearance related to the six theoretical elements (two tergal, two pleural, two sternal), traceable in the arthropod exoskeleton, of which the proctodæum and stomodæum are reflected folds.

The anus of the cockroach opens beneath the tenth tergum, and between two "podical" plates. Anal glands, such as occur in some beetles, have not been discovered in cockroaches.

The three principal appendages of the alimentary canal of the cockroach are outgrowths of the three primary divisions of the digestive tube; the salivary glands are diverticula of the stomodæum, the cæcal

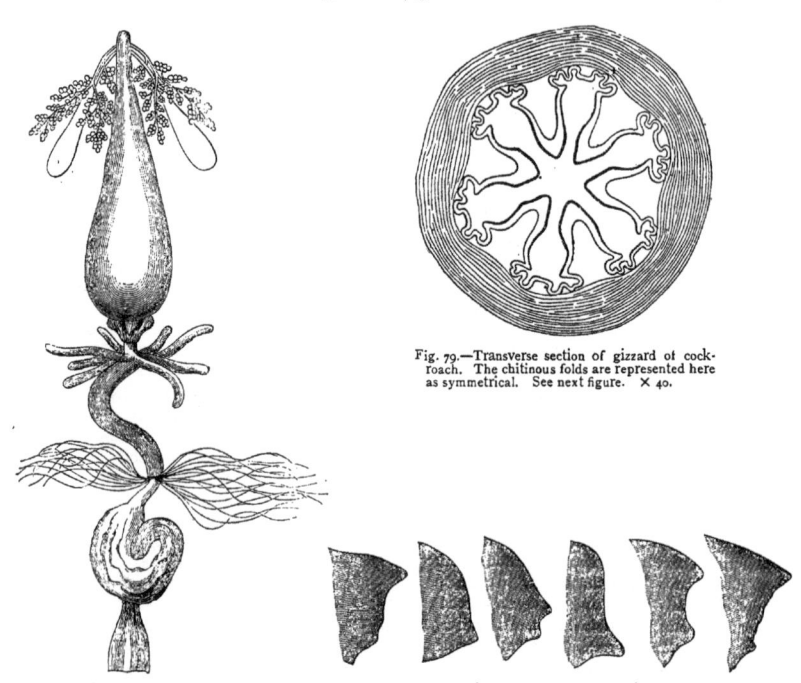

Fig. 79.—Transverse section of gizzard of cockroach. The chitinous folds are represented here as symmetrical. See next figure. × 40.

Fig. 78.—Alimentary canal of cockroach. × 2.

Fig. 80.—The six primary folds (teeth) of the gizzard, seen in profile.

of spiracles is a serious difficulty in the way of this view, as Chun has pointed out. It seems more probable that the respiratory appendages of the rectum of the dragon-fly larvæ are special adaptations to aquatic conditions of a structure which originated in terrestrial insects, and had primarily nothing to do with respiration.

The number of the rectal bands (six) is worthy of remark. We find six sets of folds in the gizzard and small intestine of the cockroach, and six longitudinal bands in the intestine of the lobster and crayfish. The tendency to produce a six-banded stomodæum and proctodæum (if not fanciful) may possibly be tubes of the mesenteron, and the malpighian tubules of the proctodæum.

A large salivary gland and reservoir lie on each side of the œsophagus and crop. The gland is a thin, foliaceous mass about ⅓ in. long, and composed of numerous acini, which are grouped into two principal lobes. The efferent ducts form a trunk, which receives a branch from a small accessory lobe, and then unites with its fellow. The common glandular duct thus formed opens into the much larger common receptacular duct, formed by the union of paired outlets from the salivary reservoirs. The common salivary duct opens beneath the lingua. Each salivary reservoir is an oval sac with transparent walls, and about half as long again as the gland.

The ducts and reservoirs have a chitinous lining, and the ducts exhibit a transverse marking like that of a tracheal tube.

There are eight (sometimes fewer) cæcal tubes arranged in a ring round the fore end of the chylific stomach; they vary in length, the longer ones, which are about equal to the length of the stomach itself, usually alternating with shorter ones, though irregularities of arrangement are common. The tubes are diverticula of the stomach and lined by a similar epithelium, but do not contain any undissolved food. In the living animal they are sometimes filled with a whitish, granular fluid. They are developed late, and are absent or of inconsiderable size in larval cockroaches.

The position of these cæca suggests a comparison with the tubular "liver" of crustacea, and with the pyloric cæca of fishes, but no adequate physiological investigation has been made as to the function of any of these organs.

The malpighian tubules mark the beginning of the small intestine, to which they properly belong. They are very numerous (60–70) in the cockroach, as in locusts, ear-wigs and dragon-flies; and unbranched, as in most insects. They are about ·8 in. in length, and ·002 in. in transverse diameter, so that they are barely visible to the naked eye as single threads. In larvæ about ⅛ of an inch long, Schindler (Zeitsch. f. wiss. Zool. Bd. xxx.) found only eight long tubules, the usual number in Thysanura, Anoplura, and Termes. In the adult cockroach the long threads wind about the abdominal cavity and its contained viscera.

In the wall of a malpighian tubule there may be distinguished (1) a connective tissue layer, with fine fibres and nuclei; within this, (2) a basement-membrane, between which and the connective tissue layer runs a delicate, unbranched tracheal tube; (3) an epithelium of relatively large, nucleated cells, in a single layer, nearly filling the tube, and leaving only a narrow, irregular central canal. Transverse sections show from four to ten of these cells at once. The tubules appear transparent or yellow-white, according as they are empty or full; sometimes they are beaded or varicose; in other cases, one half is coloured and the other clear. The opaque contents

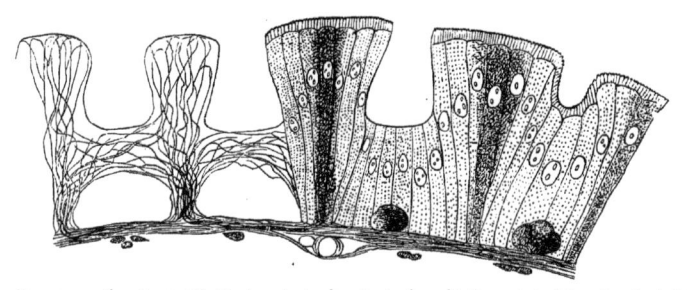

Fig. 81.—Transverse section of wall of chylific stomach of cockroach, showing epithelium and glandular cells. To the left side the epithelium is removed, to show the connective tissue skeleton. × 325.

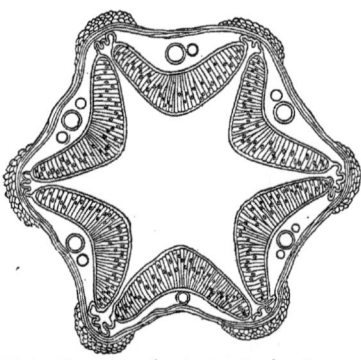

Fig. 82.—Transverse section of rectum of cockroach. × 50.

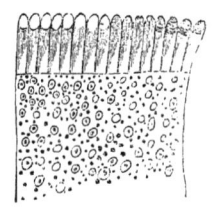

Fig. 83.—Free end of a single epithelial cell of chylific stomach, showing processes. × 3000.

consist largely of crystals, which either occur singly in the epithelial cells, or heaped up in the central canal. Occasionally, they form spherical concretions with a radiate arrangement. They contain uric acid, and probably consist of urate of soda.* In the living insect the tubules remove urates from the blood which bathes the viscera; the salts are condensed and crystallised in the epithelial cells, by whose dehiscence they pass into the central canals of the tubules, and thence into the intestine.

In the present rudimentary state of invertebrate physiology, little can be certainly affirmed of the digestive processes of any insect. The difficulties of observation and experiment are great, and the investigators are few. Plateau's memoirs on the physiology of various Arthropoda are the most laborious and comprehensive which we possess.† The observations of Basch are earlier and less complete.

Basch set out with a conviction that where a chitinous lining is present, the epithelium of the alimentary canal secretes chitin only, and that proper digestive juices are only elaborated in the chylific stomach, or in the salivary glands. The tests applied by him seemed to show that the saliva, as well as the contents of the œsophagus and crop, had an acid reaction, while the contents of the chylific stomach were neutral at the beginning of the tube and alkaline further down. From this he concluded that the deep-seated glands of the chylific stomach secreted an alkaline fluid, which neutralised the acidity of the saliva. Finding that the epithelial cells of the stomach were often loaded with oil-drops, he concluded that absorption, at least of fats, takes place here. The chylific stomach, carefully emptied of its contents, was found to convert starch into sugar at ordinary temperatures. The saliva of the cockroach gave a similar result, and when a weak solution of hydrochloric acid was added, Basch thought that the mixture could digest blood-fibrin at ordinary temperatures.

Plateau's principal conclusions as to the digestive processes of cockroaches are the following. (1) The saliva changes starch into glucose, but it is, like all the digestive secretions of insects, alkaline, and not acid. Acidity in the contents of the crop is due either to the ingestion of acid food, or to an acid decomposition of the same. The supposed solvent action of saliva upon fibrin is imaginary. (2) Absorption of glucose takes place in the crop, and glucose is neither formed nor absorbed in succeeding parts of the alimentary canal. (3) The function of the gizzard is that of a strainer; it has no triturating power. (4) The cæcal tubes pour forth a fluid which is often distinctly alkaline (never acid), and which neutralises

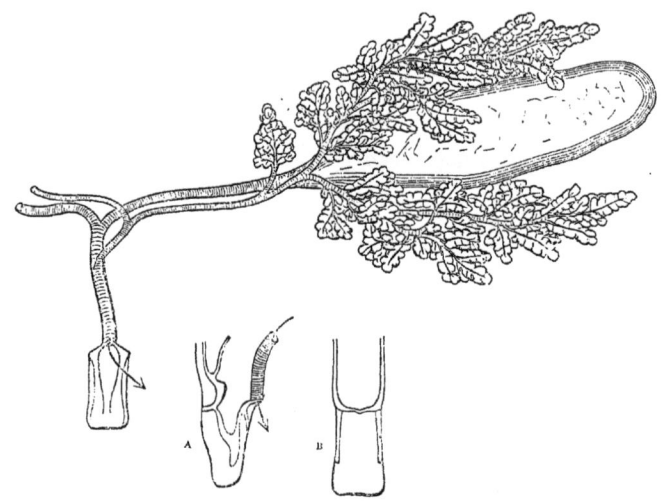

Fig. 84.—Salivary glands and receptacle of cockroach, right side. The arrow marks the opening of the common duct on the back of the lingua. A, side view of lingua; B, front view of lingua.

---

* The contents of the malpighian tubules may be examined by crushing the part in a drop of dilute acetic acid, or in dilute sulphuric acid (10 per cent.). In the first case a cover-slip is placed on the fluid, and the crystals, which consist of oblique rhombohedrons or derived forms, are usually at once apparent. If sulphuric acid is used, the fluid must be allowed to evaporate. In this case they are much more elongated, and usually clustered. The murexide reaction does not give satisfactory indications with the tubules of the cockroach.

† See his "Rech. expérimentales sur la digestion des Insectes," and various memoirs published by the Royal Belgian Academy.

any acidity in the food passed on from the crop. It emulsifies fats, which is only temporarily done by the saliva, and converts albuminoids into peptones.* The secretion of the glandular wall of the chylific stomach could not be satisfactorily isolated. (5) There are no absorbent vessels, the products of digestion passing by osmosis through the wall of the digestive tube to mingle with the blood which circulates in the perivisceral space.

GOSSIP ON CURRENT TOPICS.

IN the last and previous numbers of SCIENCE-GOSSIP are some letters on domestic aquaria, the appearance of which is refreshing, as the little tanks seem to have been snuffed out recently by the larger displays of the public aquaria which they preceded and suggested. So much interest, instruction, and elegance is afforded, even by a very small well-managed aquarium at home, that a revival of the popularity of such humble "zoological stations" is very desirable.

A remark of Mr. Walter T. Cooper, respecting the "poisonous matter" contained in Portland cement, indicates that an explanation of the chemistry of this is demanded. When the Crystal Palace aquarium was started I was consulted by my old friend Mr. Lloyd, with whom I had co-operated in his early small attempts, concerning a dilemma in which he found himself. The building was complete; his admirable duplicate arrangements for circulating the water throughout the tanks were all in order; the tanks were filled with water from Brighton, and the stocking with animals had commenced in good time for the opening as announced, but, alas! the water, at first quite clear and bright, gradually became turbid, and the animals simultaneously sickened and died.

On examination, I found it slightly alkaline, and thus at once indicated the cause of the trouble. Portland cement contains caustic lime, i.e. lime in its alkaline condition. Such lime is soluble in either river or sea-water, and when these contain, as they usually do, some carbonate of lime dissolved in excess of carbonic acid, the caustic or alkaline lime combines with this excess, and is not only itself precipitated as insoluble carbonate of lime, but also effects a further precipitation of the carbonate of lime that existed originally in the water dissolved by its free carbonic acid.

I accordingly proposed that hydrochloric acid should be carefully added, and proved its efficacy by clearing the water of an experimental tank. I also suggested that by allowing sufficient time, and keeping up a vigorous circulation of the water, the carbonic acid of the air would eventually do the work of clearing. Mr. Lloyd, having a dread of artificial chemicals, preferred this method, which was adopted with complete success, but it delayed the stocking of the tanks by some weeks.

The lecture delivered by Professor Osborne Reynolds, at the Royal Institution, on Friday, March 28th, is very interesting and instructive. It is an able popular exposition of a somewhat difficult mathematical subject, and supplies a lesson not only to those for whom it is directly intended, but also to the especially special mathematical specialists, who fall into the sin of mathematical self-righteousness by assuming, as a matter of course, that mathematical demonstration settles everything.

Professor Reynolds commenced by saying that "'in spite of the most strenuous efforts of the ablest mathematicians, the theory of fluid motion fits very ill with the actual behaviours of fluids, and this for unexplained reasons.' The theory itself appears very tolerably complete, and affords the means of calculating the results in almost every case of fluid motion, but while in many cases the theoretical results agree with those actually obtained, in other cases they are altogether different." He then proceeds to show how the theory stands experimental verification when applied to a raindrop falling through the air, but fails in the case of a large body, such as a ship moving through the water, and affords "no clue to the reason why it should apply to the one class more than to the other." We have here an example illustrating the absolute necessity of verifying by observation or experiment every theoretical conclusion that falls within the reach of such verification, without which it is merely an hypothesis and should be scrupulously described accordingly.

The calculations of Adams and Leverrier, which led to the discovery of the planet Neptune, are justly regarded as great triumphs of mathematical reasoning, but it is an error to assert, as commonly done, that these mathematicians discovered the planet. What they actually did was to formulate an hypothesis to account for certain observed perturbations of Uranus, which might be due to the gravitation of an outer planet having certain positions at given times. The actual discovery by the telescope proved the truth of the hypothesis. Subsequently Leverrier based upon similar mathematical measurements of the perturbations of Mercury the hypothesis of an inner planet, and calculated its hypothetical orbit, but it is now fairly evident that this planet Vulcan is but a mathematical fiction, though the mathematical demonstration of its existence is as good as was that of the existence of Neptune prior to its actual discovery.

Unappreciated discoveries are common enough,

---

* This was first demonstrated by Jousset de Bellesme, who, however, erroneously states that the secretion is acid; it is really alkaline, like the proteolytic secretions of the pancreas of mammalia. Plateau uses the muscles of the fly for digestion by the chylific fluid of the cockroach, and finds that the microscopic characters of the fibres render it easy to judge by simple inspection of the progress of the digestive process.

but it rarely happens that a bold innovator brings upon himself an embarrassment of over-reception. This appears to be the dilemma of M. Pasteur in reference to his antidote to hydrophobia. He has announced his belief or hopes of being able to prevent the outbreak of this dreadful malady by inoculation with modified virus; is extending his researches on animals, on dogs bitten by mad dogs, and finds himself already besieged by human applicants desiring to be mildly maddened in order to escape a possible attack of malignant madness. As ninety-nine and some odd tenths per cent. of the angry dogs who bite people, and are therefore supposed to be mad, are no more rabid than those they have bitten, the inoculation of all the latter with virus of questionable potency would constitute a very responsible series of experiments, which M. Pasteur, in spite of his enthusiasm, refuses at present to undertake in France. When he has done with the dogs, and is prepared to commence operations on human beings, he may do well to come to England, where the laws for regulating surgical practice are so stringent for the protection of frogs, but appear to empower no interference on behalf of human beings, provided the practitioner has the ordinary qualifications.

The conferences and lectures have commenced at the Health Exhibition with but small audiences on the spot, though practically extending to much larger audiences through the official reports and newspaper abstracts. I had the honour and misfortune to "open the ball," so far as the lectures are concerned, and found that only a very small percentage of the visitors knew of the existence of any such lectures. The difficulty of competing with the illustrated soap-makers, &c., in advertising them is very serious, but we may hope that as the fact of lectures being provided becomes better known, the attendance will be larger; it cannot be more appreciative than was the first and very select audience.

A good deal of flippant criticism has been applied to this "show" as regards its scientific pretensions. To those who understand no more science than is necessary for obtaining university degrees, all such popular efforts, such promenade science, mingled with military bands and bazaar business, must be shockingly frivolous; but the true philosopher, who has plunged through the superficial crust of text-book technicalities into the profound simplicities of actual science, of which these technicalities are merely the working tools; who understands the relations of science to the human mind and human progress, sees in all such efforts to intermingle science with practical daily work, and above all, with joyous recreation, the fulfilment of the great object which alone justifies the existence of science schools, universities, learned societies, original research, &c. I put recreation above all, because the elevating moral and intellectual influences of science operate most powerfully on those who make it a recreation, and least upon the specialist who uses it as a matter of business, like any other trade, or in order to obtain social status by public displays. It is even possible that these latter may be morally degraded by their scientific attainments; they certainly are so whenever they assume the pharisaical airs of a scientific priesthood, by looking down with pedantic self-complacency upon vulgar people who are not as they are.

W. MATTIEU WILLIAMS.

AT THE GATE OF NORTH WALES.

A MIDSUMMER HOLIDAY.

BY THE EDITOR.

I DO not know a single holiday place, among the many I am privileged to remember, which strikes one so much as Llangollen. As soon as we step out of the train, we feel that the place will suit. The famous river Dee rushes and brawls and seethes

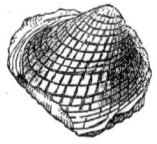

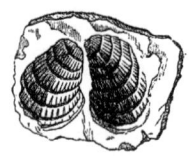

Fig. 85.—*Cardiola interrupta*, a Common Upper Silurian Fossil Bivalve.

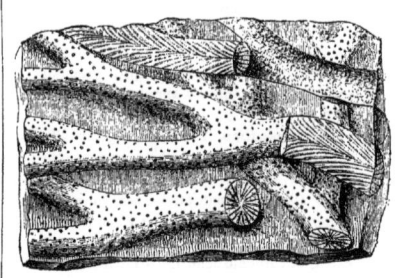

Fig. 86.—*Stenopora fibrosa*, a Common Fossil Coral at Glen Ceirog.

in and out of the huge masses of slate-rock which crop out irregularly in its bed, before it shoots through the pointed arches of its fourteenth-century bridge. The picturesque buildings of the older part of the town near the river hide the higher and newer houses. Beyond the town rise the slate hills, clad with rich foliage almost to their summits, and in front stands the bold, white, and precipitous escarpment of the Eglwyseg rocks, forming the north-western boundary of the valley. Between us and this splendid outcrop of the Carboniferous limestone stands Dinas Bran, a solitary outlier of Silurian slates, almost conical in shape, which rises to the height of 600 feet, and is crowned by an ancient castle. For delightful rambles with vasculum or hammer, or without, there are few places where one can better spend a week, or, for the matter of that, many an one, than Llangollen.

Archæologically, too, there is here much to be seen English structure, built in the year 1200, and held by the Cistercian monks. The lancet-shaped windows of the eastern end have a very telling effect, when one views these ruins against the background of the green hills. Not far away the archæologist finds the Pillar of Eliseg standing, a monument not so much to be venerated for its antiquity as for its being perhaps the last of its long race erected in the British isles. Cromwell's soldiers would have proved poor followers of the æsthetic school, for they mis-

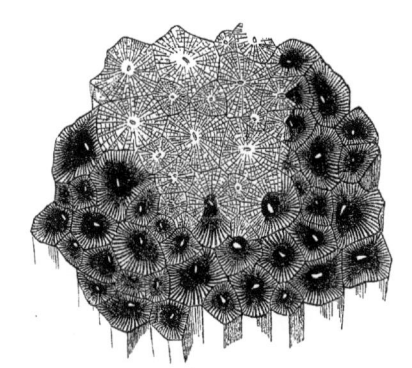

Fig. 87.—*Lithodendron basaltiforme*, a common Carb. limestone coral.

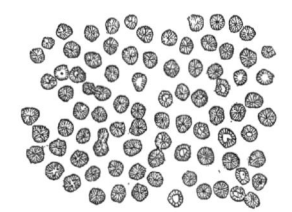

Fig. 89.—Transverse section of Fossil Corals (*Lithostrotion junceum*), Carb. limestone.

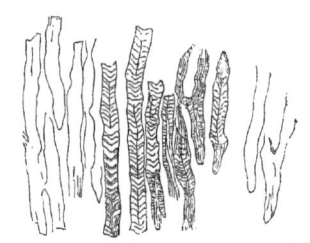

Fig. 90.—Longitudinal section of *Lithostrotion junceum*.

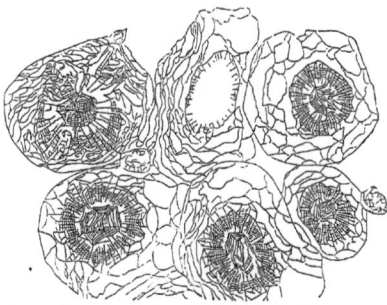

Fig. 88.—Transverse section of Fossil Coral (*Lonsdalia rugosa*), Carb. limestone.

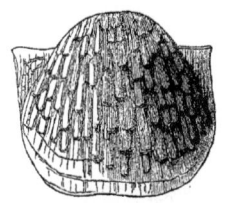

Fig. 91.—*Productus scabriculus*, Carb. limestone.

and heard of. We are on the border-land of two races and two languages. The hills around are sprinkled with tumuli, nearly covered up in the heather, which eloquently tell of the bitter feuds of days happily past, and of the fierce fights of old between Saxon and Celt. Hereabouts, all the way up the richly-wooded, close-shut-in, lovely valley of the Dee to Corwen, the placed is steeped in traditions of Owen Glendower and his fighting men.

Val Crucis Abbey is still a sight to be seen ; an Early took this Welsh monolith for a Popish Cross, and "brake it down !" It was set up again in 1779.

I cannot stay, however, to recapitulate the historical and archæological interest which hangs round the spot I selected for a brief and much-needed summer holiday. I take it for granted that my readers, like myself, are interested in the affairs of nature rather than in those of men'; at any rate, during their holidays.

The botany of the place is not so interesting

perhaps, as its geology, for the latter is unusually rich. Anyone wanting to work it out had best refer to the papers by Mr. Davis in the Proceedings of the Geologists' Association, and to Mr. G. H. Morton's handsome and useful little volume (with maps, photographs, &c.), which deals chiefly with the Carboniferous series.

All the way up the valley to Corwen, whether by the high-road or the river, nothing can surpass the loveliness of the scenery. There is a peacefulness about the spot which gets hold of the tired and weary brain at once; and the atmosphere, heather-scented from the hills, produces that delightful drowsiness which reminds us of the grounds round about Thomson's "Castle of Indolence." This gorge of the Dee, now looking so inexpressibly lovely in its midsummer dress, was once filled with moving ice, as the moraines lining the present river abundantly testify in their scratched stones. What a contrast between the slow-moving glacier which once filled this defile, perhaps up to the shoulders of these rounded hills, and the brawling salmon and trout river rushing past us, its banks fringed with forests of foxgloves, nettle-leaved campanulas, and brake ferns! The rich alluvial soil forms margins of meadowland which border the river alternately on this side or that, and which are usually the sites of old country houses, and the retreats of many a rare plant. Among others, not uncommon, may be mentioned *Inula helenium*, woodruffe (*Asperula odorata*, abundant just now in places); *Trollius Europæus* is common in sheltered localities by the river, and *Paris quadrifolia* in the woods. The very beautiful *Althæa officinalis* is abundant by the water both of the river and the canal (the latter one of the oldest in the country) and its clusters of flowers never look so pretty as when reflected in the water they love to linger by. Among other characteristic plants of the neighbourhood are the Welsh poppy (*Meconopsis Cambrica*), dwarf elder (*Sambucus Ebulus*), etc.

The Eglwyseg rocks and the gorges which here and there run up into them,—notably that beautiful *cul-de-sac* called "The World's End,"—are very rich in various plants and ferns. Here may be found

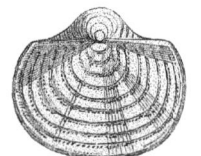

Fig. 94.—*Productus punctatus*, Carb. limestone.

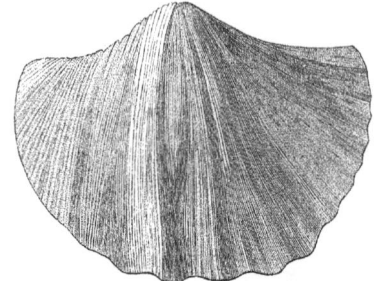

Fig. 95.—*Productus giganteus*, Carb. limestone.

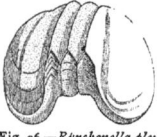

Fig. 96.—*Rynchonella pleurodon*, Carb. limestone.

Fig. 97.—*Terebratula hastata*, Carb. limestone.

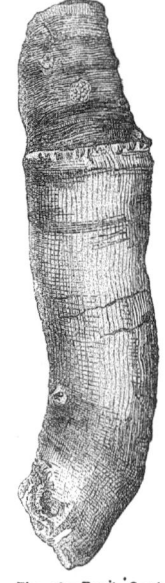

Fig. 98.—Fossil Coral (*Amplexus coralloides*) Carb. limestone.

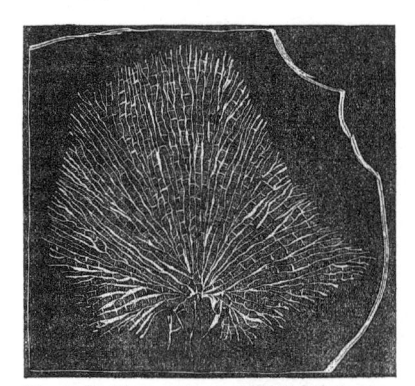

Fig. 99.—Fossil bryozoan (*Retepora plebeia*), Carb. limestone, Hafod.

the rock-rose (*Helianthemum canum*), *Thalictrum alpinum*, *Geranium sanguineum*, *Hieracium maculatum*, &c.; and the "World's End" is the habitat of *Asplenium viride*, *Cystopteris*, *Polypodium calcareum* (also to be found all over the Eglwyseg rocks), the oak-fern (*P. Dryopteris*, in the woods), the beech-fern (*P. phegopteris*), the holly ferns, &c.

On the breezy hills, on both sides the valley, the naturalist will find freedom and paradise. He may walk up Barber's Hill, and get knee deep in heather almost immediately; and, if he chooses, he can walk thus until he drops down on the other side the billowy table-land, near Bala, at the end of a long summer day. Or, should he prefer the other side he clambers up the limestone ridge behind Trevor and makes his way over wild moorland as far as Wrexham, none daring to make him afraid. On these swampy moorlands he will find the bearberry, (*Arctostaphylus uva-ursi*), the crowberry (Vaccinium); particularly on the slate-hills.

Geologically, the rambler soon finds abundant materials to fill a bag. Rock and mineral specimens abound; the fossils both of the Silurian and Lower Carboniferous formations are numerous. On climbing Dinas Bran, the loose upper Silurian slates near the top yield him Rhynchonella, &c. The small quarries opened out by the canal near Glyndyfrdwy station, abound in orthis, trilobites, and other fossils. The walls by the roadside contain abundant casts of the tapering *Orthoceras primeva*, a well-known cephalopod, always associated with those of the bivalve *Cardiola interrupta*. Should the pedestrian climb up to Moel Gamelin, or any of the near hills, he will find the refuse of the quarries abounding in these fossils; and, mayhap, he will also meet with beautiful casts of entire encrinites, such as *Actinocrinus pulcher*. The slates are seen in the quarry standing vertically, and their upper, turned over, edges, indicate the direction of the flow of the ancient ice-sheet which once masked all the hills hereabout.

On the other side, after walking up Barber's Hill, we drop into the Glen Ceirôg Valley. It is a lovely and quiet spot, little visited or invaded by civilisation, although a tramway runs up it from the quarries. The geologist soon finds evidence of ancient volcanic action in the ash-beds and traps; and, when he begins to ascend the hills on the other side, the little lime kilns will tell him where the Bala limestone crops out, and there he may settle himself down for a little hammering. The debris around are a perfect museum, full of fossils of all the kinds common to this formation; trilobites, corals, brachiopods, mollusca, &c.

But time is short, and geology is long. The Carboniferous limestone of the Eglwyseg rocks is sure to attract the geological student. All the way from beyond Trevor to "The World's End," it is worked for lime-burning, and fossils are abundant in it. The upper part is crowded with fossil corals, in a marvellous state of preservation, especially *Lonsdalia rugosa*, *Lithodendron basaltiforme*, *Lithostrotion junceum*, *Amplexus coralloides*, *Zaphrentis*, &c. The limestone generally abounds with *Productus Llangolliensis* (which appears to compose a good part of the rock), with *Productus cora*, *P. giganteus*, *P. scabriculus*, *P. punctatus*, *Terebratula hastata*, *Rhynchonella pleurodon*; Trilobites such as Phillipsia; Euomphalus, Goniatites, Nautilus, &c. Many pleasant days collecting might be spent here, as the visitor has time and opportunity.

Or, if he chooses to go further afield, he will find an outlier of the Carboniferous limestone at Hafod, two miles beyond Corwen, perhaps the richest fossiliferous spot in Great Britain, abounding in fossil corals most beautifully preserved, and some of which will tax the visitor's strength to remove, owing to their huge size. Notable among them is the pretty creeping coral Phillipsastrea. The fossil shells are coated over with bryozoa, especially with *Retepora plebeia*, in every stage of growth and development; and many others of these beautiful, lace-like fossils may be found investing shells, corals, &c., all silently testifying to the quiet life of the ancient Carboniferous seas.

How the time flies when every dewy morning ushers in a bright day's work in Nature's laboratory! A week is all too brief, but it brings with it thankful relief, and as near an approach to boyhood's joviality as one's commencing grey hairs can expect.

I was delightfully located at the Royal Hotel (so called because Queen Victoria staid there on her visit to Wales many years ago). The Iron Duke accompanied Her Majesty, and the principal dining-room still goes by the name of the "Wellington Room.". The splendid billiard room is not without archæological interest, for one of the finely carved tables was made from the timber which once formed part of Val Crucis Abbey. Under the solicitous care of the kind host and hostess (Mr. and Mrs. Hardy) the wayfarer need feel no anxiety about his inner man. A more comfortable home, away from home, he could not find throughout the length and breadth of his native land. In the pleasant gardens of the hotel he may rest his tired limbs and smoke his pipe of peace, whilst he watches the salmon and trout anglers making their casts in the rushing river which flows within a few feet where he sits; and the soothing roar of the waters is like music to a town-dweller.

NIGHTINGALES IN YORK.—It may interest readers of SCIENCE-GOSSIP to know that two nightingales have been in this neighbourhood for the past fortnight, one at Crimple, the other in the Spa grounds at this place, where they may be heard in full song after dark. It is more than twenty years since similar birds were in this district.—*H. Newman Pullan, Harrogate.*

SCIENCE-GOSSIP.

THE Report of the United States Commission of Fish and Fisheries for 1880 has been issued, and it makes a handsome and bulky volume of more than 1000 pages. It is devoted to an enquiry into the decrease of food fishes, and to the propagation of food fishes in the waters of the United States. Some of the papers are of high zoological value, such as that on "Materials for a History of the Swordfishes" (illustrated); "The Oyster and Oyster Culture," &c.

THE third annual Report of the United States Geological Survey for 1881-82, has just been published. It contains Professor Marsh's valuable paper on "Birds with Teeth," others on the "Copperbearing Rocks of Lake Superior," by R. D. Irving; "The Geological History of Lake Lahontan," by J. C. Russell; "The Geology of the Eureka district," by A. Hague; "The Terminal Moraine of the Second Glacial epoch," by T. C. Chamberlin; and "A Review of the Non-marine Fossil mollusca of North America," by Dr. C. A. White, from the Devonian strata upwards, artistically and abundantly illustrated.

MR. PRINCE, a well-known meteorologist, thinks the recent beautiful sunset glows were due to the crystallisation of saline particles from masses of seawater, ejected in the form of vapour into the upper regions of the atmosphere from Krakatoa, and that the fact of the greatest displays having been seen during the coldest weather can only be accounted for on the theory that the crystallisation of saline particles was the chief factor in their production.

THE first annual report on the Injurious and other Insects of the State of New York, by J. A. Leitner, State Entomologist, has recently appeared. It is a very interesting work, dealing chiefly with economic entomology, and describing at length the life-histories of the specific insects attacking crops and fruits.

MR. MATTIEU WILLIAMS writes in the last number of "The Gentleman's Magazine" in defence of arsenical wall papers where malaria abounds, and recommends that the hotels in the vicinity of the Maremma, the Campagna and Pontine marshes should be papered throughout with brilliant green wall papers, &c. He considers slight doses of arsenic an antidote to such fever and malarial poisoning.

AN apparatus has been employed at Harvard College, United States, to test the value of electric changes in the atmosphere, as indications of coming changes. It photographs every change in the electricity of the air, and also indicates the degree of change. The observations are said to be very promising as an aid to meteorology.

SEVEN hundred members have sent in their names from Great Britain alone to attend the approaching meeting of the British Association in Montreal.

SATURN has recently been examined carefully by Messrs. Paul and Henry, two French astronomers, who have distinguished outside the known established rings of that planet a new ring, brilliant and perfectly defined. They think that Encke's division is the result of an optical illusion, produced by the brilliant ring they have just discovered.

BRANCHES of the "Youth Scientific and Literary Society" are now in course of formation in all our large cities and towns throughout the kingdom. Information regarding this society may be obtained on application to the secretary, Mr. A. Davis, jun., High Street, Great Marlow, Bucks.

PART XIX. of Dr. Lang's now famous "Butterflies of Europe," has just appeared, illustrated as usual with very beautiful, natural-sized and coloured drawings of each species. Every species in the following genera are described and figured:—Pararge, Epinephele, and Cœnonympha.

PROFESSOR HOUGH, of the Chicago Observatory, has been engaged in measuring the companion of Sirius. He says that in a few years this interesting object will be beyond the reach of all but the very largest telescopes.

THE Edison Electric Company is trying the experiment of out-door lighting, by placing clusters of three 32-candle power lights on a couple of street corners. The lights are suspended on wires about 30 feet from the ground. It is thus intended to demonstrate that street lighting by this system is possible.

AN interesting test was made in Washington recently, by electricians and an official of the coast survey, to determine the speed of dots over a telegraph wire. A very intricate testing instrument was used, and it was finally estimated that an electric dot travels at the rate of 16,000 miles per second, or 6,000,000 miles per minute.

THREE hundred and seven Edison electric lighting plants have been sold in the United States and Canada since May 31st, 1883, aggregating 59,173 lamps.

PROFESSOR HOUGH thinks that the great "Redspot" on Jupiter, which remained until last year of a brick-red colour, but which has gradually grown paler, until it is now hardly visible, will not be seen much longer in any telescopes. A similar spot, with a diameter of about 8000 miles, was noticed in this planet in 1664, and it has reappeared and vanished many times since then. It seems to appear and reappear at regular intervals. Professor Hough thinks it is in reality the solid body of the planet, usually invisible beneath its cloudy covering.

AN optical telegraph has been successfully established between the Islands of Mauritius and Reunion, a distance of about 140 miles. Observers can read the signals without difficulty, and arrangements for announcing cyclones are nearly completed.

MR. G. H. KINAHAN, of the Irish Geological Survey, writes as follows concerning the area affected by the recent earthquake : The area of structural damages was limited, but in the country surrounding it the shock was felt for considerable distances. In the area of structural damages there were five smaller area in which the effects were greatest. Three small areas are distinctly margined at two or more sides by lines, while at one or two of these lines the maximum damage in each occurred. The greatest damage was done in the Peldon and Abberton area. In the areas of greatest effect the shock travelled in different directions, yet individually ; in all the shock appears to have occurred at the same moment. The geological formations in the area of structural damage are London clay, glacial drifts, and alluvium. In two places, Eastbridge, Colchester and Wivenhoe, damages were observed on the latter ; while from the north portion of Wivenhoe to Alresford, and in the northern portion of Colchester damages occur on the glacial drift, but elsewhere all damages, especially the greater ones, are found on or at the margin of the London clay ; while nearly invariably the destruction ceased at the margin of the young accumulations.

THE members of the Geologists' Association took their usual Whitsuntide holiday (their chief excursion of the year) to Cambridge, where they investigated the gravel pits, phosphate pits, chalk, &c., of the neighbourhood, and visited the various museums in Cambridge, under the directorship of Professor Hughes.

ON Friday, June 6th, Mr. T. V. Holmes read a paper before the above association on "Some Curious Excavations in the Isle of Portland," and Mr. C. E. De Rance one on "The Underground Waters of England and Wales."

AT the last ordinary meeting of the Essex Field Club the following papers were read :—"Report on the Flowering Plants growing in the neighbourhood of Colchester," by J. C. Shenstone, F.R.M.S., &c.; "Progress of the Report on the Recent Earthquake shock in Essex," by Raphael Meldola, F.R.A.S., and W. White; "On the Earth-subsidence at Lexden, near Colchester, in 1861," by T. V. Holmes, F.G.S., M.A.I.; "On the Occurrence of the Rhizopod, *Clathrulina elegans*, in Essex," by C. Thomas, F.G.S., F.R.M.S. On June 21st a field meeting was held in Epping Forest, under the leadership of Mr. J. H. Harting.

MR. R. KIDSTON has described the fructification of several species of fossil ferns belonging to Sphenopteris.

MR. JAMES ENGLISH announces the forthcoming publication of a work called "Flora of Epping Forest Mosses," to be issued in four parts or fasciculi of twenty species each.

THE Report of the Entomological Society of Ontario, for 1883, had just been printed by order of the Legislative Assembly. It deals with some of the noxious, beneficial, and other insects of the province.

THE May number of the Transactions of the Hertfordshire Natural History Society has been published, giving the index to papers, list of members, &c.

A RE-ISSUE of Mr. W. F. Kirby's "European Butterflies and Moths," in sixpenny parts, with coloured illustrations of the insects, their larvæ, and food-plants is being issued by Messrs. Cassell & Co.

MR. W. PHILLIPS, F.L.S., announces the speedy publication of "A Manual of the Discomycetes, with descriptions of all species of Fungi hitherto found in Britain."

AT a recent meeting of the Geological Society, Mr. H. G. Spearing stated that for the last nine years the sea has encroached at Westward Ho, North Devon, at the rate of about eighty feet annually.

MR. G. V. SMITH has communicated to the Geological Society an account of the discovery of the footprints of vertebrate animals in the Lower new red sandstone of Penrith, Cumberland. Eleven footprints were found. From their different sizes and shapes they are believed to have been made by different species of animals.

AN anthropometrical laboratory has been opened at the Health Exhibition by Mr. Francis Galton.

PROFESSOR GÖPPERT, the distinguished palæontologist and fossil botanist, has just died at Breslau at the ripe age of eighty-four.

PROFESSOR RAY LANKESTER is anxious to have the support of all lovers of zoology for the Marine Biological Association which has just been started. The subscription for membership is one guinea a-year. £10,000 is required.

HELIX PISANA.—On the 23rd of May last year I received from Jersey twenty live specimens of this snail. Four of the finest were placed in a vivarium (where several other species thrive and breed) and well supplied with food. All four died during the last week of October. A few days ago I discovered the forgotten card box in which the snails had come from Jersey, and which still imprisoned sixteen. On placing them in tepid water, all except two revived and are now feeding heartily. Eight months is a long time surely for this delicate species to remain torpid.—*K. McKean, Warham Road, Croydon.*

MICROSCOPY.

MOUNTING INFUSORIA.—I shall be glad if any one can tell me a way of killing such Infusoria as vorticellæ and stentors, and such Animalcules as rotifers with their cilia expanded, and then setting them up as permanent objects. Saville Kent, in his recent work on the Infusoria, says that a solution of osmic acid effects this perfectly. I have tried it and failed totally. At present I only have one per cent. solution (which is the strength he recommends). Is the weakness of the solution a cause of failure? It certainly kills the creatures very quickly, but yet not quickly enough. But they seem to contract more in consequence of the disturbance of the water caused by adding the acid, than from the effects of the acid itself. Is there any way of applying the acid so as not to frighten them by moving the water? Again, Mr. Kent recommends osmic acid as a preservative. I have set up half-a-dozen slides of non-contractile Infusoria in this medium. After the lapse of a fortnight, they, and the acid itself, are now the colour of ink, and opaque. I cannot help thinking that Mr. Kent must be mistaken in saying that osmic acid will preserve animalcula—such as *Euglena viridis*—in their natural colours. In SCIENCE-GOSSIP for 1879, Mr. G. du Plessis says that a saturated solution of permanganate of potash kills animalcula without contracting their cilia. This solution is a very deep violet colour, so deep as to be almost black. I have tried it, but such things as vorticellæ are quite contracted by it. A certain T. C., later on in the same volume, recommends "chromic oxydichloride" acid, combined in certain proportions with permanganate of potash. A querist asks in vain what this acid may be, and when I enquired about it of a friend of mine, a chemist, he had never heard of it. He suggested that it might be chromic acid; so I tried this, but it was not of the slightest use. The aggravating part is that these three "authorities," S. Kent, G. du Plessis, and T. C., all say how splendidly their methods succeed, whilst I follow their directions as closely as I know how, and yet fail so utterly that I feel sure that there must be some "wrinkle" in the processes which their descriptions do not make clear. If either of these people should read this note, a little further information from them, or from any one who has had better success than myself, will be exceedingly welcome to *H. M. J. Underhill, Oxford.*

IMPROVEMENTS IN MICROSCOPIC SLIDES.—I read with considerable interest the paragraph p. 136, "On Improvements in Microscopic Slides," and that Mr. B. Piffard had "obtained a patent specification under the above title." I have been experimenting upon a similar arrangement for some years past, my first attempt was made in conjunction with Mr. William Robson, about five years ago (then proprietor of the Lemington-on-Tyne Glass Works). The primary object was to grind a level cell in the ordinary glass slide with a sufficiently sunk margin to allow a glass cover to be counter-sunk and flush with the surface of the slide; hence rendering leakage impossible—simplifying, the process of mounting, as there is no cell to build, and giving a perfection and finish to the mount not otherwise attainable. He found a difficulty in levelling the bottom of the cell, as the tool used ground the edges more rapidly than the centre; this was to a considerable extent obviated by an alteration in the grinding tool, but the expense in working them became the most serious difficulty—of course this vanishes if the principle is applied to other material, vulcanite, wood, cardboard, &c., which for opaque objects would be lighter, and answer equally as good a purpose as glass. These attempts were well-known to many members of "The North of England Microscopical Society" (for which I acted as honorary secretary up to the present session), and to whom I was indebted for many useful suggestions.—*M. H. Robson, Newcastle-upon-Tyne.*

MR. COLE'S "STUDIES."—Numbers 18 and 19 of the welcome "Studies in Microscopical Science" are devoted severally to "Vascular Tissue" (illustrated by Bast, sieve-tubes, and litter-cells), and "Nerve of Horse" (the latter illustrated by an exquisitely coloured plate). No. 9 of the "Popular Microscopical Studies" treats upon the Crane fly (Tipula), its life, history and general anatomy, particularly the structure of the head. Part 10 of "The Methods of Microscopical Research" continues the subject of mounting, in which the student gets the full benefit of Mr. Cole's long and very extensive experience. The first numbers mentioned above were accompanied by slides, all of the usual neatness and high finish, but that of the head of the Crane fly is unusually beautiful, as it shows the compound eyes and mouth organs very distinctly.

"THE JOURNAL OF THE ROYAL MICROSCOPICAL SOCIETY."—The June number contains the following papers:—"On the Estimation of Aperture in the Microscope," by the late Charles Hockin, jun.; "Note on a Proper Definition of the Amplifying Power of a Lens," by Professor E. Abbe; "On Certain Filaments observed in *Surirella bifrons*," by John Badcock. Besides the above we have the usual excellent summary of current researches relating to Zoology, Botany, Microscopy, &c. All the papers are illustrated.

MR. J. E. ADY writes to say that owing to certain differences betwixt himself and his colleague, Mr. Hensoldt (which we cannot enter into here), he begs the indulgence of the subscribers until suitable arrangements can be made for continuing the "Petrological Studies."

ZOOLOGY.

FREE SWIMMING ROTIFERS—A SUGGESTION FOR A ROTIFER SOCIETY.—The rotifer which Mr. Lord has figured and described in the last number of SCIENCE-GOSSIP is very common, and if Mr. Lord will consult Pritchard again, he will find his Colurus described under the head of Lepadella. If water known to contain rotifers be left for some time, the species in question will be found to be one of the last survivors; it may be taken during the winter, under ice, and seems to act as a sort of scavenger, thriving on food which all other species disdain. Lepadella is one of the euchlainida, and bears a very close resemblance to salpina in its structure, and in the arrangement of its carapace. As a constant student of the rotifers, I have often felt the want of a satisfactory text-book, to do for them what Saville Kent has done for the Infusoria, and I am glad to be able to inform Mr. Lord that the rumour he has heard is well founded. Dr. Hudson favoured the Bristol Microscopical Society, at one of its meetings last year, with a sight of the drawings for his new Monograph on the Rotifers, and it is impossible to speak too highly of their beauty, and, which is still more important, of their accuracy. As to the probable date of publication, I am sorry to say that I can give Mr. Lord no idea. Those who have only studied the rotifers as a means of amusement can hardly be aware of the wide field for real scientific work offered by our British species alone. New ones are constantly being found, and even of the best known of all species, the *Rotifer vulgaris*, the male has yet to be discovered. I think I am well within the mark in saying that of all known species numbering about two hundred, at least seventy per cent. of the male has yet to be discovered. If a society could be formed of workers in this branch of science, the very fairy-land of microscopy, the members of which would exchange drawings and descriptions of any unfamiliar forms they might meet with, and perhaps even samples of water containing the rotifer alive, which can easily be done through the parcels post, the result would probably be a very great stride in our knowledge of the subject. Should any of your readers feel inclined to join a rotifer society, I shall be very glad to receive communications and suggestions from them, and to do anything I can to promote the formation of such a society.—*Edward C. Bousfield, L.R.C.P. Lond.*

FISH AND AIR.—The reason Mr. Finch's gold fish gulps air above the surface of the water is that it finds insufficient oxygen in the water to support life. Not being provided with lungs, of course it can't do this with impunity. The poor animal makes the best of a bad job by carrying this air to the bottom of the tank; so as to get it mixed with water and duly passed over its gills. In doing so, much of the air escapes from its mouth "in a succession of bubbles." It cannot be too often reiterated that the changing of water is fatal to aquarium keeping. You can't get your water pure enough unless you keep it exposed to sunlight and air. Water introduced into an aquarium to make up for loss by evaporation should always be kept a month or two. I have seen a small quantity of apparently pure water set up decomposition in a few days and poison a whole tank. Again, never keep the more complex or highly organised water plants for the evolution of oxygen—they are far too liable to damage and decay. The minute green confervæ, which come, as it were, spontaneously, are the only plants for your purpose. As far as I can see they elaborate infinitely more oxygen in proportion to the space they occupy and they grow on every solid object in the tank. They are welcome and longed-for guests in a newly set-up aquarium which is never ready for animals till they appear. Always look on fresh unkept water from the tap, river, or sea, and the higher order of water plants, as so much poison introduced into your tanks, and your animals will not adopt any extraordinary or fatal habits, but will confine themselves to the ordinary routine of life.—*John J. Stevens.*

GLOUCESTERSHIRE SLUGS.—Since the publication in the April number of SCIENCE-GOSSIP of a note by Mr. Dennison Roebuck, of Leeds, upon the Gloucestershire slugs, I have met with some at Brinscombe, near Stroud, which he has pronounced to be the var. *bicolor* of *Arion ater*. He also tells me that the specimens I sent him were the first he had seen from any English locality, though he had received it last year from co. Waterford, Ireland.—*E. J. Elliott.*

THE HOUSE-FLY.—Can any of your readers tell me where to find the eggs of the common house-fly, *Homalomyia canicularis* (Linn.), or whether it is possible to breed them in captivity?—*F. W. E.*

BRITISH SHELLS.—Can any reader inform me whether any marine shells have been discovered since the publication of "British Conchology"? Also whether any of the numerous species said to have been found in Britain, but considered by Dr. Jeffreys as doubtfully British, have been since found on any of the British coasts?—*T. D. A. Cockerell.*

WILD CAT IN KENT.—About a month ago a friend of mine saw a wild cat in a wood near Maidstone. He was sitting on a stile, and happening to look round, saw it sitting on the path glaring at him. Directly it saw it was observed, it made a spring through the hedge and disappeared. But what my friend saw of it was enough to prove to him that it was a true wild cat. It had a very short and thick tail, and altogether answered to the description of a true *Felis catus.*—*H. C. Brooke.*

BOTANY.

"A Synopsis of the Bacteria and Yeast-Fungi."—By W. B. Grove, B.A. (London: Chatto and Windus). Most of our readers will remember Mr. Grove's valuable papers, which appeared in this journal during 1882 and 1883, under the title of "The Schizomycetes," &c. They roused much interest at the time, for the Bacteria had never before been treated in this country in a systematic manner so that students could approach them. Consequently we have no doubt whatever that many of our readers will be pleased to hear these papers have been re-published, in an enlarged and fuller form, in a handsomely got up and attractive little volume, by the above-mentioned publishers, at 3s. 6d. To young medical students (and indeed old medical students as well) this handbook will prove most valuable, in enabling them to see at once all that has been said and done about Bacteria and other disease germs, &c. The volume is not less important to every brewer.

Œcidium Jacobææ (Grev.)—This œcidium I have demonstrated by a series of experiment cultures has nothing uncommon with either *Puccinia glomerata* or *P. compositarum*, but is a true heterœcismal uredine, the teleutospores of which occur upon *Carex arenaria*. This puccinia is quite distinct from *P. caricis*, from which it can be distinguished readily enough by the naked eye. It is more nearly allied to *P. dioica* (Magnus), but whether these two Pucciniæ be identical I can hardly at the present time say.—*Charles B. Plowright.*

Distribution of Plants in England.—I think a list of the wild flowers to be found in an ordinary country lane here, would be interesting, if similar lists from other districts, wide apart, could be published in Science-Gossip. I mean the common plants which are the particular features of the roadsides in those districts. The following are the more common plants in a lane near Maidstone. The hedges on either side are composed of hawthorn, common elder, cornel or dog-wood, common elm, hazel, and here and there a blackthorn bush. The plants under the hedges are *Ranunculus repens*, *Urtica dioica*, *Sisymbrium alliaria*, *Lamium album*, *Geum urbanum*, *Stellaria Holostea*, *Galium Aparine*, *G. Mollugo*, *Stachys sylvatica*, *Ballota nigra*, and *Potentilla reptans*.—*Henry Lamb, Maidstone.*

Cucumber-Tree.—In North America, *Magnolia acuminata*, Linn. is called the Cucumber Tree, from its fruit, which is about 3 inches long, resembling a small cucumber. The flowers of the tree are large, and the petals bluish-coloured, from which it is sometimes called the blue magnolia. The wood is of fine grain and orange-coloured.—*M. H.*

GEOLOGY, &c.

"CHALLENGER" MUSINGS.

By A Terrigenous Philosopher.

CRITICAL AREAS.*

HOW great is Nature! how divine
   The means adapted to her ends!
   The bottom of the ocean bends,
And all is part of one design.

But not in the abysmal deeps
   Of solemn ocean far from land;
   Nay round the continent and strand
This movement of the sea-bed keeps.

With such devices she entraps
   The sediment ground from the shore,
   And keeps it as a muddy store
To build up future lands, perhaps.

Thus naught is wasted in her hands;
   The balance of the globe she keeps;
   And stops deposit in the deeps;
The ocean waits her stern commands.

Outside the land she draws a line,
   The land and ocean toe the mark;
   And e'en the little islets hark,
Nor seek to disobey the sign.

Within these boundaries hard and fast,
   She garners harvests from the shore;
   Or well within the breakers' roar
Her bank-deposits are amassed.

Nor outside of the limits fixed
   Do bendings of the crust appear
   Of earth—a fact that seemeth queer
To one with notions rather mixed.

And well within this submarine
   And sunken trough around the land,
   The future strata are trepanned,
And this methinks has ever been.

So as was first of all observed,
   How great is Nature! how divine!
   That she should thus mark out the line
That continents may be preserved!

For should detritus find its way
   To ocean depths, it would be lost;
   Instead of staying tempest-tossed
Upon the shore or in the bay.

For the abyss but sinks and sinks—
   Its general tendency, they say—
   Though on occasion it may stay
To rise a little, Murray thinks.

* See "The Nomenclature, Origin and Distribution of Deep-Sea Deposits."—"Nature," June 5th.

And land, which fluctuates the more
   In average movement, still does rise ;
   A fact which need not cause surprise,
But one for brains to ponder o'er.

Since, to obey Divine command,
   The waters gathered in the deep,
   Perforce they this position keep,
And interchange not with the land.

But land is washed into the sea,
   And all around us is decay ;
   So, slowly but without delay,
The continents would wasted be.

For this Dame Nature—don't deride,
   Her labours otherwise were lost,
   Had she not counted up the cost ?—
A " critical " remedy applied,

And to a belt of sea. confined,
   The movements of the upper crust
   Of earth, and thus entrapped the dust
Of continents to be defined

When Pluto wakes from out his sleep,
   And stirs his furnace fires below,
   And land begins to grow and grow
From out the borders of the deep.

Hail, then, the savants who discerned
   The purposes of Nature, vast !
   In theory as unsurpassed
As any vouchsafed by the learned.

                     A. CONIFER.

HOW WAS COAL REALLY FORMED ?—An important paper on this subject has just been read before the Geological Society, by Mr. E. Wethered, F.G.S., who pointed out that seams of coal do not always occur in one bed, but are divided by distinct partings, some of which, as in the case of the Durham main seam, contain Stigmariæ. It was important to notice this feature for several reasons, but especially as the beds of coal, defined by the partings, showed differences both in quality and structure. In the case of the shallow seam of Cannock Chase they had at the top a bed of coal 1 foot 10 inches thick, the brown layers of which were made up of macrospores and microspores. The bright layers were of similar construction, except that wood-tissue sometimes appeared, also a brown structureless material, which the author looked upon as bitumen. He thought that hydrocarbonaceous substance would be a preferable term. What this hydrocarbonaceous material originated from, was a question for investigation. In the lower bed of the Welsh " Four Feet " seam wood-tissue undoubtedly contributed to it ; whether spores did was uncertain ; it was true they could be detected in it. In the second bed of the shallow seam they had a very different coal from the upper one. It was made up almost as a whole of hydrocarbonaceous material. Very few spores could be detected. Spores resisted decomposing influences more effectually than wood-tissue, which seemed to account for the fact that where they occur they stand out in bold relief against the other material composing the coal. Below the central bed of the shallow seam came the main division. In it was a large accumulation of spores, but hydrocarbon formed a fair proportion of the mass. The conclusions on the evidence elicited were (1) that some coals were practically made up of spores, others were not, these variations often occurring in the beds of the same seam ; (2) the so-called bituminous coals were largely made up of hydrocarbon, to which wood-tissue undoubtedly contributed. An appendix to the paper, by Professor Harker, of Cirencester, dealt with the determination of the spores seen in Mr. Wethered's microscopic sections. Taking the macrospores, the resemblance to those of Isoëtes could not fail to strike the botanist. He had procured some specimens of *Isoëtes lacustris* in fruit, and compared the spores with those from the coal. When gently crushed, the identity of the appearance presented by these forms from the coal was very striking. The triradiate markings of the latter were almost exactly like the flattened three radiating lines which mark the upper hemisphere of the macrospores of *Isoëtes lacustris*. He therefore concluded that the forms in the coal were from a group of plants having affinities with the modern genus Isoëtes, and from this Isoëtoid character he suggests the generic title of Isoëtoides pending further investigation. In the discussion which followed, Mr. Carruthers could not accept the conclusions arrived at. He remarked that there was no doubt as to the nature of the vegetation which formed the coal. The triradiate structure of the macrospores referred to was merely a superficial marking which threw very little light on the affinities of the spores. These spores had been found connected with leaves of Sigillaria and Lepidodendron. The coal itself had been a soil which supported the vegetation ; it is penetrated by the Stigmariæ, roots of Sigillaria and Lepidodendron, and by roots of other plants. Trees grew in the coal itself. Coal-seams are the remains of forests which grew upon swampy ground and were subsequently covered by clay. Spores were first noticed in coal by Professor Morris. They abound in some places, but there is no reason to attribute them to Isoëtes or to any other form of submerged vegetation. Professor W. Boyd Dawkins had never seen sporangia in coal, although macro- and microspores abounded. Coal is composed of two principal elements, carbon proper and a fossil resin, to which the blazing property of coal is due. The latter is mainly due to the spores ; but the blazing element cannot be wholly attributed to them. The carboniferous forests grew upon horizontal tracts of alluvium not far above the water-line. Mr. E. T. Newton considered this paper to be the first systematic attempt that had been made to ascertain the microscopic structure of a consecutive series of coal-beds. Certain coals were undoubtedly made up of macro- and microspores. He remarked that there are three principal kinds of layers in coal—bright, dull, and intermediate. Professor T. Rupert Jones thought that the spores would themselves supply the resinous or hydrocarbonaceous matter. Mr. Bauerman said that as plants are made up of materials varying very considerably in their resistance to decomposition, and as only the more stable ones were likely to retain their structure, the fact of such structures as those of spores being recognised in microscopic sections seems to be no proof that whole seams were made up of them. The author said there was no reason why an Isoëtoid plant should not have Stigmaria roots, and Isoëtes is the only existing plant of the group in which the macrospores show triradiate structure. His conclusions were entirely based on the evidence of the microscope.

NOTES AND QUERIES.

WHITE STARLING.—On the afternoon of the 7th of June, I was walking across some fields between the villages of Kentisbeare and Uffculme in East Devonshire, when I noticed a flock of starlings feeding on the ground with a white bird amongst them. When, on my nearer approach, they flew away it became evident that the white bird was a starling also. The size, mode of flight, &c., differed in no way from that of the rest of the flock. They settled (dark birds and the white one) all together in an adjoining field, and the white bird, in spite of its contrasted appearance, seemed to be on the best of terms with its more sombre comrades. I have never seen a white starling before, nor do I remember having heard of one.—*W. Downes, Kentisbeare, Collumpton.*

AUTHOR'S NAME WANTED.—Will some reader kindly inform me who the author of the following curious lines was, and also date of same :

The Foxe in crafty witte exceedith moste men,'
A Dogge in smelling hath no man his peere,
To foresight of weather if you looke then
Many beastes excell men, this is clere.
The witinesse of Elephantes doth letters attayne,
But what cunning doth there in the Bee remayne?
The Emmet foreseeing the hardness of winter,
Prouideth vitailes in tyme of summer.
The Nightingale, the Linet, the thrushe, the larke,
In musicall harmony passe many a Clerke.
The Hedg hogge of Astronomy seemith to know
And stoppeth his caue wher the wind doth blow.
The Spider in weauing such arte doth showe
No man can him mende, nor follow I trowe,
When a house will fall, the Myse right quick
Flee thence before, can man do the like?

—*F. A. A. Skuse,* 143 *Stepney Green, London, E.*

THE FIELDFARE (TURDUS PILARIS).—Does this bird ever breed in England or not? None of the standard works on Ornithology allow that it does. The fourth edition of Yarrell, now so slowly emerging from the publisher's hands, says that no recorded instance " seems to be free from reasonable doubt." Yet in most country districts it is asserted that nests are to be found. I have lately had a nest and two eggs brought, which I cannot attribute to any other bird. The eggs are much elongated, one is an inch and three-eighths in length ; of a blue ground colour and speckled and blotched with red brown, almost wholly at the larger end. The nest was built in an ash-tree in a wood, and is rather loose in structure ; the outer part is of dried fir twigs, inside this a quantity of coarse grass, and this in its turn is loosely lined with a few feathers. I should be glad of the opinion of any of your readers more skilled in ornithology than I am myself.—*H. Ullyett, Folkestone.*

NOTE ON HOPLOPHORA.—If Mr. J. K. Lord will carefully examine his specimen of Hoplophora, he will find that the supposed eyes are in reality the stigmates or breathing openings of the mits. None of the Oribatidæ, so far as I know, are possessed of eyes. When I wrote in 1877, I had not met with Nicolet's classical monograph. Since that time the Oribatidæ have been attentively studied in England, and most of the already described species have been found to be indigenous ; and some very curious new ones have been discovered. Several papers have been published in ".The Journal of the Royal Microscopical Society" (1879, 1880, and 1881), and my friend Mr. A. D. Michael, who has succeeded in tracing a great number of them from the egg to the perfect state is, I believe, preparing a monograph on these most interesting mites, for the Ray Society ; which will be a most complete and valuable work. The transformations which occur in the passage from the egg to the imago, in some of them, are most marvellous, and would be incredible if they had not been repeatedly confirmed.—*C. F. George.*

GEOLOGY OF TENBY.—M. L. C. would be glad if any reader would answer the following questions in April SCIENCE-GOSSIP. 1. Whether Tenby is a good centre for a collector of fossils to stay at? 2. Whether there is an inexpensive book on the geology of Tenby and its neighbourhood.

SKELETON LEAVES.—Can any of your readers tell me of some simple method of obtaining skeleton leaves ? My interest in the venation has recently been quickened by finding some good impressions of fossil leaves in a compact quartzitic deposit of Oligocene age a few miles off. The primary veins are beautifully marked, but very rarely the secondary ones. Any information will oblige—*C. A. Barber, Bonn, Germany.*

DO SWALLOWS REASON ?—As I watched a pair of swallows feeding their young last summer, I could not help propounding to myself the above question, for mere instinct seemed altogether incompetent to account for the phenomenon observed. The nest was fixed at the apex of an angle formed by the junction of two beams in a barn, and in such a position that it was necessary for extraordinary sanitary arrangements to be made to preserve the cleanliness of the nest and its surroundings. As one or other of the young birds was almost constantly excreting, a comparatively large dung heap would quickly have arisen around the nest, owing to its peculiar position, had not the sagacity or thoughtfulness of the parent birds come to the rescue ; and it was the method adopted by them that specially attracted my attention. I noticed that at about every third visit to the nest each of the old birds carried away something which I at first thought was a remnant of the food brought to the young ones. On closer observation such proved to be erroneous, and revealed the fact that it was the fæces of the young birds; and thus, notwithstanding their incessant excretions, the immediate surroundings of the nest were kept perfectly clean.

Is it not a little strange that absolute wild birds altogether destitute of reasoning powers—at least, we are asked to believe they are—should adopt extraordinary sanitary measures when necessary to preserve the cleanliness of their dwelling?—*R. S. Preston.*

MALFORMATION OF EGG OF HEDGE-SPARROW.—The outline I send is that of an egg of hedge-sparrow taken many years ago by my brother. The other eggs in the same nest were of the usual shape and size.—*K. D., Cofton.* [It was probably a double-yolked egg.—ED. SCIENCE-GOSSIP.]

WHITE FIELD-MOUSE.—I have in my possession a perfectly white specimen of the short-tailed field-mouse, which was caught near Bromsgrove. A quite young one was, I believe, caught at the same time, but it died very shortly, and the farmer who caught them did not think it worth preserving. The one I have only lived a few days, and I never saw it in the flesh. Are such albinos of common occurrence?—*K. D., Cofton.*

DEATH OF DORMICE.—I should feel obliged if some reader would kindly let me know in the next number of SCIENCE-GOSSIP, what you think was the cause of the death of my dormouse. I had had it about a twelvemonth, and it was so tame, and had such engaging little ways, that it was a very great pet. The other evening I found the little creature in one corner of its cage trembling violently, its heart beating so fast that it seemed as if every minute it would breathe its last. It was to all appearance in good health the previous evening, and it could not by any possibility have been frightened. On the second evening following, my pet died. It got weaker and weaker, and at last could not move by itself. The trembling was accompanied by a little ticking wheezy noise. The food I gave it was Spanish nuts and occasionally acorns. I should also be glad to know to what age dormice usually live. I am sorry to trouble you with such a lengthy note, but am so grieved with regard to my little pet's death, that I feel anxious to know the cause of it.—*A. M. P.*

LADY-BIRD.—I lately captured a lady-bird in my garden, that instead of being red with black spots is black with red spots. I have since diligently searched for another specimen, but although the ordinary variety abound in great numbers, I have not been successful in finding another black one. I should be glad to hear this is a rare occurrence, as I have neither read of nor met with one before.—*H. Moulton.*

MOLE CASTLES.—The following is an accurate account of a large mole castle discovered on February 17th, in Madingly Wood, near Cambridge. A large circular mound of loose clay, fairly symmetrical, stood in an open space amongst the brushwood; it measured three feet six inches in diameter, and eighteen inches in height. On removing the lumps of earth and tracing the tunnels carefully, the plan exposed to view consisted of three tiers of circular galleries one above the other, and two inches from the surface of the mound. The upper one was not completed by about an inch, but there were traces of the mole's claws, suggesting there had been recent work. These galleries were connected by three or four passages running down the sides, and then proceeding along the surface of the earth in various directions from the castle. Neither lines nor circles were geometrically true, but seemed suggestive as a rough plan throughout, and at several places the tunnels were enlarged, probably to allow moles to pass each other; the diameter of a circular section varying from two to four inches. There was only one "blind" tunnel terminating abruptly. Making random excavations afterwards, further internal galleries were exposed, on a level with the middle external one, and communicating with a large domed cavity a foot across and 6 inches high, which was on a level with the ground, and contained a number of damp brown oak leaves. Below this, extending for a foot beneath the earth, were numerous intersecting galleries from which shafts ran perpendicularly into the earth for a considerable distance, but the great difficulty of excavation, owing to the falling in of the clay, prevented further investigation. All the passages were rubbed smooth, but rather slimy. The fact of this wonderful castle being erected in an open space, and composed of firm clay, enabled us to make an almost complete examination. Can any reader inform me, (1) how this castle was made; (2) for what purpose; (3) the object of the dead leaves? Except that the outer galleries had no communication with the inner, saving on a level with the earth, this maze affords to the mole a most perfect city of refuge. The method of construction which appears most probable is the following. From the deep shafts a large quantity of clay was thrown up in the usual manner, and the loose clay was pressed to form tunnels causing enlargement and extra compactness of the heap, for there were no signs of any outlet through which to eject fresh excavated clay. But the questions as to how the domed cavity was formed, and the use of the damped dead leaves, unless as a nest for young, are still a problem, which, I trust, will be announced next month.—*A. S. E., Cambridge.*

THE DORMOUSE IN ENGLAND.—Being desirous of ascertaining as far as possible the range of the dormouse in England and Wales, I shall be extremely obliged if any of the numerous readers of SCIENCE-GOSSIP can give me any information on this subject. Well-authenticated instances of the occurrence of this little animal in any part of England or Wales will be gratefully received, especially if accompanied by a slight description of the kind of place (as regards herbage, trees, &c.) most frequently selected as a residence.—*G. T. R.*

NOTICES TO CORRESPONDENTS.

TO CORRESPONDENTS AND EXCHANGERS.—As we now publish SCIENCE-GOSSIP earlier than heretofore, we cannot possibly insert in the following number any communications which reach us later than the 8th of the previous month.

TO ANONYMOUS QUERISTS.—We receive so many queries which do not bear the writers' names that we are forced to adhere to our rule of not noticing them.

TO DEALERS AND OTHERS.—We are always glad to treat dealers in natural history objects on the same fair and general ground as amateurs, in so far as the "exchanges" offered are fair exchanges. But it is evident that, when their offers are simply disguised advertisements, for the purpose of evading the cost of advertising, an advantage is taken of our *gratuitous* insertion of "exchanges" which cannot be tolerated.

WE request that all exchanges may be signed with name (or initials) and full address at the end.

X.—We could not make anything of the mosses, &c., you sent us to name, neither could a botanical authority to whom we sent them. They are too small and imperfect.

C. C.—A vast number of insects are insectivorous. The lady-birds prey quite as voraciously on aphides as cats would on mice. Ants feed on the aphides' secretions, but lady-birds (Coccinellæ) do not.

H. J. T.—You will be able to obtain living specimens of the toad from Mr. E. W. Wilton, Northfield Villas, Leeds.

W. G.—The occurrence of four-legged chickens is by no means rare. We have seen four this summer, and the other day had a four-legged duck (a year old) brought us. The sagacious bird utilised its two hind but well-developed feet to sit upon!

A. B.—Get the following separate monographs from Messrs. Lovell Reeve & Co., 5 Henrietta Street, Covent Garden. They are from Curtis's "British Entomology."—Orthoptera (price 5s.); Hymenoptera (price £6 5s.); Neuroptera (price 13s.); Coleoptera (price £12 16s.). They contain figures of all British species. Or get Rye's "British Beetles," price 10s. 6d.; Shuckard's "British Bees," price 10s. 6d., same publisher.

EXCHANGES.

SPLENDIDLY-preserved and correctly-named Swiss Alpine plants. Price 6d. each.—Address, Dr. B., care of Editor of SCIENCE-GOSSIP, 214 Piccadilly.

OFFERED, insects from south-east of France, such as *Coleoptera diptera*, dry or in acetic acid, all fresh killed; also dry plants, from the same country, in exchange for good mounted slides.—E. Rodier, 61 Rue Mazarin, Bordeaux, France.

BEAUTIFUL fossil sand from Sancat's tertiary deposits, containing well-preserved foraminifera, shells, coral spicules (any quantity), in exchange for good mounted slides.—E. Rodier, 61 Rue Mazarin, Bordeaux, France.

WANTED, fresh specimens of stag beetle and male cricket for dissecting purposes.—John Moore, 86 Porchester Street, Birmingham.

WANTED, all classes of Reptilia, British and foreign, alive and in spirits, in large quantities: foreign correspondents especially requested. Also alive, about 500 bull frogs of America, and about 5000 common frogs of England. Also the following, alive, in large quantities: hedgehogs, moles, common snakes, tree frogs, natterjack toads, tortoises, *Helix pomatia*, blindworms, and lizards.—E. Wade Wilton, Fairfield, Buxton.

SPRINGFIELD, Bermuda, and Cambridge, U.S.A., Polycistinous earths, one oz. of either on receipt of six good 3×1 mounts on smooth edge slips.—Tylar, 20 Geach Street, Birmingham.

CLUSTER cups on *Œci. rubellum, Œci. Epilobii, Œci. Tragopogonis, Œci. urticæ, Œci. ranunculare, Œci. tussilago*, all mounted: offered in exchange for other slides.—C. Garrett, Shirley House, 30 Palmerston Road, Ipswich.

WANTED, the following birds' nests with eggs: dipper, ring ouzel, pied flycatcher, stonechat, goldfinch, and golden-crested wren; also eggs of red grouse and sea birds in quantity. Offered in exchange, other natural history specimens.—W. K. Mann, Wellington Terrace, Clifton, Bristol.

EGGS (side blown), dipper, pied flycatcher, pied and grey wagtail, tree and meadow pipits, starling, red-winged starling, rook, jackdaw, magpie, lapwing, golden-winged woodpecker, dunlin, oystercatcher, redshank, black-headed gull, sandpiper, moorhen, landrail, common tern, &c., for others not in collection.—H. Patrickson, jun., Warwick Road, Carlisle.

POWERFUL electric machine, with dial, regulator, &c., for giving shocks, by Halse; pair electric telegraph instruments, electric bell, batteries, wire, &c., in portable case. Wanted, American organ, harmonium with stops, gas engine, or boiler heated by gas, or offers.—E. R. Dale, Esq., F.S.Sc. Lond., Sherborne, Dorset.

SPLENDID double breechloader, 12 bore, central fire, rebounding locks; what offers, musical instruments, or mechanics, to value £5?—E. R. Dale, Sherborne, Dorset.

WANTED, Cooke's "Manual of British Fungi, 1883," 2 vols.; offers, other scientific books and micro slides.—W. E. Green, Cleve Dune, Belvoir Road, St. Andrew's Park, Bristol.

WELL-MOUNTED slides of British mosses, named, for other good micro slides; also micro material for exchange. Send lists to W. E. Green, Cleve Dune, Belvoir Road, St. Andrew's Park, Bristol.

SILURIAN and Carboniferous fossils for recent British marine shells. Post-Tertiary and raised beach fossils for British freshwater and land shells. Lists exchanged.—J. Smith, Kilwinning, Ayrshire.

COLLINS' "Mineralogy," vol. i.; Thorp's "Inorganic Chemistry;" "Metals;" Pennell's "Angler Naturalist." Wanted, Rimmer's "British Land and Freshwater Shells."—H. E. Craven, Bath Road, Matlock Bridge.

A DOUBLE-BENT nosepiece to fit English microscope, in exchange for one to fit Hartnack's, or for a turntable or other micro material.—E. H. Young, 18, Stoneneست Street, Tollington Park, N.

WANTED, during the summer and autumn, a few good gatherings of Desmids (exclusive of Closterium), for which first-class botanical slides are offered in exchange.—C. V. Smith, Carmarthen.

FOR exchange, Indian reptiles in spirits, and splendid specimen of death's-head hawk moth. Wanted, wing cases of diamond beetle, or micro slides, &c.—J. Boggust, Alton, Hants.

Zonites glaber offered in exchange for other British shells.— A. H. Shepherd, 4 Cathcart Street, Kentish Town, London, N.W.

CASH or useful exchange for specimen, of English and foreign ants, larvæ of ant-lion (Myrmeleon), or any curios connected with the above.—Particulars to "Manager," 3 Upper Parliament Street, Liverpool.

EGGS of nightingale, nightjar, tree creeper, and others, taken in locality, to exchange.—J. H. H. Knights, Hawthorn Cottage, Norwich Road, Ipswich.

WANTED, first volume of Darwin's "Descent of Man," or will part with second volume, containing good photo. of author have got "Imperial Lexicon," half calf, to part with, and other works, and want Lyell's "Principles of Geology."—Danie Mayor, 7 Chaddock Street, Preston.

"ENGLISH MECHANIC," fourteen vols. 18-32. Vols. 18-30 bound half calf, two vols. in one. Vols. 30-32 unbound, all clean and in first-class condition, for last edition of "Sachs' Manual," or works of Darwin and Huxley.—E. Irving, 125 Buchanan Street, Glasgow.

OFFERED, for exchange, nests and clutches of dipper, clutches of hooded crow, ringed plover, swift, coot, mute swan, teal, cormorant, gannet, herring-gull, and other sea-birds, and many other eggs, all one-holed, with data. Send list of duplicate clutches and of desiderata.

IRISH nests and eggs in clutches, side blown, of sparrowhawk, kestrel, hooded crow, land-rail, jay, ring-dove, night-jar, swallow, sand-martin, long-tailed tit, gold-crest, guillemot, puffin, razor-bill, kittiwake, cormorant, oyster-catcher, lesser black-backed and herring-gull. Wanted, British birds' eggs in clutches.—Rev. W. W. Flemyng, Clonegam Rectory, Portlaw, co. Waterford.

DUPLICATES, *Sphærium corneum, Sp. rivicola, Pisidium amnicum, P. pusillum, Anodonta cygnea, A. anatina* and *Unio pictorum*, all fine. Desiderata, British land, freshwater, and marine shells.—Robert B. Cook, 44 St. John Street, Lord Mayor's Walk, York.

MEDIA for mounting animal, vegetable, and algæ slides in, in exchange for slides.—S. R. Hallam, 22 High Street, Burton-on-Trent, Staffs.

WANTED, slides of parts of *Musca domestica* (common housefly) in exchange for media or slides.—S. R. Hallam, 22 High Street, Burton-on-Trent, Staffs.

Clausilia delicata, Cl. scalaris, &c., only found in Malta, by pairs. Desiderata, micro-slides, side-blown British eggs. Several species of European butterflies in paper, &c.—E. F. Becher, Hill House, Southwell, Notts.

WANTED, complete clutches, eggs and nests, of pied flycatcher, ring-ouzel, stonechat, rock-pipit, woodlark, twite, &c. Will give other eggs and nests, British Lepidoptera, or glass-capped boxes for preserving nests.—Thos. H. Hedworth, Dunston, Gateshead.

WANTED, land, freshwater and marine shells; will give books, slides and micro-material in exchange.—A. Alletsee, 7 Glendall, Clifton, Bristol.

SHALL be glad to exchange the exotic fungi (*Puccinia Pilocarpii*) for other micro fungi.—Geo. Ward, 26 St. Saviours, Mere Road, Leicester.

WANTED, "Nature," No. 725, September 20th, 1883, without plate. State what exchange or money to W. F., 56 Asylum Road, London, S.E.

MICROSCOPIC slide of *Hæmatopinus Canis* in exchange for other parasites.—H. F. Jolly, Stow Villa, Bath.

BOOKS, ETC., RECEIVED.

"A Synopsis of the Bacteria and Yeast Fungi," by W. B. Grove, B.A. London: Chatto and Windus.—"Petland Revisited," by the Rev. J. G. Wood, M.A. London: Longmans & Co.—"Wonders of Plant Life under the Microscope," by Sophie B. Herrick. London: W. H. Allen & Co.—"Bicycles of the Year 1884," by H. H. Griffin. London: L. Upcott Gill.—"Studies in Microscopical Science," edited by A. C. Cole.—"The Methods of Microscopical Research," edited by A. C. Cole.—"Popular Microscopical Studies." By A. C. Cole.—"The Gentleman's Magazine."—"Belgravia."—"The Journal of Microscopy."—"The Science Monthly."—"Midland Naturalist."—"Ben Brierley's Journal."—"Science Record."—"Science."—"American Naturalist."—"Medico-Legal Journal" (New York).—"Canadian Entomologist."—"American Monthly Microscopical Journal."—"Popular Science News."—"The Botanical Gazette."—"Revue de Botanique."—"La Feuille des Jeunes Naturalistes."—"Le Monde de la Science." &c. &c. &c.

COMMUNICATIONS RECEIVED UP TO 10TH ULT. FROM :— H. R.—H. D.—F. C. L.—H. M.—J. B.—H. J. W., junr.— J. Y.—E. M.—S. A. B.—J. F. R.—H. P., junr.—W. K. M.— H. N. P.—F. C. K.—Dr. O. N. W.—W. H.—J. A. D.— E. R. D.—W. J.—S. C.—W. E. G.—A. W. O.—G. S. M. F.— W. P.—E. T. D.—J. S.—H. E. C.—J. S.—J. H. G.—E. L.— J. H. D.—C. F. W. T. W.—E. A.—F. H.—   —G. H. T.— C. J. M.—Dr. W. B. K.—M. L. S.—C. DWB.—W. T.— J. S.—G. G.—A. S. W.—E. W. W.—H. G. D.—H. L.— J. H. G., junr.—J. H. H. K.—J. W. Q.—H. L.—J. H. C. K. —H. A. F.—A. P.—D. M.—M. H. R.—E. B.—H. B.— W. H. H.—T. D. A. C.—H. S.—H. A.—E. J. H.—E. H. Y. —J. B.—C. V. S.—T. M. R.—P. E.—A. B.—R. J. U.— W. F.—B. C.—W. D.—J. J. S.—C. P.—S. J. H.—J. S. E. B.—W. G.—S. R. H.—J. H.—E. A. G.—M. J. H.— W. H. H.—E. F. B.—F. W. E.—G. W.—H. F. J.—E. H. S.— M. J. J., &c.

GRAPHIC MICROSCOPY

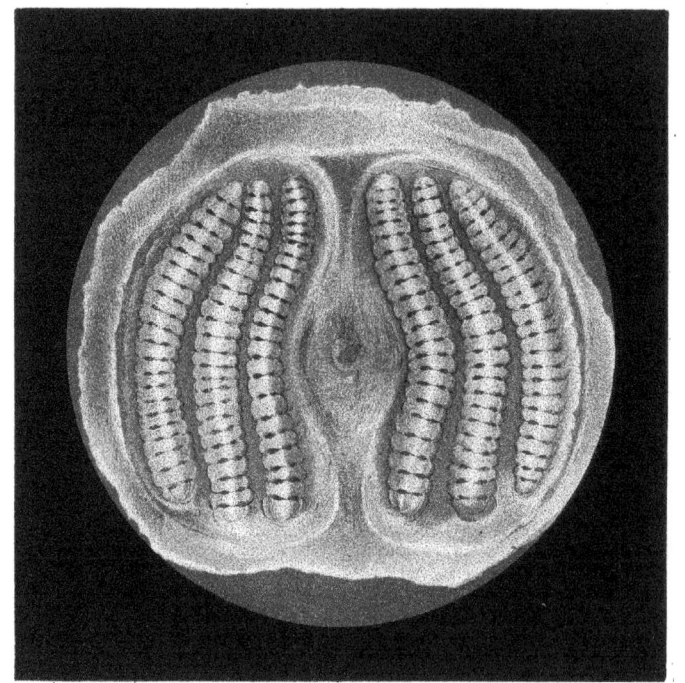

E. T. D. del, ad nat.   Vincent Brooks, Day & Son lith.

SPIRACLE OF BREEZE FLY EQUIŒSTRUS.
× 50.

GRAPHIC MICROSCOPY.

By E. T. D.

No. VIII.—Spiracle of Breeze Fly (Œstrus equi).

AN elaborate and exalted condition of the organs of respiration is, in most cases, associated with a high development of physical power and strength. This is notably seen in insects; their untiring flight, incessant locomotion, the muscular force of their various parts, nervous activity, vivid life, sense of smell, sight, reproduction, and generation of heat, are functions evidently in close connection with a capability of incessantly and perfectly incorporating air with blood through an intricate mass of respiratory vessels, more diffused than in any other living thing.

In insects there is no centralised breathing organ, involving a system of veins and arteries; instead of blood seeking the lungs for aeration as in animals, in these powerful creatures the air seeks the blood, through tracheal tubes, inosculating and permeating every part of the body, legs, wings, alimentary canal and nerve centres; and thus oxygen streams through every vital part, admitted by external apertures, "spiracles," plainly visible, on each side of the abdominal segments of any insect, either in the larval or perfect condition. These spiracles, or "stigmata," are the entrances to this elaborate respiratory system, freely admitting air. The shape and character of the opening is diverse, interesting to the microscopist, as, in distinct species they are, in shape and character singularly dissimilar; the typical condition is a contractible puncture, with filamentous processes converging from the margin to the centre, obviously to filter the air, and prevent the admission of impurities. Some are sieve-like, as in the common house-fly; in larva of many moths they take the form of a flexible slit or compressible lip; in the blow-fly, an oval opening with an elaborate screen: they are of various configurations, circular, oblong, vermiform, crescent-shaped, and all have sensitive edges expanding and contracting under the influence of respiratory action.

The spiracle of the breeze-fly of the horse (*Œstrus equi*), the subject of the plate, and a popular microscopical "slide," unlike what may be termed the ordinary form of a single aperture, has a peculiarity in a number of fine slits crossing a series of vermiform-like membranes, which, by their delicate elasticity, open and close the fissures for the admission and emission of air; in the larva of the crane-fly this condition is found arranged in a complete circle.

Spiracles are the portals of the tracheæ, a series of delicate tubes ramifying through every part of the body, and subdividing into microscopic filaments of exquisite delicacy. In large insects, expanded at points into dilatations or air-sacs, these vessels, imbedded in the tissues, are preserved from collapse by a spiral filament strong enough to resist compression and yet sufficiently elastic to maintain the required expansion, analogous to the cartilaginous rings, lining the tracheæ leading to the lungs of animals. This beautiful thread, enclosed between the tissues of the tubes, may be traced in the minutest terminations. The tracheæ are supplied with the "breath of life" by the respiratory palpitations of the insect; a rhythmical compression and distension of the abdominal parts opens the spiracles for the passing in and forcing out of air, quicker or slower, synchronal with the pulsations of the body, increased or retarded, under excitement or repose. The result of this action may be seen by immersing a large caterpillar, or beetle, in water; air bubbles will then emerge from the spiracles and cease when the contents of the tubes become exhausted; greasy substances smeared over the abdomen, blocking the apertures and preventing respiration, result in suffocation and consequent death. In the chrysalid, respiration is reduced to a passive condition, but it exists; a pupa dipped in oil is destroyed.

It is possible that sounds produced by insects during flight do not entirely emanate from the vibration of the wings, or external frictions, but may have some connection with the velocity of the air streaming through the spiracles in and out of the body. It is obvious that any unusual hum of an insect, the unmistakable sound of an angry bee, or fitful buzz of a blow-fly, cannot altogether proceed from a sudden and abrupt alteration in the motion of the wings, but is more likely to be the result of a rush of respiration under strong excitement through the trumpet-like orifices.

Curious and strange conditions exist: among others the larva of the Ichneumon, parasitic and buried in the tissues of a caterpillar, breathes through stigmata differing but little from the larvæ of other hymenoptera; they must necessarily tax the respiration of their host (and it has been observed) by piercing a tracheal tube and inspiring the air pouring from it. Such a wound might not be vitally material to the victim, as the puncture of the smallest branch would amply sustain so minute a life as the larva of an Ichneumon; caterpillars so affected are always sickly.

Aquatic insects, both in the larval and perfect condition, have varied, and curious modifications for respiratory purposes, membranes perforated with holes, prolonged gill-like flaps, in which the tracheal tubes open out and radiate into minute fibrillæ. In the larva of the day-fly, common in every pond, a most complicated organ is seen, a pronged brush-like tail; each filament is the termination of a tracheal process. In other aquatic larvæ may be found a funnel or coronal of hairy plumes capable of gathering from the surface of the water a globule of air and carrying the supply for respiration into the depths. The common pond-beetle (Dytiscus) rises to the surface and entangles air between the elytra and abdomen, where the spiracles are placed; when the captured store is exhausted, it returns for a renewal. Another large aquatic insect (*Hydrous piceus*) collects and sweeps in air with antennæ covered with fine hairs, packing and storing the minute bubbles under the thorax, which is coated with a silky pubescence, until there is accumulated a globule, often so large and buoyant as to make it difficult for the creature to sink to its home; they may be seen with their treasure cautiously creeping down the stems of plants. Water-spiders (although not in the category of insect life) excel all aquatic creatures in their respiratory resources; collecting air from the surface, they carry it down by degrees to their lairs, and, entangled in their webs, accumulate bubbles of considerable magnitude in which they reside. The arachnida, in all that concerns their condition of existence, exhibit resources, if not approaching sagacity, at least amazing powers of expedience.

For dissection, and the exhibition of the tracheal tubes and spiracles, a full-grown caterpillar, the larva of a goat moth, may be selected, and the entire system revealed. Pinned to a small plate of wax, pressed on the bottom of a porcelain or glass tray under water, an incision is made from end to end, the integuments carefully turned back and pinned down; with a sable brush, a pair of fine surgical scissors and a thick needle sharpened into a cutting edge, the respiratory organs may be traced and washed out, the tubes and spiracles removed, and the various parts prepared as permanent objects, by easy and well-known methods.

Considered from a microscopist's point of view, the insect world is endless; many of the class could creep through a pin-hole, possessing every function of a highly organised life, with respiratory machinery no less complicated than in the giants of the order. Dr. Carpenter says, "the inexhaustibility of Nature is constantly becoming more and more apparent, so that no apprehension need arise that the microscopists' researches can ever be brought to a standstill for want of an object," and it may be added, the subject of a drawing.

Crouch End.

NOTES ON THE FERNS OF THE PYRENEES.

DURING a visit of some five weeks' duration to the Pyrenees last autumn, I succeeded in finding the following species of ferns, twenty-nine in number—

Cystopteris fragilis, C. alpina, C. montana, Adiantum Capillus-Veneris, Cheilanthes fragrans, Allosorus crispus, Blechnum Spicant, Asplenium viride, A. trichomanes, A. fontanum, A. Adiantum-nigrum, A. ruta-muraria, A. septentrionale, A. Germanicum, Athyrium filix-fœmina, Pteris aquilina, Ceterach officinarum, Scolopendrium vulgare, Polystichum Lonchitis, P. aculeatum, Lastræa filix-mas, L. oreopteris, L. dilatata, Polypodium vulgare, P. Phegopteris, P. Dryopteris, P. Robertianum, Osmunda regalis and *Botrychium lunaria.*

I found several plants of *Cystopteris alpina* at the Cirque of Gavarnie, where *C. montana* occurs as well, the latter flourishing under some large rocks on the left bank of the stream, near the wooden bridge two or three hundred yards above the little hotel.

P. Lonchitis, too, grows luxuriantly in the locality, and a few plants of the moonwort, very dwarfed and scanty, were just beginning to make their appearance at the time of my visit.

Besides the Gavarnie locality *C. montana* is also to be found near Cauterets, on the right-hand side of the road between the Pont d'Espagne and Lac Gaube, but it is decidedly local in its habitats.

The maidenhair fern abounds on the cliffs at Biarritz in the direction of the lighthouse, and occurs sparingly in several localities inland, near Eaux Chaudes, by the roadside between Argelès and Cauterets, and again on a stone wall in the neighbourhood of Luchon. This fern appears to be very variable in size and form, the Biarritz specimen differing very materially in both of these respects from those picked at Luchon.

Cheilanthes fragrans I found, in some abundance, about halfway up the ascent to the little village of Cazaril above Luchon. *Gymnogramma leptophylla* and *Asplenium lanceolatum* are said to occur there as well, but I did not come across either of these species.

Asplenium fontanum seems to be pretty generally but sparingly distributed, often growing intermingled with *A. trichomanes*, from which however it differs very considerably in appearance. I cannot imagine how it can be confounded with that species, or with *A. viride*, it is so distinct from both, and I did not see any intermediate forms.

I was fortunate enough to discover three or four plants of *A. Germanicum* near Cauterets, after a careful search. They were the only specimens of this rare fern I found at all. Its near relative *A. septentrionale* is certainly much more generally distributed on these mountains than upon the Swiss Alps.

I do not suggest that the above forms anything like a complete list of the ferns of the Pyrenees, but think that, imperfect as it no doubt is, it will serve to show that a visit to that region will amply repay those who like to combine a little fern collecting with the pleasures of healthful exercise and fine mountain scenery.   T. W. B.

THE FRECKLED GOBY (*GOBIUS MINUTUS*).

THIS fish may be found in great abundance in the shallow pools left by the retiring tide where it swims about in company with the shrimps, and very much resembles these last in colour, being like them of a hue which renders it difficult to distinguish it from the sand, in which, like all the gobies, being a ground fish, it delights to lie partly buried. I find it a fish extremely well adapted for a marine aquarium, as it is very hardy and so exceedingly tamable that it soon becomes a most amusing pet. I have had the same individuals living in my aquaria for many months—I had nearly written for years—but, indeed, I have really kept them for from twenty to twenty-six months and even longer. I have had one for three years. It is a small fish, never exceeding four inches in length, but usually three inches is about the length of a mature fish. I have had numbers of hardly more than an inch, and have been much amused by the intelligence displayed by such tiny creatures, for the capacity for being tamed does certainly imply no small degree of intelligence. They seem in some sort to exercise a power of reasoning, extremely minute though their brains must be. The same may of course be said of all pet creatures, but it is most astonishing to see the familiarity displayed by such tiny creatures.

See them, for instance, wriggling their bodies up and down in the water, and flattening their mites of noses against the glass, while they do their best to attract the attention of the (to them) tremendously big being of whom they were once in mortal dread. No sooner do they catch sight of me than they make every effort to get to me, and if I just touch the surface of the water with my finger they are nibbling at it in a moment. The mystery of mind in the lower animals, but more especially in such tiny creatures as these little gobies, is to me inscrutable. It is certainly not hereditary instinct that teaches them to look for food from my hands. What is it, then? It is certainly some kind of intelligence; some exercise of reasoning; and this in creatures with brains no bigger than a pin's head.

It is true, as the poet sings, that

"There lives a soul
In all things, and that soul is God."

Or is there anything of the nature of mesmerism in the influence their owner exercises over the minds of his pets, finny or otherwise? I have had many evidences that my gobies know me and discriminate me from other persons. At the sight of strangers they will often scuttle away in alarm, and hide themselves in the nooks and crannies of the rockwork, or burrow in the sand. But I cannot discuss this matter further now; interesting—vastly interesting though the subject is, or I shall take up too much space. I may possibly revert to it again.

Although they are pretty peaceable fish, and capable of being kept for months in company, yet they will sometimes fight, and it will be found that they gradually decrease in number as the weaker get killed off by the stronger.

I feed them daily on tiny morsels of scraped beef, and sometimes, in the case of the larger sized fish, with small red worms.

I find them even more tamable and hardier than the common smooth blenny (*Blennius pholis*), also most interesting little fish, and one well adapted for a marine aquarium, of which, possibly, I may have more to say another time.

I have never had them breed, although they may be found full of spawn in the summer. They are susceptible of cold, and apt to die in severe weather.

ALBERT H. WATERS, B.A.
Cambridge.

CHAPTERS ON FOSSIL SHARKS AND RAYS.

By Arthur Smith Woodward,
OF THE BRITISH MUSEUM (NATURAL HISTORY).

I.

HOWEVER much care a palæontologist may bestow upon the naming and classifying of the relics of animals with which he is concerned as a student of fossils, it is at all times most difficult, and often impossible, for him to render his nomenclature and arrangement of equal value to that of the zoologist, who has complete animal structures to guide his determinations. Although palæontology is intimately connected with geology, and cannot well be separated from it, it still bears a close relationship to zoological science, and is, in fact, simply that branch which attempts to unravel the characters of the various faunas that have inhabited the earth during the successive ages of the past. Nevertheless, it requires but little thought to perceive that the methods of grouping living animals must generally differ greatly from those employed in the case of the extinct forms, and consequently it is obvious that palæontological species are nearly always distinct in kind from zoological species. This circumstance is chiefly brought about, as is well known, through the fragmentary nature of most fossilised organic remains ; and the imperfection in materials is due to the fact, that only the hard parts of animals have usually been preserved, while the softer tissues—so important for classificatory purposes—have almost always been completely destroyed.

Now, of all the groups of fossils, there is, perhaps, none which exhibits more clearly the difficulties of a palæontologist than that comprising the relics of sharks and rays. The larger kinds of these fishes are very rarely met with unmutilated in any rocks, and it is somewhat unusual, also, to find the smaller species in any but an exceedingly fragmentary state. Often, we discover a few teeth, or even only a single one ; sometimes, there is nothing but a fin-spine ; sometimes, we have a more or less incomplete vertebral centrum, or, it may be, a short series of centra ; at other times, we observe the faint indications of a patch of shagreen, or find a single spinous dermal plate ; and occasionally we are delighted to meet with a nearly complete mouth or head, or possibly with all the hard parts of one individual,—either a veritable "jumble," or else showing an outline of the original fish.

Such being the facts, therefore, it is not in the least surprising that the different parts of any one true species should often have been described as belonging to two or more species, or even to two or more distinct genera ; at the present time, indeed, there are very many forms of teeth and spines known by names that are merely provisional, and when their true relationships to each other have been determined, it will still be extremely difficult to relegate them to a correct place in classification.

That it is necessary for palæontologists to employ a provisional nomenclature in these puzzling instances, will be admitted by all students of fossil remains ; but, at the same time, the so-called specific and generic distinctions ought not to be based upon differences too trivial, and we can scarcely agree with those who think it advisable to assign such small limits to variations as has been frequently done, and more particularly in recent years.

In this short series of papers, it is proposed to give an outline of the palæontological history of the Selachian fishes, so far as it has been revealed through scientific research up to the present time ; and we shall attempt a zoological, rather than a stratigraphical arrangement, although it will be almost impossible to adhere strictly to either.

But before proceeding with this palæontological sketch, it will be well, perhaps, in the first place, to enter upon a few preliminary considerations, and take a general glance at the sharks and rays as represented in the living fauna of the globe. They are usually grouped with *Chimæra* and *Callorhynchus* in an order variously termed Elasmobranchii, Selachii, or Chondropterygii, and this order is further divided into the suborders of Plagiostomi and Holocephali,—the former comprising the families with which we are here concerned, and the latter the two Chimæroid genera just mentioned. Both these divisions agree in many important respects, and they are regarded by most ichthyologists as forming quite a natural order ; but others are inclined to make them much more in-

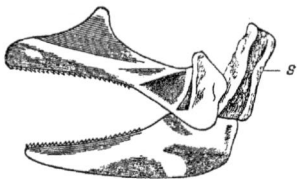

Fig. 99.—Palato-quadrato-mandibular arch (= jaws) of Spinax (nat. size). *s*, suspensorium.

Fig. 100.—Outline of Spinax (much reduced).—*a*, spiracle ; b^1 b^2, dorsal fin-spines ; *c*, gill-openings ; *d*, mouth ; *e*, nostrils ; *f*, eye.

dependent—to constitute each a distinct order—and the structure of the skull is particularly considered to justify this new classification.

The entire skeleton of the Plagiostomes is cartilaginous, and, as might be inferred from what has already been stated, the only hard parts capable of fossilisation are teeth, spines, shagreen, and some of the more fibrous and calcified cartilage. The skull is peculiar and characteristic, and specially noticeable on account of the arrangement of the jaws, which are provided with numerous teeth. Instead of the upper dentition being borne by maxillaries and premaxillaries, the palatines constitute the upper "jaw," and these cartilages do not coalesce with the rest of the skull, but, together with the mandible, form a decided "arch." This is suspended by an element

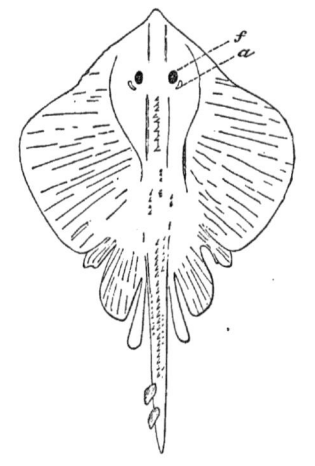

[Fig. 101.—Dorsal aspect of Ray (Raja), much reduced in size. *a*, spiracle; *f*, eye.

(*s*) that is probably homologous with the hyomandibular + symplectic of the Teleostean fishes, and is diagrammatically represented in fig. 99. Two "labial cartilages" on each side (not shown in the figure) are probably homologous with the maxillaries and premaxillaries of other fishes. The teeth vary in form considerably, not only in the different genera and species, but also in different parts of the individual mouth; sometimes they are more or less flat and smooth, adapted for crushing food; in some genera, they are only slightly flattened, and ornamented with grooves and ridges of varying fineness and complexity; while, in other genera, they are more or less conical and laterally compressed, with a sharp cutting edge, either serrated or entire, and particularly fitted for grasping or piercing or for lacerating flesh. But, however varied the teeth may be in general shape and arrangement, they all agree in possessing well-defined roots, not lodged in sockets, but enveloped in a somewhat elastic membrane, and thus attached to the jaw; those sharks with cutting teeth usually have only one row in use at a time, and, notwithstanding their loose implantation, are not able to erect or depress them at will. Among living forms, at least, the new teeth are developed behind the old, and advance forwards to take the place of those that are continually falling out, and there are only extremely few cases among fossil forms where this plan of succession appears to be altered. A little behind the head, on each side, are observed five narrow slits (fig. 100, *c*), which constitute the external gill-openings, and are not covered by any fold of skin; the "spiracle" (fig. 100, *a*), occasionally present, is also interesting,—being the remnant of a sixth (really the first) branchial cleft in the embryo,—and is the external opening of a small canal connected with the pharynx. The body of the Plagiostomes, though sometimes naked, is generally more or less protected with bony granules or "shagreen," and, in a few rays, these dermal ossifications are modified into large spinous tubercles. The dorsal fins are frequently armed in front with strong spines (fig. 100, b_1, b_2), and such spines, barbed, occur upon the tails of some genera of rays without any accompanying fin. There are still other most important characters, displayed in the internal soft parts of these fishes—brain, heart, intestine, &c.—but these are necessarily unavailable to the palæontologist, and space prevents us from considering them here.

The two sections of the Plagiostomi are easily recognised; the sharks (Selachoidei) are usually elongated in shape (fig. 100), with a cylindrical body, and the gill-openings lateral, while the rays (Batoidei) are almost all much depressed and flattened (fig. 101), with very large pectoral fins, and the gill-openings on the ventral surface. There are, however, living forms that appear to be intermediate between the two sections, and several fossil species are known that seem to be even more remarkable in this respect.

The sharks and rays of the present day are mostly marine, although some of the former often ascend large rivers to a considerable distance from the mouth, while a few species of each are found exclusively in fresh water. All (except perhaps one) are carnivorous, and those with flat crushing teeth feed upon molluscs, crustacea, and other creatures not requiring any very rapid movement for their capture, while those with sharp cutting teeth are much more agile and live chiefly upon fishes. The species of sharks have a very wide geographical range; some are true pelagic forms, but the majority exist at no great distance from the shore, and scarcely any have been met with in the oceanic depths. The species of rays have not so wide a geographical range as the sharks, and only very few venture to leave the comparatively shallow waters of the coast-line.

According to Dr. Albert Günther,* the living Plagiostomi may be divided into fifteen families,—Selachoidei nine, and Batoidei six,—and, so far as can be ascertained, all but one are represented in the fossil state. Seven others have already been established for curious extinct groups, and, as palæontological research progresses, it is not in the least improbable that the families will be still further multiplied.

Following Dr. Günther's arrangement, for the most part, and including the results of recent investigations, we may place the different fossil genera in their respective families, as in the table below. It is always difficult to determine precisely the affinities of the fragmentary relics preserved in nature's stony record, and hence this classification cannot be regarded as more than provisional, although it is perhaps the best that the present state of science will allow. Only the genera that are now quite extinct are printed in italics, and more attention will be given to these in the sequel than to those which still exist and inhabit recent seas.

LIST OF FAMILIES OF PLAGIOSTOMI
(WITH THEIR PROBABLE FOSSIL REPRESENTATIVES).

1. CARCHARIIDÆ. [Dorsal fin-spines absent; teeth, in most genera, conical or triangular, and laniary.] *Corax*, *Galeocerdo*, Hemipristis, Zygæna (Sphyrna). Range, Cretaceous—Recent.
2. LAMNIDÆ. [Mostly pelagic, fishes of large size. Dorsal fin-spines absent; teeth conical or triangular, and laniary.] Lamna, *Odontaspis*, *Otodus*, *Oxyrhina*, *Carcharodon*, *Sphenodus*. Range, Jurassic—Recent.
3. PETALODONTIDÆ. [No fossil forms known.]
4. NOTIDANIDÆ. [Only one genus, Notidanus, living. Dorsal fin-spine absent; teeth laniary.] Notidanus. Range, Jurassic—Recent.
5. SCYLLIIDÆ. ["Dog-fishes."] Dorsal fin-spines absent; teeth very small, and more than one row at a time in use.] *Thyellina*, *Scylliodus*, Scyllium. Range, Cretaceous—Recent.
6. HYBODONTIDÆ. [An extinct family.] *Ctenacanthus* (*Cladodus*), *Tristychius*, *Pristicladodus*, *Hybodus*. Range, Carboniferous Limestone—Chalk.
7. ORODONTIDÆ. [An extinct family.] *Orodus*, *Lophodus*. Range, Carboniferous.
8. COCHLIODONTIDÆ. [An extinct family.] *Psephodus*, *Cochliodus*, *Streblodus*, *Deltodus*, *Sandalodus*, *Tomodus*, *Pœcilodus*, *Pleurodus*. Range, Carboniferous.
9. COPODONTIDÆ. [An extinct group, possibly a family.] *Copodus*, *Labodus*, *Pinacodus*. Range, Lower Carboniferous.
10. PSAMMODONTIDÆ. [An extinct group, possibly a family.] *Psammodus*. Range, Lower Carboniferous.
11. CESTRACIONTIDÆ. [Only one genus living,—the "Port Jackson Shark" of Australian coasts. Dorsal fin-spines present in Cestracion and *Acrodus*; teeth obtuse, adapted for crushing.] *Acrodus*, *Strophodus*, *Ptychodus*. Range, Trias—Recent.
12. SPINACIDÆ. Two dorsal fin-spines generally present; teeth numerous, and very variable in form.] *Palæospinax*, *Drepanephorus*, Spinax. Range, Lias—Recent.
13. RHINIDÆ. ["Angel-fishes" and "Monk-fishes." This family, and the three following, may be regarded as intermediate between the Selachoidei proper and the true Batoidei.] *Thaumas*, Squatina. Range, Jurassic—Recent.
14. XENACANTHIDÆ. [An extinct family.] *Pleuracanthus*. Range, Carboniferous—Permian.
15. PETALODONTIDÆ. [An extinct family.] *Ctenoptychius*, *Petalodus*, *Petalorhynchus*, *Polyrhizodus*, *Janassa*. Range, Carboniferous—Permian.
16. PRISTIOPHORIDÆ. [Snout produced into flat rostrum, generally with lateral teeth.] *Squaloraja*. Range, Lias—Recent.
17. PRISTIDÆ. ["Saw-fishes."] Toothed rostrum proportionally larger and better developed than in the last family.] Pristis. Range, Eocene—Recent.

* "Catal. Fishes Brit. Mus.," vol. viii. (1870). See also Günther's "Study of Fishes," pp. 312-348.

18. RHINOBATIDÆ. [Body not remarkably depressed, tail long. Teeth numerous and obtuse. Dorsal spines absent.] *Spathobatis*, Rhinobatus. Range, Jurassic—Recent.
19. TORPEDINIDÆ. [Body naked, much depressed, and gently rounded in front. Provided with an electric organ.] *Cyclobatis*. Range, Cretaceous—Recent.
20. RAJIDÆ. ["Rays" proper. Body generally protected with rows of spinous dermal tubercles, which are often found fossil.] Raja. Range, Pliocene—Recent.
21. TRYGONIDÆ. ["Sting-Rays."] Teeth usually flattened. Tail often armed with barbed spine.] Trygon, Urolophus. Range, Eocene—Recent.
22. MYLIOBATIDÆ, ["Eagle-Rays."] Often attain to a large size, and sometimes live in open sea far from the coast. Teeth flattened, and forming a kind of tessellated pavement in both jaws. Occasionally a barbed spine behind the dorsal fin.] Myliobatis, Ætobatis, *Zygobatis*. Range, Eocene—Recent.

GOSSIP ON CURRENT TOPICS.

NOT long ago the geological conundrum of the origin of the chalk appeared to be answered; "the continuity of the chalk" was an established phrase, but its endurance was very brief. The idea that our characteristic national sea-walls are the ancient representatives of the globigerina ooze, still in course of deposition, and that there has been no break of time between, appears now to be officially replaced by the statement that chalk "must be regarded as having been laid down rather along the border of a continent than in a true oceanic area."

It is rather perplexing that these opposite inductions are both based on the results of the great national yachting expeditions. Once upon a time, before "The General Results of the Dredging Cruises of H.M.SS. Porcupine and Lightning," were published "under the scientific direction of Dr. Carpenter," &c., I could spend a holiday on the south coast, bask on the beach, and peacefully contemplate the chalk cliffs and their horizontal lines of flint nodules without suffering any brain-splitting perplexities concerning their origin. Somehow and somewhere I had imbibed a very simple theory concerning the formation of chalk. I supposed it to be a redeposited limestone; that an older rock, such as the mountain limestone of the carboniferous period, had been exposed over a large area and denuded; that this denudation had formed a calcareous mud in the sea, which ultimately consolidated; that sponges abounded there, and as they died and their sarcode decomposed, their spicules were rolled over and aggregated into rounded lumps by tidal or other currents, and thus the flints were formed. Whether this theory was derived from Jamieson's lectures, which supplied my first geological notions, when both myself and geology were young, or whether from some old geological treatise, or whether I dreamed it while dozing on the beach, I know not; but after struggling through Wyville Thompson's book on the above-named expeditions, and the more recent contradictions supplied by the Challenger expedition, I am strongly disposed to revert to the old notions and look upon those cliffs again as peacefully as of yore.

The absence of meteoric particles of glassy spicules

of volcanic dust, disturb not this simple theory, and no miraculous "segregation" of siliceous molecules is demanded to explain the abounding flints.

The International Forestry Exhibition of Edinburgh is much needed. To most Englishmen, Scotchmen, and Irishmen, the idea of the existence of forestry as a branch of science is quite novel, but in other countries it is publicly acknowledged and accepted, with its national schools receiving state endowment. The abundance of our coal supplies is not an unmixed blessing. It renders all our cities grimy and hideous, and by averting the necessity of obtaining wood-fuel has left vast tracts of country bare. With a humid and equable climate like ours, every acre of ground that is not available for pasture or tillage is (unless absolutely bare rock) capable of being restored to its ancient condition of woodland, and should be so restored, for every reason. Every year of current extravagance brings us nearer to the approaching time when our *easily obtainable* coal will be exhausted; —note the italics—I do not say when all our coal shall be exhausted, as that time will never arrive, but when by the exhaustion of our best and most accessible seams we shall be no better off than our neighbours are. Then the miserable poverty of our wood supplies will press crushingly upon us, unless something is done at once to anticipate the coming demand. In Scotland and Wales there are five to six millions of acres now lying waste which are available for timber growing, and would yield enormous wealth if properly planted. In Ireland, where the climate is especially favourable to foliage, there are above two millions of acres in a similar condition.

The configuration of Ireland affords further advantage. Speaking generally, it is a flat bottomed basin, the flat bottom being an area of mountain limestone, the rim of the basin an irregular ridge or ridges of mountains, notched and furrowed on the outside to form innumerable estuaries. The shipment of timber from the slopes thus terminating in these fjords or "loughs" is so very easy, that a respectable degree of enterprise on the part of the landlords would supply more prosperity to Ireland than if all the coal seams that have been denuded from above the inner plains of carboniferous limestone were restored.

Connemara, Mayo, and Donegal, or the whole of the north-west mountain region would become one of the most picturesque in the world if it were amply wooded. That labyrinth of little lakes that complicate the singular promontory projecting on the north of Galway Bay would become a tourist's paradise if it were but richly wooded. At present it is practically unknown, beyond the drive from Galway to Clifden and Leenane, which only skirts the curious region, which, orientally speaking, I may call the land of a thousand and one lakes. When I have an afternoon to spare, I will count them as laid down on the ordnance map. I tried to count the number I passed in the course of a drive from Roundstone to Recess, but gave it up somewhere in the thirties. Miserable barrenness prevails throughout, where the loveliest of woodland should be.

An example of the trustworthiness of the anti-vivisection agitators is afforded by some expressions of Mrs. A. Kingsford, M.D., quoted in last month's number of the "Journal of Science." At an international congress of these peculiar people, she exclaimed that M. Pasteur was not justified in "torturing thousands of animals," with the object of abolishing so "very rare" a disease as hydrophobia. As the editor observes, "the thousands when translated into the language of sober reason shrunk to forty!" and "as for the rarity, we must remember that twenty-one persons died of hydrophobia in the department of the Seine within twelve months."

An observation made by Mr. J. B. Armstrong of the botanic garden, Christchurch, New Zealand, is very interesting, and suggestive of further research. The red clover of that island is becoming modified in structure, and he believes that this occurs in such wise as to admit its fecundation by insects that differ from those which visit it in England. If this is confirmed, and the progress of the change carefully watched, we may have, within a reasonable period, an interesting display of the full history of modification by natural selection, by isolating specimens of plants that show the first traces of favourable modification, and watching the development of their offspring.

The transplantation of sponges promises to become a commercial enterprise of some magnitude, in the course of which we may fairly hope that biological science will be enriched. The French are already at work; the field selected for the first experiments is the coast of Algiers, and the species to be acclimatised are the choicer kinds of "Turkey" sponges, found in the Archipelago, and on the Syrian coast. With modern diving appliances fine specimens can be deliberately selected and carefully removed with a sufficient amount of adherent rock to render their transplantation a fairly hopeful enterprise. Thus raised, they will be placed in perforated boxes and towed through the water to their new homes, where, if the experiment succeeds, they will multiply in the course of the following year and yield a good crop in the course of three years. If these sponges increase with anything like the rapidity of our own coast sponges, this crop will be very abundant. When the Crystal Palace aquarium was newly stocked some of the tanks became infested with a growth of sponges which must have originated in germs carried in the water that conveyed the fishes, &c. They spread like cobwebs; destroying the beauty of tanks, and demanding much trouble for their clearance. Mr. Lloyd told me that the nuisance was still greater at the Hamburg aquarium, where they appeared in like manner with the first supplies of animals.

Now that the authorship of the "Vestiges" is fully revealed, there is no reason in gossiping about the condition of the secret. Long ago when the book was new, and many of the large family of the author were little children, I spent many bright evenings in Doune Terrace, in the midst of the circle of his intimate friends, which included all the eminent men and women of Edinburgh; for Robert Chambers, although distinctly on the liberal and progressive side in everything, was so free from mere partisanship that representatives of every social set in Edinburgh, excepting the extreme Calvinists, were present at his larger gatherings.

Somehow, but nobody could say definitely, how or why, we generally believed that Robert Chambers was the author of the book; nobody asserted it, simply because nobody positively knew; all respected the secret and understood well enough why it should be kept, and therefore scrupulously avoided any discussion of the subject in the presence of the supposed author. To have asked him anything about it would have been regarded by all as a gross impertinence. The chief confirmation of my own suspicion was supplied by the fact that I never heard him allude to the book, then so fruitful a subject of conversation, and rarely to any of the topics discussed in it, though I knew that he was a self-taught geologist and naturalist, with that freedom from text-book trammels and scholastic orthodoxy that leads to the philosophical breadth and bold originality of thought that characterise the work, as well as to the gaps in the knowledge of detail that it also betrays. As I heard George Combe remark, he never denied the authorship.

W. MATTIEU WILLIAMS.

THE TEETH OF THE HOUSE-FLY.

BY W. H. HARRIS.

IN the year 1878 there appeared in SCIENCE-GOSSIP, at page 147, a very interesting paper on "The Teeth of the Blow-fly," accompanied by an accurate illustration enabling anyone to recognise these organs in a properly prepared object.

Having prepared some lips of the blow-fly by the method therein indicated, in order to make myself thoroughly assured of the presence of these organisms, I was induced to seek further among the diptera, with the view of ascertaining if other members of this large order presented a similar development. In very many cases I have found this to be so. The teeth in different species vary not only in number but also in form, so far as their free ends are concerned, but in one particular they all appear to agree, viz., the bands of chitine of which the teeth are composed, are all, for a certain portion of their length, turned in on each margin, and thus form a split tube from their point of attachment to about two-thirds of the length of each tooth, when they begin to expand, and continue to do so until the extremity of the tooth is reached. Usually the end presents a V-shape, but there are modifications of this form.

The illustration represents the position and number of teeth in one lip of the house-fly (*Musca domestica*). The teeth are six in number, there being three in each lip. The free ends are trifid and serrated; the two lateral expansions each bear two small saw-like processes, while the central one has seven similar denticulations, the central apex carrying one with three on either side, there being eleven denticles or serrations on each tooth.

Fig. 102.—Teeth of the house-fly (*Musca domestica*), showing position in one lip.

In nearly all creatures possessing teeth these organs afford a ready and reliable means of identification; and after some little study, the outline of which is indicated in these few remarks, I have little doubt the same would apply to this order of insects, and might possibly prove of advantage in determining species apparently slightly removed from each other by outward appearances.

As an improvement on drying the objects, let me suggest the use of one drop of carbolic acid; this clears the object and prepares it for mounting in Canada balsam. The whole operation can be completed under five minutes, and a permanent mount secured in every respect equal to one obtained by more lengthened treatment.

PROFESSOR ROSCOE, the distinguished chemist, has just received the honour of knighthood.

CHOLERA BACILLI.

A FEW weeks back, I furnished a short note to SCIENCE-GOSSIP, on the discovery of the specific bacillus which Professor Koch believes to be the cause of cholera. Through the kindness of a friend, I am now able to send you a sketch made from a pure cultivation of the specific bacillus, drawn from a preparation mounted by the Professor while in Calcutta. *b* in the sketch is the cholera-germ; and for the purpose of comparison I have also drawn *a*, a bacillus which presented itself in a solution of

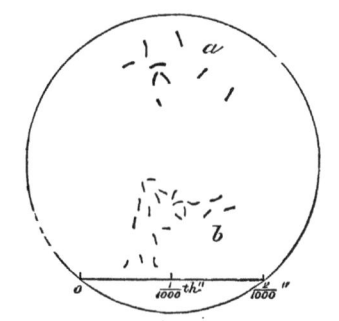

Fig. 103.—Bacillus of Cholera.

gelatine in my book-case, and which was wind-sown, and I have subjoined a micrometric scale of two one-thousandths of an inch drawn with the eye-piece at the same distance from the paper as *a* and *b*, in fact from a scale which simply replaced the above objects on the stage without disturbing the microscope. Professor Koch's Sixth Report to the German Government appears translated in the Indian Medical Gazette for May current, and will be of great interest to mycologists.

Calcutta.   W. J. S.

NOTES ON THE NATURAL HISTORY OF JERSEY.

By EDWARD LOVETT.

SOME time ago I wrote a few chapters in SCIENCE-GOSSIP on this subject, and the present contribution is not intended so much as a continuation of them, as a brief sketch of what use can be made of a very brief visit to a favourable locality.

My stay was a short one, all too short from a naturalist's point of view: for every time I revisit my favourite island, I am fortunate enough to find the home of some rare or local form, and become more than ever convinced that the word "rare" is only a term of comparison, and that the rarest species exist in quantities, if we only knew where to find them.

I am also convinced that the shores of Jersey are comparatively unworked, and that it would be no light task to investigate their natural history thoroughly, owing to their vast and rugged expanse.

Although I have done a good deal of dredging about the island, I have not found it yield such good results as shore hunting during a low tide. On this occasion I experienced strong N. E. winds, which made dredging not only unpleasant but dangerous; still, I obtained fairly good results by going under very easy sail. Our dredge brought up great numbers of *Aplysia punctata*, several *Aphrodita hystrix*. Shells, some dead, of *Venus verrucosa*, *Lutraria oblonga*, *Cardium norvegicum*, *Anomia ephippium*, *Pecten maximus*, and many others; in one spot *Pectunculus glycimeris* came up in abundance.

Of Crustacea I did not obtain much with the dredge; *Pagurus prideauxii*, *P. cuanensis*, *P. bernhardus*, *P. hyndmanni*, *Stenorhynchus longirostris*, *Eurynome aspera*, *Maia squinado*, being about all. The common hermit and *P. prideauxii* were in shells on the outside of which was *Sagartia parasitica*, in many instances several animals on one shell. Of small objects, the wind prevented my saving much, for, as soon as we managed to turn the dredge out on our little deck, a good deal of the contents got blown overboard again; still, I secured the ova of *Nassa* attached to Zostera, and the curious ova of the *Aplysia*. I had almost forgotten to mention that I got a few specimens of the star-fishes *Cribella* and *Ophiocoma rosula*.

Part of the ground dredged was in the course of the steamers, and I was struck with the effect produced by the cinders thrown overboard by them. It is well known what enormous quantities of this material finds its way to the bottom of the sea in the track of steamers, and I have often thought that the effect must be injurious to marine life, but I did not find it so. Our dredge brought up quantities of these cinders at each haul, and they being somewhat porous and rough seemed to offer suitable anchorage for many things. Every specimen of *Anomia* that I obtained was anchored to a cinder, and many of the best shells and Crustaceæ were living on and among them; so that, after all, these cinder banks, as they must become in time, may prove of actual service to the life of our seas.

Whilst in Jersey the breaming season was in full swing, and I obtained several things which the fishermen took on their long lines. One morning a fine star-fish (*Uraster glacialis*), about one foot eight inches in diameter, was brought me, and another (*Uraster rubens*), about ten inches in diameter and very robust; this latter, the common star-fish, is called by the fishermen the five-finger Jack. One man brought up on his line two very large *Bunodes crassicornis*, attached to good-sized stones; they had each absorbed a hook baited with a whelk, and I was told that they often came up on the long lines. Octopus was also "in season," and many of the fishermen I visited had the walls of their cottages hung with dried "cats"—"cat-o'-nine-tails" being the Jersey fisherman's logical name for a mollusc possessing eight tentacles. When properly dressed and cooked, the octopus is by no means bad eating, a large one skinned and grilled being really palatable and unlike anything else in flavour.

I was fortunate to be in Jersey during one of the best spring tides of the year, and was thus enabled to walk over part of the sea bed very seldom exposed to view; but as this paper has reached its limit I will reserve my further observations for another article.

Croydon. June 4, 1884.

MY WINDOW PETS.

TO the true naturalist there are few things more pleasing than the confidence exhibited by the lower creatures, when encouraged to become familiar by a long course of kindness. This trustfulness is, perhaps, to be expected when the creature is a domestic pet—one familiarised with him whose hand supplies the daily food—but doubly pleasing does it become when the pet is one enjoying its wild freedom—this being more particularly true of birds.

What true ornithologist would cage the gladsome bird rejoicing in its freedom; who fetter the joyous spirits of creatures so exquisitely fitted for facility of motion? Some, indeed, there are who would—nay, do—but of their number am not I, far preferring to see them exhibiting their manifold beauties, and graceful airy movements as near to my person as if they were caged, and yet, as free to come and go as if such an individual as myself were not in existence.

How accomplish this? methinks I hear a reader ask. Simply thus.

Without my sitting-room window I have placed across from wall to wall, and about midway up the lowest panes, a slender, unbarked stick, and daily, both summer and winter, at the hour of one, or thereabout, I spread a portion of food. Summer, bread crumbs alone; all the rest of the year, bread crumbs and chopped up mutton fat mixed. So nice is their perception of time that, half an hour beforehand, the garden near the house teems with my pensioners, who, flitting from spray to spray, await the scattering of the daily portion to alight and feast. One of the first to descend is that active little fellow the cole titmouse (*Parus ater*), who, poised upon the blossom of a giant wallflower, which scarcely bends beneath his tiny weight, for scarce an instant surveys the spread feast. Alighting upon the perch on his way, he snatches from the sill the most alluring morsel of fat, and lo! as quickly as he came he goes, to return almost immediately for another scrap, and so, till all is gone. Most delightfully indifferent is he to our presence, and, when not disturbed by other sharers in the feast, eats his morsels upon the perch, casting many a friendly glance at his purveyors. He is a busy little fellow, and has a large place in my affections, on account of his trustful character. No match is he for that courageous, azure-capped little bird, the blue titmouse (*Parus cæruleus*), who, unless too busy laying in his own stock of provisions, drives him from the scene. A most pugnacious little rascal is the blue tit, and although, at times, as many as five or six are quietly feeding side by side, more often than not one more combative than his fellows loses a good portion of his own meal, whilst he tries hard to keep them from theirs. He, too, is a very trustful little fellow, and seldom indeed is it that two or three are not busily employed picking up the microscopic crumbs, left after the bulk of the feast has been devoured; and many a tap against the window panes may be heard, as, standing on tiptoe, some blue cap peeps in, seeming to say as plainly as words could do, "Nothing to eat—have you nothing more for us?"

Foremost for beauty of plumage, if not for sprightliness, next comes our largest titmouse, that active and restless bird the titmouse, the greater titmouse, or oxeye (*Parus major*). It is a beautiful bird, the yellow and rich blue-black of its plumage presenting a contrast that the most unobservant could scarcely overlook.

A great benefactor is he to the gardener, being most active in his search for insects, which I, as an apiarian, know to my cost.

I have somewhere seen it stated, that this bird will not eat bread crumbs; a statement not in accord with my own almost daily observations, during the past four years. Undoubtedly he is very partial to mutton fat—a partiality shared in common with many other birds—but when of *that* there is none, he certainly does not disdain the humble crumb. His call note so closely resembles the "twink, twink" of the chaffinch as occasionally to deceive the most practised ear.

Who does not listen with delight to the soft and joyous strains of our confiding, homely little friend, robin redbreast (*Erythaca rubecula*)? Of course he never fails to put in an appearance. Even whilst I write he trills his sweet lay from the perch, literally singing for the meal he so patiently waits for, and casting, ever and anon, his most lovely eye towards the room, to see if there are any signs of its coming. At this time of the year two pairs occasionally alight on the window-sill at the same moment, it is for an instant or two only, and more frequently one pair makes a quick departure as the other appears, the boldest seeming to consider it a point of duty to keep the coast clear of all intruders. I am sorry to say that Robin has a most unamiable trait in his character, being decidedly quarrelsome, and even bloodthirsty, but he atones for this fault by the exhibition of many virtues.

Another bird most familiar with me is that lovely,

soft-plumaged bird, the chaffinch (*Fringilla cælebs*), whose graceful motions and sweet and varied song make it a general favourite. "From rosy morn till dewy eve" is either this song or call note to be heard, and, to me, there are few country sounds so sweet and suggestive of peace. Were I compelled to banish from my garden and orchard every bird but one, I think I should elect to preserve the chaffinch.

And yet I should look with lingering fondness after his first cousin, that merry songster the goldfinch (*Carduelis elegans*), for it is a docile and gentle bird, delighting the eye as much by the beauty of its plumage as the ear by the sweetness of its song. She does not visit my window so frequently as many other birds, and almost invariably waits until the flock has dwindled down to ones and twos, when, with her friend the female chaffinch, they together gather up the almost microscopic fragments of the feast. I say *she*, for, singularly enough, although one pair always nests in my garden, or orchard, the male bird never favours me with a visit; but sings to cheer his mate.

One day last summer, when she was leading off her scarcely fledged nestlings, a playful puppy in my orchard contrived to kill one in the long grass, a second barely escaped, and taking the panting little downy ball in my warm hands, I kept it there until its throbbings were stilled. I fed it with biscuit, soaked in milk, and beaten into a soft pulp, and, after keeping it for 24 hours, placed it upon the lower branch of an apple-tree, then watched to see if the parents would recognise their lost one. To my delight they at once answered its chirp, and set to work to feed it, and soon I had the satisfaction of seeing it led to the side of its three brethren. A few days after, it set up life on its own account.

There is yet another most innocent little bird ever near my window, although not often upon the sill. I allude to the soberly-dressed hedge warbler (*Accentor modularis*). Most social and trustful is this bird, almost never, either summer or winter, absent from my small lawn, or gathering crumbs beneath the window—never molesting nor being molested, since he never thrusts himself into the midst of the chirping, twittering crowd. Very active in his movements, although less so than the titmice, he is to be seen searching for insects from early morn till darkness shrouds every distant object. Yesterday, whilst I was raking my garden, he was pouring out his full heart in a strain of sweet melody from a shrub within a yard of me, following me as I moved, simply, as it seemed, from a desire to be near my person.

I have yet another constant visitor to my window; but I do not encourage him, seeing that his voracious appetite leads him to clear off in the course of a few minutes, more than sufficient to feed a host of smaller birds. I speak of that bird whose flute-like notes charm every woodland wanderer, the blackbird (*Turdus merula*). He pecks, then peeps in at the window, pecks again and again, and then another peep before he flies off with his usually somewhat discordant cry, mostly uttered when disturbed. Last year one built a nest and laid five eggs within two feet of a beehive where I was daily feeding the bees, but forsook it when I injudiciously swept from the lowermost branches the snails that had collected there—shortly after building another in the branches of a pear-tree trained against the house.

This bird seems to entertain a particular antipathy to the common thrush (*Turdus musicus*), and chases him viciously whenever he comes to claim his share of the good things I provide. Owing to this persecution thrushes are not often to be seen upon the window-sill, generally picking up stray crumbs that have been scattered beneath. For these he repays me with his sweet song, from break of day till nearly all other birds have gone to rest. Last Christmas his strains were remarkably strong and full, and during the month of February his song seems to have acquired its full strength.

I must not forget my noisy, combative, ubiquitous friend the common sparrow (*Passer domesticus*); of course he is always in full force, always wary, always obtrusive and full of self-assertion, and always able to hold, not only his own, but also what properly belongs to others. Both the courageous blue tit and robin, however, contrive to keep him at a respectful distance.

I have by no means exhausted the list of visitors to my window. The bullfinch (*Pyrrhula vulgaris*), the tiny wren (*Troglodytes Europæus*), the starling, and the water wagtail (*Motacilla yarrellii*) may be included, but they are not frequent visitors; the first named being far too busy with my fruit buds to do more than hop down for a brief survey, just to see what can possibly have attracted so many of his feathered neighbours; and the wren too much engaged in prying into every crack and cranny in search of insects, to often venture into the midst of a feathered crowd.

Well, it is a source of pure delight to me to see the uncaged birds so full of trust, seemingly regarding me as a friend, and many a charming sight have I seen within the compass of half-a-dozen window panes. A lovely picture is formed when side by side appears a chaffinch in full plumage, a greater titmouse, and a robin; the contrast of colours is remarkably striking.

During the severe winter of 1880–1 the snow had been swept into a ridge, some four or five feet from my window, and upon the summit of this ridge the live-long day there perched a small regiment of birds, awaiting the supply of food which was served three times per diem. I sometimes counted together as many as nine male chaffinches, beside a host of blue tits, sparrows, robins, thrushes, starlings, blackbirds, &c. 'Twas a pretty sight, and although never since repeated, the daily smaller gathering of our window pets is, to us, an unfailing source of pleasure.

"What about your fruit?" methinks I hear some reader ask. Let me at once frankly admit that my pets help me to get rid of it—it is, however, little enough they take, if I except standard cherries, nor do I begrudge them their share, when they contribute so largely to my enjoyment; but, of course, I take proper precautions to preserve it in the shape of cotton stretched from bush to bush, which affords ample protection, and, being almost invisible, is no disfigurement to the garden.

EDWARD H. ROBERTSON.

OUR BRITISH MARINE ANNELIDA.

By Dr. P. Q. KEEGAN.

AN unobtrusive tribe of animals, of an annular conformation and bilateral symmetry, of no great dimensions, rather sluggish in habit, and of a structure sometimes singularly beautiful, yet in itself delicate and fragile and needing the shelter of a shell, &c. A very remarkable tribe of creatures, mostly tenants of the profound depths of the ocean; yet some prefer the quietude of the seashore or the oozy slime of the matted sea-plants. Some are denizens of chambers hollowed in the soft yielding deserts of sand between tide-levels; others tunnel through the water-penetrated sandstone rocks of the seaboard of an inland tidal harbour. Many of them roam and prowl about, free, naked, and unattached, in search of prey or in wantonness of animal activity; others are sedentary, and move in the restricted sphere of a shell or habitation permanently affixed to a rock, loose stone, or a sea-post. Nearly the whole tribe are embellished with beautiful colouring, and exhibit microscopic structures of extreme loveliness and unparalleled constructive ingenuity.

Such are the principal features of the Annelida—a class of animals which stand on an equal rank with the crabs, lobsters, &c. Generally speaking, they may be characterised as having elongated, rounded, segmented bodies, often divided by many superficial transverse constrictions into membranous rings, which frequently develop lateral rudimentary limbs usually provided with locomotive organs called setæ (bristles and hooks); there is also frequently a distinct "head," and, in some instances, auxiliary branchial organs called "elytra" are distributed over the back. The nervous system consists of a chain of ganglia extending throughout the length of the body and connected by longitudinal and transverse bands of varying lengths; the eyes are very simple, consisting of "an expansion of the extremity of the optic nerve imbedded in pigment, and provided occasionally with transparent spheroids or cones;" the ears have sometimes many otoliths; the digestive apparatus extends the whole length of the body, being straight (except in Pectinaria, &c., where it is convoluted, and in Polynoe, Aphrodita, and Sigalion, where long sacs or processes are developed on each side thereof); there is, except in the tube-building species, a muscular proboscis armed with strong teeth and knobs; the functions of a liver are discharged by glands opening into the digestive canal, and a sort of renal organ (the segmental organs) has been observed in some instances. The gills vary much in development, sometimes they are not specialised, while in other species large and exquisite organs are distinctly referable to aeration. Between the walls of the body and the stomach a corpusculated fluid moves to and fro, while special close trunks or canals with varying branches spread over the body, contain what has been denominated the "blood proper," or the "pseudhæmal system."

A child wandering by the seashore is attracted by a variety of interesting marine objects cast up from the deep by the ever-chiding waves. He muses over these forms, he gathers and closely inspects them. We ourselves remember being specially attracted in this manner by the curious aspect of a series of white convoluted tubes investing the surface of various sea-pebbles, old shells, &c. These tubes are tenantless habitations of a curious and interesting sea-worm of the genus Serpula. Several species of this genus are found on our shores. Sometimes we observe two or

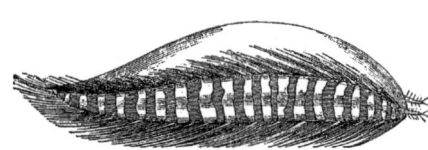

Fig. 104.—Sea mouse (*Aphrodite aculeata*).

Fig. 106.—*Serpula contortuplicata*.

Fig. 105.—*Sigalion boa*.

three long, round, pinkish, almost smooth tubes about a quarter of an inch in diameter, twisted together slightly, and incrusted with sponges, &c. This is *S. contortuplicata* (Cuv.). At the entrance of the tube we observe a neatly-made circular structure of a turkey red colour, and serving the purpose of a door. When the shell is popped into a jar of sea-water in a dark situation, this membranous cover or operculum is seen to be pushed forth, and a fan-shaped expansion stealthily emerges, consisting of two large scarlet tufts of about thirty filaments, each furnished with arge and energetic cilia. These are the gills of the

lets. *S. triquetra* (Mont.) is another species which is found in considerable profusion adherent to shells, stones, &c. Its tube bears a sharp keel all along its length, and a knob above the aperture; the operculum is calcareous and conical, and fixed between two soft processes of its stalk; the gills have about eight filaments in each tuft; the bristles are similar in shape to the last species, but I am unaware of any hairs or toothlets having ever been detected at their extremity; the hooks are broader and more triangular than those of *S. contortuplicata*, and are cut into about nine teeth, and there are no mop-like bristles..

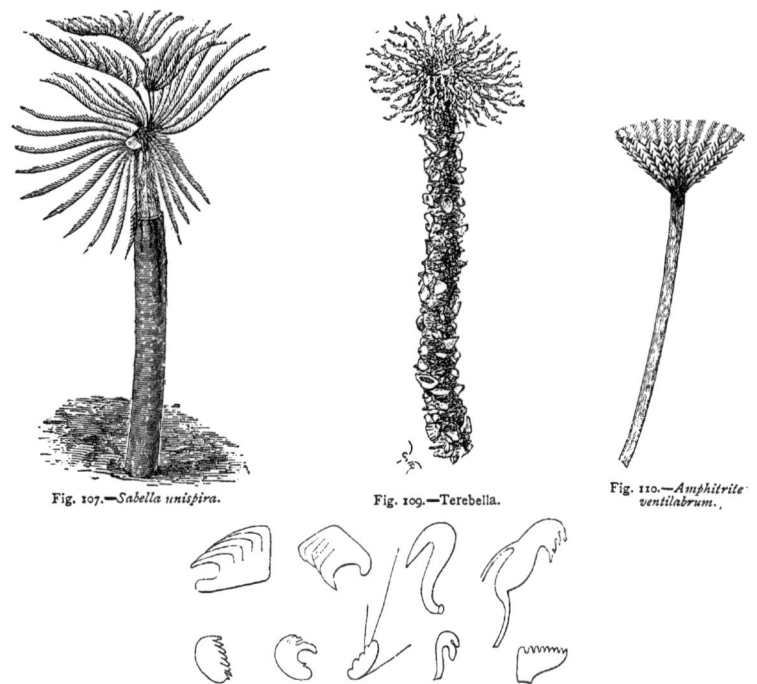

Fig. 107.—*Sabella unispira.*

Fig. 109.—Terebella.

Fig. 110.—*Amphitrite ventilabrum.*

Fig. 108.—Various forms of hooks (*uncini*) used by the *Tubicolor annelida*.

animal, and right vigorously do they operate as aerators of the true blood (not the peritoneal fluid). The body is divided into a great number of segments; the upper branch of the feet carries a number of bristles which are long and acicular, and when viewed under high powers are seen to be fringed at the extremity with a series of very fine hairs or toothlets; on the lower branch of the feet there is a singularly curious ribbon or string of bent hooks, each cut at the edge into five teeth; there is also a third set of bristles of a mop-like shape, the broad flat end being cut into a series of beautifully fine and delicate tooth-

S. vermicularis is a third species which constructs a cylindrical tube without a keel, and with a plain aperture; the operculum is a very elegant structure, having a double funnel, a shallow one planted above another deeper one, both being neatly sliced on the rims into a number of saw-like teeth; the branchiæ have numerous filaments in each tuft, and the bristles and hooks are of similar structure to the last, differing only slightly in shape and in the number of teeth.

The function of these beautiful hooks and bristles in the tube-tenanting sea-worms may be indicated as follows. The tube is smooth and hard. By means of the bristles, which may be regarded as long levers or leaping-poles, the animal slowly pushes its lungs and body forth into the water, using the toothed end as a sort of fulcrum; and by the hooks, to which long tendons are attached, and which are arranged in transverse rows on the under-side of the body, it can withdraw its soft slimy body into the tube with lightning-like celerity on the approach of alarm, &c. Some of the species can also revolve in the tube in a curious manner.

In the genus Sabella the tube is soft, tenacious, and flexible, and is usually made up of a fine membrane coated with a layer of smooth fine mud on the exterior. The branchiæ have an almost similar structure, situation and function to those of the last genus; there is no operculum, and sometimes the two small branchial tentacles are absent; the abdomen as well as the thorax carries both bristles and hooks, which on the former are situated above, and on the latter below the body. *S. vesiculosa* is about six inches long, and the branchiæ bear on their filaments dark-coloured solid globules filled with coloured granules supposed to be eyes; there are eight pairs of setæ-bearing feet on the thorax; the bristles are thickened and serrated near the end, while the hooks are bent at both ends with an entire undivided point; the tube is leathery, about ten inches long, and coated with sand and fragments of shells. We need merely refer to a few other species of Sabella, such as *S. penicillus*—a most beautiful and extraordinary animal over one foot long, and living some ten fathoms deep in an indiarubber-like tube, some eighteen inches or two feet in length. The creature is provided with very efficient masonic tools, in the shape of a couple of trowels and a scoop, whereby, aided by a secretion from its own body, it prepares the mud as cement for the composition of its tube, shaping the material, laying it into position, and smoothing and polishing the whole with the finish and dexterity of an accomplished plasterer; the branchiæ are very large, and exquisitely beautiful in shape and colouring. There is also *S. bombyx*, notable for its talent in constructing a tube of a silken texture, which it spontaneously exudes from its own body.

In the genus Terebella we encounter structures different in many features from what we have hitherto noticed. The tube is membranous, and lavishly coated with rather large pieces of shell, gravel, or sand, is open at both ends, and unattached to any foreign body. The anterior ganglia are fused, while the posterior ones are separate and distinct; the posterior end of the body bears hooks only, not bristles, while from the fourth segment for a certain specific number we have feet-bearing bristles, with or without hooks. The tentacles are very conspicuous, being very long, extensile, and ciliated organs of touch, prehension, and pulling, penetrated by the peritoneal fluid; the branchiæ, which aerate and propel the true blood, are comparatively small and without cilia, are of a red colour, and are erected on two or three of the anterior segments; there are strong and muscular lips at the opening of the alimentary canal. *T. littoralis* is perhaps the commonest and the most numerous of all the British marine annelida. When traversing the sandy area between tide levels, we can hardly fail to observe a curious tuft of fibrous threads coated with numerous sandy or gravelly particles, and crowning the orifice of a tube similarly composed, that stands buried erect in the sand for about a foot deep or so. The animal during the ebb-tide remains at or near the bottom of this case, and he is an exceedingly wily, vigilant, and wary customer. It is about as difficult to catch him as it is to shoot a curlew. No matter how cautiously, dexterously, and swiftly you bury your trowel, and try to shovel him up, it is all no-go, except under very favourable circumstances indeed. This Terebella, which is about four inches long, bears tentacula that vary in number from 60 to about 100, and are of a carnation colour; the gills are most elegantly branched; the skin is smooth and of a peach-blossom colour, with a bright red stripe down the belly; the bristles are arranged in sixteen pairs of bundles, and have a smooth double-edged sharp point like a lancet, while the hooks are furnished with one long large tooth with two smaller ones above it. *T. nebulosa* is about six inches long, and bears twenty-three setigerous feet, which are of service more for trowelling, plastering, and polishing the tube, than for locomotion; there are thick muscular ridges on the under part of the body; there are three pairs of branchiæ; and the hooks are cut into three teeth. There is also *T. textrix* (the weaver), which manufactures a real cobweb as a covering seemingly for its spawn; and *T. figulus* (the potter), which constructs a tube of soft mud or clay. The genus Amphitrite comprises a number of very remarkable forms. Herein the branchiæ are usually distributed over the back, and are long prominent appendages bearing on their surface a spirally arranged series of large vibratile cilia, which are remarkably beautiful when seen in motion under the microscope. A single blood-vessel containing the peritoneal fluid penetrates the interior of the organ, and at the extremity thereof returns upon itself. The tentacles of this group are fleshy, unciliated filaments clustered about the mouth, and penetrated by the same fluid; the feet bear hooks (which appear on ridges extending round the body), and also bristles, and long tactile warts or cirri; and the tail is provided with bristles, or fleshy appendages. Frequently, when walking by the margin of the sea after a gale, we may observe amongst the multitudinous rejectamenta of the tumbling billows a number of very straight, neatly and trimly made, smooth tubes of the shape of a cheroot. On peering into the broader end of one of these sand-woven cases, we are surprised to observe a sort of tiny·hair-comb whose teeth are of a bright lustrous gold colour. This animal is the golden-haired Amphitrite (*A. auricoma*). Its branchiæ have a leaf-like structure, and sprout from the sides of the third and fourth segments; the alimentary canal is convoluted.; the tentacles are capable of being shortened or extended, and they are the instruments whereby the tube is constructed ; and in addition to the thirty golden bristles that garnish the head, there are tufts of bristles ranged along the flanks of the body. Under the microscope the anterior bristles are seen to be grooved longitudinally on the surface, and taper to a fine smooth point ; the lateral bristles are more serpentine in outline, and some of them are very finely serrated at the extremity. Below the feet there is a long winding file of minute hooks marshalled closely together, each hook (as seen × 400) closely resembles a pigmy human hand, with a big thumb, and eight fingers of an equal height. The honey-comb sea-worm (*A. alveolata*) is one of the most extraordinary animals of the sea-shore. It exhibits eminent social proclivities, so that a small colony is generally found in one spot, such as the surface of a soft rock between tide-levels, or a root of laminaria, or an old shell or stone from deep water. The tube is fragile, and built up of a dirty dark sand and mud ; the branchiæ are of the typical form and arrangement, being long and narrow, and distributed in pairs on all the segments of the body ; the head is decorated with three rows of very singular striated bristles—those of the outer row are shaped like a hand of six or seven fingers or prongs, those of the middle circle have a large triangular or spear-shaped head supported on a narrow stalk, while those of the inner rank are like holdfasts, being bent at the end analogously to a foot in relation to the leg ; the sides of the thorax are armed with three pairs of bristles, while the abdominal segments carry tufts and very slender capillary bristles ; the ridges of hooks are composed of minute narrow structures cut into six teeth, and attached to very slender threads. *A. ostrearia* is the tenant of a small, neatly drilled hole, which it bores in the thickest part of an old oyster shell, &c. ; it bears three or four pairs of branchiæ, and a pair of tentacles in front; there are bristles and hooks along each side.

(*To be continued.*)

At the last meeting of the Geologists' Association, a paper was read by Professor Blake on "The North-West Highlands and their Teachings," and another was read by Mr. W. A. E. Ussher, on "The Geology of South Devon, with special reference to the Long Excursion." The latter commenced on July 21st, and occupied the entire week, the directors being Messrs. Champernoune, Pengelley, and R. N. Worth.

SCIENCE-GOSSIP.

The Lambeth Field Club has started a new line of departure. Mr. A. Ramsay, F.G.S., has been nominated recorder in Physical Geography, and Mr. W. E. Bowers in Meteorology. Everything that by any possibility can be grouped under these two elastic terms, in the county of Surrey, will henceforth be recorded.

A proposal is popular in Spain to cut a canal from the Bay of Biscay to the Mediterranean. The plan suggested is to deepen the Gironde for some distance, and reach the open sea at Narbonne. The proposed work would be about 250 miles, and if carried out, it would save a distance of about 2000 miles between London and Suez.

Count von Stein has found a new place to obtain cilioflagellate protozoa. He has examined the stomachs of alcoholic ascidians, echinoderms, and worms, and found multitudes of specimens.

The plan of using the enormous water power of the Alps for working electric railways in Switzerland is about to take a definite shape, the idea being to connect the towns of St. Moritz and Pontresina by an electric railway $4\frac{3}{4}$ miles long, the motive power to be supplied by the mountain streams ; the line, in case the plan proves a success, to be extended a considerable distance.

M. de Fonvielle has suggested the following method of detecting infernal machines. All luggage to be placed on wooden tables supported by iron feet but not nailed to them. A microphone to be placed on each of the tables, when any ticking or other noise proceeding from the luggage would at once become audible.

The city of Brussels is going to try the experiment of using electricity to drive its street cars. One line —that of the Rue de la Loi—is to be equipped with motors, and separate accounts are to be kept, in order to ascertain definitely the cost of running, as compared with the use of horses. The test is to last for one entire year, and then, should the result warrant it, electricity will be employed exclusively on the street railways of Brussels.

The Naturalists' World is about to present, in an early number, a series of facsimile autographs of eminent naturalists and scientific men of the day. Among them are the following : Sir John Lubbock, M.P., F.R.S., Professor T. H. Huxley, F.R.S., Richard Jefferies, the Rev. J. G. Wood, M.A., Dr. J. E. Taylor, F.L.S., and many others.

Many readers will be sorry to hear of the death of Mr. Henry Watts, F.R.S., editor of the well-known "Dictionary of Chemistry," and for many years editor of the "Journal of the Chemical Society."

MR. MONTIGNY has recently drawn attention to the influence of the atmosphere in the apparition of colours seen in the scintillation of stars, and he shows that there is some connection between the colours and the coming weather. Thus, when rain is approaching, there is a great predominance of blue in the scintillating colours.

IN the last volume issued of the "Encyclopædia Britannica," Professor Ray Lankester, F.R.S., has a most important article, in which he gives a new classification of the Mollusca. Of late years all naturalists have felt how very slim and unsatisfactory the existing scheme is, and all such will be delighted with this able instalment to what we hope will, ere long, form part of a larger work by the same author.

MR. THOMAS LAURIE, 31 Paternoster Row, has published a letter, entitled "Suggestions for establishing cheap popular and Educational Museums of Scientific and Art Collections and Industrial Productions and Inventions in all towns and villages, and for their arrangement and classification on a new plan," &c.

MESSRS. LE TALL AND A. R. WALLER have edited and issued a new edition of Mr. Henry Ibbotson's "Ferns of York, including also Nidderdale, and the districts around Thirsk, Scarborough, and Whitby." (York : H. Sessions.) The price is only sixpence, but we hope the zealous efforts of the editors will result in realising their wishes that its sale may benefit the original author.

MR. T. ROBERTS, F.G.S., has just described a new species of Conoceras from the Lllanvirn beds, Abereiddy, Pembrokeshire.

FEW fossils have been more debated than the Receptaculites. They have been called fossil pine-cones, foraminifera, corals, cystidians, &c. Dr. Hinde, in a paper just read before the Geological Society, concludes that they constitute a distinct family of siliceous hexactinellid sponges, whose nearest relationships are to Protospongia, Dictyophyton, &c.

WE regret to hear that Mr. Edmund Wheeler, so well known for many years to microscopists, has, through continued ill-health, been obliged to retire from business. No man ever pursued it with more devotion or intelligence. During the many years he was in business, Mr. Wheeler accumulated a large and carefully-selected stock of microscopic material, especially prepared objects. We understand that Messrs. Watsons & Sons, of Holborn, have purchased the whole of this valuable stock, so that it is still available to microscopists.

THE Brazilian diamonds have been traced to a bed of white clay, evidently the result of the felspathic decomposition of the granite mountains, and into this the diamonds seem to have been washed.

WE are pleased to see our best biologists strongly denouncing the folly and cowardice of the "mackerel" scare. Professor Huxley has held it up to scorn and ridicule, and Dr. Spencer Cobbold has shown that the "worm" so much talked about (*Filaria piscium*) is perfectly harmless to man, whether swallowed alive or dead.

MR. LAMEY has been measuring the height of certain mountains on Venus. He found a perfect protuberance in the southern hemisphere, which may be a volcano, but it must be seventy miles high ! He does not think this height incompatible with the volcanic nature of Venus.

IN Italy, oil is now being extracted from the seeds of grapes. Young grapes yield most, and black kinds more than white.

WE have received a specimen of a skeleton of the common frog, from Mr. Edward Wilton, Fairfield, Buxton. It is admirably dissected and mounted, and we are glad to draw the attention of science teachers to the capital preparations now being issued from Mr. Wilton's laboratory.

MR. GURLEY, of the Marietta Observatory, United States, has succeeded in photographing a flash of lightning, during a thunderstorm which occurred five miles away. Professor Wheatstone, many years ago, showed that a flash of lightning could not occupy more than the millionth part of a second. Mr. Gurley took the flash by means of the Bromogelatine process.

PROFESSOR MARSH, the distinguished American palæontologist, has recently named an almost complete reptilian skeleton, of oolitic age, Ceratosaurus. It was seventeen feet long, and had a large horn on its skull (whence its name). The vertebræ are of a peculiar type, and the reptile had a pelvis in which all the bones were ossified as they are in birds.

DR. TRONCIN'S experiments with oxygen in cholera are attracting much attention. His system appears to be efficacious in preventing the deadly chills which are the usual accompaniments of this dread disease. When the patient inhales oxygen his bodily composition and heat are intensified, and the chills may be thus averted.

DR. SPENCER COBBOLD, in a lecture on Parasites, recently given at the Health Exhibition, said that vegetarians who flattered themselves that by abstention from meat they escaped the ills of parasitism, only jumped out of the frying-pan into the fire, for the most common parasite known in England is almost entirely nourished by the cellulose and protoplasm of vegetables; and that it was by means of vegetables and the careless use of unfiltered water, employed in the washing of salads and other herbs, that certain parasites were introduced into the human body.

M. BERTHOLET has been making experiments with nitrates and their origin and transformations. They are partly derived from the soil, and partly from the atmosphere, and are found chiefly in the stems of plants.

WE have received a copy of Dr. Lindorme's paper read before the National Eclective Medical Association at its annual meeting at Topeka, Kansas, on the "Scientific Basis of Eclecticism in Medicine."

MR. WILLIAM MAWER, F.G.S., has published a charming little contribution to Lincolnshire geology, (illustrated), under the title of "How the River Lud cut through Hubbard's Hills."

THE Seventh Annual Report of the Hackney Microscopical and Natural History Society has appeared, giving the address of the President, Dr. M. C. Cooke, abstracts of the papers read last winter, the Report of the Council, &c., from which we are pleased to learn the society is flourishing every way.

DR. M. C. Cooke has given an account, in "Grevillea," of the occurrence of a microscopical fungus (*Tilletia sphærococca*) in the ovaries of a grass (*Agrostis pumila*) from Glen Cluny.

THE Marine Biological Association has now been incorporated under the Act relating to associations not aiming at commercial profits. Its primary object is to establish a thoroughly well-organised laboratory on the English coast, where the study of marine zoology and botany may be carried on by naturalists —as at Naples, at Roscoff, in France, and at Beaufort and other institutions in the United States. These studies will be especially directed to such questions as oyster-breeding, and the spawning, food, and habits of sea fish, so as to provide knowledge which is urgently needed by our various fishing industries. Ten thousand pounds are needed to establish the laboratory. Those who desire to aid scientific men in this really national enterprise should communicate with the secretary of the association, Professor Ray Lankester, of University College, London. Messrs. Clowes and Sons, Limited, the well-known printers, have generously contributed to the preliminary expenses of the association, by undertaking the gratuitous printing of some of its circulars.

MR. A. BALDING notes the fact, that the round-leaved sun-dew catches insects as large as dragon-flies.

AN Exhibition of Forestry is now being held in Edinburgh. The catalogue contains much valuable information.

MR. A. J. DOHERTY requests us to inform our readers that he has withdrawn from his connection with Mr. J. E. Ady, and that none of the slides accompanying the "Popular Studies in Comparative Histology," with the exception of the sections of *Physcia stellaris* and *Rosa canina*, will be prepared by him.

A MOST interesting paper on "Scottish Galls," by the editor, Professor J. W. H. Trail, appears in the last number of the "Scottish Naturalist." We are glad to see it is "to be continued."

MICROSCOPY.

DIFFICULTIES IN MOUNTING.—I should be very much obliged if some readers would help me out of a difficulty I have in mounting objects for the microscope in Glycerine jelly or Deane's medium. My objects require cells, say $\frac{1}{10}$ of an inch deep; they have been well soaked and contain no air. I heat the medium till it just melts, and mount the object without any air bubbles; when cold, I varnish with gold size. Some slides remain perfect; others, perhaps the next day, perhaps a fortnight afterwards, are quite spoiled by air-bubbles or vacua, which spread in a branching manner just under the covering glass. None of the books on mounting which I have consulted allude to this in any way. What mistake do I make? In mounting in balsam I have great difficulty in making the balsam (dissolved in benzole, or Remington's depurated balsam) harden in cells. I have heated the slides for six hours a day for a week over a gas stove, the slides being quite hot to touch. What ought I to do?—*G. S. S.*

MOUNTING INFUSORIA.—Writing in SCIENCE-GOSSIP for July, under the above head, H. M. J. Underhill, Oxford, states that a friend of his, a chemist, said he never heard of a compound called "chromic oxydichloride" acid. Perhaps he would know it better under the name of "Chlorochromic acid," the deep red liquid obtained by distilling a mixture of common salt (or some other chloride), potassic bichromate, and strong sulphuric acid.—*H. L. E., Widnes.*

PETROLOGICAL STUDIES.—Mr. J. D. Ady has issued another part of these Studies, dealing with the Dolorite of Whitwick, the Olivine-Serpentine of Saxony, and the Luxulyanite of Cornwall. It is illustrated by three admirably drawn plates, and the text is turned out with the usual neatness and finish.

COLE'S MICROSCOPICAL STUDIES.—The various parts of the different departments of microscopical research in which Mr. Cole is engaged, appear with marvellous regularity. These various departments are well suited for the numerous classes of students who use the microscope as a tool to work with. Part ii. of "The Methods of Microscopical Research" continues the subject of "Mounting," wherein the learner gets the benefit of the full experience of the most able mounter of the day. No. 10 of the "Popular Microscopical Studies" has appeared, dealing with "Sponge," and accompanied by a plate showing the various types of sponge structure. Nos. 20 and 21 of the original "Studies in Microscopical Science" are devoted, severally, to blood-vessels and cells, and the human cerebellum, so that both departments of biology get equal attention. The coloured plate of the latter is one of the best yet struck off. The slides which accompany these various "Studies," we hardly need say, keep up their high histological and artistic character.

"THE JOURNAL OF MICROSCOPY," &c.—This now well-known periodical, edited by Mr. Alfred Allen, grows both in importance and interest. The last part issued, contains the following papers: "Some new Infusoria from Bristol," by J. G. Grenfell (illustrated); "On the Collection and Preparation of the Diatomaceæ," by A. W. Griffin; "Further Researches on Tubifex," by A. Hammond (illustrated); "The Action of Ammonium Molybdate on the Tissues of Plants," by Dr. T. S. Ralph; "The Microscope in Palæontology," by Dr. M. Poignand (illustrated); "Diamonds and their History," by J. A. Forster; "Hydrozoa and Medusæ" (illustrated), by J. B. Jeafferson; "The Larval Forms of the Crustaceæ," by Edward Lovett; "Examination of the External Air of Washington," by Dr. J. H. Kidda. In addition we have the selected notes from the Postal Society's book, reviews, current notes, &c.

MOUNTING THE ANTHERS OF FLOWERS.—In the Bulletin of the Belgian Microscopical Society, M. Ratabone gives the following method for preparing the cells of anthers for examination. He places the latter in alcohol of $\frac{95}{100}$, for about five minutes, gently rubs them about, and then transfers them to distilled water. The cells then open in a remarkable manner; the pollen-grains are easily detached, and no trouble ensues from air-bubbles. The specimens are then mounted in glycerine.

MOUNTING INSECTS.—A writer in the "American Monthly Microscopical Journal" says, that a mounting needle, bent like a hook at the end, and dipped in alcohol, is the best way for capturing small insects on windows, stones, &c. The insect is always drawn into the drop included in the hook. Dipping the needle into the alcohol frees the insect from the drop and loads the needle again. The editor suggests concentrated carbolic acid as better to dip the needles in, owing to the stiffening effects of alcohol.

THE MILDNESS OF LAST SEASON.—As an instance of the mildness of last season, on the 26th of January I took a piece of sallow in full bloom. The branch contained six catkins, all bearing the pollen, as in April; the other portion of the shrub was very forward. Is not this an exceptional case?—*Wm. P. Ellis, Enfield Chase.*

ZOOLOGY.

THE HORNED APHIS (*Cerataphis lataniæ*).—The author of the article on this aphis, in last November's SCIENCE-GOSSIP, was doubtless unaware when he wrote it that this insect had been previously described and figured in the "Garden" of the 10th of July, 1880, by myself, from specimens which I found in some orchid houses belonging to my uncle, Mr. Joshua Saunders, at Clifton, in 1879. It was described under the name of *Boisduvallia lataniæ*, which was the name it was then known by. Unfortunately, I mistook the winged form for the males, foolishly taking it for granted that they were so, as the insect belonged (as it was then supposed to do) to the Coccidæ, though the neuration of the wings puzzled me. I found a considerable number of winged specimens among the apterous ones which were abundant on various plants in April and May in 1879, and again in May and June in 1880. I cannot agree with Mr. Richter whom Mr. Anderson quotes, as stating that "they never occur on the leaves as do the apterous forms, but only on the stems of the plants, hidden under the leaf-stalk which embraces them," as I have found them on the leaves of various kinds of plants, though generally near the basis, nor are "they very difficult to catch or even to see." I took them easily with the wet point of a camel's hair brush, for they are very sluggish in their movements. I never saw one even attempt to fly and, as soon as I knew what I was looking for, had no difficulty in seeing them. One day I found among some apterous specimens I was examining, one which was evidently a pupa of a winged form, as it had rudimentary wings. I at once carefully searched all the plants infested by this insect and found several more pupæ. The next day I was delighted to find a winged specimen, and subsequently I found a dozen or more. My figure shows the neuration of the lower wings, which Mr. Buckston's does not. I did not know, until the last volume of his monograph was published, that he intended to include this insect, or I would have provided him with a perfect specimen.—*George S. Saunders.*

THE CRYSTAL PALACE INSECTARIUM. — Mr. William Watkins is delighting the frequenters of the Sydenham Palace with a most interesting display of beautiful and curious tropical and indigenous insect life. I visited the exhibition on Saturday last, and, although Mr. Watkins was not there, I found no difficulty in understanding the inhabitants of each cage from the lucidly-written description of each species exhibited. Entomologically speaking there are some fine specimens, notably cocoons of *Castnia eudesmia* from Chili, one actually fourteen inches long, another nearly a foot, from which a fine moth has recently emerged and deposited some eggs which are as large as grains of rice. A very interesting display of the larva, pupa and imagos of *Aponia Cratægi* and their ichneumons, the cocoons of which Mr. Watkins tells his visitors are frequently called by gardeners "caterpillars' eggs." In another cage a number of ant lions which are very lively, puffing the sand from their curiously-constructed traps. In different cages are also to be seen in all their stages, *Attacus Cecropia, Attacus Cynthia, Attacus Pernyi, Saturnia pyri,* and *Actias luna.* Some forcing must have been necessary to show, as Mr. Watkins does, many of these species spinning up at this early period of the year; and to those who know but little of these creatures these cages are highly popular, and as this is evidently the aim of the Insectarium I think it must be pronounced successful,—*George D. Harvey.*

BORING SPONGES.—A recent number of the "Science Record" (a new American scientific journal of great ability and promise) has the following :—In the seas of both Europe and America, are a group of sponges which possess the power of boring into shells and into limestone. On our coast *Cliona sulphurea* causes the rapid disintegration of the dead shells of the molluscs. It excavates tortuous channels in the shells, which soon result in their complete destruction. How these sponges bore is still a mystery. Some have supposed that it was by chemical means, and some by mechanical agencies. Nassonow has recently attacked the problem again, and describes the process. He is, however, uncertain as to the way in which it is accomplished, though inclined to accept both views. He claims that the spicules take no part in the boring, a view which would appear to be well founded, for he found that the young sponge begins its excavations before any spicules are formed.

THE REPORTED SCARCITY OF BUTTERFLIES.—I am afraid the increasing dearth of butterflies results, in a great degree, from causes which it would be a difficult matter for naturalists' societies to interfere with. The raids of collectors do not extend to this side of the Channel, but for some years we have had singularly few butterflies. The red admiral and peacock have become quite scarce, and the common blue is no longer to be seen in fields where I remember it abundant. Two of our more local insects, the brown hairstreak and dingy skipper, will probably become extinct in a few years, in consequence of the recent cutting down of the remnant of Killoughram Forest; a picturesque old wood of oak and birch, the resort of innumerable butterflies, and in whose glades the two last-named species seemed to have established their peculiar home. The purple hairstreak, too, though enjoying a more general distribution than its brown congener, was to be met with in profusion in Killoughram only. I suppose we must now count it as one of our rarer kinds. A very remarkable non-appearance during the present season is that of the marsh fritillary (*Melitæa Artemis*). Of this insect I find it recorded in Mr. Coleman's "British Butterflies," that it has one known locality in Ireland, viz. "Ardrahan Castle, county Galway." A quarter of a mile or so from where I write, there is a little boggy field, where, until the present year, it might have been observed flitting about in quantities among the orchids on any sunny morning of June or in the beginning of July. I first noticed it in 1877, and have since naturally been to the spot many times in each succeeding summer. For no very obvious reason, it had prescribed to itself an area embracing scarcely more than a third part of the field; so that should this piece of ground ever be drained, or, as seems much more probable, totally overgrown with alders, the marsh fritillary would become extinct, perhaps from here to the county Galway! In the meantime, having visited us regularly, sometimes before the end of May, for seven successive seasons, it has this year up to the present date (July 1st) altogether failed to appear. I should like to know whether it has been at all scarcer than usual in its English haunts. It is five years since I have seen a clouded yellow; 1876 was the only year it was really common in Wexford, but it came out sparingly in the three following summers, and then was seen no more. On the other hand, the painted lady is commoner just now than she has been since 1879. Of course, the specimens are hibernated, which makes their sudden reappearance all the more perplexing.—*C. B. Moffat, Ballyhyland, Enniscorthy.*

FREE-SWIMMING ROTIFER.—I cannot accept Mr. Bousfield's statement that I shall find my rotifer described in "Pritchard" under the genus Lepadella, fam. Euchlanidota. In two species of that genus the lorica is depressed; the lorica of my specimen is compressed. The third species is described as having a "lorica oblong, prismatic, obtusely triangular, back crested, denticulated." This description will certainly not fit. Then again the absence of eyes is one characteristic of the genus Lepadella; in every specimen I examined there was a small, but well-defined eye. This character, however, for reasons which I need not here specify, is not considered by some authorities to be of great value; but my specimen had one character to which no such objection can be taken, viz. the lorica being open on the ventral side; this at once shows us that it belongs to the genus in which I placed it. It differs however from every species of Colurus, mentioned in "Pritchard," by its one eye, by its lorica being pointed anteriorly, and by the long, turned-up posterior spines. Since my note appeared, I have received numerous letters from correspondents, referring my rotifer to widely different genera, which goes to prove that the descriptions and figures given of many of the more minute free-swimming forms are very imperfect, and give a student little help towards identifying many of his finds. In another journal, "Microscopical News," I am contributing a series of articles upon many doubtful, and one or two new forms of the free-swimming rotifera. Since my last note was written I have become fully aware of the good things in store for all enthusiastic lovers of these charming animals, and I hope that Dr. Hudson will soon be in a position to make some announcement as to its appearance. The first portion of Mr. Bousfield's suggestion, in the concluding part of his communication, is very good, and I shall be happy to further the object in view, the more so as I have a large number of drawings of the less-frequently recurring forms. The last portion of this suggestion is, I am afraid, impracticable There would be little difficulty in sending specimens of the fixed Rotifers, but the roving fraternity would be difficult to manage, as we seldom get pure gatherings, as a diatomist would say; that is, we generally find numerous species, representing probably several genera, all together in one gathering, and if the particular specimen of interest was only present in limited numbers, there would be a great difficulty in indicating which you meant; and if this was got over, by good drawings and precise descriptions, the receiver might never come across it in the wilderness of weeds. This difficulty being got over, I predict a large field of usefulness to the "Rotifer Exchange Club."—*J. E. Lord, Rawtenstall.*

THE ESSEX FIELD CLUB.—Part 8 of vol. ii. of the Transactions of this flourishing society has just appeared. It contains the conclusion of Mr. R. M. Christy's valuable paper on "The Species of the genus Primula in Essex; with observations on their variation and distribution, and the relative number and fertility in Nature of the two forms of Flowers" (illustrated); a Report of the Committee appointed to investigate the ancient earthwork in Epping Forest, known as "Loughton" or "Cowper's" Camp, (illustrated); "Notes on the London Clay, and Bagshot Beds at Oakhill Quarry, Epping Forest," by the hon. sec. Mr. William Cole. Besides the above, there are abstracts of short papers read, a Journal of Proceedings, &c.

BOTANY.

MIMICRY OF MINT BY DOG'S MERCURY.—We have several pots of common lamb mint in our garden, and a few days ago we made a curious discovery in connection with one of them. One pot had only one genuine plant of mint in it, the rest of the vessel being filled with dog's mercury (*Mercurialis perennis*), in colour, shape, and habit closely resembling mint. To a botanist, of course, the deception would be evident at once, but a casual observer would, I know, find it difficult to detect the difference, unless he knew exactly what to look for. The imitating plant, which had probably been conveyed there in the first instance by birds, had almost crowded the unfortunate mint out of existence, for although in a fairly flourishing state last year, there was now, as I have said, only one plant left, and that in a somewhat sickly state. I do not think that in a wild state one could possibly mistake the two plants, but here, placed close together as they were, and under the same conditions, except that the dog's mercury was in more luxuriant quarters than it usually inhabits, the resemblance was, at a glance, almost perfect.—*P. S. Taylor.*

SKELETON LEAVES.—These may easily be made, in the case of most leaves at any rate, by macerating for a few days in rain-water. The leaves should be placed in a plate or dish, and the water changed daily, or every two days. In a short time, varying according to the leaf operated on, the outer tissue will become soft; when it should be carefully removed with a camel's hair brush. It will occasionally be necessary to use the fingers as well as the brush, but this should be avoided if possible. The skeleton may be bleached by several methods; but for scientific purposes they are perhaps better in their natural state.—*H. Snowden Ward.*

SKELETON LEAVES.—A simple method of obtaining skeleton leaves is to make a solution of chloride of lime (one oz. chloride of lime to a quart of water), and then to steep the leaves in it for from four to six or seven hours, according to their size. On taking them out, the skin will be found to peel off easily, by the aid of a small camel-hair brush.—*W. G. H. Taylor.*

SENECIO VULGARIS, VAR. RADIATUS.—It may interest your botanical readers to know that I gathered two plants of *Senecio vulgaris*, variety *radiatus*, on the sand-hills between Wallasey and Leasowe, in Wirral, Cheshire. I gathered them on the 23rd of May, and I went again about a month later, and found several plants which were just recognisable as the same variety, although they were almost over and dying down.—*S. Slater.*

SAGITTARIA SAGITTIFOLIA.—Last autumn, I planted two strong specimens of this plant in a pot, and sunk it in a shallow tank, wishing to have it under my observation. Early this spring, intending to examine its roots, I turned the pot over, and, to my surprise, at first sight it seemed to have entirely disappeared. After a careful search in the mud, I found two small tubers about the size and shape of snowdrop bulbs, hard and very compact, and covered with a thin, duck-green skin. Is this the manner in which this plant usually hybernates? and is this fact generally known?—*B. P.*

A "SPORT" IN HIBISCUS.—My attention was called by my sisters one morning lately, to a strange freak on the part of our buff-coloured, double-flowered hibiscus. On a branch on one of the principal stems and surrounded by double flowers of the normal buff colour, was a solitary red single flower. There was no mistake about it, and no cause such as grafting or budding to account for it. The single flower was perfect, and as bright and pure in colour as if it had grown on a red hibiscus shrub. I have tied a thread round the stem which bore it, to see if any more single or abnormal coloured flowers will appear on it.—*W. J. S., Calcutta.*

GEOLOGY, &C.

THE GEOLOGY AND MINERALOGY OF MADAGASCAR.—Dr. G. W. Parker has just contributed a paper on this subject to the Geological Society. A central plateau from 4000 to 5000 feet high occupies about half the island, rising above the lowlands that skirt the coasts, and from this plateau rise in turn a number of volcanic cones, the highest, Ankaratra, being 8950 feet above the sea. The known volcanic cones extend from the northern extremity of the island to the twentieth parallel of south latitude. Beyond this granite and other primitive rocks occur as far as lat. 22°, south of which the central parts of Madagascar are practically unknown to Europeans. Only a single trap-dyke is known near Antanarivo. The hills around this city are of varieties of granite. The general direction of the strata is parallel to the long axis of the island. Marine fossils have been found, by Rev. J. Richardson and Mons. Grandidier, in the south-west part of the central plateau. These fossils are referred by the last-named traveller to the Jurassic system. Remains of hippopotami, large tortoises, and an extinct ostrich-like bird have also been recorded. North and north-west of the fossiliferous rocks, between them and the volcanic district of Ankaratra, sandstone and slate occur. North of this volcanic district again is a tract of country in which silver lead (mixed with zinc) and copper are found. Near the north-western edge of the central plateau are granitic escarpments facing northwards and about 500 feet high. Some details were also given of valleys through the central plateau and of lagoons within the coral reefs on the coasts.

THE OOLITIC ROCKS UNDER LONDON.—Professor Judd has recently made an additional communication on this interesting subject to the Geological Society. The well-boring at Richmond has now been carried down to a depth of more than 1360 feet, 220 feet deeper than has been reached by any other boring in the London Basin. A temporary cessation of the work has permitted Mr. Collett Homersham to make a more exact determination of the underground temperature at Richmond. At a depth of 1337 feet from the surface, this was found to be $75\frac{1}{2}°$ F., corresponding to a rise of temperature of 1° F. for every 52·43 feet of descent. The boring is still being carried on in the same red sandstones and "marls," exhibiting much false-bedding. The Rev. H. H. Winwood, of Bath, has found the original fossils obtained by the late Mr. C. Moore from the oolitic limestone in the boring at Meux's Brewery in 1878. A careful study of these proves that though less numerous and in a far less perfect state of preservation than the fossils from the Richmond well, they in many cases belong to the same species, and demonstrate the Great Oolite age of the strata in which they occurred. Dr. Hinde has described five new species of Calcispongia from the materials brought up by the boring. One species is closely allied to *Blastinia costata* from the lower Jurassic strata at Streitburg. Professor Rupert Jones also described the Foraminifera and Ostracoda, and Mr. G. R. Vine the polyzoa found at the Richmond well-boring. The polyzoa included fourteen different forms, most of which are characteristic of the oolite, and some of which are new.

THE UNDERGROUND GEOLOGY OF NORTHAMPTON. —At a recent meeting of the Geological Society, Mr. H. J. Eunson, F.G.S., read a very important paper on the "Range of the Palæozoic Rocks beneath Northampton." In two borings made near the town by the local water company, after passing through 738 feet of the upper, middle, and lower lias, a series of conglomerate sandstone and marls were found resting on an eroded surface of carboniferous dolomite, passing into the usual fossil-crowded limestone. Forty-six feet of carboniferous strata were drilled, and the boring was discontinued at 851 feet. A second boring at a place called Gayton, after passing through various strata, came upon an eroded surface of carboniferous limestone at a depth of 699 feet. In this fossils were found down to a depth of 889 feet, the boring being continued to a total depth of 944 feet. At Orton, near Kettering, a boring was made through white lias, rhætic, sandstone, and breccia, into quartz-felsite in a futile search for coal. As none was found down to a depth of 789 feet, the boring was discontinued.

THE GEOLOGISTS' ASSOCIATION. — No. 6 of vol. viii. of the "Proceedings" has just been published. It contains the following papers (besides notices of meetings): "Fossil Plants," by J. S. Gardner, F.G.S.; "Fossil Plants from various Formations," by W. Fawcett, F.L.S.; "Notes on the Krakatoa Eruption," by Grenville A. J. Cole, F.G.S.; "The Implementiferous Gravels of North-east London," by J. E. Greenhill; and another on the "Implementiferous Gravels near London," by Professor T. Rupert Jones, F.R.S., &c.

FASCIATED STEM IN WHITE BROOM.—There is now growing in my garden a plant of white broom, one branch of which is fasciated till it is three-quarters of an inch wide.—*K. D., Cofton.*

NOTES AND QUERIES.

PIED LAPWING.—Your correspondent, G. Bristow, asks whether any cases of pied lapwings have been noted other than the one he mentions. I am able to give the following four instances of this strange occurrence. (1) In the "Zoologist," October, 1881, a Dublin correspondent inserted a note mentioning a white-heron swallow and lapwing which had been shot. (2) The editor of the "Zoologist" (J. E. Harting) appended the following note to the above : "If these varieties had been secured alive and kept in confinement they would in all probability have assumed their colours on moulting. Such at least has been the case with a cream-coloured lapwing which has been for some time in the Western Aviary at the Zoological Gardens, Regent's Park, and we have known the same thing to occur in the case of a pied blackbird." (3) On October 20, 1881, I examined a lapwing in the shop of the late Mr. Young, bird-stuffer, York, which was so curiously pied and streaked with white that it looked almost as if it had been dipped into whitewash. (4) In the "Natural History Journal" (Feb. 1883) a lapwing is mentioned having the head cream-coloured and a light band across the breast, this was exhibited for sale in a poulterer's shop in York on Jan. 29, 1883. I am not aware of any other instances having occurred, although I do not doubt that many such have passed unrecorded in the pages of any scientific journal. I shall be obliged to Mr. Bristow if he will kindly let me know the locality and the date on which his specimen was procured. As I am collecting all the instances of pied and albino birds, &c. which I can hear of, I shall be extremely obliged to any one who may be able to send me notes on this very interesting subject. (See SCIENCE-GOSSIP, vol. xix. p. 117).—*Edward J. Gibbins, Neath, Glamorganshire.*

WILD FLOWERS IN BLOOM, JANUARY 1884, IN WEST NORFOLK.—It seems to my limited knowledge of botany a list of plants of this description cannot fail to have some interest. On January 1st, in a short walk from Castleacre, I gathered *Ranunculus bulbosus, Caltha palustris, Capsella bursa-pastoris, Sisymbrium officinale, Viola odorata, V. tricolor, Arenaria,* sp.? *Veronica agrestis, Euphorbia Helioscopia, Lamium album, L. purpureum, Salvia verbenaca, Urtica urens, Bellis perennis, Senecio Jacobæa* and *vulgaris, Chærophyllum sylvestris, Leontodon taraxacum, Picris virens? Poa annua, Dactylis glomerata.* Until the 7th I had little or no time to work among plants, but on that date I found *Carduus lanceolatus, Ulex Europæus,* and *Hieracium pilosella.* Next day, in another direction from that before taken, I gathered *Brassica oleracea, Ranunculus reptans, Corylus avellana,* and *Primula vulgaris,* and on the 11th added *Vinca minor* and *Veronica chamædrys* to my list. Till the 15th I found nothing new, but on that day I noticed the alder in full flower as well as chickweed. Next day I found *Cheiranthus cheiri* and *Daphne laureola,* and on the 18th added my best find, *Helleborus fœtidus.*—*J. Harvey Bloom, Westbury House School, Worthing.*

PARIS QUADRIFOLIA.—Last summer I found a large patch of the above in a wood in this neighbourhood ; many of the plants had five leaves below the flower instead of four, and one had six. Is not this last unusual ?—*K. D., Cofton.*

L. PEREGER, VAR. PICTA.—Perhaps some readers can inform me whether *L. pereger* var. *picta* is common throughout the British Isles or otherwise, as Mr. Rimmer in his "Land and Fresh Water Shells" only mentions one locality, viz. Ulva Island in the Hebrides. I have found it on the east coast, but not common. I have also submitted specimens to parties well able to be sure of the variety.—*Wm. Duncan.*

CORONULA DIADEMA.—I have noticed in a late specimen of *Coronula diadema* (from the Tay whale) that the enlargement of the shell appears to me to have been by lines of growth at both the apex and base. I shall be pleased to learn whether my surmise is correct or not.—*Wm. Duncan.*

CLIMBING MICE.—Those in Texas have many opportunities of noticing the "climbing power" of mice. The mouse which comes into houses here is very like the English dormouse. It has large ears and very large eyes, and is a pretty creature, but very mischievous. The houses here are lined with a thin calico called "domestic," and the mice run up and down the walls as nimbly as they run over the floor. On one corner of the ceiling of my bedroom, I noticed that the domestic had a hole in it. Looking through it I could see creatures moving at the other side, and soon found they were large red wasps, which had suspended a nest from the rafter just over the hole. Said hole was not their work, it was done by the mice, which made nightly raids upon the nest, eating the under part where the young wasps were stored away in their sealed up cells. I thought the matter over with "a house-that-Jack-built" feeling. The wasps have built a house, and they are a nuisance. The mice come to eat the wasps, but they are a nuisance. The cat comes to the mice, but the cat is a nuisance, because there are fleas on the cat, and fleas are a nuisance. How am I to get rid of all these nuisances ? No other way than to take the nest. So when the wasps were asleep, I made the attack. They fell into a tin of water, and their beautiful paper house came to the ground. But I ought to be talking of mice. The climbing propensity makes these little depredators doubly dangerous, for nothing can be protected from them. They get into any box not lined with metal, gnaw the clothes or papers into little pills, in the centre of which they repose until disturbed. A mouse about to begin life ran up the tail of my room, ate the shoulder out of a new cloth overcoat hanging up in a recess, and with the fluffy wool made a snug little nest in a box below, where he and his nest met the doom they deserved. As the mice run up the outside walls and posts of the house as easily as the inside, it is not the domestic that assists them. I fancy English house mice are climbers too.—*Mrs. C. Brent, Kerrville, Kerr County, U.S. America.*

THE DORMOUSE IN ENGLAND.—When a small boy, I frequently had dormice given me. These were obtained from Freshford Wood, near here, and close to a village of that name ; and also from a wood at Eastcott, near Devizes. Sometimes I have had a mother and two or three young given me. Though every care was taken, they seldom lived long in confinement. I am informed that dormice are still frequently met with, not only in the above-named woods, but also in various places near this city.—*Charles F. W. T. Williams, Bath.*

HELIX PISANA.—I can fully corroborate Mr. McKean s statement as to the vitality of the above, after a prolonged period of hibernation. On September 5th of last year, I received from a friend in South Wales, a number of *H. pisana* taken at Tenby. I placed a few of them in a vivarium, together with a number of other species of Helix.

They soon sealed themselves in with the usual epiphragm, and remained in that condition all the winter, in spite of various attempts to induce them to come out by offers of fresh food. Examining them a few months ago I found that, with the exception of one specimen, all had perished. On reading Mr. McKean's note, I immersed the survivor in tepid water, and, after an interval of a few minutes, I had the satisfaction of seeing it gradually emerge from its retirement. It soon became quite active, and appeared to be quite as healthy as when first received, in spite of its having sustained a prolonged fast of just ten months.—*A. Jenkins, New Cross.*

NOTES ON BIRDS.—There is nothing unusual in the "strange habit of a bullfinch" mentioned by H. B. R. in a recent SCIENCE-GOSSIP. I have had both cock and hen bullfinches in my aviary cage, and it seems to be one of their peculiarities to collect a quantity of seed in the mandible and retire to some quiet place to crack and eat it. The bullfinch I have now has the same habit with all small seeds, especially with maw seed, of which he is particularly fond. It may interest some of your readers to learn that last spring (1883) I had a robin (cock) which I allowed to have liberty in the room, and sometimes of the whole house, and which quickly became tame enough to feed from the hand. One afternoon, one of the bedroom windows having been left open, he flew away, but a couple of hours afterwards we were surprised to see him sitting on the yard wall looking at the birds inside, and making that peculiar "cricking" call with which we are so familiar. On opening the door and calling "bobbie" he flew in, and on to the hand to be fed. After this experience, we allowed him to fly in and out of the house at his pleasure, and this he did a dozen times a day during several of the summer months, always coming to us when we called him in the garden; and some of my friends have been much surprised to see him fly down to us from a tree or the housetop when called. Once or twice he stayed away all night, and came home in the morning. On one occasion, after being away one day and two nights he returned with his head, breast, and legs covered with pitch, though how he got it I do not know. We cleansed as much of it off as we could, but he pulled nearly all the feathers out of his breast, and this and the pitch he must have swallowed, made him so ill that he died a few days afterwards. Since then I have been robinless. I may add that when confined to the aviary he did not molest the other birds (canaries, linnets, goldfinches, and bullfinch) in any way, as I have heard robins in confinement will do.—*Mark L. Sykes, Pendleton.*

TAMING WILD HUMMING-BIRDS.—A lady residing at San Rafael, one of the many pleasant health resorts of California, has sent to friends in London an account of the taming of two free wild hummingbirds by her daughter, who, under medical direction, has for some months passed several hours daily reclining on rugs spread on the garden lawn. "E. has a new source of interest," her mother writes. "The humming-birds have claimed her companionship and manifested their curiosity by inspecting her with their wise little heads turned to one side at a safe distance, watching her movements, evidently wishing to become acquainted. To entice them to a nearer approach E. plucked a fuchsia, attached it to a branch of a tree over her head, and filled it with sweetened water. The intelligent little creatures soon had their slender bills thrust into the flower, from which they took long draughts. Then E. took honey, thinking they might prefer it, and filled a fresh flower each day. They would sometimes become so impatient as scarcely to wait for her to leave before they were into the sweets, and, finally, while she held a flower in one hand and filled it with drops from a spoon, the now tame little pets would catch the drops as they fell, and dart into the honey cup their silvery threadlike tongues. E. is delighted, and so fascinated with them that she passes hours each day of her resting-time talking to them and watching their quick lively movements. Although these tiny birds are humming all day among the flowers, two only have monopolised the honey-filled flower, and these are both males, consequently there are constant squabbles as to which shall take possession. They will not permit a wasp or a bee to come near their honey flower, and not only drive them away, but chase them some distance, uttering a shrill note or protest against all intruders." Referring to them again, at the close of the rainless Californian summer, in a letter dated October 26, this lady writes : "We have had threatening clouds for two days, and a heavy rainfall to-day. E. has continued her devotion to her little humming-birds. Since the change of weather she has tried to coax them to the parlour windows. They appeared to think there must be some mistake, and would hum about the window where she stood with the honey flower and spoonful of honey, or they would sit on a branch and watch every movement, yet not daring to take a sip until to-day, when at her peculiar call, which they always recognise, one ventured repeatedly to take the honey from her hand."—*Times.*

SWALLOWS IN CHURCH.—On Sunday, July 6th, a somewhat unusual lesson was given in Rhylstone church during the morning service. Soon after the commencement of the sermon, two old swallows and one young one flew screaming through the open porch into the body of the church. Perching together on a rafter, the birds paused a few seconds, and then the old ones commenced to give their offspring a flying lesson, in which they were only partially successful, owing perhaps to the unusual surroundings, and the preacher's voice.—*H. Snowden Ward.*

NOTICES TO CORRESPONDENTS.

TO CORRESPONDENTS AND EXCHANGERS.—As we now publish SCIENCE-GOSSIP earlier than heretofore, we cannot possibly insert in the following number any communications which reach us later than the 8th of the previous month.
TO ANONYMOUS QUERISTS.—We receive so many queries which do not bear the writers' names that we are forced to adhere to our rule of not noticing them.
TO DEALERS AND OTHERS.—We are always glad to treat dealers in natural history objects on the same fair and general ground as amateurs, in so far as the "exchanges" offered are fair exchanges. But it is evident that, when their offers are simply disguised advertisements, for the purpose of evading the cost of advertising, an advantage is taken of our *gratuitous* insertion of "exchanges" which cannot be tolerated.
WE request that all exchanges may be signed with name (or initials) and full address at the end.

H. ABBOTT.—Your zoophytes are: No. 1, *Sertularia rosea*; No. 2, *Sertularia operculata.* See Taylor's "Half-hours at the Seaside," page 97.
D. BRADLEY.—Many thanks for your interesting monstrosity of *Campanula medium.* It is due, as you say, to petalody, but it is the most singular form we have seen.
F. H. A.—Your plants are *Crepis biennis* and *Lepidium draba.*
J. BOGGS, jun.—The specimens are those of the hair-worm (*Gordius aquaticus*).
J. STEWART.—We have no doubt that the calcareous plates you found in the stomach of the catfish, are those of a Holothurian, but it is impossible to identify them from your sketches.
C. MORGAN.—The specimens go by the name of "oak spangles." They are galls, produced by a species of Cynips. See Taylor's "Half-hours in the Green Lanes."

W. MILLS.—You will find Pascoe's "Zoological Classification" a capital handy book of reference. It gives tables of the sub-kingdoms, classes, orders, &c., of the entire animal kingdom, together with their characters, and lists of the families and principal genera. (London: Van Voorst; price 7s. 6d.)

MR. DIXON.—It is impossible to tell the insects from the sketches sent. They appear to be some species of Podura. Can you not send a specimen?

T. JONES.—Get Norman Lockyer's "Lessons in Astronomy," price 5s. 6d. (London: Macmillan), or Plummer's "Elementary Astronomy," price 1s. (Glasgow: W. Collins & Sons).

A. O. (Winsdar).—The grass you send is not the Avena as you suspect, but the *Arrhenatherum avenaceum*. The work you name is not very reliable.

J. H. C. R. (Switzerland).—Thanks for specimens of *Ophrys arachnitis*. You are quite right, it takes two or more years for the bulb to flower.

W. G. R. (Liverpool).—It is one of the Planes, or, *Platanus acerifolia*, a tree generally grown in shrubberies and plantations for *P. occidentalis*.

J. C. S. (Penrith).—No. 1 is the *Orchis incarnata*, L. No. 2 *O. maculata*, or No. 1 may prove to be the *O. divaricata*, Reich., a form closely related to *O. maculata*; it is however difficult to tell in dried specimens.

S. J. H.—Apparently they are all blights. No. 2 is Hazelblight (*Phyllactinia guttata*). If you procure "Microscopic Fungi," by Dr. Cooke, it will make your researches delightful.

N. F. D. (Romford).—Your paper on bifurcation will have our attention shortly.

J. S. (Bolton).—We are sorry to say your specimens are too small and imperfect to name accurately. No. 4 is *Bryum argenteum*.

A. M. P. (Newport)—The small insects which issued from your chrysalid are Ichneumon flies. Their mother pierced the body of the caterpillar, and deposited her eggs there; so that when the caterpillar passed into the chrysalis state it had the young parasites in its body. There they hatched and developed.

EXCHANGES.

SPLENDIDLY-preserved and correctly-named Swiss Alpine plants. Price 6d. each.—Address, Dr. B., care of Editor of SCIENCE-GOSSIP, 214 Piccadilly.

WANTED, March number of SCIENCE-GOSSIP; six well-mounted and interesting micro-slides offered in exchange.—R. L. Hawkins, 24 Baker Street, W.

Well-mounted slide of diatoms *Synedra capitata* and *Cocconema cymbiforme* in exchange for other objects of interest.—T. B. Forty, Market Square, Buckingham.

SEVERAL specimens of *Ceratodus polymorphus, Rhætic, Aust*; also large collection of Rhætic fossils from Aust. What offers? H. B. Capell, Great Easton, Dunmow, Essex.

WANTED, to exchange splendid double-barrel rifle, muzzle loader, of Charles Lancaster's make, very finely engraved and little used, cost £80 and has case and all fittings, for a good binocular microscope by good maker; free particulars to be given, references exchanged.—Mark L. Syker, Pendleton, Manchester.

H. lapicida, H. rupestris, C. dubia or *C. laminata*, a number of any of the above for single specimens of *H. pomatia, H. pisana, H. Carthusiana* or *P. secale*.—W. Webster, National School, Lofthouse, Wakefield.

FOR exchange, eggs of common guillemot, herring-gull, mallard, blackhead gull, &c.—John Murray, 10 St. Paul's Street, Aberdeen.

OFFERED, nests and clutches of dipper, clutches of sparrowhawk, kestrel, hooded crow, magpie, swift, ringed plover, coot, cormorant, herring-gull, kittiwake, 11 cl. mute swan eggs of gannet, puffin, varieties of razor-bill, and many others, one holed with data. Send list of duplicate clutches and desiderata. R. J. Ussher, Cappah, Lismore, Ireland.

FOR slide of diatomaceæ from Mackintosh Lake, Canada, send any good slide of tropical gatherings to Alfred W. Griffin, Saville Row, Bath.

WILL send one dozen of fine young plants of Vallisneria for half-a-dozen sections well-mounted for the microscope.—John Simm, West Cramlington, Northumberland.

A FEW well-mounted miscellaneous slides in exchange for other well-mounted slides of interest.—H. Abbott, 40 St. Peter's Street, Lowestoft.

FRESH plants of sundew (*Drosera rotundifolia*) sent for good slides or material.—W. Sim, Gourdas, Fyvie, N.B.

WANTED, specimens of stag beetle and male cricket. Will give micro-slices or cash in exchange.—J. Moore, 86 Porchester Street, Birmingham.

NEW slide for polariscope, transparent crystals of sulphur. Other micro-slides for sale or exchange. Wanted botanical works or fret saw.—A. Wire, Kreochyle Co., Leytonstone, E.

WANTED, unmounted pathological or anatomical material, slides, Northern Microscopist, 1881-3, or turn-table, in exchange for slides mounted of the material, or "Science Monthly," vol. i. new, in parts, "Boys' Own Paper," vols. v. (and vi. not yet finished), monthly parts (except vol. i.), Cassell's "Sea," vol. i., perfect condition, clean.—V. A. Latham, F.M.S., 15 Thorncliff Grove, Oxford Road, Manchester.

OFFERED, first-class slides of rare diatoms (selected), also pieces of *Hyalonema Sieboldii* and *Synapta Bessellii* in exchange for unmounted diatomaceous gatherings.—Dr. Otto N. Witt. E. 8. 13. Mannheim, Germany.

SIX plants of *Drosera rotundifolia* free by post, offered in exchange for well-mounted micro-slide.—Thos. Richardson, Rose Cottage, Heath End, Farnham, Surrey.

FINE *L. stagnalis* and *P. corneus, P. spirorbis, P. vivipara*, also number of foreign postage stamps. Desiderata, British or foreign shells.—W. Hewett, 26 Clarence Street, York.

PATHOLOGICAL material wanted: also Stirling's "Practical Histology," and Woodhead's "Pathology." First-class slides in exchange—A. J. Doherty, 33 Burlington Street, Oxford Road, Manchester.

OFFERED L. C., 7th edit., Nos. 45, 121, 133, 147, 161, 197, 275, 280, 326, 398, 406, 859, 914, 924, 1008, 1349, 1361, 1379, 1422, 1447, 1448, 1458, 1494, 1501, 1504, 1515, 1577, send lists.—H. J. Wilkinson, 17, Ogleforth, York.

A couple of young long-tailed field mice (*Mus sylvaticus*) in exchange for a pair of harvest mice, or books on natural history.—F. Hayward Parrott, Walton House, Aylesbury.

EXCHANGE Transactions of the Hudd. Nat. Soc., containing complete catalogue of the Lepidoptera of the district, for similar publications of other societies.—S. L. Mosley, Hon. Sec., Beaumont Park Museum, Huddersfield.

TO COLLECTORS.—I will give the first four half-crown parts of my "Varieties of British Lepidoptera" to the person who, before Dec. 1st, shall send me the largest number of different species of insects of all orders (except Lepidoptera) especially Diptera. Usual specimens not objected to, but date and locality indispensable.—S. L. Mosley, Beaumont Park Museum, Huddersfield.

WANTED, fresh specimens of stag beetle and male cricket for dissecting purposes.—Wilton's Zoological Station, Fairfield, Buxton.

WANTED, all classes of Reptilia, British and foreign, alive and in spirits, in large quantities: foreign correspondents especially requested. Also alive, about 500 bull frogs of America, and about 5000 common frogs of England. Also the following, alive in large quantities: hedgehogs, moles, common snakes, tree frogs, natterjack toads, tortoises, *Helix pomatia*, blind worms, and lizards.—E, Wade Wilson, Fairfield, Buxton.

WANTED marine animals for dissection, specially octopus and sepia, also any other type specimens.—Zoological Station, Fairfield, Buxton.

EGGS of snipe, grebe, plover, black-headed gull, teal, coot, and waterhen (all Yorkshire specimens); also the following shells: *Limnea stagnalis* and *Planorbis corneus*, very fine; also *Limnea glabra, L. peregra*, var. *fragilis* of *L. stagnalis, P. hypnorum, P. polymorpha*, and *P. vivipara*; desiderata very numerous eggs and shells.—W. Hewitt, 26 Clarence Street, York.

SHELLS.—*Helix pisana*, several varieties. *H. aspersa*, var. *minor* and *conoidea*. Crustaceans: *Porcellana platycheles, P. longicornis, Portumnus latipes* (rare): *Pagurus Bernhardus, Corystes Cassivelaunus*, &c. What offers in exchange?—C. Jefferys, 15, Warren Street, Tenby.

BOOKS, ETC., RECEIVED.

"The Life of the Fields," by Richard Jeffrey. London: Chatto & Windus. "Practical Taxidermy," by Montagu Brown, 2nd ed. London: Gill. "The Blowpipe in Chemistry, Mineralogy and Geology," by Colonel Ross, F.G.S. London: Crosby Lockwood & Co. Annual Report Manchester Microscopical Society, 1883-84.—"Studies in Microscopical Science," edited by A. C. Cole.—"The Methods of Microscopical Research," edited by A. C. Cole.—"Popular Microscopical Studies." By A. C. Cole.—"The Gentleman's Magazine." —"Belgravia."—"The Journal of Microscopy."—"The Science Monthly."—"Midland Naturalist."—"Ben Brierley's Journal."—"Science Record."—"Science."—"American Naturalist."—"Naturalist's Leisure Hour"—The Electrician and Electrical Engineer."—"Canadian Entomologist."—"American Monthly Microscopical Journal."—"Popular Science News."—"The Botanical Gazette."—"Revue de Botanique."—"La Feuille des Jeunes Naturalistes."—"Le Monde de la Science."—"Cosmos, les Mondes," &c. &c. &c.

COMMUNICATIONS RECEIVED UP TO 11TH ULT. FROM:—
W. J. S.—Dr. P. Q. K.—G. D. H.—W. H. H.—T. A. W.—P. S. T.—M. L. H.—J. G.—C. M.—T. J.—L. E. A.—W. W.—C. D. B.—Dr. O. A. W.—M. L. S.—B. P.—C. F.—W. T. W.—W. G. H, T.—J. P. G.—R. F. Z.—Mrs. G. F. H.—W. E. B.—W. R.—D. B.—Dr. S. A.—H. B. C.—T. B. F.—J. E. L.—H. L. E.—C. B. M.—M. L. S.—S. J. H. P.—F. A. J.—W. W.—R. J. U.—J. M.—A. W. G.—H. A.—J. S.—H. F.—V. A. L.—G. W. S.—A. P. W.—J. M.—T. D. C.—S. W.—A. J. D.—A. O.—S. L. M.—H. S. W.—S. S.—W. H.—H. J. W.—C. O.—S. T. D.—F. H. P.—F. L. E.—T. W.—H. G. G.—G. H. K.—L. M.—A. D.—Dr. J. H.—W. A. P.—Dr. C. P. C.—J. C.—R. A. F.—T. R.—W. G. R.—A. J. D.—F. B.—W. H. H.—H. B., &c.

GRAPHIC MICROSCOPY

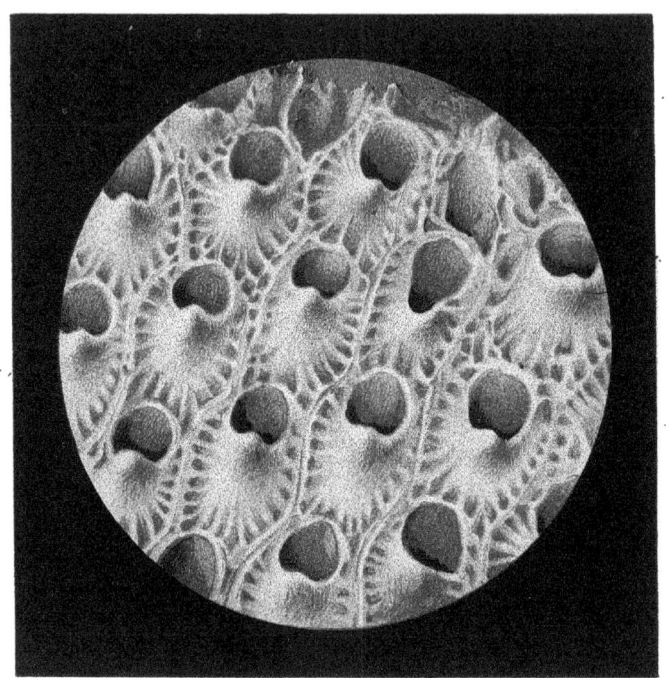

POLYPIDOM OF LEPRATIA NITIDA.
× 50.

GRAPHIC MICROSCOPY.

By E. T. D.

No. IX.—Polypidom of Lepralia nitida.

THE holiday month, September, affords the microscopist the long anticipated pleasure of reaching the sea—to luxuriate in all its associations, the terraces and plateaux of rocks, dredging, and its mysterious lottery of attractions, and the enchanting solitude. For fine broken ground, certainly for successful explorations, attractive to the naturalist, and especially the geologist, perhaps no locality within easy distance can equal the Dorset coast, in that part stretching from Studland and Swanage Bays towards St. Alban's Head. With Mr. Gosse's "Tenby," "Devonshire Coast," and invaluable manuals, "Marine Zoology," as companions, much accurate knowledge may be secured. At low water, in pellucid pools, fringed with delicate algæ of many hues, and replete with quaint and curious forms, are found specimens of the polyzoa, polypes inhabiting cells, creatures allied to the mollusca, but living under many strange and diverse forms and conditions. In the class under consideration they are delicately spread over, or commensal, on sea-weeds, shells, or submerged substances, forming crusts, or attached as minute plant-like tufts to rocks. Many of these forms, so frondose is their appearance, appear, to a casual observer, as minute plants. Upon closer examination they are found to be horny, or calcareous congeries of cells; polypidoms, domiciles, in each a polype, in a berth of its own, not only in association and in intimate contact with countless neighbours, but individually aiding in, and contributing to, the general construction and exquisite architecture of the settlement.

The solitary polype (Hydra), and its more beautiful fresh-water allies, Lophopus, Plumatella, and Cristatella—the latter a community of individuals, with power of locomotion, exhibiting a colony incessantly seeking fresh pastures—are well known to every possessor of a microscope. These forms are merely imbedded in a gelatinous substratum; they do not found a permanent establishment in well-constructed and castellated homes, but in the marine forms, with some exceptions, the polype is discovered in a separate lodgment of a substantial character, formed by its capability of secreting carbonates of lime and other substances, packing itself into a cavity, and by united efforts forming an aggregated building, a general association, a caucus. And these residences—polyzoaries or polypidoms ("cells for retreat in danger")—are varied in substance, form, and character. Space does not admit the enumeration of all the varieties; but those most generally found, and typical, are horny and flexible (Flustra), arborescent and plant-like (Sertularia), stony and calcareous (Madrepores and Corallines), filamentous and tubular (Anguinaria), and greater varieties forming crusts on various substances, resulting in a pavement of cells.

The subject of the plate, the dried polypidom of *Lepralia nitida* (?) (or *verrucosa*), of the family Celleporidæ, order Ascidioda (Johnston), is a congery of such cells. From the mode of increase, there is no limit to their number or security of preservation. The polypes die, but the solid walls remain in accurate connection with each other. New individuals increase at the margin. It appears that, when an original or seminal cell is set up and completed, another begins to form at its side. This process can be seen in the Flustræ (Sea-mats), a common form, where round the edge of the crust cells may be observed in every stage; some beginning, others half formed, and many nearly completed. Although the polypidoms of the order Ascidioda assume many diversities of form, the habit and structure of the polypes have a general uniformity. When in repose, the creature lies doubled up in its cradle, and, expanded, exhibits a crown of ciliated tentacles; in the centre a mouth. When extended, and the cilia in full play, the vortices produced in the water attract and carry into the mouth the minute particles of food. This lovely microscopical sight is well displayed in the fresh-water species; but the marine forms are no less interesting. The little flosculous heads peep out from the opening of the cells, cautiously emerge, and suddenly expand in regular order, like living blooms and flowers of inconceivable delicacy.

The individual polype, and its domicile, are, in almost every instance, microscopic; but the united and accumulative efforts of these minute creatures attain gigantic results, as seen in the madrepores, and arborescent species, notably the reef-forming corals, capable of producing even geological changes. Although not quite germane to the present subject, reference may be made to another species, the red coral of commerce. Here the polypary, or foundation of the colony, is branched like a tree, centralised into a solid axis, throughout covered with a living crust or "cortex;" studding the surface of this fleshy substance are the innumerable polype centres, appearing as minute rayed stars; the stem is hard, and, as well known, especially to ladies, an exquisite ornament, as it takes a fine polish. In the Gorgonia, or Sea-fans, the axis of the branches is formed of a substance not clearly understood; it appears like a concreted albumen, flexible, horny. A beautiful microscopic object is a transverse section through one of the thicker stems. Its solidity is seen to be made up of layers of calcareous deposits, imbedded in some connective tissue, undulating round a centre; the dried crust, or desiccated remains of the living envelope, furnishes the well-known "Spicules of Gorgonia."

These homes of the Ascidioda, when dried and empty (the "deserted village" of departed polypes), beautiful as they are, and exciting wonder that such structural results could be potential in a creature so simple and elementary, afford no idea whatever of the sight presented under the microscope, with dark spot illumination, of any of the class in a living condition, protruding and unfolding a coronal of supremely beautiful tentacles, which when disturbed are instantly gathered into a parallel band, and sucked into the recesses of their cradles; but when dead, and the tenant disappeared, a domicile of such rare beauty is worth a place in the cabinet, especially as the configuration of the cells is an element in classification. Any of these forms, after soaking in fresh water (to remove saline incrustations) and carefully dried, may be easily "mounted" as opaque objects, and good reflected light reveals all the salient points, as seen in the picture.

The pith and terseness of purely scientific words are well exemplified in the description of species Dr. Johnston gives in "British Zoophytes." Our younger readers should cultivate the art of word-painting. A fair lesson might be to describe a glass tumbler in precise language. Many works of purely technical character abound in fine and graphic description: often the accuracy of phraseology demanded by science touches the poetical, even the humorous: The works of the late Edward Forbes, the most genial of naturalists, reveal many instances of curiously quaint and elegant description.

The Polyzoa may be secured as permanent objects, with the polypes extruded, *in situ*. Such marine forms as Sertularia or the Corynidæ may be arranged in a cell, in their native fluid, under a covering glass; when fully expanded, and in vigorous action, pure alcohol from the tip of a sable pencil is allowed to run under the cover. It immediately kills the creature, and, although not successful in every case, generally they are paralysed and die before retraction; the cell is then closed with the ordinary cements—for experiment, the fresh-water polyzoon Lophopus is always available, and easily managed; it takes to alcohol very kindly. A group of three or four should be arranged in a ring of glass under cover; and when the horse-shoe tentacles are fully expanded, the stimulant is cautiously admitted, and generally a fair and permanent preparation secured.

Crouch End.

GOSSIP ON CURRENT TOPICS.

MANY and loud are the complaints concerning the condition of the lower Thames. All who can see beyond the present moment must be thankful for the hot dry weather for its friendly demonstration of the dirty truth. So long as the poison remains hidden we may go on converting our rivers into sewers, by the barbarous practice of throwing into them the material demanded for the restoration of the fertility of the soil; a practice which, if general and continued long enough, would simply exterminate the human race, and with them all the rest of the higher animals that live on land and cannot obtain their food supplies in the form of fish and sea-weeds. The agricultural desolation of England would have been practically effected ere this but for our importation of guano and other foreign manure. I am sorry to find that Dr. Andrew Wilson, in his valuable magazine "Health," says, "The only clear solution of the question seems to be in the direction of the free and absolute conveyance of all sewage in closed aqueducts or sewers to the sea." Such an expedient is only a device for hiding the evil. For my own part, I prefer that this and all other consequences of human sin, whether sins of commission or sins of omission, should be nakedly exposed, and the more they stink the better. What we require is that all the sewage of all our towns, all our villages, all our farm-houses, and all other human habitations shall be deodorised and restored to the land. This certainly can be done, and would be done if the doing of it were compulsory, and therefore the sooner we are compelled to do it, by the urgency of intolerable stench and wholesale pestilence, the better.

In a letter from "Our Correspondent," dated "Marseilles, Friday, August 1," and printed in *The Times* of August 2, is the following: "The municipality is washing the lower parts of the houses, the drain openings, and the kerbstones with sulphate of copper. The red colour of our streets may be imagined." This is curious, seeing that sulphate of copper is not red, but very deeply green or nearly blue. It is quite true that the red colour "may be imagined," but certainly it does not exist as a consequence of using sulphate of copper. Sulphate of iron is sometimes called "green copperas," and this, if applied to limestone, would be decomposed into sulphate of lime and iron-carbonate, which carbonate by exposure to the air would presently become red sesquioxide of iron. I suspect that "O. C." has been a victim of the word "copperas," though I am not aware of its use in France. Sulphate of copper applied to limestone would produce a malachite-green stain.

The old subject of the influence of gun-firing on rainfall has been revived at the Antipodes, in a paper read at the Royal Society of New South Wales by Edwin Lowe, who advocates the practical use of explosions for the purpose of effecting precipitation. The subject is one of such vital importance in Australia, where so many millions of sheep have been lost during the last five years for lack of rain, that something more than mere speculative paper-reading is demanded there. Experiments on a scale of sufficient magnitude to settle the question one way or the other should be made; and it is idle, feeble, and unjust to leave such researches to private enterprise. The value of the sheep lost in a single month would, if judiciously expended, settle the question for ever.

The climate of Australia is specially suitable for the experiment. The uniform sweep of the upper currents over its vast area, and its normal uniformity of climate, permit the testing of the vexed question by promoting a local disturbance and observing the local effect.

During the recent sultry weather, I have made some observations that strikingly illustrate the folly of one of our customs, which is blindly followed by people who ought to know better, and would know better, if at all addicted to thinking about common things. I refer to the practice of opening doors and windows in hot weather for the purpose of keeping the house cool. Seeing that the heat of a summer day comes from without, the absurdity of such a practice is evident enough. I placed thermometers in two rooms; one a large room extending from front to back of the house, the other just half the size, with windows only on one side, facing north-west. Both windows of the larger room were thrown open to secure the "thorough draught" so much esteemed by the majority in hot weather. The window and door of the smaller room were closed. I found a difference varying from five to nine degrees on different days, between two and three P.M. The superior coolness of the closed room was evident at once, on passing from the open room; it appeared even greater than the thermometer indicated. To keep a house as cool as possible in summer time, all doors and windows should be closed from ten A.M. to five P.M., and the blinds as well as the windows closed on the sunny side. The more open the better during the early morning, the evening, and night. Of course, in very small houses, where the inmates are crowded, the hot air from outside must be endured for oxygen sake.

School holidays should be arranged on meteorological principles. Having determined the number of weeks of the summer vacation, tables of average day temperature should be consulted, and the hottest five, six, or other number of weeks should be chosen. A crowded school-room cannot be kept cool by excluding the outer air.

Various devices have been suggested for the diffusion of popular scientific knowledge, the latest being one that has come about by a process of natural selection. Professor Milne, of Tokio, in Japan, has constructed a pair of pendulum seismographs, instruments for the automatic delineation and registration of the earthquake disturbances so frequent in that country. He wrote describing them as "conical pendulums," each consisting "of a heavy mass suspended by a string." The printer of a local paper improved the original by describing them as "comical pendulums," consisting of "a heavy man suspended at the end of a string." This rendered the apparatus extremely interesting to the general reader, and led to many inquiries that have diffused extensive knowledge of the subject, so much so, that in the interest of popular education, the typographical errors have not been corrected.

It appears from the testimony of several correspondents in "Health," that the crowing and clucking of a neighbour's fowls is a serious nuisance. The complaints of the aforesaid correspondents are very loud and bitter.

I have heard much of this before, and have deduced a very curious psychological law, by collating the facts connected therewith. This law is that people generally are kept awake and seriously annoyed by their neighbours' fowls, but not by their own. This law is so universal that if A., who lives in a semi-detached villa and keeps fowls which have destroyed

K 2

the rest of B., his co-semi, he may, by skilfully and delicately offering to B. a setting of his choice silver-pencilled Rotterdams, and the loan of a setting hen, terminate the trouble by securing B.'s acceptance and thus making him a fowl owner ; for when the cocks are crowing and the hens are cackling in the back premises of both houses the harmony is so perfect that nobody is disturbed.

Where no such unity of action is attainable another remedy might, I think, be found. Let prizes be offered at Agricultural shows for a breed of dumb cocks. By artificial selection, such as Darwin describes to have effected the wondrous modifications of pigeons, this desideratum may surely be obtained, and presently extended to hens by only breeding from those which lay their eggs modestly, without any vociferous proclamation of the achievement.

In the "Journal of Science" of June last is an account by Henry H. Higgins of the piercing of five lawn handkerchiefs by the grass of a closely-mowed lawn upon which the handkerchiefs were spread for bleaching. All were pierced : some of the blades had grown two inches above the handkerchief without destroying the texture of the lawn. The experiment was repeated with partial success under less favourable circumstances. Many other curious experiments of this kind may be made. At the present moment I have in my garden an example of vegetable perseverance, an attribute that might supply an additional chapter in the next edition of Mr. Taylor's "Sagacity and Morality of Plants." Last year I grew some artichokes on a nearly worthless piece of ground, and reaped the usual crop. Late in the autumn, I placed a bee-hive on a stand made up of old packing-cases over this same ground. Some of the small tubers that had escaped the fork sprouted, as they are wont to do, but under the inverted packing-case, and, after struggling for a while in their gloomy prison, discovered a chink through which they have thrust themselves, horizontally at first, and since have bent and grown upwards, like ordinary plants.

The old subject of the utilisation of the Niagara Falls is again in course of agitation over the water. The American Association of Civil Engineers have had it under discussion quite lately. Mr. Benjamin Rhodes estimates the total horse-power at work night and day at seven millions, and that to utilize this by means of water wheels generating electric currents, and to transmit these to cities within five hundred miles' radius, would require an outlay of 5000 millions of dollars. This latter and practical part of the estimate is curious when applied to power which has been described as "running to waste."

* Electrical dreamers are much addicted to financial fallacies ; and when the electrical transmission of this vast supply of power was first suggested, the notion prevailed that it might be conveyed to New York, &c., for "next to nothing," by merely laying a wire to carry it. These projectors had not considered the fact that a wire of given length must have a thickness proportionate to the quantity of electricity it has to carry, and that the longer the wire the greater the thickness demanded for carrying a given supply. Taking this and all the sources of dispersion as well as the cost of the primary dynamos into consideration, I think it will be found practically that the conversion of the mechanical power of this or any other waterfall to a distance of a few hundred miles will cost about as much as a tubular aqueduct that would carry the water itself. In other words, the Niagara might be tapped and $\frac{1}{100}$ part of its waters be carried to New York for about the same cost as the electrical transmission of the mechanical power of the falls, of which at least 99 per cent. would be lost by dissipation and conversion. The cost of either would be monstrous.

W. MATTIEU WILLIAMS.

ON OUR BRITISH SEA-WORMS.

By DR. P. Q. KEEGAN.

(*Continued from page* 183.)

FREQUENTLY, at low-water mark, or hurled ashore by the waves, we may observe pieces of fucus seaweed studded with a number of little white shells, twisted into a spiral. If we break one of these shells, extract the contents thereof, and place it under the microscope, we shall observe an organism with branchiæ, bristles, and files of hooks, very similar to those of the Serpula already described. These are specimens of Spirorbis, of which *S. nautiloides* is the commonest form. There are about seventeen British species of this genus, distinguished from one another by such features as a dull or glassy aspect of the shell, its more or less cylindrical form, its being ridged or smooth, being pierced in the centre or not, and so forth.

We now advance to a group of Annelids that may be considered intermediate between the tube-constructing forms we have already described, and those which are free and unconfined to the fixed tenure of a case. Among this intermediate group, the common lugworm or lòbworm is placed, and it is so common and so familiarly known, that we need not particularly delineate it.

It may be sufficient to observe that therein the blood-system is more centralised than in any other annelid ; there are a series of feathered bristles and also hooks, and a large proboscis eminently adapted for enabling the animal to burrow dexterously in the soft watery sand wherein it dwells. There is also another anomalous worm called *Trophonia plumosa* and a number of other names. Its habits are sedentary and porcine ; it grovels in the dirtiest crevices of a muddy shore, &c., and it is a very queer-looking customer, having all round its circumference long, mud-smeared bristles, seen under the microscope to resemble, with uncommon closeness, a series of tiny bamboo canes of great pellucidness.

Cirratulus borealis is exceedingly abundant in certain localities of the shore where soft, porous sandstone segments of the body as well as from the head: they are organs of touch, have no cilia, and serve to aërate the peritoneal fluid; there is a curved black line on each side of the head; the bristles are not very inter-

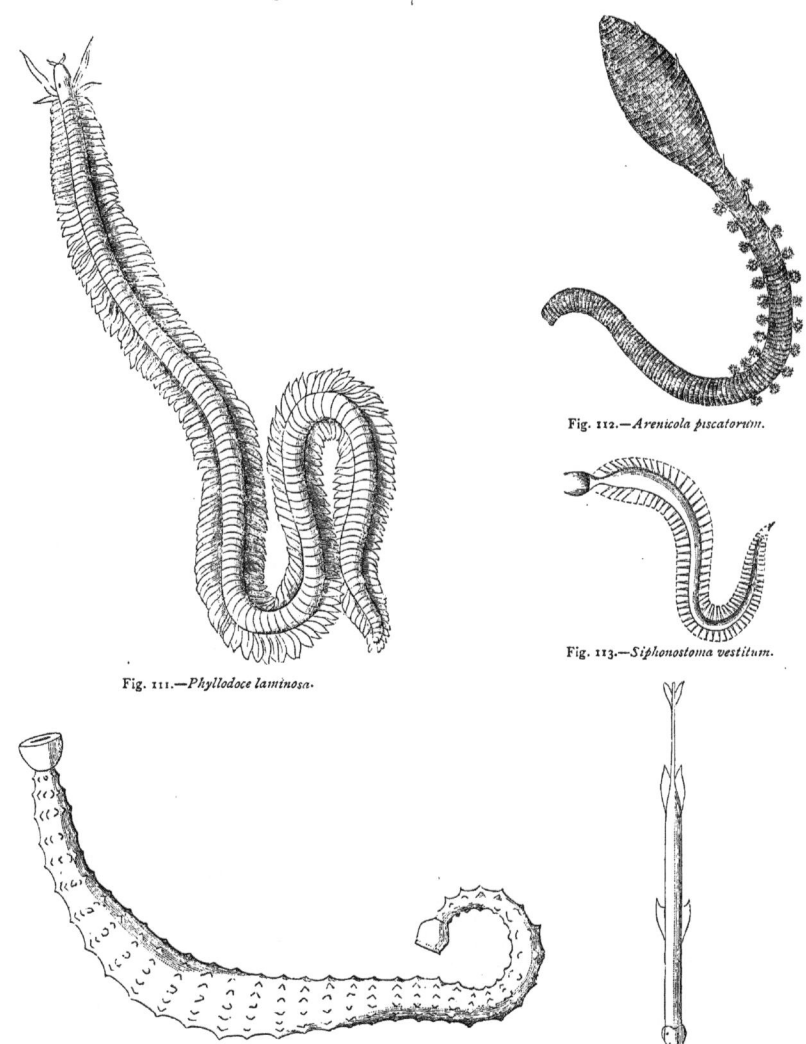

Fig. 111.—*Phyllodoce laminosa.*

Fig. 112.—*Arenicola piscatorum.*

Fig. 113.—*Siphonostoma vestitum.*

Fig. 114.—*Pontobdella muricata.*

Fig. 115.—*Sagitta bipunctata.*

abounds. It is somewhat like a Terebella, but, save in the presence of bundles of filaments near the head, it differs from that genus in most other respects. The branchial filaments spring irregularly from several esting, some being curved like an S, while the rest are long and slender. *Leucodore ciliatus* is another rock-boring worm of a small size, and distinguished chiefly by having flattened conical branchiæ, of a blood-red colour and richly ciliated in a spiral manner; the hooks are bisected at the top, and the bristles are like a flat turn-bolt.

The genus Spio is distinguished by the unique length and development of the tentacles of the head, and by the extreme beauty of the branchiæ, which are ciliated, and serve to aërate the peritoneal fluid. *S. seticornis* inhabits a tube built up of grains of sand; the antennæ are extraordinarily developed, and there are four eyes arranged in a square: here also do we observe well-developed feet with an upper and a lower branchial thread or cirrus, and two tubercles carrying bristles and hooks—the bristles are long, slender, and sharp at the point; the hooks are short, stout, and divided at the apex into sharp claws.

Nerine vulgaris is a long, spare animal of a reddish-brown colour, darker behind, and marked with red cross lines over the back, indicating the branchiæ folding over thuswise; the antennæ are about half an inch long, and incline forwards, and there is a long tapering sort of snout to the head; the eyes seem very small or rudimentary; the tail is a broad, horizontal semicircular fin, and assists the cirri and bristles in enabling the creature to move with vigorous agility through sand or shingle, although in water its progress is rather dilatory. *N. coniocephala* is distinguished by the distinctly conical shape of the head, while the antennæ spring from behind it and incline backwards; it exhibits a starfish-like talent for shuffling off its antennæ, and breaking itself to pieces.

The genus Syllis comprises a company of seaworms that manifest a special predilection for creeping about and browsing upon seaweeds, the shape of their feet and bristles being eminently adapted for this sort of life. The head bears some curious lobes projecting in front, and has four eye-like dots; a necklace-like structure of long antennæ and cirri encompasses the body; there is a long proboscis, quite destitute of jaws or glandules, but the stomach exhibits a highly-organised glandular organ at the rear of the throat; the branchiæ are furnished with cilia at the base of the feet, whereby the peritoneal fluid only is aërated. *S. armillaris* frequents the deep sea; it has two lobes in front of the head, and three antennæ longer than the lobes; the upper cirri exhibit a necklace-beaded pattern, and are four times longer than the width of the body; the bristles are jointed and curved near the pointed apex. *S. prolifera* is a small species, half an inch long, and has no head lobes; the antennæ are very long, ciliated and unjointed, and are invariably seen curled or twisted; the eyes are set in a square; the upper cirri are only about twice as long as the width of the body, and are merely slightly wrinkled; the bristles are simple and unjointed.

Glycera alba is a very peculiar form. It is generally found buried in the sand or mud heaped between the crevices of the rocks and stones that lie scattered about between tide-levels. Shovel the worm out of its oozy tenement, and immerse it forthwith in a bowl of sea-water, and you will witness a lithe and vigorous wriggling on the part of the creature, as if it had "gone daft," winding up the performance by gracefully twisting itself like a serpent into a neat spiral. There are bright red lines on each side, and a large proboscis, which perpetually protrudes, and then vanishes with admirable facility. This proboscis is thickly villose with papillary glandules, and bears four brownish-black teeth which are curved at the point, and have a number of processes fore and aft, whereby it presses the sand and sowise moves onwards; the feet are much less complicated than in Nereis, consisting only of one wedge-shaped piece with very short cirri; the peritoneal fluid is abundantly supplied with red corpuscles, and is aërated in hollow cylindrical branchiæ, which bear on their inner wall vibratile cilia; the bristles are very protrusile and dovetail-jointed, the terminal piece being bayonet-shaped and very sharp.

(*To be continued.*)

STAINER AND OTHERS *v.* BACILLUS.

THE case for the plaintiffs having concluded, Counsel for the defendant rose.

May it please your lordship, and gentlemen of the jury, this action is an attempt to dislodge my client from ancient right of domicile, user, and easement, involving an issue charging him as a common malefactor and disseminator of disease. In combating these allegations, I shall show absolutely a divarication, and contradiction of evidence. Assuming the defendant does not bear the best of characters, I submit the plaintiffs have no case. The actual existence of the defendant must be accepted: whether he be "a vegetable germ, not originating from the world outside," or the result of a "decomposition or degradation of other life," he has some "functional object," possibly creating new combinations from effete matter. It has been said, the defendant could not be supported "except under abnormal conditions, and an enormous power of fecundity." This is a grave charge, but I submit does not justify the persecution and annoyance to which he has been subjected. Take a case. The actuality of "splenic fever;" it does not emanate from him; he necessarily must have some habitat, and such a condition is his *raison d'être*. At all points meeting antagonism, his disposition has naturally become soured and obstinate. No wonder! When I present before you the story of a persecution, an oppression, aimed at existence itself; who could tolerate the ferocity of opponents attempting to reveal your (so-called) wickedness and vice with a battery of chemical obscurities, aided by magnifications of 1800 to "5000" diameters?

The defendant, gentlemen, is (said to be) a member of the family of "*Schizomycetous* fungi,"—I suppose coeval with all things; but even his origin has been questioned. To quote the delicious ambiguity of a great authority, he is an organism which requires "tendency of thought;" certainly, delicate handling. Conceive an inch (a popular measure) divided into "one hundred thousand parts;" one part would be more than the dimensions of my client round the waist. Do not accept this from me; one of his greatest enemies, Billroth, in 1874 measured him,—how, does not appear, but at that time, under the name of Coccobacteria, he was accused of producing "Anthrax, Charbon, Mitzbrand, and Frustules, the wool-sorters' disease." This allegation seems somewhat vague, and insusceptible of proof; however, it was accepted, even supported, in 1875, by Kjøbenhavn, Mangin, Archer, Ewart, Dallinger, Cienkowski, Roberts, Waldstein, Tieghem, and others. The cases are before me; you may take my statements unchallenged. About this time the defendant found a friend in Livon, who pronounced he was not malicious, poisonous, or even infective. In 1877 Klebs impeached him, evidently with the idea of accepting, sifting, and consolidating all opinions in one indictment,—to wit, mischievously concerned with lesions of the lungs, glandular and scrofulous conditions, comprised in the general term "tuberculosis." Klebs virtually pulled the whole thing together, and went so far as to "cultivate" the defendant by "crushing tubercles" (an obvious act of trespass) and inoculating fluids. I admit the result was an undoubted re-discovery. Klebs was naturally delighted, and then and there gave my client his fatal name, "Bacillus." A complete triumph was not approached. Klebs attempted too much; for in trying to "cultivate from the juice of a human tubercle," where my client was then in residence, and plant him "through a guinea-pig, into the abdominal cavity of a cat," the experiment (as may be imagined) was not successful; at all events, the result has never been disclosed.

Following Klebs, in 1880, Schuller filtered liquid from a tubercular lung, and inoculated flasks containing a "solution of Bergmani" (the recipe is not given); but both had to confess their experiments came to nothing, on the plea that the defendant, consequent on the trying ordeal of so many "changes of locality," might have been "weakened and exhausted in infective power." Klebs gave up; Schuller persisted, and eventually succeeded in localising "Micrococci" (the generic term of all the members of the defendant's family) in the "synovial membranes of joints,"—not the quietest place to select!

At this time, gentlemen, enemies were fast and furious, and consequently opinion differed. In 1881, Touissant entered for the "cultivation" stakes; at the same period Aufrecht found the defendant's family in a "centre of tubercles," and attempted a classification of three species. How they were discovered, or in what respect they differed, has never yet been disclosed; but in a burst of confidence, Aufrecht, wandering from, and entirely dropping his tubercle, stated that "all varieties" may be seen in sputum by staining it in a mixture entitled "a half per thousand watery solution of fuchsin." This may appear facetious, but I assure you, gentlemen, every statement hitherto or hereafter to be made can be supported by documentary evidence.

Now approaches a decided phase in the history of my client. In 1882, Professor Koch caught the idea that the defendant was not only "always present in tubercle in all animals," but that similar conditions might be produced by inoculation. Baumgarten, at this time, made independent observations, corroborating Koch. At this point Touissant revives, and cultivates infusions of " blood serum, taken direct from the heart, with heated scalpels and sealed pipettes," operating also with "phthisical sputum," and with these fluids inoculating and infecting guinea-pigs and kittens, marking them by "clipping their ears." Koch, Baumgarten, and Touissant formed a triumvirate, determined to face the difficulty of absolutely detecting the *then* pale and ghost-like form of the defendant in nearly every tissue.

At this critical period the plaintiff Stainer, a common detective with no scientific ability, a colourable impostor, although he had been previously employed in a tentative way, now proposed to disclose the defendant by a series of experiments unparalleled in the history of persecution. Conceive, gentlemen, any one of *you*, after being hardened in alcohol, immersed for twenty-four hours in "an aniline dye," an "alkaline solution of methylen blue," or "a watery solution of vesuvin." Even then hope gleamed, the defendant might have escaped! It occurred to Baumgarten that the mycelia of any fungus—in fact, any vegetative or protoplasmic globule, thread or cell, however minute, or wherever placed—would accept a stain, reveal itself, and under high powers materially interfere with the identity of the defendant; but after savage persistency he was at last found in tuberculosis, *stained*, and "spotted," comfortably imbedded round the edges of a bronchus. This was effected after the frightful expedient of using "caustic potash, hardening for twenty-four hours in absolute alcohol, and staining with a solution of safranin."

Ziehl now made an effort to improve Baumgarten's method, but at this interesting point, when "all the talents" were engaged, the processes employed were unable to again subdue or overcome the defendant's obstinacy or coax him from obscurity. Ehrlich in the effort started a new formula, never yet explained, although, he goes on to say, "so well known as to be universally adopted." In fact, gentlemen, the plaintiff Stainer is nothing more than a picturesque medium. He has created a profound impression, and his fascinations seem to have monopolised all consideration, to the detriment of the main issue ; but he is not everybody, his imperfections are manifest in his power of staining generally ; he detects too much. Intoxicated by success, in his endeavour to apprehend my client, he includes a mob of lookers-on, especially in sputum.

Employed by Ehrlich, Stainer certainly (as I have admitted) revealed the defendant, but how ? To quote an official précis recorded in authoritative pages, by the use of "solutions, watery and alcoholic, of a basic aniline dye—fuchsin, crysoidin, vesuvin, methyl violet, and gentian violet." Weigart now made a singular discovery, the difference between a "saturated alcoholic and a saturated watery solution of aniline incorporated with the absolute necessity of alkalinity." Ziehl opposed the alkaline theory altogether, declaring the defendant could only be stained and discovered when "acetic acid" was employed. "Experiments were then made with phenolptalein," a substance of the "aromatic series" combined with carbolic, resorcin, and pyrogallic acid. An unexpected difficulty approached: aniline, the founder of the Stainer family, was found to be an imperfect medium, could not be relied upon; containing toluidine, nitrobenzole, and paraniline, his purity was questioned ; the alcohols differed in quality and character (they generally do) : "methylic, propylic, butylic, amylic," were tried, and did not give satisfaction. "Phloroglucine was better; boracic was somewhat feeble compared with salicylic." Gentlemen, under this trying ordeal the defendant became captious, and showed his resentment by sometimes taking up one stain, then another, and exhibiting a curious acumen in detecting a difference between aniline and fuchsin, and being especially stubborn and sulky under the influence of "phenolphthalein." Differences of temperature naturally affected him ; locality also. I admit, a distinguished member of his family resides in leprosy, and accepts a stain with amiable readiness, but obstinacy seems to be the family failing, for Lichtheim found a similar micrococcus that "would not;" another in "psorospermiæ" behaved very handsomely, "coming out beautifully reddened ; " "those who (sic) dwell in elastic tissue and cheesy matter" display the same readiness. To crown these discoveries, it was found the various washings and soakings made the sections of the tissues "shrivel up." It might end here, but the hue and cry after "Bacillus" is severely eager ! He had to meet the exhaustive researches of Laulanie, Eimer, Aufrecht, Ponfick, Weigert and others, especially Mügge; beyond this foreign research, there is a perfect chain of English authority pressing hard. Is the defendant a factor ? That is the real issue. A cause, or an effect ? Essays have been written, experiments made, but no reliable standpoint reached. Conceive in all seriousness this : Cohnheim endeavouring to produce his development by introducing "infected cork, gutta-percha, and other inert materials into the abdomen of animals ; " Fraenkel failing, in repeating these experiments ; Lebert and Waldenburg opposing inoculative power, Koch swearing by it ; Tappeimer and Schottelius trying "inhalation of tuberculous matter," the result obtained being only "inflammatory ; " Carl Solomonsen attempting to infect the "anterior chamber of the eyes of rabbits," with no result. But no longer to trespass on your patience, I claim a verdict on the testimony of an antagonistic witness.

Baumgarten with astounding patience tried to inoculate, without in any one case producing tuberculosis or infection, organic "foreign" bodies. To quote the instances in the exact words: "carcinoma, sarcoma, lymphona, chancres, lupus, typhus, glands, actinomycosis, crupous and diphtheritic masses, granular tissue, scars, pus, gregarinæ, cocci, all sorts of fungi, and cheesy infarcts." Nothing as affecting the character of the defendant can excel the sublime bathos of such experiments. In tracing the history of obscure organisms, different minds, however alert, cultivated, and conscientious, may be so warped by preconceptions or an eagerness to establish foregone conclusions, as to give accepted facts any interpretation chosen. Opinions may actually be supported or opposed by the same observations. If the allegations against the defendant could be incontestably proved, no living creature would be exempt from rapid and fatal disease.

E. T. D.

Crouch End.

DREDGING IN THE FRITH OF CLYDE.

By S. P. ALEXANDER.

THAT the Frith of Clyde affords a rich field for the study of Marine Zoology, is, I feel, a fact not very generally known or appreciated.

Seaside visitors and many young naturalists have doubtless often roamed along the sands and rocks of our coasts at low tide ; have peered into the rock-pools ; watched the habits of, collected, and admired the beautiful creatures of the ocean left by the receding waters.

In this way they will have picked up many specimens of our littoral species, and occasionally, even, of those more oceanic in distribution.

It is in this manner, I say, that the lover of Nature has probably made his first acquaintance with Marine Zoology, and by the products of which he may have stocked an aquarium.

But it is to the more fruitful and higher pleasures of dredging for marine objects, that I would now call his attention.

A great stride was made towards the advancement

of science when Dr. Robert Ball invented the naturalist's dredge. By its use, a great part of our present knowledge concerning the varieties and numbers of the inhabitants of the ocean's depths was gained.

Previously to this, naturalists availed themselves of the huge ungainly apparatus which fishermen used for obtaining oysters and scallops. This consisted of a large, heavy, single-bladed frame, and a bag formed of iron rings, after the fashion of the ancient chain-mail. The meshes of the net were large, in order that only oysters of a marketable size might be retained; thus other animals of considerable dimensions alone could be obtained by its use.

Ball's dredge is a much smaller and more portable affair, the scraping blades being double, and the bag composed of strongly-netted twine.

It was with such dredges that the late Sir Wyville Thomson did so much in the Natural History department of the famous "Challenger" Expedition.

Owing to the enormous depths at which he used them, his dredges were necessarily very heavy and powerful ones. The weight of the smallest was 20 lbs.

Such a heavy dredge, it is apparent, could only be used from steamships or large vessels, and thus would hardly meet the requirements of the amateur marine zoologist. What the latter needs is a dredge at once so portable, that it may be with ease carried under the arm; and suitable for the lugsail, or ordinary rowing boat. These requirements are, I think, fully met by a dredge of the following dimensions:—

Galvanised iron frame, 18 inches by 4½; double bridles on each side, 2 feet long; scraping blades, 2 inches broad; netted bag, about 2 feet deep. The bridles, attached to the cross-bars of the frame by means of eyes, are movable, thus allowing them to be folded down upon one another, when not in use. The meshes of the bag may be half an inch in diameter, except at the bottom, where it is well to have them considerably smaller.

The whole apparatus weighs only 5½ lbs.; and this at a depth of twenty or thirty fathoms, with forty fathoms of rope, is a load quite as heavy as the ordinary single rower can manage.

When dredging at greater depths, the weight may be increased if desirable, by attaching heavy sinkers to the rope, a short distance in front of the bridles.

The general appearance of such a dredge is seen from the sketch below.

It was with such a dredge that Mr. P. H. Gosse, the author of the Manual of "Marine Zoology for the British Isles," made so many discoveries in that subject, and to the productions of which, doubtless, we owe the writing of his numerous popular works on marine fauna.

Of all pastimes open to the young naturalist, that of dredging ranks pre-eminent. This is so, not only on account of the exercise involved in, or the health gained by its pursuit, but from the very beautiful and rare nature of its production. During the summer months, much amusement and instruction is to be gained by the dredge, at any of our watering places, and by those in the vicinity of the Frith of Clyde.

Let the reader accompany us on a dredging

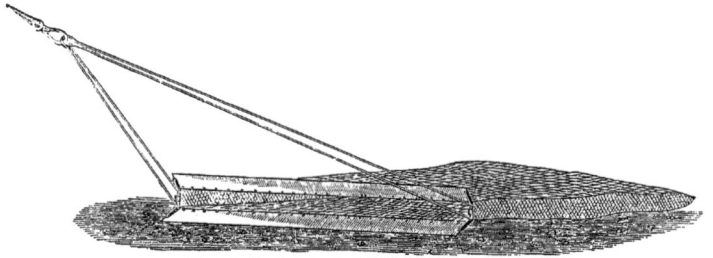

Fig. 116.—Dredge, showing its position on the ground.

excursion. Let us see what is to be scraped up from the bottom of the Frith of Clyde.

A calm day being chosen, we start away in a rowing boat—let us say from Dunoon.

At about a hundred and fifty yards from the shore we throw out the dredge. The depth here is from eight to ten fathoms; so, to ensure the dredge biting properly, about twenty fathoms of rope must be let out. The longer the rope, the deeper the blade of the dredge scrapes, due to the angle it makes with the bottom being more an acute one.

To dredge properly the net should glide along the bottom, and not dip too deeply into it. The bottom most easy for dredging is one of gravel; a sandy one is more difficult, and a muddy or clayey bed almost out of the question. That most productive of good specimens is probably a mixture of gravel and sand with patches of seaweed. The distance traversed in a given time will of course vary with the nature of the bottom. Having rowed the boat for about ten or fifteen minutes, let us pull our dredge on board. Up it comes, with the mouth of the net clogged with masses of dripping seaweed. Chief among these are to be noticed the long brown ribbon-like bands of the Laminaria.

These seaweeds should be carefully examined, not only for good specimens of Algæ, but because clinging to them, especially to their roots, one often finds many rare creatures.

Masses of jelly-like spawn, white, yellow, and delicate green (the future progeny of many species of animals), are frequently to be obtained in this way.

Sprouting from the seaweed also, are constantly to be noticed colonies of those beautiful plant-like animals belonging to the orders Sertularida and Campanularida. I would also mention those genera of Polyzoa, Diastophora and Membranipora, colonies of which are found encrusting the surface of the seaweeds, like the grey lichen the surface of the rock, and often assuming a shape which may be said to resemble that of the prothallus of a fern. Having now examined and cleared away the seaweed, let us take a look at the contents of the dredge. If the haul has been a successful one, we will here find quite enough employment for some time.

On the surface of the mass of stones and sand, we find quite a wriggling maze of beautifully coloured brittle-stars (Ophiocoma) of many different species; and in fewer numbers the Sand-star (Ophiura) with its long and more flexible arms.

Preserving a few of these as specimens, we turn out the contents of the net. Among the many commoner shore crabs, here are one or two little fellows with brown or white testa, in which we recognise the nut-crab (Ebalia).

The common hermit-crab (*Pagurus Bernhardus*), which is but seldom found at low tide, we see before us in considerable numbers. Of these, the larger specimens are to be noticed occupying the shells of the whelk (Buccinum); the smaller, those of the top (Trochus), or Turritella.

Beside sand-stars and brittle-stars, the class Echinodermata is here represented by many specimens of the common Sea-urchin (*Echinus sphæra*), varying in size from the smallest pea to that of a pomegranate; also the common star-fish (*Uraster rubens*), the *Asterias aurantiaca*, the Goniaster, and *Luidia fragilissima*.

If close enough into shore, we may obtain specimens of the "cobbler" (Cottus), fifteen-spined stickleback (*Gasterosteus spinachia*), and sucker (Lepidogaster).

If very fortunate, a large yellow sea-lemon (*Doris tuberculata*) may be obtained. This is sometimes seen at low tide clinging to the rocks, presenting in this situation a no very attractive sight; looking like, to use a homely simile, "a half potato." Put it in its native element, however, and its real beauties will manifest themselves. In a few minutes the creature flattens itself out, protruding its two fleshy tentacles. At the same time the purple mark on its dorsal surface will appear in its true character, as a cluster of delicate feathery branchiæ.

Other examples of Nudibranchiata we have here in those little papose-crowned Eolis (*Eolis coronata*).

(*To be continued.*)

NOTES ON THE NATURAL HISTORY OF JERSEY.

By Edward Lovett.

IN a recent short paper in SCIENCE-GOSSIP, I described a little experience in dredgings off the coast of Jersey. I will now say a little about shore collecting. During my visit I was favoured by an extra good spring tide, nearly forty feet; and, knowing the value of such a tide for shore work, I was anxious to get a fine day. In this, too, I was fortunate, and we started for La Rocque on one of the most charming of May mornings. La Rocque is, as its name suggests, an exceedingly favourable part of the coast for the marine zoologist. The tide goes out for an enormous distance (five to seven miles at a tide like this one), the result being that many thousands of acres of rocks, shelly or shingly banks, "graps," pools, and sandy lagoons are approachable and workable.

Of course we obtained, or could have done, a large quantity of common mollusca, but we also succeeded in obtaining some not so common. In the big sand and shingle banks we dug out *Donax politus*, *Venus verrucosa*, *Venus fasciata*, *Cardium Norvegicum*; but the prize was a fine series of *Lutraria oblonga*: the tips of their siphons were just visible, and, upon alarm, they ejected a small jet of water. By rapidly "grubbing" up the firm wet shingle, we got a them about nine inches above the surface.

On the zostera, in the lower pools, we obtained *Trochus exiguus*, and in spots where there was a kind of muddy deposit, probably formed by the decomposed felspar of the syenite, tainted with decomposed algæ and animal matter, we observed large numbers of *Trochus magnus*, many of which were very fine.

Few of these were clean; indeed, this species seems to be a very dirty feeder, and certainly those I saw were all on these muddy flats, where no other shells were to be seen.

Tapes aurea, *T. virginea*, and *T. pullastra* were in considerable numbers, and some beautiful varieties were obtained.

We found a solitary but very large valve of Pholas. This was interesting, inasmuch as I have never seen this genus on these shores. The beautiful *Psammobia vespertina* was in considerable number, and *Nassa reticulata* were of fine size and remarkably bright and clean.

The season has been a fine one for *Octopus vulgaris*, and this part of the coast was their favourite haunt: this was clearly seen by the great scarcity of crustacea. The arrival of the cuttle is the signal for the departure or death of a host of things, but especially of crustacea, small heaps of the shells of which testify to their rapacious appetite; nor are mollusca free from their embrace, although they are not so easily made a prey as crustaceans. We only found a couple of this cephalopod. I was much amused on turning over a large stone to see a pair of remarkably bright eyes staring at me, and to see eight leathery arms scraping together stones and shells, in order to screen the creature from observation; however I picked him up and transferred him to my fish basket, and, in doing so, experienced the curious sensation caused by the suckers of the skin. Being a somewhat young one and out of water, his hold was very slight, but, having experimented with a large one on a former occasion, I am inclined to think that their power is somewhat overrated. But to return to our captive, who by the way is now before me in a large jar of spirit. Whenever I opened the basket he commenced swarming over the side most quickly, and I nearly lost him more than once. Their mode of progression, by means of their arms, out of the water, is by no means slow. Of crustacea we did but little in the way of collecting, for reasons already given, the only species I observed being *Cancer pagurus, Carcinus mænas, Maia squinado, Porcellana platycheles* and *P. longicornis, Palæmon squilla*, a few of the Hippolytes, and perhaps one or two others; in fact a very poor series indeed for a spot usually so rich. But a few days' later, after it had been blowing a bit, I found at high-water mark several dead specimens of *Pirimela denticulata*, a very uncommon and pretty crustacean.

Unfortunately a tide is not long enough to do much work, and it is very seldom anyone is able to penetrate so far as we did on this occasion, besides the fact of its being unpleasant to be five miles from high-water mark with the tide flowing.

Again, no dredge could possibly work for five minutes amongst such a tangle of outcropping rocks, with gullies and stretches of sand between, so it will be understood why I believe it almost impossible to arrive at anything more than a scanty knowledge of the Fauna of the coasts of an island so close to us as Jersey.

THE Fitzroy weather-glass is composed, I believe, roughly speaking, of spirits of wine, camphor, water, &c., and is enclosed in a glass bottle which is hermetically sealed (or not). But in the former case, how can the atmosphere act on the contents of the bottle, which is reputed to cloud and clear for wet and fine respectively? I have never heard this conclusively answered.—*L. G. F.*

THE ORGANS OF RESPIRATION AND CIRCULATION IN THE COCKROACH.*

By Professor L. C. MIALL and ALFRED DENNY.

THE respiratory organs of insects consist of ramified tracheal tubes, which communicate with the external air by stigmata or spiracles. Of these spiracles the cockroach has ten pairs; eight in the abdomen, and two in the thorax. The first thoracic spiracle lies in front of the mesothorax, beneath the edge of the tergum; the second is similarly placed in front of the metathorax. The eight abdominal spiracles belong to the first eight somites; each lies in the fore part of its segment, and hence, apparently, in the interspace between two terga and two sterna; the first abdominal spiracle is distinctly dorsal in position.

The disposition of the spiracles observed in the cockroach is common in insects, and, of all the recorded arrangements, this approaches nearest to the plan of the primitive respiratory system of Tracheata, in which there may be supposed to be as many spiracles as somites. The head never carries spiracles, except in Smynthurus, one of the Collembola (Lubbock). Many larvæ possess only the first of the three possible thoracic spiracles; in perfect insects this is rarely or never met with (Pulicidæ?), but either the second, or both the second and third are commonly developed. Of the abdominal somites, only the first eight ever bear spiracles, and these may be reduced in burrowing or aquatic larvæ to one pair (the eighth), while all disappear in the aquatic larva of Ephemera.

From the spiracles short, wide air-tubes pass inwards, and break up into branches, which supply the walls of the body and all the viscera. Dorsal branches ascend towards the heart in the intervals between the alary muscles; each bifurcates above, and its divisions join those of the preceding and succeeding segments, thus forming loops or arches. The principal ventral branches take a transverse direction, and are usually connected by large longitudinal trunks, which pass along the sides of the body; the cockroach, in addition to these, possesses smaller longitudinal vessels, which lie close to the middle line, on either side of the nerve-cord.† The ultimate branches form an intricate network of extremely delicate tubes, which penetrates or overlies every tissue.

The accompanying figures sufficiently explain the chief features of the tracheal system of the cockroach, so far as it can be explored by simple dissection. Leaving them to tell their own tale, we shall pass on

---

* For the natural history, the outer skeleton, and the alimentary canal of the cockroach, see this Journal, March, May, and July, 1884.
† The longitudinal air-tubes are characteristic of the more specialised Tracheata. In Araneidæ, many Julidæ, and Peripatus, each spiracle has a separate tracheal system of its own.

to the minute structure of the air-tubes and the physiology of insect respiration.

The tracheal wall is a folding in of the integument, and agrees with it in general structure. Its inner lining, the intima, is chitinous, and continuous with the outer cuticle. It is secreted by an epithelium of nucleated, chitinogenous cells, and outside this is a thin and homogeneous basement membrane. The integument, the tracheal wall, and the inner layers of nearly the whole alimentary canal are continuous and equivalent structures. The lining of the larger tracheal tubes at least is shed at every moult, like that of the stomodæum and proctodæum.

In the finest tracheal tubes (·0001 in. and under) the intima is to all appearance homogeneous. In wider tubes it is strengthened by a spiral thread, which is denser, more refractive, and more flexible than the intervening membrane. The thread projects slightly into the lumen of the tube, and is often branched. It is interrupted frequently, each length making but a few turns round the tube, and ending in a point. The thread of a branch is never continued into a main trunk. Both the thread and the intervening membrane become invisible or faint when the tissue is soaked with a transparent fluid, so as to expel the air. Both, but especially the thread,

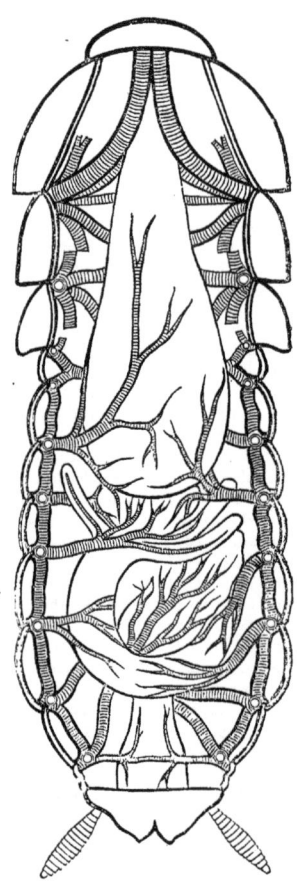

Fig. 117.—Tracheal System of Cockroach. The dorsal integument removed and the viscera in place. × 5.

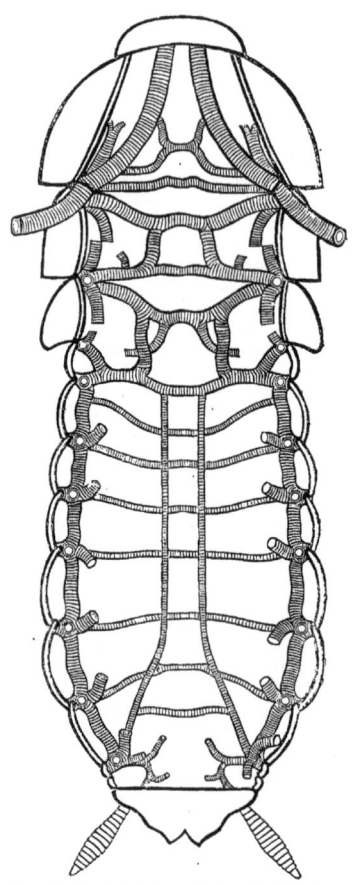

Fig. 118.—Tracheal System of Cockroach. The viscera removed to show ventral tracheal communications. × 5.

absorb colouring matter with difficulty. The thread, from its greater thickness, offers a longer resistance to solvents, such as caustic alkalies, and also to mechanical force. It can therefore be readily divided into polygonal areas, each of which is occupied by a reticulation of very fine threads. This structure may be traced for a short distance between the turns of the spiral thread.

Fig. 120.—Tracheal System of Cockroach. Top and front of head seen from without. × 15.

Fig. 119.—Tracheal System of Cockroach. The ventral integument and viscera removed to show dorsal tracheal communications. × 5.

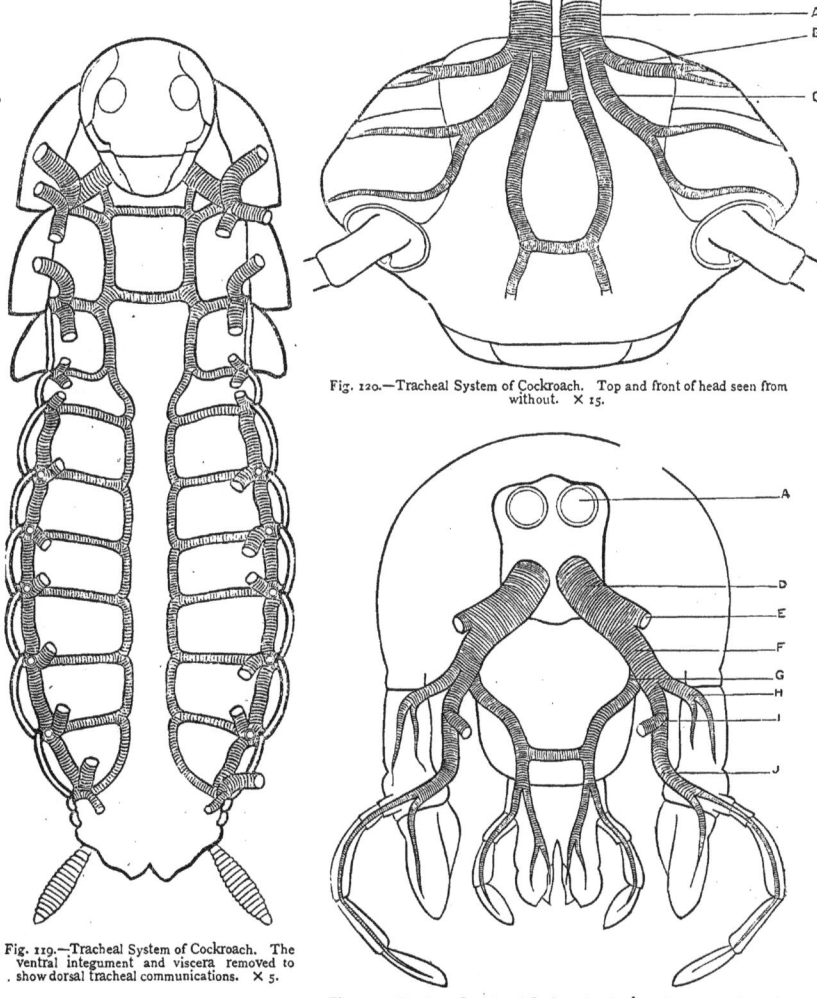

Fig. 121.—Tracheal System of Cockroach. Back of head, seen from the front, the fore half being removed. × 5. The letters A—J indicate corresponding branches in figs. 120, 121, and 122.

unrolled, and often projects as a loose spiral from the end of a torn tube, while the membrane breaks up or crumbles away.

The large tracheal tubes close to the spiracles are without spiral thread, and the intima is here sub-

The chitinogenous layer of the tracheal tubes is single, and consists of polygonal, nucleated cells, forming a mosaic pattern, but becoming irregular and even branched in the finest branches. The cell-walls are hardly to be made out without staining.

Externally, the chitinogenous cells rest upon a delicate basement-membrane.

Where a number of branches are given off together the tracheal tube may be dilated. Fine branches, such as accompany nerves, are often sinuous. In the very finest branches the tube loses its thread, the chitinogenous cells become irregular, and the intima is lost in the nucleated protoplasmic mass which replaces the regular epithelium of the wider tubes.*

The spiracles of the cockroach offer much intricate detail, and it is far from easy to master their structure and mode of action. The first thoracic spiracle is the largest in the body. It is closed externally by a large, slightly two-lobed valve, attached by its lower border, which protects the opening. The air-tube divides into two primary trunks immediately within the orifice. The second thoracic spiracle is simpler and smaller; its valve is nearly semicircular, and the orifice, when slightly open, has the figure of a horse-shoe. The abdominal spiracles are obliquely truncated papillæ without valves and directed backwards; their openings are vertical, and oval or elliptical.

We have already pointed out that the wall of the air-tube for a short distance from the spiracular orifice has a tesselated instead of a spiral marking. The tesselated border is most regular in the case of the second thoracic spiracle. The chitinous cuticle within the opening is crowded with fine setæ, while the setæ of the external integument are often arranged so as to form a fringe on one side of the aperture.

Fig. 123.—First Thoracic Spiracle of the Cockroach (left side). The arrows indicate the external openings. × about 35 times.

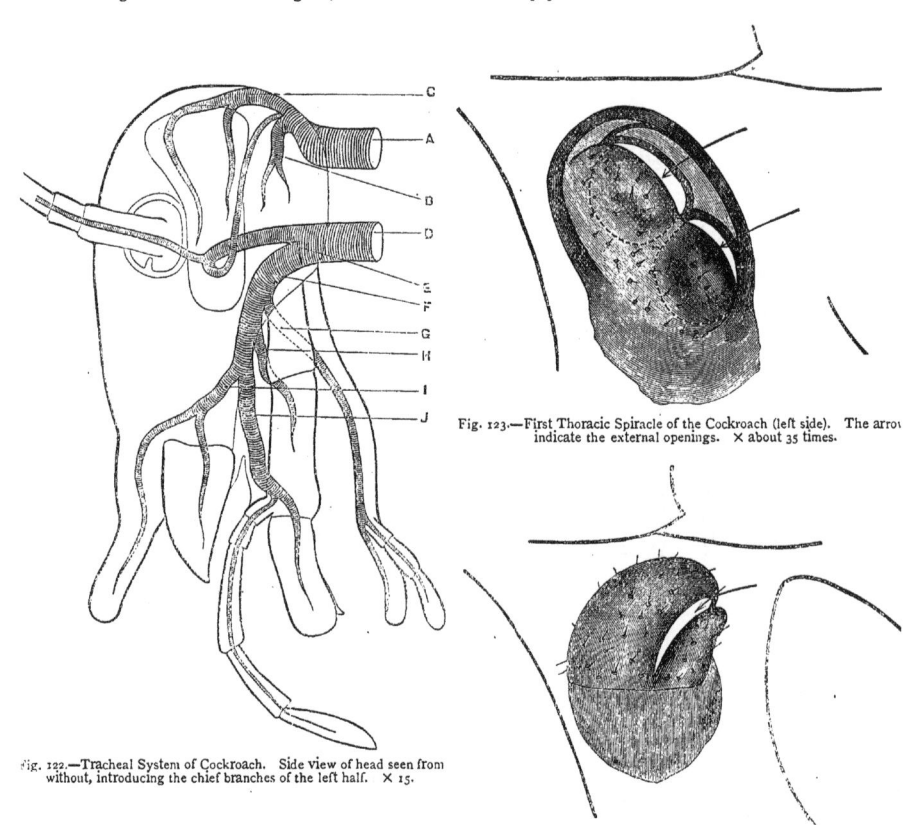

Fig. 122.—Tracheal System of Cockroach. Side view of head seen from without, introducing the chief branches of the left half. × 15.

Fig. 124.—Second Thoracic Spiracle of Cockroach (left side). × about 35.

---

* It has been supposed that these irregular cells of the tracheal endings pass into those of the fat-body, but the latter can always be distinguished by their larger and more spherical nuclei.

In animals with a complete circulation, aërated blood is diffused throughout the body by means of arteries and capillaries, which deliver it under pressure at all points. Such animals usually possess a special aërating chamber (lung or gill), where oxygen is made to combine with the hæmoglobin of the blood. It is otherwise with insects. Their blood escapes into great lacunæ, where it stagnates, or flows and ebbs sluggishly, and a diffuse form of the internal organs becomes necessary for their free exposure to the nutritive fluid. The blood is not injected into the tissues, but they are bathed by it, and the compact kidney or salivary gland is represented in insects by tubules, or a thin sheet of finely divided lobules. By a separate mechanism, air is carried along ramified passages to all the tissues. Every organ is its own lung.

We must now consider in more detail how air is made to enter and leave the body of an insect. The spiracles and the air-tubes have been described, but these are not furnished with any means of creating suction or pressure; and the tubes themselves, though highly elastic, are non-contractile, and must be distended or emptied by some external force. Many insects, especially such as fly rapidly, exhibit rhythmical movements of the abdomen. There is an alternate contraction and dilatation which may be supposed to be as capable of setting up expirations and inspirations as the rise and fall of the diaphragm of a Mammal. In many insects, two sets of muscles serve to contract the abdomen,—viz., muscles which compress or flatten, and muscles which approximate or telescope the segments. In the cockroach the second set is feebly developed, but the first are more powerful, and cause the terga and sterna alternately to approach and separate with a slow, rhythmical movement; in a dragon-fly or humble-bee the motion is much more conspicuous, and it is easy to see that the abdomen is bent as well as depressed at each contraction. No special muscles exist for dilating the abdomen, and this seems to depend entirely upon the elasticity of the parts. It was long supposed that, when the abdomen contracted, air was expelled from the body and the air-passages emptied; that when the abdomen expanded again by its own elasticity, the air-passages were refilled, and that no other mechanism was needed. Landois pointed out, however, that this was not enough. Air must be forced into the furthest recesses of the tracheal system, where the exchange of oxygen and carbonic acid is effected more readily than in tubes lined by a dense intima. But in these fine and intricate passages the resistance to the passage of air is enormous, and the renewal of the air could, to all appearance, hardly be effected at all, if the inlets remained open. Landois accordingly searched for some means of closing the outlets, and found an elastic ring or spiral, which surrounds the tracheal tube within the spiracle. By means of a special muscle, this can be made to compress the tube, like a spring-clip upon a flexible gas-pipe. When the muscle contracts, the passage is closed, and the abdominal muscles can then, it is supposed, bring any needful pressure to bear upon the tracheal tubes, much in the same way as with ourselves, when we close the mouth and nostrils, and then, by forcible contraction of the diaphragm and abdominal walls, distend the cheeks or pharynx. Landois describes the occluding apparatus of the cockroach as completely united with the spiracle. It consists of two curved rods, the "bow" and the "band," one of which forms each lip of the orifice. From the middle of the band projects a blunt process for the attachment of the occlusor muscle, which passes thence to the extremity of the bow. The concave side of each rod is fringed with setæ, and turned towards the opening, which lies between the two.[*]

The mechanism described by Landois undoubtedly exists, and his explanation of its action has been generally accepted. It may be doubted, nevertheless, whether he has taken sufficient account of the difficulty of forcing air by mechanical means into tortuous and incredibly minute tubules. Inexpert dissectors know well the annoyance caused by the presence of an air-bubble in the injection syringe. The air will not pass along the vessels, though collapsed and empty, but blocks them up, and, if

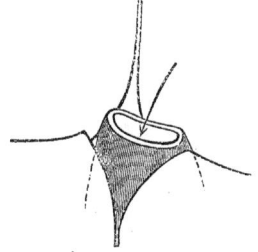

Fig. 125.—Abdominal Spiracle of Cockroach (left side). × about 35.

further pressure be applied, bursts them. It is equally difficult to withdraw air from fine passages. If the body of an insect be opened, and placed in the receiver of an air-pump, the trachea cannot be completely emptied of air without much time and exertion. The tenacity with which air adheres to closely approximated surfaces is so great as to constitute an apparently insuperable difficulty in the way of distension of the tracheal system by any muscular arrangement whatsoever. If unlimited force could be applied, the air could not be made to fill the delicate, branched tubes; it would either shrink to a small portion of its original volume, or burst the tissues.

It is worthy of remark that in the Chilopoda, which breathe by tracheal tubes, and whose respiratory

---

[*] Landois und Thelen, Zeitsch. f. wiss. Zool. Bd. xvii. (1867).

activity is probably as great as that of most insects, there appears to be no occluding apparatus.

If there is no current of air in the finer tracheæ, how is the substitution of oxygen for carbonic acid effected? The late Professor Graham showed how by a diffusion-process the carbonic acid produced in the remote cavities would be moved along the smaller tubes and emptied into wider tubes, from which it could be expelled by muscular action. The carbonic acid is not merely exchanged for oxygen, but for a larger volume of oxygen ($O\ 95 : CO_1\ 81$); and there is consequently a tendency to accumulation within the tubes, which is counteracted by the elasticity of the air-vessels, as well as by special muscular contractions.*

The occluding apparatus has, no doubt, its proper function, which is probably that of excluding water, dust, and noxious gases from the tracheal system. It may be found possible to apply a direct test to Landois' explanation by observing whether in living insects the occluding apparatus works rhythmically, as, upon his hypothesis, it should do.

The respiratory activity of insects varies greatly. Warmth, feeding, and movement are found to increase the frequency of their respirations and also the quantity of carbonic acid exhaled. In Liebe's† experiments a Carabus produced ·24 mgr. of carbonic acid per hour in September, but only ·09 mgr. per hour in December. A rise of temperature raised the product temporarily to twice its previous amount; but when the same insect was kept under experiment for several days without food, the amount fell in spite of increased warmth. Treviranus‡ gives the carbonic acid exhaled by a humble-bee as varying from 22 to 174, according as the temperature varied from 56° to 74° F.

Larvæ often breathe little, especially such as lie buried in wood, earth, or the bodies of other animals. The respiration of pupæ is also sluggish, and not a few are buried beneath the ground or shrouded in a dense cocoon or pupa-case. Muscular activity originates the chief demand for oxygen, and accordingly insects of powerful flight are most energetic in respiration.

The question has been raised whether the respiratory movements are voluntary, or not. They continue after decapitation, and may be observed even in the detached abdomen of insects whose abdominal nervous centres are not concentrated and fused with those of the thorax. Plateau§ observes that in the abdomen after severance the respiratory movements are stimulated or restrained by the same external agents as in the uninjured animal. This points to a nervous control independent of will or consciousness. The voluntary and conscious life may nevertheless modify actions which it does not originate; and that this is the case with the respiratory movements of insects may be inferred from the fact that they can be checked, stopped, or restricted to a single segment, apparently at the pleasure of the uninjured animal. The respiratory centres seem to be the ganglia of the respiratory segments.

A rise of temperature proportionate to respiratory activity has been observed in many insects. Newport* tells how the female humble-bee places herself on the cells of pupæ ready to emerge, and accelerates her inspirations to 120 or 130 per minute. During these observations he found, in some instances, that the temperature of a single bee was more than 20° above that of the outer air.

Some insects can remain long without breathing. They survive for many hours when placed in an exhausted receiver, or in certain irrespirable gases. Cockroaches in carbonic acid speedily become insensible, but after twelve hours' exposure to the pure gas they revive, and appear none the worse. H. Müller† says that an insect, placed in a small, confined space, absorbs *all* the oxygen. In Sir Humphry Davy's "Consolations in Travel"‡ is a description of the Lago dei Tartari, near Tivoli, a small lake whose waters are warm and saturated with carbonic acid. Insects abound on its floating islands; though water-birds, attracted by the abundance of food, are obliged to confine themselves to the banks, as the carbonic acid disengaged from the surface would be fatal to them, if they ventured to swim upon it when tranquil.

Kowalewsky, Bütschli, and Hatschek have described the first stages of development of the tracheal system. Lateral pouches form in the integument; these send out anterior and posterior extensions, which anastomose and form the longitudinal trunks. The tracheal ramifications are not formed directly by a process of invagination, but by the separation of chitinogenous cells, which cohere into strings, and then form irregular tubules. The cells secrete a chitinous lining, and afterwards lose their distinct contours, fusing to a continuous tissue, in which the individual cells are indicated only by their nuclei, though by appropriate reagents the cell-boundaries can be defined.

It has been held that the spiracles correspond morphologically with tracheal gills, and that they are the scars or broken ends of tubes, which in other cases project beyond the body as respiratory appendages. Palmén seems, however, to have proved that there is no constant relation of spiracles to

---

* "Phil. Mag.," 1833. Reprinted in "Researches," p. 44. Graham expressly applies the law of diffusion of gases to explain the respiration of insects. Sir John Lubbock quotes and comments upon the passage in his paper on the Distribution of the Tracheæ in Insects (Linn. Trans. vol. xxiii.), but it has dropped out of sight in the more recent discussions.
† Ueb. d. Respiration der Tracheaten. Chemnitz (1872).
‡ See Table in Burmeister's "Manual," Eng. trans., p. 398.
§ Rech. exp. sur les mouvements respiratoires des Insectes (1882).

* Art. "Insecta," Cyc. Anat. and Phys., p. 989.
† Pogg. Ann. 1872, Hft. 3.
‡ Works, vol. ix. p. 287. This passage has been cited by Rathke.

tracheal gills, either with respect to their number or position; both may occur in the same segment; and the tracheal gills themselves are not everywhere homodynamous or equivalent, but have merely a physiological correspondence.

A very long chapter might be written upon the views advanced by different writers as to the circulation of insects. Malpighi first discovered the heart or dorsal vessel in the young silkworm, and observed its progressive contraction; he regarded it as a large pulsating vein.[*] Swammerdam thought that his injections ascertained the existence of lateral vessels, but this proved to be a mistake. Lyonnet added many details of interest to what was previously known, but came to the conclusion that there was no system of vessels connected with the heart, and even doubted whether the organ so named was in effect a heart at all. Marcel de Serres maintained that it was merely the secreting organ of the fat-body. Cuvier and Dufour doubted whether any circulation, except of air, existed in insects. This was the extreme point of scepticism, and naturalists were drawn back from it by Herold,[†] who repeated and confirmed the views held by the seventeenth century anatomists, and insisted upon the demonstrable fact that the dorsal vessel of an insect does actually pulsate and impel a current of fluid. Carus, in 1826, saw the blood flowing in definite channels in the wings, antennæ, and legs. Strauss-Durckheim followed up this discovery by demonstrating the contractile and valvular structures of the dorsal vessel. Blanchard affirmed that a complex system of vessels accompanied the air-tubes throughout the body, occupying peritracheal spaces supposed to exist between the inner and outer walls of the tracheæ. This peritracheal circulation has not withstood critical inquiry, and it might be pronounced wholly imaginary, except for the fact that air-tubes and nerves are found here and there within the veins of the wings of insects.

The insect heart is a chambered tube occupying the middle line of the dorsal surface of the abdomen. The chambers may correspond, as in the cockroach, to the eight foremost abdominal segments, or they may be fewer; they increase regularly in length and width as they approach the thorax. From the anterior chamber a simple tube, the aorta, passes forwards to the head. If the abdomen is marked off by a decided constriction, the aorta is bent down at its commencement. In the thorax it lies above the œsophagus and crop, sunk among the muscles of the dorsal wall. The fore end of the aorta is usually a trumpet-shaped orifice, from which distinct vessels have in a few and doubtful cases been seen to proceed; it is possible that the blood escapes freely from the aorta, but the course of the circulation has not been distinctly and continuously traced beyond this point. A sinus resembling a pericardium, but containing blood, can be plainly seen to surround the heart in the cockroach and other large insects. Other sinuses or lacunæ occupy the interspaces of the viscera, and one large ventral sinus envelopes the nerve-cord in some insects. The wall of the heart is muscular, and the fibres are wound about the tube in opposite spirals. A pair of lateral valvular inlets lead from the pericardial sinus into the hinder end of each cardiac chamber, and every chamber communicates with the next in front by another valvular opening. In the living insect a wave of contraction passes rapidly along the heart from behind forwards; and the blood may under favourable circumstances be seen to flow in a steady, backward stream along the pericardial sinus, to enter the lateral aperture of the heart. The peristaltic movement of the dorsal vessel may often be observed to set in at the hinder end of the tube before the preceding wave has reached the aorta. In white cockroaches which had just moulted the pulsations were counted by Cornelius to eighty per minute.

Lyonnet, in his famous memoir on the larva of the goat moth, describes what he terms the wings of the heart—paired lateral muscles, which radiate from opposite points of the dorsal integument, and unite by their broad bases beneath the heart, so as to form transversely elongate lozenges. These alary muscles form thin sheets of loosely connected fibres, the interspaces being occupied by granular matter. It has often been explained that the muscles by their contraction dilate the heart, but this cannot be true. A pull from opposite sides upon a flexible, cylindrical tube would narrow, and not expand its cavity; moreover, the muscles are not inserted directly into the wall of the heart, though a muscle passes upwards from their junction to the ventral surface of that organ. They form a transverse diaphragm which, whenever it contracts, depresses the heart and enlarges the pericardial space. The same action compresses the abdominal viscera, and may be supposed to force out the blood from the surrounding lacunæ, impelling it towards the pericardial sinus. It should be explained that direct observation of this and similar points is nearly impossible in large and opaque insects. We are therefore largely dependent upon inference from minute structure, and upon the microscopic examination of transparent aquatic larvæ.

No satisfactory injections of the vessels of insects have been made; and the large lacunæ, or cavities without proper wall, which lie in the course of the circulation, render complete injections impracticable. Reticulated blood-vessels can however be traced in the wings and other transparent organs. The course of the blood is in general forwards along the anterior, backwards along the posterior, side of the appendage. The direction of the current is not absolutely constant in all the branches, but the same

---

[*] Dissert. de Bombyce, 1669.
[†] Schrift. d. Marburg. Naturf. Gesellschaft, 1823.

cross-branch may at one time convey blood towards, and at another time away, from the heart.

The blood of the cockroach is a clear coagulable fluid, containing large nucleated corpuscles. A drop of blood deposits on drying a number of colourless crystals, which often form radiate clusters. The quantity varies greatly according to the nutrition of the individual; after a few days' starvation, nearly all the blood is absorbed. Larvæ contain much more blood in proportion to their weight than adult insects.

SCIENCE-GOSSIP.

MR. E. J. BEAUMONT showed at a meeting of the Pathological Society, at the Royal Berkshire Hospital, Reading, a newly-invented automatic arm and hand. A man who had lost his arms was fitted with the artificial limb, and showed the audience that he could pick up a pin and move it about, and still retain the grasp; after which he ate a piece of cake and drank a cup of coffee, and then wrote his name. Mr. Beaumont showed that with these patent arms a person can pick up things off the ground, and carry a parcel or basket of considerable weight.

A BOOK-WORM is described in the dictionaries as "a great reader or student of books," and also as "a worm that eats holes in books." The "Publishers' Circular" says: "We confess that, although quite familiar with the little circular tunnel to be met with in bound books as well as in 'quires,' we have never seen the engineer that so scientifically performs this destructive kind of work, until Mr. Bowden sent a specimen." This is figured in the last number of the "Circular." It is a white wax-like grub, exactly resembling the little white maggots seen in a well-decayed "stilton." Mr. Bowden says: "Booksellers are often made aware, in a manner that is more painful than pleasant, that there are such things as book-worms in existence. However, it is not many booksellers that have ever seen one; for despite its large ravages, the worm itself is very rare. Mr. G. Suckling discovered three at Messrs. Sotheran's Strand house a few days ago. They were half-way through a bundle of quires, and were evidently on their second or third journey, judging from the number of perforations made in the paper. Mr. Blades devotes, in his 'Enemies of Books,' some space to a description of this destructive, but withal interesting species of worm."

PROFESSORS AYRTON AND PERRY have brought out their electric tricycle. Its driving-wheel is forty-four inches, the electro-motor is placed beneath the seat, and the battery acts directly upon its cogged spur wheel. The battery is equal to two horse-power, and can be regulated with the utmost nicety.

As the result of six years' experiments, M. de Cyon has arrived at the conclusion that borax may be introduced in any required quantities into the system to preserve it from all contagions caused by parasites or germs. For cholera, M. de Cyon recommends boric acid or borax, to be applied to all the external mucous membranes, and about six grains to be taken every four hours with the food and drink.

IN a paper read before the Paris Academy of Sciences, Mr. Sace gave a full account of the immense deposits of saltpetre which are found in Bolivia. He says they contain over 60 per cent. of nitrate of potash, sufficient to supply the whole world, and over 30 per cent. of borax. Mr. Sace believes that the saltpetre is the result of the decomposition of an enormous quantity of fossil animal bones.

THE period of the double star β Delphina, as calculated by Dr. Dubjajo, is 26·07 years, and its periastron passage 1882·29.

A FRENCH electrician, M. Reynier, has brought out a new accumulator, stated to be the lightest yet introduced; an important factor in its application to tram-carriages, tricycles, &c. In practice, its actual weight of seventeen kilogrammes gives 7,600 kilo-grammeters' storage per kilo, much greater than has yet been yielded.

AN almost new industry promises to be developed by the discovery that the hitherto intractable metal iridium can be easily melted by the addition of phosphorus.

IN the midst of the controversy about overhead wires comes a practical testimony in their favour. The new drill hall at the State University, Minneapolis, was struck by lightning, when there were above four thousand people in it. All at once a series of electric lamps were lighted, and as suddenly extinguished. A loud report followed, and balls of fire were seen following the electric wire away from the building. The fine wire circuit feeding the lamps was fused by the intensity of the current. This is not the first time that overhead wires have proved protective during a thunderstorm.

THE French Northern Railway Company have commenced a series of experiments bearing on the transformation of electrical into motor force. An electric lift has been constructed at the Chapelle Station, with two Siemens' electro-magnetic machines, one for elevating the weight, and the other for moving the machinery alongside the railway.

WHILST the Australians are complaining of the rabbit pest, another has started in Victoria and New South Wales,—a breed of semi-wild dogs. Great slaughter of sheep takes place through them, and the Government has offered rewards for their destruction.

BARON HOCHSTETTER, distinguished for his geological explorations of New Zealand, has just died at the comparatively early age of fifty-five years.

MR. EDISON is showing a thirty-ton dynamo at the Philadelphia Electrical Exhibition. It is the largest yet constructed. Visitors have to be warned not to approach too near.

A PATENT has been granted for a process of facilitating submarine exploration, in which the oxygen required for breathing purposes is produced by electrically decomposing sea-water.

THE Eruption of Krakatoa is on its defence. The sunset glows have, if possible, been more conspicuous and beautiful than ever during the last month. Surely, there cannot be so much dust in the upper regions of the atmosphere now as there was twelve months ago!

AN important new departure in railway lighting is reported from the district where the passenger railway had its birth—the Liverpool and Manchester line of the London and North-Western Company. This is the utilisation of electricity for lighting the carriages by the help of Swan's incandescent 20-candle power lamps and Brotherhood's patent engine, stationed on the tender, and fed with steam from the locomotive boiler, locomotives being specially fitted for this service. The electric current passes from the engine through the train, and back to the locomotive, where, fixed to the footplate, is a regulator fitted with an electric burner, showing the driver the power of light in the train. Each compartment of the train is fitted with a duplicate lamp, the arrangement securing the instant lighting of one lamp if the other should become extinguished.

THE oldest inhabitant in the zoological collection in the Regent's Park died the other day. This interesting individual was a specimen of the black parrot from Madagascar (*Coracopsis vasa*). It was presented to the Society by the late Mr. Charles Telfair, a corresponding member, so far back as July 1830, just two years after the gardens were opened. This bird has therefore lived for 54 years in the gardens. How old the parrot was when it arrived we cannot learn, beyond the fact that it was represented as an "adult bird."

MR. ANDREW CARNEGIE, of New York, has recognised the importance of the microscrope in medical practice. He has recently made an absolute gift of fifty thousand dollars to Bellevue Hospital Medical College of that city, to be expended in the erection of a building and an apparatus to be devoted to laboratories for practical work and teaching in medicine. It is the design of Mr. Carnegie to establish a laboratory for the conduct of microscopical investigation.

AN illustration of the perfection to which lip-reading can be brought was given by a deaf girl before delegates to the recent convention of the teachers of the deaf and dumb. By the movement of a speaker's lips outlined in shadow on a wall she was enabled to decipher the words uttered.

DR. ALFRED WRIGHT says that the house-fly is a most dangerous agent in the propagation of disease, and he insists on the need for doing all that is possible to keep this insect at bay. He draws attention to a popular fallacy that flies will not pass over a barrier of geraniums or calceolarias, a quite erroneous idea, but he points out that flies cannot endure the eucalyptus and its products, and he wisely commends an eucalyptal preparation as a pleasant and valuable disinfectant in any room whence flies are to be excluded.

PROFESSOR MACLEOD, of Cooper's Hill Engineering College, has invented an ingenious sunshine recorder. He places a water lens or a globular bottle of water in front of a camera obscura, in such a position that the ray of light falls on a sensitive piece of paper spread on the bottom of the camera box. As the sun revolves a curved band is produced on the paper, which stops when the sun is obscured.

IT is our sad duty to chronicle the death of an old and valued friend, and an ardent field-worker in Lancashire geology. Mr. John Aitkin, F.G.S., was for many years president of the Manchester Geological Society, where the chief of his papers were read. Of late years he worked hard with the microscope. He was a true scientific student to the end—without a particle of that scientific dogmatism now only too threatening—always willing to learn, and thankful for being taught. No man ever lived a more blameless or upright life, or was more widely respected.

THE Fourth Annual Report of the Walthamstow Natural History and Microscopical Society has just been published, from which we gather that the Society is in a flourishing condition. The curator's Report is excellent.

MR. J. B. SUTTON, the lecturer on comparative anatomy at Middlesex Hospital, has shown that the commonly-accepted notion that monkeys brought to this country die of tuberculosis is a wide-spread error.

MM. DEPIERRE AND CLOUET have communicated to the Industrial Society of Mulhouse some experiments upon the bleaching action of rays of solar and electric light upon colours printed upon calico. The electric light bleaches as does the solar light. All colours of rays bleach, but not equally. The bleaching takes place either in air or in vacuum. The yellow rays are the least active, and the red rays the most active. Of all artificial lights the electric light is the most active.

IT will be some time before the great heat we experienced in England in the earlier part of August is forgotten. On the 11th, the thermometer at Greenwich was 94·8 in the shade, and 157·5 in the sun. This is the highest temperature we have experienced for twenty years past.

MR. A. IRVING states that the deleterious effects of nitric acid burns may be quite easily prevented. Mr. Irving severely burnt his face with concentrated nitric acid while making some electrical experiments, and, reasoning that the effect was one of oxidation, concluded that dilute sulphuric acid should alleviate the suffering. In a few minutes after the application was made, he says, "The blister was reduced, and the oxidation completely arrested; the painful irritation disappeared, and the wound was comparatively cured."

DR. RICHARDSON, in the last number of "The Asclepiad," pleads for "Euthanasia for the lower creation," and suggests a lethal chamber for animals which have to be destroyed from disease, old age, injury, or other causes.

DR. RICHARDSON also contends for "A sea atmosphere in a sick room." A spray consisting of a solution of peroxide of hydrogen (10 volumes' strength), containing 1 per cent. of ozonic ether, iodine to saturation, and 2·50 per cent. of sea-water.

THE subject of poisoning by canned meats has been thoroughly discussed before the Medico-Legal Society of New York. It was proved that the so-called fatal cases failed, on endeavouring to trace them to poisoning by muriatic acid or otherwise. Mr. Barrett said 60 million dozen tins were exported annually, so that many more people ought to be killed, if canned provisions are dangerous.

MR. CHARLES MANBY, F.R.S., who was so long associated as Secretary with the Institution of Civil Engineers, has just died at the ripe old age of eighty years.

THE Essex Naturalists' Field Club held a capital field meeting at Colchester on Bank holiday, and went over the recently visited earthquake district, under the leadership of Mr. R. Meldola. At Colchester the party were led by the Rev. C. L. Ackland and Mr. Henry Laver, F.L.S. Mr. H. Stopes, F.G.S., addressed the members on the "Salting Mounds," or "Red Hills."

A NEW comet has made its appearance. It made its perihelion passage on August 17th, but it will be observable in this country for some time. Mr. Chandler calculates its orbit as follows:

12 h. G.M.T.	R.A. h. m.	N.P.D. °	Distance from Earth.	Intensity of Light.
September 3	18 27·2	123	8 0·682	1·06
,, 7	18 44·2	122	6 0·701	0·98
,, 11	19 0·2	120	58 0·722	0·91
,, 15	19 15·9	119	46 0·747	0·83

VISITORS to the Health Exhibition continue to flock in at the rate of nearly a quarter of a million persons a week.

THE Marine Biological Association is now an established fact, thanks to the energy of Professor Lankester and Professor Moseley. Plymouth has been selected for the site of a laboratory and experimental aquarium. Friends are still required to successfully carry out this truly national project, and we hope many of our readers will see their way to helping it, either by donations or otherwise. We should be happy to receive subscriptions for the purpose.

A COMMITTEE appointed by the Paris Academy has examined and rejected no fewer than 240 "infallible nostrums" for curing and arresting the progress of cholera!

MR. MARES, a French geologist, states that he has discovered in Tunisia a regular superposition of the upper cretaceous, eocene, and miocene formations.

WE are sorry to note the death of Professor Sir Erasmus Wilson.

NEW YORK and the district have recently experienced an earthquake shock, very similar to that which occurred in East Anglia last April. No lives were lost, and little property was destroyed, but the incident has been welcomely received by the "Earthquake prophets."

MICROSCOPY.

DIFFICULTIES IN MOUNTING.—If G. S. S. will varnish twice, at intervals of a couple of hours, with a solution of shellac in alcohol, and then finish off with ordinary bitumen, I don't think he will be further annoyed with air-bubbles in glycerine cell-mounting; at least, that is my experience, and also that of Professor Rothrock, from whom I heard of it. As to balsam mounting, I find that the method which hardens quickest is to keep a stock bottle of balsam, which is quite hard. Cut off a piece as required, and dissolve in chloroform: this hardens up again quickly by merely laying the slides on a tray in a window exposed to the sun, keeping the covers in position either with a clamp or an elongated rifle bullet.—*B. Sc., Plymouth.*

THE QUEKETT CLUB.—The May number of the "Journal of the Quekett Microscopical Club" contains articles "On the Florideæ, and some newly-found Antheridia," by F. H. Buffham, and "On Parasitic Vegetable Organisms in the Gabbard and Galloper Sands," by J. G. Waller.

COLE'S MICROSCOPICAL STUDIES.—The various numbers of these now widely-known "Studies" appear with remarkable regularity. Part 12, of the "Methods of Microscopical Research," has for its subject "Microscopical Art," by E. T. D. No. 11, of "Popular Microscopical Studies," treats on "Starch," and is accompanied by a beautiful plate, showing the starch grains *in situ* of *Sarsaparilla officinalis*, × 400. Nos. 22 and 23 of "Studies in Microscopical Science" deal with "Fundamental Tissue" (illustrated by wood-vessels and cells, and a transverse section of the petiole of *Limnanthemum*), and "The Human Cerebrum." The slides sent out with these numbers, and on which they treat, are all in Mr. Cole's best and neatest style of mounting and finish.

ZOOLOGY.

LOCAL SCIENTIFIC SOCIETIES.—The Fourth Annual Report of the Hampstead Naturalists' Club, containing the Presidential Address, list of members, &c., is to hand. The president, Mr. William Boulting, L.R.C.P. Lond., deals with "The Correspondence of Body with Mind." A valuable address on "The Natural History of the Diamond," by Professor Rudler, F.G.S., is embodied in the "Proceedings." The "Proceedings of the Bristol Naturalists' Society" deserves more than a passing notice. Part VI. of the "Catalogue of the Lepidoptera of the Bristol District," by Mr. Alfred E. Hudd, M.F.S., appears in this issue, and, among others, are the following: "Report on Wells sunk at Lockes, Somerset, to test the alleged power of the Divining Rod," by Professor W. J. Sollas ; "On an Ergometer for small Electromotors," by the Rev. F. J. Smith ; "Recent Researches on Dynamo-Electric Generators," by Professor Silvanus P. Thomson ; "On the Primary Divisions and Geographical Distribution of Mankind," by James Dallas, F.G.S., F.R. Hist. Soc.; "Fungi of the British District," by C. Bucknell, Mus. Bac. The "Papers and Proceedings of the Royal Society of Tasmania" contains, in addition to others, the following papers : "Description of New Tasmanian Animals," by E. T. Higgins, M.R.C.S., and W. F. Pettard, C.M.F.S., and "Notice of Recent Additions to the List of Tasmanian Fishes," by R. M. Johnston, F.L.S., &c. Among the "Transactions of the Chichester and West Sussex Natural History Society," are the following : "A New Aphis," by J. Anderson, Jun.; "Vaseline," by A. Lloyd; "The Three Oldest Fossils," by Rev. H. Housman ; "Germs," by Dr. Dutton. The following papers, in addition to the Address of the President, Mr. G. E. Davis, F.R.M.S., F.C.S., appear in the "Annual Report" of the Manchester Microscopical Society : "The Oval Organs of the Gad-Fly," by J. B. Pettegrew ; "*Gyrodactylus elegans*," by Herbert C. Chadwick,

F.R.M.S.; "The Hatching of Rotifers in *Volvox globator*," by James Fleming, F.R.M.S.; "Penetration in Objectives," by G. E. Davis, F.R.M.S., F.C.S. ; "The Hydra, its anatomy and development," by T. W. Dunkerley, F.R.M.S.; "Spontaneous Fusion in an aquatic worm," by W. Blackburn, F.R.M.S.; "The Ginger-beer Plant," by G. E. Davis, F.R.M.S., F.C.S., &c.; "The Forms, Origin, and Development of the Teeth," by Parsons Shaw, D.D.S.; "On a Small Collection of Hydroid Zoophytes and Polyzoa from the Menai Straits," by Herbert C. Chadwick, F.R.M.S.; "Selaginella : Alternation of Generations," by W. Stanley, F.R.M.S.; and "The Application of Quantitative Methods to the Study of certain Biological Questions," by H. C. Sorby, LL.D., F.R.S.

REVERSED HELICES.—The early part of this week I had a very pretty banded *Helix nemoralis*, var. *hybrida*, var. *minor*, var. *sinistrorsum* (reversed), sent me from a village near Bristol. The shell is nearly an adult, and the rich pink lip adds greatly to its beauty. I am keeping it in company with some other *Helix hortensis*. It is healthy, which is not often the condition of the snails inhabiting reversed shells. This specimen adds another to our Bristol list of reversed helices.—*Fanny M. Hele.*

PALUDINA VIVIPARA NEAR MANCHESTER.—Mr. Dyson, in his little work entitled "Land and Freshwater Shells of the Manchester District," does not include either this species or *Planorbis corneus* in his list. I have within the last week, however, discovered both species inhabiting a pond in Baguley, near Manchester, *P. vivipara* being especially plentiful. In addition to these two species, I took specimens of the following from the same pond : *Limnea pereger*, *L. auricularia*, *Bithinia tentaculata*, *Planorbis carinatus*, *Cyclas lacustris*, and *Anodon cygneus*, making in all eight species in one small pond.—*C. Oldham.*

BOTANY.

CAREX OVALIS, VAR. BRACTEATA.—In a walk from Thirlmere to Keswick (Cumberland) on the 8th of July, I found among a number of patches of *Carex ovalis* growing by the roadside one in which the bracts of the lower spikelets were mostly leafy and longer than the spike. On examining the tuft, I found four different spikes (now in my herbarium) as follows :—1st, lower bract leafy, one inch longer than the spike ; 2nd, lower bract half inch longer than the spike ; 3rd, lower bract still leafy, but about ¼ inch shorter than the spike—reaching in fact to the base of the terminal spikelet ; and lastly, one of the ordinary form of *C. ovalis*, with the lower bract short and membranous like the other bracts. The variety with the lower bract longer than the spike has been distinguished as var. *bracteata*, and is recorded from "Castle Morton Common, Worcestershire." Besides the one tuft mentioned above, none of the other tufts appeared to contain long bracted specimens, so that these observations would tend to show that the long bracted form was a chance variety and not a permanent one.—*G. H. Bryan.*

DISEASES OF FRUIT.—In the "Popular Science Monthly," Mr. D. P. Penhallow remarks that, as a result of the peculiar methods of cultivation employed by fruit-growers, various diseases have appeared from time to time in several important fruits, and indeed such has been their extent within the last ten or fifteen years, that in some places the fruit industry has been completely destroyed. "Blight" has been known in pear-trees for the last hundred years, and the disease has undoubtedly increased during that period. "Yellows" in peaches again has been constantly developing since first heard of eighty years ago. The disease is so thoroughly established, and has become so much a matter of inheritance, that the life of the tree is greatly determined by it. The peach is naturally a long-lived tree, and has in many cases survived for a hundred or more years, and, with proper care, bears fruit for a long period. Now, however, about Delaware and New Jersey, owing to the certainty of the disease appearing, or the inherently weak constitution, the trees are rooted up as worthless at the end of nine years.

GEOLOGY, &c.

SONGS FOR THE TIMES.

WHICH IS THE OLD ARCHÆAN LAND?

[To be sung by the "Brotherhood" in Section C at the forthcoming meeting of the British Association at Montreal.]

AIR.—*Was ist des Deutschen Vaterland?*

Which is the old Archæan rock?
Who can the mystery unlock?
Is't found on bold St. David's Head,
Where Hicks has fought and ever led?
    Oh, no! Oh, no! Oh, no!
    Says Geikie, We must wider go.

Which is the old Archæan rock,
That does the Survey's feelings shock?
Does't cover much of Scotland's land,
Where works that single-minded band?
    Oh no! Oh no! Oh no!
    As Murchison does plainly show.

Which is the old Archæan rock,
'Gainst which our shins so often knock?
In Ireland is it to be seen,
Set like a gem in emerald green?
    Oh no! Oh no! Oh no!
    The Kinahan does whisper low.

Which is this old confounded rock
That seems our intellects to mock?
Is it within the English land?
Or shows it on wild Cambria's strand?
    Oh no! Oh no! Oh no!
    Sir Andrew Ramsay well does know.

Which is this hard outrageous rock,
Our Wednesdays' meetings born to block;
When oaths are sworn with graspèd hand,
And few their feelings can command?
    Oh no! Oh no! Oh no!
    Our sentiments we plainly show.

Which is that old Azoic rock?
Canst pick it out in crumpled block?
'Tis where we find the gnarlèd schist,
And fossils do not now exist.
    That let it be! That shall it be!
    Archæan, it belongs to thee.

Whenever, on whatever strand,
We meet with schist and gneissic land,
That is Archæan land we see,
And that our fatherland shall be.
    On that we go. That shall be so.
    We'll claim it from whatever foe.

Nor e'en except the lunar land,
That left those gashes broad and grand
To which the deeper waters drew,
As from the earth it skyward flew.
    Oh no! Oh no! Oh no!
    For Fisher saw some left below.

So in a firm united band
At Montreal we take our stand;
In Alpine schist our land we meet,
We claim the rocks beneath our feet
    Oh yes! Oh yes! Oh yes!
    No claim so opportune as this.

That is the wild Archæan's land,
And he possession will demand;
While in his eye clear truth does shine,
His native rocks he'll not resign.
    Gneiss let it be; schist shall it be;
    The whole of schistose land it be.
                    A. CONIFER.

NOTES AND QUERIES.

CURIOUS TRAP FOR A BIRD.—Whilst waiting in the train at Ashchurch station in Gloucestershire, on the 30th June, my attention was attracted by a bird fluttering in one of the lamps. A young fellow presently passed by, and I asked him to let it out. He put his hand into the lamp, caught it, and brought it over to me, when I was much surprised to see it was a wryneck (*Yunx torquilla*), one of our summer visitants. The only way I can account for it being in so peculiar a situation is, that being an insectivorous bird, it was attracted by the flies which usually abound inside lamps. This is the first time such a thing has come under my notice, and I thought it might be interesting to some of the readers of SCIENCE-GOSSIP. —*P. T. Deakin.*

MISCELLANEOUS NOTES.—Last year I described several species of Chrysomela which I had not been able to identify. I have now ascertained the names of the British species, which are as follows :—*C. polita, C. staphyloca, C. hyperici,* and *C. gœttingensis.* The foreign species I believe to be *C. menthrastri.* The lady-bird described by Mr. Moulton is a very common one: I have often taken it myself in company with the ordinary red form. In answer to G. T. R., I may say that the dormouse (*Myoxus avellanarius*) is by no means uncommon in the woods near Battle in Sussex, and is sometimes taken in the hollows of trees. Mr. Elliot mentions the occurrence of *Arion ater,* var. *bicolor,* near Stroud, but he gives us no description of it, nor do I remember having ever seen it described in any British journal. *Vertigo tumida* of Westerlund is included in the Conchological Society's list of British shells. Can any reader inform me where it has been found in Britain? As far as I can ascertain, it is confined to Scandinavia.—*T. D. A. Cockerell.*

HYDROGEN GAS GENERATED, BUT NOT CONSUMED.—A clergyman, scientifically ignorant on the subject, would feel greatly obliged by information needed for a special object on which he is engaged. It is assumed that about 150 millions of tons of coal are now annually raised in the British Isles, and that probably about the same quantity is supplied by all the other coal-producing countries of the earth. It is estimated that at least one-third of the coal used for household fires and for the furnaces of factory and other engines is unconsumed, through the thoughtlessness of servants, in loading the fires with too much coal at one feeding; so that the gases evolved from such portion do not combine with the oxygen of the air to produce combustion and heat, but ascend through the chimney, and float in the surrounding air perceptibly as smoke. If this be in any degree the case, the inquirer is anxious to ascertain what becomes of the immense volume of hydrogen thus generated, being about 70 per cent. of the coal elements? Does the hydrogen, as the lightest of all known elements, rise in the earth's atmosphere, working its sinuous course upward, and forming a stratum above the atmosphere, remaining superincumbent upon it, and surrounding it, or does the hydrogen combine with the atmosphere, at a low level, and become in any way absorbed by other substances, into contact with which it may be drawn, or otherwise chemically united?—*S. C.*

DOES THE SPARROW-HAWK ATTACK TOADS?—A short time ago, while driving to Needham Market, I met a sparrow-hawk, flying exactly in the opposite direction at a slow pace, as if heavily ballasted, and about fifteen feet above my head. I stopped to watch him; he alighted in the middle of the road, about sixty yards behind me, and began plucking at some object which he had carried in his claws. As he pulled no feathers, I drove back to find what he had got, when he flew away, leaving behind him a toad so knocked about that it did not attempt to crawl. I had never before supposed that a hawk would prey on toads.—*Henry Ridley, Ipswich.*

EARLY DRAGON AND LACE FLIES.—On Sunday, May 11th, I captured two specimens of *A. minium,* both males. They were very weak, and appeared as if they had just left the pupa case, their flight being slow. I also captured at the same place a lacefly (*H. pula*), on a piece of floating wood. The vicinity was Penylam, and the time four in the afternoon. The sun was very brilliant, the thermometer registering 64 degrees. Is it not early for them? The water-beetle (*G. natator*) was very abundant, as well as the larvæ of the caddis fly, in the pond I captured the dragon flies.—*H. J. Wheeler, Jun., Cardiff.*

LADY-BIRD.—Lady-birds with red spots on black ground, though not nearly so common as the red ones with black spots, cannot be called uncommon in the neighbourhood of Birmingham, and are sometimes very beautiful.

THE DORMOUSE IN ENGLAND.—At Longbridge, Northfield, seven miles from Birmingham, on the Worcestershire side, I have twice seen the dormouse within the last ten or twelve years; the first time in an ordinary country garden, running across a lawn and disappearing under a wall thickly covered with ivy. The soil is loam, and there are many oak-trees about. The second specimen (or it may have been the same individual some years older) was brought in by the cat, who was very fond of the field-vole, of which she caught many large specimens and of the long-tailed field-mouse. The colour of these dormice was a rich chestnut brown, darker by several shades than the yellowish-brown individuals mostly seen in confinement.—*Benjamin Scott.*

PIED LAPWING.—The pied lapwing was shot at Pevensey on the 14th of September, 1883. I might also state that a cream-coloured starling was caught at Crowhurst, near St. Leonards, last May, and is in my possession. On the 1st of August, I saw a white martin at Rye; it was being chased by several other martins, who no doubt were astonished by the unusual colour of their mate.—*G. Bristow.*

NOTICES TO CORRESPONDENTS.

TO CORRESPONDENTS AND EXCHANGERS.—As we now publish SCIENCE-GOSSIP earlier than heretofore, we cannot possibly insert in the following number any communications which reach us later than the 8th of the previous month.
TO ANONYMOUS QUERISTS.—We receive so many queries which do not bear the writers' names that we are forced to adhere to our rule of not noticing them.
TO DEALERS AND OTHERS.—We are always glad to treat dealers in natural history objects on the same fair and general ground as amateurs, in so far as the " exchanges " offered are fair exchanges. But it is evident that, when their offers are simply disguised advertisements, for the purpose of evading the cost of advertising, an advantage is taken of our *gratuitous* insertion of "exchanges" which cannot be tolerated.
WE request that all exchanges may be signed with name (or initials) and full address at the end.

H. MOULTON may be informed that his ladybird was no doubt *Coccinella variabilis*, a species quite as common as the more familiar *C. septem-punctata*.—W. C. HEY.
J. O. B.—The fragments of spider sent belong to *Dysdera erythrina*. Get Staveley's "British Spiders," published by Lovell Reeve.
J. DRABY.—The piece of rock from quarry near Sunderland is covered with the dendritic crystallisation of oxide of manganese. It looks very much like a fossil moss, but it is not organic at all.
W. HAMBROUGH.—Many thanks for sending us the curious and beautiful specimen of the double feather of a pigeon. It is very remarkable; we have not seen anything like it before.
E. A. DENNIS.—Thanks for the paper containing Mr. Cooper's lecture on Darwin. The lecturer evidently understands as much of the Darwinian theory as he does of Sanscrit.

A. M. P.—In the Rev. J. G. Wood's recently published book "Petland Revisited," you will find a chapter on Tame Chameleons, which will supply you with all the information you need.

R. RIDINGS.—You will find it difficult to get a book on Marine Conchology, with illustrations, at the price you name. There is a nice little book, with plenty of woodcuts, by the Rev. J. G. Wood, published by Messrs. Warne at 1s., called "Common Shells of the Sea Shore." Get this first.

T. S. K.—The gold-fish do not need much feeding. One reason for their destruction in aquaria is over feeding. All foods liable to ferment should be avoided. Small pearl biscuits are best. One broken up now and then, say every month, is sufficient. The vegetation induces hosts of Infusoria to breed. Don't keep the fish in too bright a place.

A. J. F.—Your grubs are the larvæ of the stag beetle (*Lucanus cervus*). Watering the ground with a solution of sulphate of ammonia might get rid of them without injuring the plants. You will get this solution from any chemist.

EXCHANGES.

SPLENDIDLY-preserved and correctly-named Swiss Alpine plants. Price 6d. each.—Address, Dr. B., care of Editor of SCIENCE-GOSSIP, 214 Piccadilly.

A PHOTOGRAPH of the interior of the celebrated Wren's Nest Cavern, Dudley, in exchange for one good micro slide of interest.—W. Tylar, 20 Geach Street, Birmingham.

SHELLS for exchange: *Paludina vivipara*, *Zonites glaber*, *Helix Cartusiana*, *Unio tumidus*, *Achatina acicula*, and many others. Wanted, well-mounted slides; or other shells, of which send lists of duplicates and desiderata to—S. C. Cockerell, 51 Woodstock Road, Bedford Park, Chiswick, W.

SPONGE (*Halichondria panicea*) and other material (quantity if wanted) for non-botanical slides. Send lists to—G. H. Bryan, Thornlea, Trumpington Road, Cambridge.

WANTED, condenser on stand; will give in exchange one dozen well-mounted slides.—E. G. Adams, Holmleigh, Parkhurst Road, Bexley, S.E.

WANTED, "Practical Microscopy," by G. E. Davis; will give in exchange eight well-mounted objects.—E. G. Adams, Holmleigh, Parkhurst Road, Bexley, S.E.

To pharmaceutical students: 35 officinal indigenous plants, mounted on card, 8 × 10, equal to new; small agate mortar or microscopic slides accepted in exchange.—H. E. Ebbage, 34 Queen Square, Wolverhampton.

"NATURE," 3 Vols., 1882–3; SCIENCE-GOSSIP, 1880, 81, 83; "The Zoologist," 1883; "Science Monthly," Vol. I. ; all clean, unbound; will exchange for biological or geological books, or Quarterly Journals Geological Society.—H. E. Quilter, 55 Earl Howe Street, Leicester.

WANTED, living specimens of *Dytiscus marginalis*, mounted insect slides, and also wing cases of "diamond beetles," in exchange for first-class anatomical and pathological sections.—Henry Vial, Crediton, Devon.

A COLLECTION of well-stuffed British birds, in or out of cases, for breech-loading gun. List sent on application to—G. Bristow, Danbury House, Silchester Road, St. Leonards.

OFFERS wanted in micro slides for entomological apparatus, and 400 specimens, good and bad, named and unnamed. Also unmounted micro material, and slides of botanical, histological, and general specimens. Write for list to—A. Downes, Kentisbeare, Cullompton, Devon.

WANTED, Lowne on the Blowfly, or any literature relating to the Diptera; state requirements.—W. H. Harris, 44 Partridge Road, Cardiff.

DUPLICATES: *Limnæa stagnalis* and *P. corneus* (very fine), *L. glabra*, *L. peregra*, *P. spirorbis*, *L. palustris*, *Z. alliarius*, &c. Desiderata.—W. Hewett, 26 Clarence Street, York.

WANTED, eggs of the smaller British birds, and rare British shells; many varieties of marine shells to exchange.—C. D. S., Maplewell, Loughborough.

MAGIC lantern slides, books, and microscopical slides; exchange other microscopical slides.—G. Harrison, 12 Dalmain Road, Forest Hill, London, S.E.

To exchange: 1. "Intellectual Observer," February and March, 1862, for odd numbers of other natural history periodicals. 2. A few Maltese marine and land shells for British land and freshwater shells (varieties), insects, not Lepidoptera, &c.—E. F. Beecher, Hill House, Southwell, Notts.

WANTED, Darwin's "Insectivorous Plants." I have for exchange a lot of microscope slides, a fine collection of foreign postage stamps, and a lot back numbers of "English Mechanic" (consecutive). Offers to—Kilgour, 63 Dallfield Walk, Dundee, N.B.

BUTTERFLIES for exchange: *P. alsus*, *A, galathea*, *E. hyperanthus*. Wanted, local British butterflies, especially Argynnis or Melitæa.—F. Rutt, 7 Connaught Road, Folkestone.

Unio pictorum, 4½ inches long, a few specimens for exchange. Desiderata : *Helix revelata*, *H. villosa*, *H. lamellata*, *Bulimus montanus*, *B. Goodallii*, *Succinea oblonga*, *Paludina Zisteri*, *Limnæa involuta*, *Planorbis dilatatus*, the rarer species of Vertigo and Pisidium, and the rare varieties of other shells; a great variety for exchange.—W. Gain, Tuxford, Newark.

OFFERED, well-preserved specimens of British plants for British fossils, shells, or minerals. Send lists of duplicates and desiderata to—F. C. King, 2 Clarendon Street, Preston, Lancashire.

WANTED, British, Devonian and carboniferous fossils, particularly corals, fish remains, and Brachiopoda, in exchange for rare Devonian and carboniferous fossils of Iowa. Send as full lists as possible.—W. R. Lighton, Ottumwa, Wapello Co., Iowa, U.S.A.

WELL-MOUNTED slides of foraminifera from soundings taken by H.M.S. 'Porcupine,' for well-mounted material containing foraminifera from silts at Sutton Bridge, Dee, Clyde districts, &c., or gatherings from Ireland or the coast of Cornwall.—G. Bailey, 1 South Vale, Upper Norwood, S.E.

WANTED, good microscope or scientific books in exchange for Cassell's "History of England," 9 volumes, half-bound.—John B. Blakeley, New Street, Horbury, near Wakefield.

WANTED, in exchange for fossils from chalk and Thanet sand, &c., medals, English or Roman coins or war medals, also rubbings of monumental brasses.—A. Leonards, 6 Clifton Gardens, Margate, Kent.

OFFERED, L.C., 7th ed., Nos. 45, 148, 197, 406, 859, 1008, 1333, 1349, 1412, 1422, 1447, 1448, 1458, 1494, 1501, 1504, 1515, 1577, &c. Send lists.—H. J. Wilkinson, 17, Ogleforth, York.

A BEAUTIFUL slide of ruby sand, well mounted, in exchange for grouped diatoms or well-mounted foram slides.—W. Aldridge, Upper Norwood, London, S.E.

WANTED, eggs in exchange for chiffchaff, jackdaw, magpie, ringdove, turtledove, coot, partridge, red-legged partridge, pheasant, black-headed gull, herring gull, &c.—F. Fenn, 20, Woodstock Road, Bedford Park, Chiswick.

WILL exchange this year's eggs of lesser redpole and winchat for eggs of rook, crow, or others.—J. Ellison, Steeton, near Leeds.

GOOD ⅘ in. object-glass (French triplet, in English mounts). Wanted, Dana's "System of Mineralogy," with the Appendices, or lantern slides.—C. R. L., 69, West Worsley Street, Salford.

WILL send living budding specimens of *Hydra viridis* on receipt of good mounted object or for specimens of Chara, Nitella, Volvox, or marine microscopic life. Wanted, a first-class ⅛ inch objective.—Thomas W. Lockwood, Lobley Street, Heckmondwike, Yorkshire.

A FEW duplicates of each of the following for exchange:—*Helix lapicida*, *H. rufescens*, v. *alba*, *H. arbustorum*, the new var. *pallida*, *Physa hypnorum*, *Bulimus acutus*, *Clausilia laminata*, *Cyclostoma elegans*, *Neritina fluviatilis*. Wanted, *Helix pisana*, *H. lamellata*, *H. revelata*, *H. obvoluta*, *H. fusca*, *H. concinna*.—George Roberts, Lofthouse, near Wakefield.

MICRO SLIDES. Offered, stained sections, as petiole of *Nuphar lutea*, &c., also named mosses and Hepatics, and many other good slides in exchange for insect mounts or others of interest.—Lists to W. E. Green, Cleve Dune, Belvoir Road, St. Andrew's Park, Bristol.

BOOKS, ETC., RECEIVED.

"Manual of the Mosses of North America," by Leo Lesquereux and T. P. James. Boston: S. E. Cassino & Co.—United States Geological Survey, Second Annual Report, 1880–81.—"Photography for Amateurs," by T. C. Hepworth. London : Cassell & Co.—"Proceedings of the Liverpool Naturalists' Field Club, 1883–84."—"Studies in Microscopical Science," edited by A. C. Cole.—"The Methods of Microscopical Research," edited by A. C. Cole.—"Popular Microscopical Studies." By A. C. Cole.—"The Gentleman's Magazine." —"Belgravia."—"The Asclepiad," No. 3.—"The Journal of Microscopy."—"The Science Monthly."—"Midland Naturalist."—"The Journal of Conchology," July.—"Ben Brierley's Journal." — "Science Record." — "Science." — "American Naturalist."—"The Electrician and Electrical Engineer." — "Canadian Entomologist."—"American Monthly Microscopical Journal."—"Popular Science News." — "The Botanical Gazette."—"The Medico-Legal Journal" (New York)—"Revue de Botanique."—"La Feuille des Jeunes Naturalistes." —"Le Monde de la Science."—"Cosmos, les Mondes."—"Revista," &c. &c. &c.

COMMUNICATIONS RECEIVED UP TO 12TH ULT. FROM :—R. P.—T. D. A. C.—W. R. L.—S. C. C.—E. J. B.—H. E. E. —H. L.—H. K. Q.—W. H. H.—F. G. S.—O. P. C.—J. A.—E. G. A.—E. A. D.—A. C.—S. B. A.—J. D. B.—J. B.—H. R. —H. J. W., jun.—W. S.—W. H.—B. S.—A. O.—R. R.—A. F. —Miss F. G.—C. F. G.—J. A. W.—L. G. F.—G. H. B.—W. C.—W. C. H.—D. C.—W. M. G.—H. I. T.—A. D.—C. B. —W. H. H.—H. T. T.—T. L.—W. H.—H. V.—D. B.—W. G. —E. L.—S. H.—C. D. S.—A. M. P.—P. K.—A. H. B.—W. D. —F. M. H.—E. F. B.—A. D. C.—H. J. M.—G. W. B.—S. B. B.—A. L.—H. J. W.—W. A.—F.—F. J. E.—C. R. L.—T. W. L.—G. B.—G. R.—T. B., jun.—F. C. K.—W. E. G.—W. M. W.—J. S.—H. & Co.—A. J. F.—G. W. R.—A. C. C. —G. W.—C. D., jun.—E. J. H., &c.

GRAPHIC MICROSCOPY.

E. T. D. del. ad nat. Vincent Brooks Day & Son lith

EGGS OF HOUSE FLY.
× 75.

GRAPHIC MICROSCOPY.

By E. T. D.

No. X.—Eggs of House-Fly.

IN form, colour, opalescence, and as exhibiting elegant-sculptured markings, the eggs of insects afford a class of microscopic objects of engaging interest.

Numerous as they must be, they are rather difficult to find and identify; a knowledge of the habits of the insect and its food-supply are the best clues, but even then they often evade detection.

The common house-fly (*Musca domestica*), ubiquitous, residentiary, and numerous as it may seem in the dwelling-house, even invading the sacred drawing-room, is found in far greater numbers in gardens and the surroundings of domestic offices; and it is only in obscure and neglected places where their elegantly-winged eggs are discovered, under circumstances depending on the economy of the creature, where the larvæ, when hatched, find sustenance, generally in moist putrefying substances; under such conditions, 60 to 80 eggs are deposited, in groups of a few; the laying is repeated after intervals, the eggs hatch in two or three days, and the perfect insect so quickly arrives at maturity that it has been calculated and placed on authoritative record, a single female in one season, through four generations, may produce two million descendants. To check this enormous fecundity, eggs are often placed in unfavourable places, when hatched, food failing, the larvæ again become the prey of enemies; the eggs may be sought for in neglected places, rarely inside an orderly dwelling. The fly only enters into the domesticity of the house when in perfect condition of trimness and cleanliness; occasionally its instincts may lead it into a pantry, lured by the odour of its attractions; the eggs may be cultivated under favourable circumstances, but the conditions should be watched, as they hatch quickly.

The general anatomy of the insect is not touched upon; past volumes of this journal are replete with accurate information, and the subject is ably exhausted in Mr. Samuelson's book, "The Earthworm and the House-fly."

Serious accusations have been made against the house-fly. A writer in the "Times" of the 8th of August last suggested that "savants," to detect infective germs, ought to "examine the feet of flies," as by such contacts, microbes or bacilli might be disseminated, and, if proved, we could "protect ourselves from disease by excluding flies from everything we eat or drink." This provisional elimination of the diptera (and to be effective it would include the whole group) from the equilibrium of existences is a bold idea, even in these days of bacillimania, and would establish a condition of things of startling interest.

Consider the *one* friendly fly, which we all know and have seen in the quietude of evening marching over the book, paper, or drawing, under the full glare of the lamp: revisiting at the same time, under the same conditions, evening after evening —a singular instance of sociability. He parades and jerks along in a somewhat weak and inane manner (the season being over), and with a familiarity bordering upon insolence, settling down within an inch of your pen or pencil to dust himself with scrupulous care, sweeping his legs over his head, and skimming his wings, finally washing with "imperceptible soap, in invisible water," and shuffling the result on to the paper. Shall he be accused of thus depositing the germs of disease?

Flies, generally, are unpopular; every one dislikes them; inventions have been devised to "keep them down;" they have been severely censured in the writings of the early ecclesiastics, who have accused them of "immorality," as "enemies to sleep," "satellites of man," "importunate dependents of

No. 238.—October 1884.

humanity," and "as emissaries of the devil, and the ghosts of heretics." Flies have always been distinguished for evil ; they were one of the plagues of Egypt, now afflicted with irritations of another Order.

The fly has curious habits ; it is persistent in attack, and repels interference. Most insects seek safety when meddled with, but the more a fly is buffeted, the more eager he is to return, and generally to the same spot. The perseverance of their attack is extraordinary. An individual "forest fly" will follow a horse for miles, and they have been known to travel to London in cattle vans containing New Forest ponies, and causing the direst excitement at Waterloo Station among the town acclimatised cab-horses.

It is somewhat surprising, considering the numbers deposited by each individual insect, how rarely, unless specially sought for, the eggs are discovered. Doubtless insectivorous birds clear off those most prominently placed, and it is possible that many rare eggs of the lepidoptera might be found in their crops and intestines—an idea supported by a very unique observation lately made by Mr. Tegetmeier, and graphically described by him in the "Field" of the 30th of August last. The rare incident of a cuckoo captured in the gardens of the old Charterhouse, in the very centre of London, afforded the opportunity of examining the contents of its gizzard. It was found full of a "mass of dark fragments composed of the wings and bodies of insects, skins of caterpillars mingled with an infinite number of small white granules." The writer had the opportunity of first seeing these white granules under the microscope. They proved to be the eggs of a moth—afterwards identified as those of the Vapourer (*Orgyia antiqua*). When washed and carefully dried, they were as intact as when laid, and, notwithstanding the ordeal they had undergone, with care, and under favourable circumstances, might be hatched. Mr. Tegetmeier is under the impression the bird fed freely on fertile insects, the eggs retaining a vitality which prevented their digestion, but it is possible the passive moth, lingering over her deposit, was swallowed with her treasure in one "bonne bouche." However this may be, it is a curious and interesting observation, and may lead to the examination of gizzards and intestines of birds for the discovery of objects of this character, otherwise unattainable.

Specimens are easily prepared and mounted for the cabinet, the only difficulty is to destroy vitality, without spoiling their beauty ; immersion in alcohol touching each with a hot needle, has been suggested, but a dip in water just under boiling-point is effectual, and in some cases improves their appearance. Before mounted, they must be carefully dried. Opalescent or iridescent eggs are seen to the best advantage empty, as mere shells, after the larvæ have escaped ; this especially applies to the eggs of parasites attached to hairs, or feathers.

Crouch End.

THE ORIGIN OF DOUBLE FLOWERS.

MR. MOTT'S article, following the suggestive one by Mr. Gibbs, has elicited my admiration as an effort of *à priori* reasoning. His preliminary observations command the assent of the understanding, and his law of the wave-form or force-wave is simply Herbert Spencer's law of rhythm as laid down in the First Principles. But I fail to see by what process, inductive or deductive, he arrives at the conclusion that "the double flower should indicate the climax of a species." In the previous sentence he says, "But as the climacteric of the species approached the individuals attaining more complete development, those organics at the terminal growing-points which formed the reproductive element, would now unfold into perfect petals, and reproduction would gradually cease." This sentence implies, to my mind, that Mr. Mott regards a change from reproductive organs to perfect petals as indicative of more complete development. Much depends upon the meaning of the word "development." I take it that an advance in development means more complete differentiation so as to secure a more thorough subdivision of labour, and I believe that is the view entertained by most biologists. If that is so, the presence of leaves, petals, and stamens (for example) in the same plant would represent a greater differentiation than the presence of petals and leaves only, and consequently a higher stage of development. If petals and stamens are specialised leaves, one would regard the petals as being less altered from the leaf-form than the stamens, and therefore an approach of the latter to the petaloid state would signify an approach to homogeneity in the organs of the plant—in other words an approach to the reverse of differentiation and of the complete development referred to by Mr. Mott. The stamens must be regarded as being of more direct importance to the plant than the petal which is merely a protection to the reproductive organs, or a means to fertilisation. In the absence of the stamens the petal is without function. What then can be the benefit to a plant when its stamens are metamorphosed into petals? Can a plant be said to be more highly developed because it has produced organs which are of no use to it? Such a change, as of stamens into petals, I should say, is a decided retrograde movement, and occurs rather when the species has passed its climax than at the time it actually attains that point. One may perceive how that owing to the easy circumstances of a species that has been successful in the competition for existence, the energy requisite for producing reproductive organs may be lacking, though sufficient to produce these petaloid representatives. But by the time this stage is reached the species must already be on its downward career.

In arriving at the probable cause of variability, Mr. Mott says, "Plants which are perpetually self-fertilised should vary very little, while cross-fertilisation will occasionally produce oscillation so extreme as to become the starting points of new specific waves." A similar view was put forward by a writer in the "Science Monthly" for April. My opinion, if it is worth anything, is just the opposite, and I gathered from the "Origin of Species" that Mr. Darwin was of the same mind. Shortly stated, I regard cross-fertilization as a check to the law of variation. Every individual is subject to the modifying influence of its environments acting in antagonism to hereditary forces. As a consequence no two individuals are exactly alike. Every difference is either of advantage or disadvantage to the species, and as the tendency to vary, acting unchecked, would operate more quickly than the rate of change in the environments, it is exceedingly probable that the disadvantageous points of difference would vastly out-number the advantageous ones. Such a state of things would be ruinous to the race, and it would be perpetuated by self-fertilization, because this process would simply exaggerate and make more positive the disadvantageous character thus introduced. Cross-fertilization, however, steps in, and by this means the departures from the normal type are reined in, as it were, by the influence of other plants which do not manifest the same eccentricity; equilibrium is restored, an average of character is maintained, and the fixity of the species is secured as we find it. Nevertheless a species must not be too rigid. To be successful it must retain a certain amount of plasticity, because, like the individuals mentioned above, no two sets of environments are exactly the same. To allow for this, self-fertilization is frequently permissible, and indeed, we find special arrangements to secure it, but naturalists generally admit that it plays its part only in a limited degree or during limited periods in the life history of the species. At any rate this affords an opening for the slow rate of modification necessary to keep pace with the slowly changing conditions of its existence. Thus, this view accounts for both the fixity of some and the variability of other species.

J. HAMSON.
Bedford.

MR. THOMAS FLETCHER, of Warrington, has succeeded in making an elastic rubber tube perfectly gas-tight and free from smell. The tubing just patented by Mr. Fletcher is made of two layers of rubber, with pure soft tin-foil vulcanized between. It is perfectly and permanently gas-tight under any pressure, and free from the slightest trace of smell after long-continued use, whilst it retains the flexibility and elasticity of an ordinary rubber tube. The tube has been in use for some time, and has been thoroughly tested for months under continuous and heavy pressures.

MINERALOGICAL STUDIES IN THE COUNTY OF DUBLIN.

No. II.

THE southern shore of Dublin Bay is remarkable for a fine exposure of granite rock, extending from the town of Blackrock, to a point near the hill of Killiney whence a beautiful coast view is obtained. The granite is flanked on one side by a portion of the upper carboniferous limestone formation which stretches north and westwards, and on the other by a narrow strip of mica schist. Beyond this, lies a series of rocks of lower Silurian age, which is in turn succeeded by a splendid section of the Cambrian period, composing the headland of Bray. These are all easily reached, and in fact can be well observed in the course of a few hours' excursion. At the inland side of Killiney rises a smaller eminence, known as Rochestown Hill, where for some time a granite quarry has been extensively worked. This stone is of excellent quality, another quarry in the same neighbourhood having afforded material for the harbour and breakwater of Kingstown, well known as an attractive watering-place, and packet station for the mail steamers running to and from Holyhead.

Starting from Kingstown one afternoon with specimen bag, hammer and chisel, a walk of less than half an hour brought me within sight of the Rochestown quarry, which presented the usual appearance of such places. There were cranes, a wooden house or two for repairing tools, and appliances for weighing, whilst a number of heavy crowbars were lying about. The quarrymen had just ceased working, which afforded a better opportunity for examining the rocks. The granite appears on a close inspection to present variations in quality, that is, there are portions which may be considered coarse in texture, the constituent minerals being all prominently defined. Large foliated masses of mica are frequently found standing out conspicuously. On the other hand, the quarry is traversed here and there by veins of a much finer quality, known to the geologist as eurite, in which the mica is not nearly so evident. This latter variety is remarkably clean-looking and attractive in appearance, being highly crystalline, and appearing to scintillate under artificial light. The stone of the quarry generally gives good examples of the characters both of felspar and mica. Taking up a specimen in which felspar predominates, the eye is first attracted by its distinct cleavage and pearly lustre. Viewed obliquely, the cleavage plane presents a silky appearance. When opaque quartz occurs with felspar in the mass, they may often be mistaken for each other by an inexperienced observer, particularly in a dubious light, but cleavage being obtained, the character of felspar is evident, as this feature is entirely absent in quartz. The felspars have been ascertained to be principally orthoclase, but albite

L 2

is also a very general constituent of these granites, as is also microcline. There are both white and black micas the former being described as margarodite, by the Rev. Dr. Haughton, who, judging from the mean derived from several analyses, believes that the amount of water of crystallization present differentiates it from the variety known as muscovite. The isolated masses before referred to, give good illustrations of the hexagonal or lozenge-shaped plates, and may be studied with interest, as the form in which mica usually appears in granite does not afford the information to a beginner. A peculiarity in the felspathic and eurite portions of this granite, consists in there being a number of crystals of a reddish-yellow colour scattered through the mass. These are very minute, none of them exceeding the size of a small pin's head, but they are translucent, with a high vitreous lustre. Closer inspection reveals certain

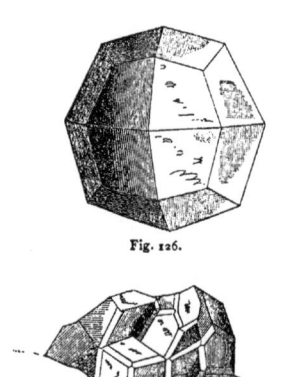

Fig. 126.

Fig. 127.

faces more or less imperfectly developed, owing to the way they are imbedded in the matrix. The application, however, of an inch power to the microscope gives a fuller insight into the forms. One of the more developed crystals then appears as follows, shewing the icositetrahedron—the $m\ Om$ of Naumann, the specific variety of the crystal being apparently 2O2. Fig. 126.

The colour also appears more distinct, and suggests a substance used in some descriptions of jewellery known as cinnamon stone or essonite. The essonites belong to the garnet family, which is divided into many sections; the essential constituent of all, with few, if any exceptions, being silicate of alumina. This composition is accompanied by certain oxides, which, in different specimens, determine the class to which the various garnets belong. Iron is, of course, present in varying proportions, and in essonite, is believed to impart the colouring matter. Lime is another constant element which has caused essonite (and also another variety known as grossularite) to be classed by Naumann and others as lime-alumina garnets. Essonite appears to be widely distributed, as it is found abundantly in the United Kingdom and also in different localities in Europe and the United States. In a good specimen in the Museum of the College of Science here, the following combination of the rhombic dodecahedron with the icositetrahedron ($\infty\ O.\ m\ O\ m$) is well developed. Fig. 127.

Crossing over the top of the hill, the shore is reached by a steep road. At the base of the hill and stretching away to the right as far as Bray, is a wide margin of strand, sloping gently to the water's edge. Here the observer cannot fail to be struck with the manner in which the flowing and ebbing tides have sorted the sands into different sizes, from coarse shingle to the very finest particles. Picking up a handful at random, it is not difficult to distinguish at least half-a-dozen distinct minerals, besides many others, the exact nature of which requires further investigation. A pocket magnet reveals the presence of quantities of magnetic iron, some of it being in small lumps, quite large enough to enable the density to be ascertained, and also to give the characteristic black streak on a piece of porcelain. Filling my bag with a sufficient quantity of the sand, I washed it on my return home. This is an interesting process, and for the benefit of any readers who may not be acquainted with mineral work, I shall describe it. A circular wooden dish is used for the purpose. It should be of hard wood bevelled smoothly from the rim to the centre, and not more than ten inches in diameter for convenient use. Portions of the sand are placed in this with a little water, and after a rotatory movement, the water with the lighter particles of the sand is poured off, leaving the heavier matter behind, which then may be examined as desired. A small portion being put into a test tube and treated with strong hydrochloric acid, effervescence shews the presence of limestone, whilst the yellow colour of the solution indicates that iron in combination is being dissolved. In all sands, particularly those in the vicinity of the primitive rocks or their derivatives, siliceous particles, such as pieces of quartz and felspar constitute by far the largest portion. Proportions, of course, vary according to localities, but there will hardly be much error in stating, that, in the sand under examination, these minerals reached at least 70 per cent. When the sand has been carefully sized by passing it through a fine sieve beforehand, nearly all the minerals whose specific gravity only reaches about two, or a little beyond, pass off, leaving the denser bodies behind. Amongst these latter we find a large number of the essonite stones before described disengaged from the matrix in which they were formed, and now appearing in their perfect crystalline form. Their specific gravity is about 3·5, and under the blow-pipe, reactions both for iron and manganese are obtained, but that for alumina is masked, owing to the presence of the other oxides. There is also a large quantity of a black mineral in small amorphous fragments with a semi-metallic lustre, and whose density appears to be about five. It is not easy to apply a specific name to the mineral substance of which these fragments must have formed part, but they contain oxides of iron and manganese, and another reaction is given which would appear to indicate the presence of tin, a not improbable assumption, considering the proximity in the neighbourhood of granitic rock with which that metal is often associated. There are some small particles of an olive-green colour, and appearing translucent, but they are so few, and their size is so minute that it is impossible to approximate with accuracy to their density, unless in a chemical balance of the finest construction, but it is just possible they may belong to the class of minerals known as olivines. These, however, are much more abundant in the neighbourhood of basalts or augitic lavas, and on the north-east coast of Ireland, they form a considerable part of the sands. Beside these olive-coloured particles there are a few isolated fragments of a brighter tint, approaching that of the emerald, and also translucent, which would appear to be small pieces of apatite. Density and streak being difficult to ascertain with such very minute particles one cannot do more than surmise the presence of the mineral, judging from the greenish coloration of the flame obtained with sulphuric acid. It is, however, quite possible that a few might be minute pieces of beryl detached from a larger mass.

There are other stones of a dull greenish colour scattered abundantly through this sand, though distinct in appearance from the supposed olivines or apatites. They would appear to belong to the rock known as greenstone porphyry. Many large masses of this substance, more or less rounded by attrition, may be found on the strand at Bray, a mile or two off, where a number of interesting stones may be seen at any time thrown into bands of various sizes. It may be stated that the island of Lambay, lying to the north of Dublin Bay, is known to consist in a great measure of greenstone porphyry, which would tend to warrant the conclusion as to these fragments. It need hardly be stated, that none of the minerals last described give any reaction with hydrochloric acid, but in nearly all of them there is distinct effervescence, when fused with carbonate of soda on charcoal, shewing them to be silicates, whilst the reaction for iron, whether in the state of protoxide or sesquioxide is well defined.

Space will not permit of all the various minerals appearing in the sand being enumerated, but one or two others may be briefly referred to. In turning over the siliceous particles, a number of transparent globules distinctly rounded, and having a peculiar pearly lustre, appeared amongst the darker portions. They were different in form from the silica before referred to which was amorphous, and the probability of their being topaz might at first be suggested. However, on picking out a sufficient number to determine the specific gravity, it was found after careful experiment to be 2·4, and as they gave no reaction with the blow-pipe beyond that for silica, it is probable they may belong to the opal group, and be classed as hyaline quartz. They were certainly inferior in hardness and density to the quartz usually found in abundance in every description of sand. A few isolated masses of a dark red colour subtranslucent and with a vitreous lustre also occurred, and were at first somewhat likely to be confounded with the essonites. They were, however, without crystalline form, and on testing with soda on platinum foil, a yellowish colour appeared, indicating chromium, from which it would probably not be far from the truth to consider them as pyropes, or garnets of the magnesia-alumina class, which often shew this reaction.

When the nature of sand is taken into consideration and the difficulty of discriminating correctly as to form and colour in very minute bodies, it will be seen that a description of any kind of sand which has undergone the process above described, must be made with more or less uncertainty. The subject, however, is well worthy the attention of students, as shewing more accurately the constituent elements of rocks, particularly those which appear near the coast. I purpose, however, making experiments on other sands on different parts of the coast in this county during the coming season.

W. McC. O'NEILL.
Dublin.

DREDGING IN THE FRITH OF FORTH.

By S. P. ALEXANDER.

(*Continued from page* 202.)

SUCH then, are a few of the principal specimens obtained by the dredge, at this depth. Let us see what can be brought up from still greater depths. We throw in our net again, at a distance of two hundred yards or so, out from the Gantocks yonder. Here the depth is about twenty fathoms, so we let out forty fathoms of rope. At this depth, and with such a length of rope, one cannot propel a rowing-boat very far. If in a sailing-boat, however, we will do more execution. When the dredge is brought up, perhaps our attention is at once riveted by a reddish-yellow object hopping about at a great rate in the bottom of the net. When examining this more closely, we see a delicate yellow fleshy substance covered with scarlet spots; and, protruding from it, long delicate pink filaments rapidly twining about, like those hair worms (Gordius) most of us have noticed in fresh water.

This lovely object, which at first glance is puzzling, is the cloaklet anemone (*Adamsia palliata*) adhering to, and completely covering, the shell inhabited by the large and powerful purple hermit crab (*Pagurus Prideauxii*).

The filaments above mentioned are the "acontia," which the cloaklet has thrown out, on finding itself removed from its natural element.

Place it in a glass jar of sea-water and allow to rest. In a few minutes the "acontia" will be withdrawn through the minute apertures or "cinclides" in the body of the anemone, and the crown of beautiful white tentacles will be protruded. At the same time the crab puts out its head and expands its antennæ, when we can admire its rich purple colouring, and contrast it with its less beautiful ally, the common brown hermit crab.

That these two creatures, so widely separated in the animal kingdom, and surely so opposed to one another in habits, should so invariably be found in association with one another, is, indeed, a wonderful fact; and that the permanent separation of the two animals means certain death to both, is alike wonderful. That this is so, I can testify from my own observation.

Among the sand and débris, we notice some cylindrical rod-like tubes composed of grains of sand, little pebbles, small shells, &c. These are the dwelling houses of that most complex annelide, the terebella. In some of the tubes the perfect animal is to be found.

This, when placed in water, unfolds before our eyes its delicate pink tentacles and branchiæ, the latter more deeply pigmented, from the contained respiratory or pseudohæmal fluid. That these terebellæ, as it were, grow perpendicularly out of the sand, is, I think, manifest, from the cluster of little roots with which the bottoms of their tubes are provided.

We have here other specimens of the Tubicola in the *Serpula contortuplicata*, and *Spirorbis communis*. These we find adhering to those stones and shells, and with a mass of barnacles (*Balanus balanoides*) to the back of that veteran old crab. The blue and green iridescent worm crawling and twining itself about in the net, we recognise as one of the sea centipedes (Nereis); and here is a smaller brown species of the same genus, the pearly nereis (*Nereis margaritacea*). Clinging to the stones, are a good many little red, brown, and speckled Chitonidæ (*Chiton ruber* and *Chiton cinereus*).

In throwing overboard the debris, care must be taken not to loose those little yellow straws, attached to the shell of that pretty little scallop (Pecten). They look insignificant, but are in reality specimens of the *Tubularia indivisa*, a member of the order Corynida. The small red filament protruding from the end of the straw-like tube, will, when placed in water, expand into a beautiful disc surrounded by tentacles.

There are very few young naturalists, I imagine, who have seen this little organism in the living condition.

It is, I think, only to be obtained at considerable depths, by means of the dredge.

We have here, among many other specimens, one or two spiny spider crabs (*Maia squinada*); the Torbay bonnet (*Pileopsis Hungaricus*); a frond of a species of delicate white sponge, and one or two little shrimp-like crustaceans, of the genus Arcturus. These crustaceans are rather rare. I have obtained them also in the Holy Loch.

Those pearly white tusk-like shells are not to be mistaken for Tubicola, which they somewhat resemble. They are specimens of the tusk shell (*Dentalium entalis*), a species of gasteropoda.

The animals now considered, are, I think, a few of those most commonly met with along the north coast of the Frith.

Sand stars and brittle stars seem to be very widely distributed over the whole Frith of Clyde. Off Snellan, especially, they exist in enormous numbers. I have there brought up my net literally living with them.

The beautiful Holy Loch as a field for dredging is very poor. The bottom there seems to be almost entirely composed of a thin muddy sand, in which the dredge sinks, and is at once clogged up. In this are to be found the thick leathery tubes of that genus of Tubicolous annelide, the Sabellidæ.

In the shallows at the head of the loch, one often finds the pipe fish (Syngnathus). Here also I have seen several hermit crabs, to the shells of which colonies of little pink "polypites" were attached, resembling somewhat in appearance pink plush velvet. As to the exact nature of these organisms, I am not quite certain whether they were Polyzoa or Hydrozoa.

However, I have used the term "polypite," as it is more than probable that they belong to the latter class, owing to the circumstance that a few little medusiform gonophores were to be discovered by a low power of the microscope in the water in which I had preserved them.

On the south side of the Frith, opposite the Cloch Lighthouse, beautiful large specimens of the dead man's fingers (*Alcyonium digitatum*) may be obtained. Here also we find the twelve-rayed sun star (*Solaster rubens*), and varieties of spongida, &c.

As we pass down the Frith, the marine fauna becomes more varied and abundant.

Brodick and Lamlash Bays on the east coast of the Island of Arran, are probably the most fruitful of all the dredging grounds in that neighbourhood. Here, among crustaceans, the hermit crab, spiny spider, nut crab, olive and red squat lobster crabs (*Galathea squamifera* and *nexa*) abound. Beautiful sea urchins, ranging through all tints of colour, from purple to almost pure white.

Among many others, I would simply mention: the purple (Purpura); numerous varieties of tops (Trochus) and Patellæ; the sea cucumber (Holothuria); the sea mouse (Aphrodite); errant annelides (Errantia); wrasses (Labridæ); Chitonidæ, the genera Gammarus and Talitrus; Asterinæ; Palmipes; Solasters and Goniasters.

Here too, we find in large numbers, those little Tunicates, of the family Ascidia, with glassy, perfectly transparent tests, through which their delicate and perfect anatomy may be seen. Larger and coarser varieties of Ascidians are also to be found.

Among the bivalve Mollusca in this locality, are the banded venus (*Venus fasciata*), and numerous varieties of scallops (Pecten).

These gorgeous Pectenidæ, with the delicate and rich colouring of their shell and interior; with their fringing branchiæ; their mantle lobes bejewelled with emerald-like ocelli; and with the activity of their graceful motions, form I think, a sight as beautiful as any in Nature; and one too, which surely make us think of our adorable Maker, and of His wisdom as manifested in these His matchless creations.

But in speaking of the fauna of this part of our coast, we must not forget to mention that most graceful species of Crinoid, the only British member of its genus, the rosy feather star (*Comatula rosacea*).

This beautiful creature, which is only to be obtained by means of the dredge, is present in considerable numbers in Lamlash Bay.

Off the north end of the Holy Island they especially abound; thus affording an example of the tendency there is to localisation in the distribution of certain animals. This may be due, in the present instance, to the fact that the bottom here is composed of a bed of coral spicules.

The *Comatula rosacea* is also to be obtained in the Sound between the two Islands of Cambrae, off Millport. In the other districts of the Frith of Clyde before considered, I have not once met with the Comatula. This however, may have been owing to the time of year, which no doubt influences the distribution of such marine animals.

That certain creatures are only to be met with at certain periods of year, is undoubtedly the case; others again, I am confident, may be found at any time, and throughout the four seasons.

Rivaling in beauty, the various species of Pecten, we have the *Lima hians*. It is to be found in considerable numbers in most parts of the Frith, especially off Snellan and the Great Cumbrae.

To picture the loveliness of this creature from a mere description, would, I think, be vain. One can only see it in its natural element to admire it, and to be enchanted with it.

Having now given the reader some little insight into the nature of the fauna of the Frith of Clyde, I hope the effect of it will be to encourage him to make the inhabitants of the ocean's depths in this way a subject of practical study.

Let the young naturalist, who does not already possess, procure a dredge, and see what he can do with it. Let him not only collect, but dissect, and draw every creature he obtains. In this way, I think, he will soon come to agree with me, that the study of Marine Zoology is perhaps the most charming of all the departments of Natural History.

MATERNAL INSTINCT IN ANIMALS.

THE marvellous and beautiful sentiment of self-denying love of offspring which pervades all animated nature, from the human mother down to the poor insect that we tread upon, is a Heaven-sent instinct. So much has already been written about the maternal instinct of mammals and birds, that the anecdotes we propose translating from M. Ernest Menault's "Amour maternel chez les Animaux" will, at least at present, be confined to those creatures we have been in the habit of considering too low in the scale of beings to be capable of affection. Like Creation itself, M. Menault commences his work from the "creeping thing" and the fish, rising gradually up the scale to the animals most approaching man. We will begin with some of his anecdotes respecting insects.

SPIDERS.

"It is in the month of July that one sees in the middle of fields the numerous dwellings of those minute spiders which are known by the name of *Clubiories*. M. Emile Blanchard, professor at the Museum, coming on one occasion to pay me a visit, I pointed out to him the pretty eggs of the spider, artistically installed among the stalks of oats. He admired them, and then caught sight of the spider herself. Half-hidden in her nest, and watching over her eggs, she was seated, hatching a brood, and surrounded by her young ones, which she seemed to contemplate with anxious solicitude.

"Since then I have made a more careful study of the nests of *Clubiories*, and this year have observed that that species of spider generally takes its station upon two or three stalks of oats, and there weaves its fine silky web, white as swansdown, and of the same consistency as what is denominated silk paper. . . . At the close of several days, the young spiders make their appearance out of the eggs, and find themselves on the web which their careful mother has stretched over the entrance of their nest. It is there that they begin to exercise their little paws, beginning very early to spin, and to feed themselves upon the provisions which their provident mother has carefully stored up round the cradle of her offspring. . . . The spider which builds her nest among the oats is very small; in colour of a sort of greyish-yellow, with a longitudinal dark-brown stripe over the back. She has six paws, of which the two hind and the two fore ones are much more developed than the others; the head, which is almost as large as the rest of the body, is of a transparent greyish-yellow; it is armed with two strong mandibles, surmounted by from seven to eight little points, black but very luminous, which constitute the eyes. At the back of the head, and forming, as it were, two little paws, are the antennæ, which are constantly in motion. It is by the aid of these organs of touch that the spider is able to make herself acquainted with everything she comes across on the road—the antennæ enabling her to distinguish what would be of use to her from what would be hurtful. Such are these charming little creatures, which are all sensibility, all intelligence, all heart, and which shew an astonishing affection for their off-spring. On one occasion, tempted by curiosity into forgetting my duties as a member of the Society for Preventing Cruelty to Animals, I had the barbarity to tear open one of these spider's nests; I was like a child who wishes to see what is inside. I beheld issuing from it an immense number of little eggs, each smaller than a grain of semolina. I counted as many as a hundred and fifty. Some of these appearing rather deformed, I examined them with the microscope, and ascertained that these eggs were in process of transformation. I could already trace, though somewhat indistinctly, the form of a nascent spider. But whilst I was engaged in making my observations, the poor mother, frightened and distressed, rushed to see what had become of her beloved eggs. She endeavoured to gather them together again, but it was in vain, they were scattered far and wide, and she was compelled to resign herself to her unhappy fate. Another time, dare I avow it? I amused myself by tearing the silky envelope that covered the nest, but the diligent mother soon set to work spinning a patch to cover up exactly the breach which I had made. I had the cruelty again several times to attempt breaking into the domicile of this innocent creature, but each time she set to work afresh to repair the mischief which I had caused. Ever since then I have entertained the highest respect for these mothers, so devoted to their progeny, and I proclaim everywhere the maternal affection of spiders.

"But it is not only the *Clubiories* which show so much solicitude for their young; the *lycosa* is equally courageous in defending her eggs. As soon as they are laid, she gathers them together in such a manner as to form a little ball, which she then wraps in a covering of silky tissue, not thick, but compact and solid. The cocoon is of the shape and size of a slightly flattened pea, and its smooth surface is usually of a whitish grey. As this species of spider is somewhat vagrant in her habits, instead of remaining assiduously watching over her cocoon by resting beside it, as do other spiders, she sticks it to her web, drags it after her, and never quits it for an instant during the chase, or even in the face of danger. When pursued, she runs as quickly as the weight of her precious burden permits of her doing, but if any attempt be made to seize hold of the cocoon, she stops suddenly and tries to get it back. Berthoud has well described the agitation of this poor mother. She first of all turns herself slowly round the robber, then approaches nearer and nearer to him by a series of jerky movements, and finally throws herself upon him and combats him with fury. But if the cocoon has been destroyed, the *lycosa* retires into a corner, and dies in a short time of sorrow and of numbness, for from that time forth she takes no exercise. After a month, at the outside, the germs become hatched and issue from their prison, but are still too feeble to obtain food for themselves, or even construct a web; they would inevitably perish, were their mother to abandon them. From that time forward her maternal devotion becomes redoubled. Obliged, in order to obtain nourishment, to be incessantly on the watch, and unwilling to be separated from her progeny, she places her little ones upon her back, and, charged with this beloved burden, pursues her way over hill and dale.

"It is impossible to behold without emotion this little creature, naturally so quick and jerky in all her movements, acquire a motion so much gentler when carrying her treasures. She carefully avoids all dangers, only attacks easily-won prey, and abandons all chance of obtaining such as would necessitate a combat which might cause her to drop the young ones, which press and move by hundreds round her body.

"Attention must have long been directed to the habits of the *lycosa* in this respect, for the ancients believed this species of spider to nourish, and even suckle their young. Bonnet witnessed on one occasion a touching and decisive proof of the marvellous attachment borne by the *lycosa* to her offspring. He threw one with its cocoons into the den of a large ant-lion. The spider endeavoured to escape, but was not sufficiently active to prevent the ant-lion from getting possession of her bag of eggs, which he tried to cover with sand. She made the most violent efforts to counteract those of her invisible enemy; but her resistance was of no avail, the gluten which held the sack gave way, and the sack became loosened. The spider snatched it up with her mandibles, but only to have it seized from her again by the ant-lion. The unfortunate mother, vanquished in the struggle, could still have saved her own life: she had only to abandon the sack and escape from the fatal den, but she preferred being buried alive with the treasure which was dearer to her than her own existence. It was by force that Bonnet at last took her away, but the sack of eggs remained in the robber's possession. In vain did Bonnet draw the spider away several times on a small piece of wood. She persisted in remaining in the scene of danger. Life had no longer any charms for her; she preferred remaining to be swallowed up in the tomb where she had left the germs of her progeny." J. Y.

TEETH OF FLIES.

HOUSE-FLY No. (2). (*Musca domestica.*)

By W. D. HARRIS, CARDIFF.

No. II.

MESSRS. KIRBY & SPENCER in their introduction to Entomology mention in the chapter devoted to "direct injuries caused by insects" a species of fly described by De Geer as "*Musca domestica minor*," and the same authorities assert that he was "one of the most accurate observers that ever existed." We may therefore safely conclude that the flies which pass current as house-flies differ in kind, notwithstanding their great similarity. It was not, however, until I undertook the investigations which form the subject of the present notes that I became so thoroughly convinced of this fact.

The sketch of the teeth of the house-fly which formed the subject of number one of this series, should, I believe, have been described as *Musca domestica minor*, while the illustration which accompanies these notes refers to the common house-fly ordinarily accepted as *Musca domestica*. In general appearance they are very much alike, but the latter is slightly more robust in structure. *Musca minor* has a trifle more white on the face; with this exception the latter might be easily mistaken for a stunted or dwarfed individual of the former species.

When, however, we come to examine the teeth, a very decided distinction is immediately apparent, and is a striking instance of the usefulness of these organs as a means of separating species so nearly resembling each other.

In the present example there are six teeth in each lobe of the proboscis, these are of two different types. The two marginal teeth are distinctly of the blow-fly pattern, while the four intermediate ones are serrated throughout, and terminate in a simple wave-like form quite unlike *Musca minor*, which possesses a distinct central apex in each of the three teeth, in addition to being similarly serrated.

Both figures are drawn to same scale, viz. one inch, representing three thousandths of an inch.

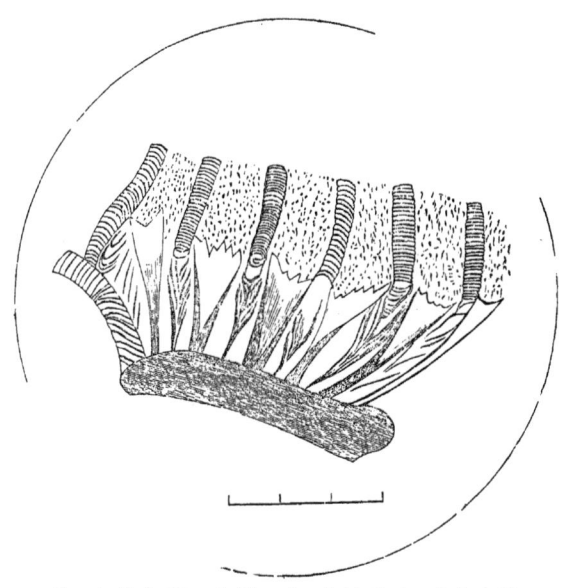

Fig. 128.—Teeth of House-fly (*Musca domestica*) (scale of 1000th of an inch).

GOSSIP ON CURRENT TOPICS.

By W. MATTIEU WILLIAMS, F.R.A.S.

MUCH encouragement to the promoters of technological education is derivable from the experience of the Swiss watch trade which has been artificially fostered with wonderful success, not by protectional tariffs, nor by exportation bounties, but by educating the workmen in such wise that they shall be able to compete successfully with their rivals in spite of serious natural disadvantages. Horological colleges have long been established as public institutions. The first of these was founded in 1824 and became a municipal institution in 1843 directed by a committee of watchmakers. The course of instruction for the students corresponds in its relation to their business to that which we supply to medical, law and divinity students. Like our usual medical curriculum it extends over four years. There are now four more in the Canton of Neufchatel, and two in the Canton Berne. The "Journal Suisse d'Horologie" has been issued monthly since 1870. Public observatories in Geneva and Neufchatel—that of Geneva built in 1773 and restored in 1829—regulate the time for the chief watch-making centres, which are now in electrical communication with the observatories. It must not be forgotten in reference to these that the whole population of Switzerland is under two and three-quarter millions—little more than half of the population of London—and that the watch-making district is but a corner where three of the twenty-two cantons meet.

When shall we have trade universities as public institutions directed by committees of practical artizans? I suppose we shall wait until our commercial advantages have passed, and the trades demanding such education are ruined by the stupidity of our "practical" men. Then, when it is too late, these "practical" men will make a spasmodic effort and fail. The Swiss began in 1824. We are just beginning to talk about beginning. Since 1824 the Swiss watch-trade has steadily advanced and now commands all the markets of the civilised world. Since 1824 or thereabouts, the Coventry watch trade has steadily declined, and is now ruined.

My old friend the late William Bragge has struggled heroically to compete with the American machine-made watches, and told me when I went over the works of the English Watch Company in Birmingham, that his greatest difficulty was in finding workmen who had brains as well as fingers; men who understood the principles of their trade sufficiently to be able to apply their mechanical skill under new and improved conditions.

The controversy to which I alluded in June (page 125) has received from Russia a contribution which will doubtless be accepted and freely used by those who are interested in the continuance of our existing sources of water supply for the metropolis. They cannot of course deny that the sewage from the large population of the Thames valley above the intake of the water companies enters the river, but they contend that its evil things are oxidized away by the agitation of the water in its downward course. This has been much discussed, more discussed than investigated, hy the contending parties. Dr. Pehl, of St. Petersburg, has counted the number of bacteria in cubic centimetres of different specimens of water, and finds that the canals of St. Petersburg contain as many as 110,000 to the centimetre in good weather, while the Neva has only 300. The conduits supplying the city which are fed by the Neva, contain 70,000 against 300 in the river itself. There is but little chemical difference between these waters. He attributes this difference to the motion of the water in the river, and finds by experiment that when water is agitated for an hour by means of a centrifugal machine, 90 per cent. of the bacteria disappeared. This is a result of no small practical importance; it indicates that agitation is more effective than filtering. If further investigation confirms Dr. Pehl's results, the threatened doom of the existing water companies may be averted by the skilful application of a little steam power to the work of agitation. The contradictions brought out by the Society of Arts debate and other controversies demonstrate pretty clearly the worthlessness of the chemical analyses and the reports periodically put forth. It is not the percentage of organic nitrogen, nor the total quantity of organic matter, but the quantity and character of living organisms in which we are vitally interested.

M. Minard, in a paper read a few months since before the French Academy of Sciences, proposed to diminish the violence of storms by attaching a large number of lightning rods to tall telegraph posts, and connecting them with the metals of railways. Others have proposed to do the like by the discharge of artillery, while at a recent meeting of the Academy M. Xamber referred to the deplorable custom of ringing church bells during thunderstorms, which still prevails in certain parts of France. If there is any truth in the theory which assumes that the aerial agitation of sound waves promotes a gradual discharge between oppositely electrical strata of air, or between the air and the solid surface of the earth, this bell-ringing may have a better foundation than the superstition which promotes the practice. The sum total of disturbance effected by an hour's continuous ringing of a large church bell would equal that obtainable by the explosion of a very large quantity of gunpowder.

All who desire the intellectual advancement of the nation must be pleased to see that the reaction against our exaggerated system of examinations is progressing. Its progress would be still more decided if a satisfactory substitute for it were practically set on foot. As regards science teaching in the universities there is no difficulty in doing this. It is already carried out in Germany, and the method is so well and concisely described in the "Journal of Science" of May last that I quote the passage as it stands. "A young man enters a university, and attends the class-room and laboratory of Professor M. or N. If he shows zeal, industry and intelligence, the professor, who keeps a watchful eye on every student, gives him some idea to work out experimentally, and assists him with advice and suggestions. When the investigation is completed it is sent for publication to one of the scientific journals, and the youth sees his name bracketed with that of a.

Hoffmann, a Baerjer, a Kolbe, &c. More and more difficult problems are placed in his hands, and the assistance of the professor is gradually withdrawn till he feels himself fully capable of original research, whether in speculative or applied chemistry. It is the interest of the professor to detect, train, and bring out ability. The researches and discoveries made in his class are his 'results.' The more numerous and important such researches, the more students flock to his laboratory and his lectures. Rival universities contend for his service, and Government awards him public honours."

Here we see a community of interest between the professor and the student. Not so with our examiners and their victims. The examiner, who is paid on the piecework priniciple, so much per gross of papers examined, is interested in the non-success of the candidate. If "plucked" he comes up again, pays another fee, contributes another paper, and thus improves the examiner's income. If he is passed, the examiner gets no more out of him. The drudgery of toiling through a multitude of examination papers is so irksome that nobody undertakes it for any other than mere pot-boiling motives, excepting in the case where the teacher examines his own pupils for the legitimate and necessary purpose of determining his progress, and for filling up the gaps of knowledge revealed by the examination. When I was a student in Edinburgh the only examiners were the professors; the ablest of these had weekly or monthly class examinations, and kept a record of the status of each student, so that the formal examination for his degree was only one of a series. In such cases the grinder could do nothing more than assist the student in the legitimate recapitulation of the work he had already gone over. The professor himself should be a trustworthy man and a teacher. The only use of outside examiners in such cases is the examination of outside students, those who may have acquired the requisite amount of knowledge without entering the university classes. Such examiners should receive a fixed salary or honorarium, not be paid by piecework, and all examinations should be both written and oral. By cross-questioning on the answers given on a written paper, mere verbal cramming may easily be detected.

An experiment recently made on feeding the horses of the 7th Cuirassiers with a mixture of oats and cocoanut meal, is worthy of the attention of our commissariat officers who are on duty in the tropics. We are told that the condition of the horses was much improved, and that the reduction in the cost of horse-keep amounted to 50 francs per annum.

On the 6th of November, 1883, the marble memorial statue of Liebig erected at Munich was found to be covered with a number of black spots and stripes. These were examined by Pettenkoffer, Baeyer, and Zimmermann. The stains were found to consist of a mixture of silver with a little hydrated manganese dioxide, from which it is inferred that the liquid used by the contemptible defiler was a solution of silve nitrate and potassium permanganate. The stains have been removed by converting the metals into sulphides, and then dissolving these sulphides with potassium cyanide. A paste was made by moistening porcelain clay with ammonium sulphide. This was laid over the stained surface and renewed after twenty-four hours. After the lapse of another day it was carefully washed off. The silver and manganese compounds were thus converted into sulphides, and still black. Two applications of a paste of porcelain clay moistened with a saturated solution of potassium cyanide, restored the marble to its original whiteness.

CHAPTERS ON FOSSIL SHARKS AND RAYS.

By Arthur Smith Woodward.

II.

AS the genera of sharks belonging to the first four families possess no dorsal spines, they are only represented in the fossil state by teeth, vertebral centra, and shagreen. They are all very rare in strata earlier than the Cretaceous, and do not appear to have flourished in considerable numbers until the old Hybodonts and Cestraciontoid sharks were on the verge of extinction.

CARCHARIIDÆ.

The earliest fossils that can be definitely referred to this family are the teeth included by Agassiz in his genus, *Corax*. Two species occur abundantly in the English Chalk, and the smallest of these (*C. falcatus*) has also been met with lower in the Cretaceous series in the rocks of the Continent. The teeth, which are triangular in shape, with regularly serrated edges, are solid throughout, having no internal cavity; and in external form they agree so closely with the dentition of the living *Carcharias*, that some palæontologist are inclined to believe in their generic identity. Fig. 134 represents a tooth of *C. falcatus*,* and fig. 132, one of the larger species, *C. pristodontus*. In the latter, it will be observed, the root is relatively much larger than in the former, and the edges of the crown describe curious curves that are characteristic and unmistakeable. The existing genus, *Galeocerdo*, seems to replace *Corax* in the Tertiary formations. The teeth of this shark are also serrated on the edges of the cone, but not quite to the summit, and the denticulations are irregular, those at the basal portion being much less pronounced than those somewhat higher. Only one species of *Galeocerdo* has been

---
* Some of the smaller teeth referred to this species are destitute of serrations on the edges; such may possibly have belonged to young individuals, for it is known that the dentition of the young *Carcharias* consists of teeth without serrated margins.

recorded from British strata; this is Agassiz' *G. latidens* (fig. 130), from the Middle Eocene of Bracklesham. Another species (*G. aduncus*) is found in Continental Miocene formations, and the Tertiary strata of North America have yielded some others (*G. contortus*, &c.). The extinct *Hemipristis* appears to have been intermediate between *Galeocerdo* and *Carcharias*, if dental characters are safe guides to such conclusions; no remains of this genus have been discovered in British rocks, and only one species (*H. serra*) is at present known. Fig. 133 will give a good idea of the curious tooth of *Hemipristis*. It occurs in the Tertiaries of both Europe and America, being found in the Miocene of Malta and Würtemberg, and also in the Eocene and Miocene of the United States.

The remarkable Selachoidei, commonly known as "Hammerheads," are classed with the Carchariidæ,

also been made to show that the detached vertebræ are capable of at least generic determination.*

Although marked differences can be observed between the living forms of each, *Lamna* and *Odontaspis* are scarcely separable palæontologically, and hence it will be necessary here to consider the latter as a sub-genus only of the former. The teeth of both are of the type shown in figs. 129 and 137, with the crown long, slender, and sharp-pointed, with distinct lateral denticles, and with the root deeply cleft, and the radicles much elongated. *Lamna* proper is generally much compressed and flattened anteriorly, while *Odontaspis* is nearly cylindrical in section, and often has a sigmoidal curvature (fig. 129, *a*). The teeth of the latter kind appear before those of the former, and the earliest hitherto met with in English strata are from the Gault. *L.* (*Odontaspis*) *rhaphiodon* occurs in the Chalk, and this tooth (fig. 129) is remarkable for the

Fig. 129.—*Lamna* (*Odontaspis*).

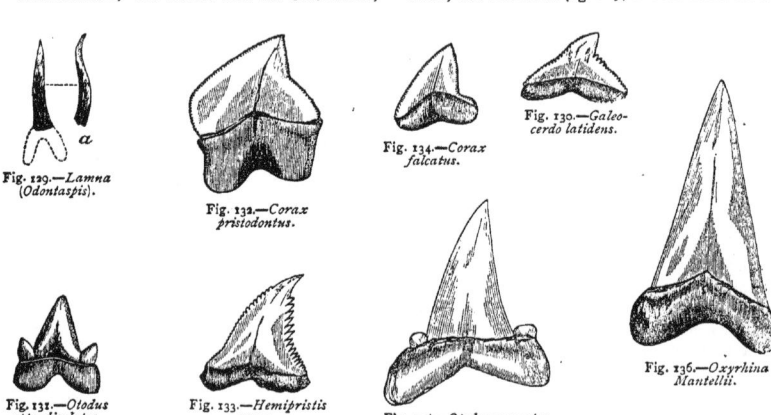

Fig. 132.—*Corax pristodontus*.

Fig. 134.—*Corax falcatus*.

Fig. 130.—*Galeocerdo latidens*.

Fig. 131.—*Otodus appendiculatus*.

Fig. 133.—*Hemipristis serra*.

Fig. 135.—*Otodus macrotus*.

Fig. 136.—*Oxyrhina Mantellii*.

and appear to be represented in the fossil state by certain detached teeth. Palæontologists, however, in this instance, find themselves greatly perplexed, for the dentition of these sharks is much less distinctive than the well-marked characters available to zoologists studying living species would lead us to expect. Teeth of *Sphyrna* (*Zygæna*) *prisca* occur in the Miocene rocks of Malta, and the Eocenes of South Carolina, and one species is known from the Chalk Marl, near Dresden.

LAMNIDÆ.

The majority of the Selachian fossils found in Cretaceous and Tertiary formations are referable to the Lamnidæ, and belong to the familiar genera, *Lamna*, *Odontaspis*, *Otodus*, *Oxyrhina*, and *Carcharodon*. The teeth of all these, except *Odontaspis*, are readily distinguished one from another, and attempts have

curious ornamentation produced by the folds of enamel (gano-dentine) on the back of the crown; the structure of the radical portion is so delicate, that specimens are rarely found perfect, and their condition is usually that of the fossil represented in the figure. *L.* (*Odontaspis*) *subulata* is a second Cretaceous species, ranging from the Gault to the Chalk, and destitute of surface ornament on either side. The Chalk examples usually assigned to this species are notably smaller than those from the lower horizons. Only one form in the British Cretaceous formations has been referred to *Lamna* proper (*L. acuminata*, Ag., from the Chalk); but this differs so much from the ordinary type of tooth characterising the genus, that there is reason to doubt the correctness of the identification, and evidence is not wanting to suggest its being a variety of *Oxyrhina Mantellii*. *L. elegans* (fig. 137)

---

* C. Hasse, "Das Natürliche System der Elasmobranchier" (Jena, Gustav Fischer, 1879–1882).

is represented abundantly in the Eocene strata of Sheppey, Barton, and Bracklesham, and the teeth may always be easily recognised by the delicate longitudinal striæ on the hinder side. *L. (Odontaspis) Hopei*, of the same stratigraphical position, much resembles the last species, but is distinguished by the absence of striæ, and by the more cylindrical shape of the tooth. Another species from the Eocene is *L. compressa*, which appears to be almost intermediate between *Lamna* and *Otodus*, and is remarkably flattened.

Otodus is an entirely extinct genus, and ranges from the Lower Cretaceous strata to the Upper Eocene. The teeth are distinguished from those of *Lamna* by the more compressed character of the crown, and the much less elongated and branched condition of the root; the lateral denticles, too, which are always present, are large in proportion to the central cone.

from those of *Otodus* in possessing no lateral denticles, but are very similar in other respects. The earliest species occur in the Gault, and the genus survives to the present day. *O. macrorhiza* (fig. 138), from the Gault and Greensand, is characterised by its very depressed form, and the consequent great breadth of the root. *O. Mantellii* (fig. 136) is a species abundant in the Chalk, and sometimes occurring in the Greensand. No teeth of this genus appear to have been recorded hitherto from the English Eocene formations, but *O. xiphodon* and *O. hastalis* are characteristic fossils of the Pliocene Crags, and these are also met with in Eocene and Miocene strata on the Continent.

The only remaining genus of Lamnidæ of much palæontological importance, is *Carcharodon*, of which but one species, *C. Rondeletii*, survives in our present seas. It first appears in early Eocene times,

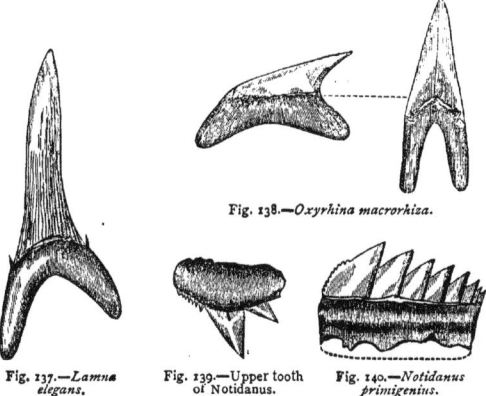

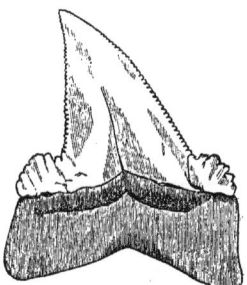

Fig. 138.—*Oxyrhina macrorhiza.*

Fig. 141.—*Carcharodon angustidens.*

Fig. 137.—*Lamna elegans.*

Fig. 139.—Upper tooth of Notidanus.

Fig. 140.—*Notidanus primigenius.*

Fig. 142.—*Notidanus microdon.*

Fig. 143.—*N. serratissimus* (all figs. nat. size).

Of the earlier species, *O. appendiculatus* (fig. 131) is the most important, and occurs in Cretaceous rocks almost wherever they are developed, being found in Britain, many localities on the Continent, and even in the New World. The root of this tooth is very thick and depressed, and both crown and denticles are likewise particularly stout. *O. obliquus* is the largest known species, and is met with in Eocene strata; in form, the teeth are closely similar to the *Carcharodon* shown in fig. 141, but there are no traces of serrations on the edges, and the size is occasionally much greater. Some are oblique, as in the figure, and some straight, the former being generally regarded as belonging to the upper jaw, and the latter to the lower. Another Eocene species is *O. macrotus* (fig. 135), remarkable for the large development of the denticles and the compressed form of the tooth, and, like the previously mentioned species, having a wide geographical range.

The teeth of *Oxyrhina* (figs. 136 and 138) differ

attains its maximum development in the Miocene, and has since been gradually approaching extinction. The teeth are flattened, triangular, with or without lateral denticles at the base of the principal cone, and in all cases serrated on the edges; although very similar to the teeth of *Carcharias* in external appearance (except as regards size), they are easily distinguished by being solid throughout, and the dentition of the lower jaw is much more like that of the upper than in the latter genus. The English Eocene species are *C. angustidens* (with which Agassiz' *C. heterodon* is now incorporated) and *C. subserratus*,— the former (fig. 141) with lateral denticles, from the Bracklesham and Barton beds, and the latter, without lateral denticles, from the London Clay of Sheppey. But the most important species of the genus is *C. megalodon*, a form whose enormous teeth are not unfrequently met with six inches long and nearly five inches across the base, and occur in Miocene (and perhaps later) deposits almost all over the world. Among other regions, they have been discovered in Belgium, France, Spain, Italy, Malta, Arabia, the West Indies, South Carolina, Central America, and New Zealand, and derived specimens occur in the British Crags. It is interesting to note, too, that numerous examples of equal size, and probably the same species, were dredged from the bed of the Pacific Ocean by the "Challenger" expedition,—a fact indicating the comparatively recent extinction of the huge sharks whose dentition they constituted. *C. sulcidens* is a smaller species, with the teeth much flattened, wrinkled longitudinally at the base, and destitute of lateral denticles, occurring in the Pliocene Crags of England and the Lower Tertiaries of South Carolina.

Two other genera, *Sphenodus* and *Meristodon*, from European Jurassic strata, are regarded by Agassiz as the forerunners of the Lamnidæ, but are very imperfectly known, having been founded merely upon the crowns of broken teeth.

NOTIDANIDÆ.

Teeth of *Notidanus*, the single living genus of this family, are met with in strata so early as the Jurassic, and occur in most of the marine formations of later date. *N. Münsteri* is Jurassic; *N. microdon* and *N. pectinatus*, Cretaceous; *N. serratissimus*, Eocene; and *N. primigenius*, Eocene, Miocene, and Pliocene. The first of these species has not been recorded from any British formation, but the four latter are known from several localities, and are not unfrequently met with. It is singular, however, that the teeth of the upper jaw, which differ much from those of the lower in the living species, have very rarely been recognised: fig. 139 represents one of these uncommon specimens from the Middle Eocene of Hampshire. The ordinary fossil—the mandibular tooth—may be described as consisting of a series of sharp, compressed cones, more or less oblique, and placed one behind the other upon a well-developed root, the foremost being the largest, and the rest diminishing in size as they approach the hinder end. Figs. 140, 142, 143, are sketches of the three most important British species, and indicate their characters better than any description: it will be observed that, in each case, the length of the crown of the tooth is much greater than the height,—the ratio being frequently more than 2 : 1.

SCYLLIIDÆ.

Clusters of minute pointed teeth, in association with more or less fragmentary pieces of calcified cartilage, are sometimes found in the English Chalk, and most of these are distinctly referable to Selachians of the family of "Dog-fishes." In the table at the end of our last article (p. 174) we alluded to the fact, that in the Scylliidæ several rows of teeth are generally in function at the same time, and an examination of some of the more perfect specimens of the chalk fossils, just referred to, can leave no doubt but that such was the case in the ancient fishes of which these are the dilapidated relics. The fossil teeth are frequently not much more than $\frac{1}{10}$ inch in length, and consist of a central sharp-pointed cone, with a slightly diverging denticle on each side. Agassiz, who had excellent examples for his study in the cabinets of the Earl of Enniskillen and Sir Philip Egerton, founded upon them the genus *Scylliodus*, and described a single species, *S. antiquus*. These type-specimens, now in the British Museum, are sufficiently perfect to reveal other characteristics of the shark besides the dentition, and exhibit remarkably well a portion of the vertebral column. No other dog-fishes have been described from British strata, but one species of the living genus, *Scyllium*, is known to occur in the Cretaceous rocks of the Lebanon.

ON OUR BRITISH SEA-WORMS.

By Dr. P. Q. KEEGAN.

OCCASIONALLY, when exploring the fauna of the extreme limits of the tide-abandoned shore, we turn up a stone whereon we perceive a very long, slim worm of a peculiar purplish colour fringed with brown, and furnished with a lavish amount of feet that move in a regular, waving, rythmical manner adown the flanks. This is *Phyllodoce lamelligera*, a very beautiful form enrolled in a genus characterised by the deep and well-marked separation of the body segments, and by the series of beautiful broad leaf-like ciliated branchiæ on the sides for the aeration of the peritoneal fluid (the blood-proper being scanty); the proboscis is beset with glandular structures filled with oleous cells and globules; the blood is colourless, but the bile is dark green. *P. lamelligera* is sometimes two feet long; the head is roundish, with a number of rather conspicuous tentacles on or near it; the branchiæ are of two sizes, one half the dimensions of the other, arranged alternately along the sides, and penetrated with an exquisite network of blood-vessels. The bristles project between these laminæ, and are slender with a dovetail, elastic, flexible joint, and a single yellow spine in the middle of each brush. *P. viridis* is a smaller species, and is immediately recognised by the bright emerald green colour which seems to deeply tincture its entire structure; there are five antennæ; the branchiæ are lanceolate, being narrower and longer than in the last species; the bristles are jointed, and very slender and sharp at the point.

When trowelling about the loose, damp, clean sand near low water, you are pretty sure to disentomb a lithe and supple worm tapered at the end, and of a bright silvery iridescent hue. This is a species of rag-worm (Nephthys), a member of a genus which has the two series of its nerve ganglia fused into a single chord; the tail is exceedingly mobile and flexible; there is a huge proboscis cleft in two at the top, and bearing a fringe of papillæ there, and also a little way down; the feet are large and carry a leaf-like expansion in front of each branch, &c. *N. cæca* or *margaritacea* is a large and brilliantly-coloured sea-worm, the central space adown the back and belly exhibiting a remarkable silvery iridescent play of colour; the upper lamella of the foot is twice as large as the lobe of the foot itself, while the lower lamella is larger and much broader than the upper one; the bristles are arranged in two distinct tufts, and do not extend much farther than the edges of the lamellæ, they are very slender and pointed, and apparently smooth and simple. *N. longisetosa* is very similar to the foregoing species, but it is smaller and more common; the lamellæ are oblong, and some of the bristles are very long indeed, and exceedingly flexible; there are three sorts of these bristles —some are bulged near the middle and minutely denticulated thence to the long sharp point, some are of a similar structure but jointed, while the third kind are very short and of a flat triangular shape, with the upper flat edge grooved crosswise like a forceps blade, so that the narrow edge seems cut into minute teeth.

The genus *Nereis,* although perhaps not the most elaborately formed or marvellously constructed of the marine annelids, is in many respects the most important. The body is always slender, linear, more or less cylindrical, and deeply divided into a large number of segments; the head is conspicuously and abundantly supplied with antennæ or tentacles; and there is a powerful proboscis armed with strong jaws bearing various shapes of notched teeth; the feet are exceedingly well developed, protrude freely from the sides, and are amply furnished with branchiæ, cirri, and jointed bristles; the nervous system has very numerous ganglia; the branchiæ are penetrated by the peritoneal fluid, and are without cilia; the blood-proper is elaborately developed, and is distributed in a plexus embracing the circumference of the conical branchiæ; the skin is more vascular than that of any other annelid; two lateral pouches communicate with the throat, and the wall of the stomach is embraced by a dense tissue of blood-vessels; the habits are carnivorous. There are some twenty-two species of Nereis, of which *N. pelagica* is perhaps the most common. If you fumble about amongst the loose shingly sand and wave-worn stones on a low-lying shore, you are pretty sure to unearth a lively, wriggling, acrobatic worm of a dark-greenish colour, and a not particularly attractive aspect. It may be observed that the distinguishing marks of the various species of *Nereis* have not been always satisfactorily indicated; but naturalists usually assign as peculiar features of *N. pelagica* the following characters—the segment immediately behind the head is about twice as long as the following segment; the branchiæ lobes are conical and round; the superior cirrus is twice as long as the pedal lobe to which it is attached; the proboscis is a very remarkable organ, its orifice being furnished with two curved jaws, the inner edges of which are cut into ten teeth, while below these there are patches of curious darkish horny prickles; the bristles have a sort of dovetail joint of great flexibility, the end piece of the superior ones is short, curved, and toothed at the edge, that of the inferior ones being slender and prolonged like a fine French bayonet. *N. Dumerilii* is distinguished by the post-occipital segment being only very slightly longer than the second one; the tentacular cirri are nearly three times longer than those of the last species; the dorsal cirrus considerably overreaches the apex of the foot lobe, and the branchial lobes are short and obtuse though conical and round; the jaws have about .12 denticles; there are two brown or yellow spots at the base of the dorsal lobe of every foot; the bristles are smooth and slender. *N. fucata* is generally to be found lodged in some old whelk or spindle shell as the co-tenant and messmate of that marine curiosity known as the hermit crab. It is three or four inches long, of an orange colour, and the tentacular cirri are about the same length as the width of the head; the dorsal lobe of the feet has a strong hump near the base; the lower cirrus of the foot reaches beyond its lobe; the jaws show four or five rather shallow notches and a number of minute prickles. *N. renalis* differs in some rather important features from the aforesaid species; the hinder feet are not of the same pattern as the fore feet, for above the base of the upper lobe there is a flat crest; the branchiæ are large, flat, and kidney-shaped; there is a curved lobule at the base of the lower cirrus; there are five denticulations in the jaw; the bristles are very beautiful, those of the upper feet bearing a sharp fang in the centre, while those of the lower feet present the usual dovetail-joint, the end-piece being broad flat, and serrated very beautifully on the edges. *N. longissima* is allied to *Phyllodoce*, and is rather a lengthy customer of some eighteen inches or more in extent; the feet are similar to those of the last species, the proboscis has no prickles, and the jaws are very slightly serrated.

Euphrosyne foliosa is a curious species, resembling a sea-mouse with shrub-like gills behind the double-branched feet and reaching from one branch to the other; the bristles are slender and unequally forked at the end.

Sigalion boa is, in many particulars, an unique and interesting form. It is the only species of its family that is furnished with external organs of respiration— these consisting of threads beneath the scales; and it is the only annelid wherein superior cirri and shield-plates exist on the same foot; these shield-plates are fringed with hairs on the outer edge, and occur on alternate segments as far as the twenty-seventh, whence they continue on every segment as far as the tail; there are three antennæ, and two large palpi; the proboscis carries two pairs of sharp horny teeth which work vertically, its aperture bristles with an array of warty tentacles, and its inner surface is villose and highly vascular; there are three bunches of bristles on each foot, and at least four shapes of bristles, some quite smooth, some serrated on the upper half or on both sides for a short way down, while others are double-jointed, or bear a kind of claw at the end, &c.

The genus Polynoe comprises a series of forms, some of which are among the commonest of shore-haunting fauna. Overturn or displace any stone about low-water mark, and you will discern one of these scale-clad creatures wriggling away into a place of protection and obscurity. The two longitudinal nerve chords run close together, but are not so fused as in Nephthys, &c. The peritoneal fluid is very abundant, yet but little corpusculated, while the blood-proper is comparatively insignificant, there being no proper vascular system. We must here indicate two important anatomical features about the genera of annelids we are now reviewing:—(1) The conspicuous scales are not true gills, but only organs specially designed to generate branchial currents for the oxygenation of the peritoneal fluid within the body; (2) the stomach dilates on each side into a number of pouches or blind sacks, with muscular orifices, and always filled with a dark-green chyme, and bearing on their exterior the oil-filled glandules of the true liver. It has been supposed that the oxygen, infused by the action of the scales into the peritoneal fluid acts on this stored-up chyme, and by its medium replenishes the true blood itself. The proboscis of Polynoe is exceedingly well developed, and is of the structure already sketched under the preceding species; there are any amount of antennæ, palpi and cirri along the body, all being organs of touch, the latter specially developed and solid and unciliated; there are one or two pairs of eyes; the bristles and spines are strong, very large, and beautifully constructed. *P. squamata* frequents the shore, but is common in deep water; it has twelve pairs of fixed scales which are ciliated on their outer margin, and the two last scales are excavated for the anus; the tentacles and their cirri are thickened below the point; the bristles are very beautiful, those on the upper branch of the foot are ranged in two rows, or kinds of somewhat similar structure, each having a thick round blade tapering to a fine point and transverse serrated ridges on its convex side, those of the lower division having a long handle and a short-pointed slightly curved blade bearing eight or nine transverse serrated ridges on its lower half; this species is sluggish and dilatory in movement. *P. cirrata* is one of the commonest sea-worms of the shore; it is more vivacious than the preceding species, it is also larger, darker in colour, and has more prominent feet, &c.; there are fifteen pairs of scales; the tentacles are ringed with black; the upper bristles are slightly bent near the minutely serrated end, the lower ones have the blade cut into two teeth at the end, and are armed with spinous denticles on the convex side. *P. scolopendrina* is about four inches long, and is flattened in shape; there are pairs of small, deciduous scales only on the fifteen anterior segments, the remaining divisions, although dotted with tubercles, being quite naked; the antennæ, palpi, &c., are only slightly thickened below the point, and are covered with fine hairs; there are only two eyes; the upper bristles are blunt and roughish on one side, the lower bristles are larger, two of them having a large triangular lance-head (this is very characteristic), and the other one being like a hedge-knife with two sharp teeth at the end of the minutely-toothed blade.

The genus Aphrodita embraces forms that may be regarded as among the wonders of the animal creation. Here we have the sea-mouse which frequents the mud and sand of the deep sea bottom, and is frequently hurled ashore by the tide. What a wondrous structure of organism, what a brilliant glow of colour! What an exquisite combination of green and gold and reddish-brown, all supremely lustrous and beautiful! Whence is the origin of all this splendour, of what utility is it, what function does it discharge in the life-economy of the animal? It is a sluggish creature, it crawls by the apparently rhythmical, alternate sheathing and projecting of its hard stiff bristles; but its life-energy, its individual force must be remarkably powerful, its affluence of beauty must spring from and be sustained by animal or vital energies pre-eminently vigorous and efficient. The anatomy of this genus exhibits manifold features of interest. The nerve chords are contiguous; the peritoneal fluid is very voluminous, and although it is highly charged with oxygen, it contains but few corpuscles, while the blood-proper is almost obsolete. The scales on the back are supplied with numerous muscles, and are coated with a layer of felt which is permeable by the water, while the internal cavities of the body are shut out by a membranous partition from the spacious exterior enclosure beneath the scales called the peritoneal chamber. The stomach floats as it were in this chamber, and is a straight tube with a series of lateral pouches invested with functions similar to those of the preceding genus, and there are segmental organs, which are branched tubuli filled with the reproductive products, and having a renal function; the proboscis has no or only rudimentary teeth, but its orifice is encircled with a fringe of filaments, each being a short stalk crowned with a tuft of forked papillæ, the inner membrane is highly vascular, and on the exterior of this fringe there are four fleshy tubercles. The feet are of two kinds, squamiferous and cirrigerous, and are divided into two branches. *A. aculeata* is from three to eight inches long, and is profusely decorated on the back and sides with a splendid raiment of green and gold, and lustrous red-brown hair and bristles. The upper branch of the foot carries long, flexible bristles; the lower branch bears three rows of stout short bristles. These hairs and bristles are seen under the microscope to be very finely grooved longitudinally. *A. hystrix* is found in deep water; it is only about two inches long, and has about thirty pairs of feet; the lower branch of the foot has a yellow spine, and four or five brown bristles, the upper branch has two bundles of bristles arranged in a fan-shaped fashion, and curved with minute granulations on their upper half; the bristles of the brush placed between these two branches are very remarkable. They are stout and long, of a rich dark-brown colour, and are straight with a lanceolate point notched on each side into four alternate reverted barbs, and enclosed in a sheath which opens and shuts upon them, and protects them from injury in a way that all teleologists must regard with exceeding wonder and delight.

SCIENCE=GOSSIP.

FEW of the many visitors to the Health Exhibition find their way to the unpretentious and out-of-the-way room known as the "Biological Laboratory," under the care of Mr. Watson Cheyne, M.B., F.R.C.S. It affords a tolerably complete exposition of the method of the cultivation and examination of the micro-organisms known as microbes. Shortly indicated, the process is as follows. Some sterilised gelatine meat infusion is exposed to the air for a short time, and then enclosed in a tube whose opening is plugged with cotton wool. The germs which have settled from the atmosphere proceed to develop, and a number of coloured or colourless spots appearing on the surface of the infusion result. These are the colours of the various microbes. If, now, some sterilised infusion be inoculated by pricking it with a needle whose point has previously touched one of these spots, a fine growth of the microbe results, and is allowed to develop undisturbed, and out of immediate contact with the atmosphere. Examples of sterilisers and incubators are shown. The use of the incubators is chiefly to keep the temperature of the infusion in which the microbes are growing constant. A number of test tubes, about one-fourth full of gelatine meat infusion, from Dr. Koch's laboratory at Berlin, containing pure cultivation of various organisms, are shown. In addition to the above, several cases of microscopes and of apparatus are exhibited by the manufacturers, and a number of microtomes are on view.

THE Rev. T. E. Espin has published a catalogue of the magnitudes of 500 stars in Auriga, Gemini, and Leo Minor, which have been determined from photographs taken by means of the equatorial stellar camera at the Liverpool Astronomical Society's Observatory.

IN Japan, where earthquakes are very common, a house has been invented not to be affected by the movements of the earth. The building is of wood, with plaster walls and ceiling, supported upon iron bales resting in hollow, saucer-like plates, which method of support, it is claimed, prevents momentum in a horizontal position from being communicated from the ground to the house, and there is just sufficient friction at the points of support to destroy the slight motion that might otherwise take place. It might naturally be supposed that people who are always being shaken would get used to earthquakes, but Professor Morse says that, far from this being the case, upon the first going there, one thinks lightly of such a visitation, but terror increases with every recurrence, until life becomes miserable from a constant state of dread.

IN order to obtain reliable information concerning the upland wilds of Iceland, the Government of that country commissioned Mr. Thoroddsen to undertake systematic explorations, in order to establish the geology of the country on a sound basis, and to correct its geography when necessary. In the course of last summer he explored the peninsula of Reykjanes and its upland connections, and determined the existence and site of no less than thirty volcanoes, and at least seven hundred craters, although, up to then, it was only supposed that there were two volcanoes in these parts which had been active within historic times. In other localities volcanoes of colossal size are found in addition to numbers of hot springs, solfataras, and boiling clay-pits. Mr. Thoroddsen maintains that this peninsula must be one of the most thoroughly burnt spots on the face of the globe, and a most instructive tract for geologists wishing to make a special study of volcanic phenomena.

THE first volume of a work on British Fungi, by the Rev. John Stevenson, illustrated by Worthington G. Smith, F.L.S., is announced. It is to contain full descriptions of all British Hymenomycetes, with habitats, seasons of growth, &c., and all genera and sub-genera will be figured.

AN interesting case of mimicry is recorded in the last issue of the "Canadian Entomologist." While examining the flowers of a bed of May apples (*Podophyllum peltatum*) the writer found a specimen of the Phalænid moth (*Tetracis lorata*) adhering to the stamens of a flower, its head towards the centre and the wings easily mistaken for petals. A little search discovered another in exactly the same position.

THE high temperatures observed in manures are due to the oxidation of organic matter by free oxygen. This oxidisation is partly induced by a bacillus. The disengagement of marsh-gas in manure deprived of oxygen is exclusively due to a bacillus.

THE second volume of "Topography of Lofthouse, and Rural Notes," by an old contributor to our columns, Mr. George Roberts, is in the press. It will contain a continuation of the Natural History Diary, forming in both volumes, a series of twenty-two consecutive years; a Memoir of Charles Forrest the antiquary (the discoverer of rock sculptures on Rombald's Moor); and additional lists of the mollusks and plants of the Wakefield district.

THE Americans are turning to practical account the discovery that the Utricularia is a devourer of young fishes. There are from twelve to fifteen species in the United States waters, and Professor Baird thinks the discovery has an important bearing on the abundance of the food fishes. Utricularia is therefore to be destroyed in spite of its acquired ingenuity.

SHEEP'S horn is being employed at Lyons for making horse-shoes. It is said to be particularly adapted to horses employed in towns, and known not to have a steady foot on the pavement.

IN the Geographical Section of the recent British Association meeting, Lieutenant Greely was present, and delivered an address. He said not a word about his sufferings; and the details he gave of his observations, especially on Grinnell Land, and the comparatively warm tide coming from the Pole, are a real contribution to science. Here is a land well worthy of an expedition to itself, and of the special study of geologists. It is an Arctic land, a neighbour of North Greenland, with a great double ice-cap and immense glaciers, yet with abounding vegetation, willows, saxifrages, grass sufficient for herds of musk oxen, freshwater lakes and rivers, old moraines, and apparently receding ice.

THE most remarkable statement in Lieut. Greely's address at the British Association Meeting was the discovery that when the tide was flowing out from the North Pole it was found that the water was warmer than when flowing in the opposite direction. He made an elaborate set of observations (which will shortly be published), showing this wonderful phenomenon.

PROFESSOR MOSELEY, in a paper read at the British Association on the presence of eyes in the shells of chitonidæ, said he discovered the eyes during the present summer. No other mollusca have any sense-organs in their shells. In such chitonidæ there are 11,000 eyes in the shell of a single animal. Each eye has a calcareous cornea or bicornea, a lens or soft tissue, with retina composed like the eye of the common snail. New eyes are constantly being formed at the margin of the shell by the growth of the latter during the animal's life. Besides eyes, elaborate organs of touch permeate the shell; their end organs can be protruded at its surface by means of pores. The shells of the chitonidæ thus differ fundamentally from those of other mollusca.

AT the same meeting, Messrs. R. Law and James Horsfall gave an account of small flint implements found beneath the peat on the Pennine hills of Lancashire and Yorkshire.

BEARINGS made of glass are being experimented with in the rolling-stock of certain American railroads, in regard to their frictionless quality.

Mr. W. W. COLLINS is giving a series of clever expositions of Spencer's "Principles of Biology" in the "Midland Naturalist."

DR. BARCENA, Director of the Department of Geology in the National Museum of Mexico, has recently discovered the facial and mandibular bones of a human skull in a hard rock not far from the city of Mexico. He will shortly describe the specimen fully.

MR. J. H. GURNEY, JUN., has some remarks in the "Ibis" on the occurrence of the Egyptian nightjar in Nottinghamshire.

MR. W. M. MASKELL has published, in the Transactions of the Philosophical Institute of Canterbury, an additional paper (illustrated) on the Coccidæ in New Zealand.

Mr. T. M. READE'S important paper, entitled, "Experiments on the Circulation of Water in Sandstone," appears in the last number of the Proceedings of the Liverpool Geological Society.

FORMIC ACID is recommended as the quickest means of destroying Bacteria.

SOME very interesting and highly readable Miscellaneæ have been contributed by Mr. M. H. Robson, hon. sec. of the North of England Microscopical Society, to the Transactions of the Natural History Society of Northumberland, Durham, and Newcastle-on Tyne, &c.

THE Meeting of the German Society of Naturalists and Physicians was held at Magdeburg on Sept. 18–23.

AT the British Association meeting, Mr. Wethered stated that an expert in the microscopical examination of coal could judge of its nature from an examination of a piece with a pocket lens.

MR. MAIRET has expressed his belief that phosphoric acid is intimately connected with the nutrition and action of the brain.

THE French railway companies are about to adopt an electric gate-opener.

M. D'ABBADIE, a well-known French astronomer, in a paper read before the Paris Academy of Sciences, proposes the adoption of 10,000 kilometres as a unit for the measurement of celestial spaces, and that this unit shall be termed a *megiste*, from the Greek μέγιστον.

The sums awarded in money grants for scientific purposes, at the conclusion of the recent meeting of the British Association Meeting at Montreal, amounted to £1525.

The Iron and Steel Institute held their annual meeting at Chester on September 23rd and the three following days.

The museum recently opened by the Prince of Wales at Newcastle-on-Tyne will cost £42,000, of which £38,000 has already been raised.

Warehouses for the storage of cold air are now in operation in New York, and from these cold air will be served through pipes to any part of the city. In the new Washington market a network of pipes is fixed running through the building, and cold air will be served to any of the stalls furnished with perishable articles.

The French Association for the Advancement of Science commenced its meetings at Blois on the 3rd of September. The president was M. Bouquet de la Grye, who gave an address on Oceanic Hydrography. He expressed his belief that the level of the sea presents many variations, owing to the quantity of salt in the water.

The brilliant sunset glows continue to be almost, if not quite, as strikingly beautiful as ever.

Drs. Maurier and Lange, who have been working at Marseilles ever since the departure of the German Commission, report that they have found a Mucor which they believe to be the actual agent in the propagation of cholera. They state the mucor is the mature form of which Dr. Koch's bacillus is only an earlier and lower stage. The mucor appears only on the fourth or fifth day. It consists of a mycelium which bears sporangia, the latter containing myriads of spores. When the latter come into contact with putrid organic matter, they develop into a mucor of another form which produces cholera, and originates the bacillus.

A laboratory devoted to special researches in Bacteria has been established at Munich.

Captain Dutton is engaged in the study of the extinct volcanoes of the Rocky Mountains.

Another Polar expedition is to start next autumn, to attempt to reach the Pole by way of Franz Josef Land!

Mr. H. H. Johnston writes from Mount Kilimanjaro, in Equatorial Africa, at the altitude of 5000 feet. He describes it as one of the loveliest sites in the world.

M. Olszewski has succeeded in liquifying hydrogen under a pressure of 190 atmospheres, cooling with oxygen boiling in a vacuum.

A balloon centenary was held on September 15th on the grounds of the Honourable Artillery Company. Five thousand invitations were sent out. The balloon steering competition for £1000 was the chief event.

The late Lord Lytton's Vril, described in "The Coming Race," was no such wild dream as many supposed! A French wild-beast tamer has just patented an electrical apparatus, shaped like a stick of about 3 feet long, with which he can inflict severe shocks on the animals. On receiving it, the latter are said to crouch down in terror, all except the bear, which appears to withstand Vril the best!

A wild and improbable story is tramping the round of the newspapers concerning a "solid mountain of alum" in Gila River country — of course, in America!

Recent experiments show that oats contain a substance easily soluble in alcohol, which has an irritant action on the motor cells of the nervous system. It has been called Avenin by Sanson. It is a nitrogenous substance, apparently of an alkaloid character. The quantity present varies according to the quality of the grain and soil on which it is grown. The darker varieties contain more than the light. Its composition is given as $C_{56}H_{21}NO_{18}$. The bruising and milling of oats diminishes the quantity of this substance very rapidly, but it is quicker in its action.

H. Baubigny has determined the atomic weight of chromium to be 52·16.

Mr. Edward Lovett has read an important and exhaustive paper before the Croydon Microscopical and Natural History Club, on "The Edible Mollusca of the British Islands."

Mr. David Houston, F.L.S., who will be well known to our readers as the writer of the Botanical portions of Cole's "Microscopical Studies," is commencing a class at the Birkbeck Institution, on Elementary and Advanced Botany, and in General and Vegetable Biology, with systematic laboratory work.

MICROSCOPY.

The Journal of the Royal Microscopical Society.—The August number of this highly interesting serial, besides the usual summary of current researches relating to zoology, botany, microscopy, &c., contains the following original papers (illustrated) :—" Researches on the Structure of the Cell-Walls of Diatoms," by Dr. J. H. L. Flögel; "On a New Microtome," by C. Hilten Golding-Bird; "On Some Appearances in the Blood of Vertebrate Animals with reference to the occurrence of Bacteria therein," by G. F. Dowdeswell; "On *Protospongia pedicellata*, a New Compound Infusorium," by Frederick Oakley; and "On a New Form of Polarising Prism," by C. D. Ahrens.

THE QUEKETT MICROSCOPICAL CLUB.—The Journal of this Society for July (edited by Mr. H. F. Hailes) contains the following original papers :— "On an Undescribed Species of Myobia," by A. D. Michael; "On the Hexactinellidæ," by B. W. Priest; "On some New Diatomaceæ from the Stomachs of Japanese Oysters," by F. Kitton; "Note on *Mermis nigrescens*," by R. T. Lewis; and List of the Objects obtained during the excursions, as well as Proceedings of the Meetings, &c.

COLE'S MICROSCOPICAL STUDIES.—We understand that these valuable papers are to be continued, and that it is intended to put fresh interest in them, so as to render them still more widely known. There will be four sections, as follows : "Animal Histology," by F. Greening ; "Botanical Histology," by D. Houston ; "Pathological Histology," by Mr. Fearnley ; and the "Popular Studies," by A. J. Cole ; the entire work being issued under the editorship of the latter.

WE are pleased to note that Mr. Martin J. Cole has been appointed lecturer at the Birkbeck Institution, and that he is about to deliver a course of lectures on Practical Microscopy in the Biological Laboratory on Saturdays.

ZOOLOGY.

RARE SHELLS NEAR LONDON.—Some time ago my brother wrote a list of shells found near London; since then many new forms have been met with, and many new localities found for those included in the list. It would occupy too much space to give a full list, but it may be worth while to notice a few of the rarer or more interesting forms, which are as follows :—*Anodonta anatina*, not uncommon in the ornamental waters in Regent's Park, living specimens may be obtained ; the ornamental waters also contain *Sphærium corneum*, *Bythinia tentaculata*, *Planorbis albus*, *Limnæa auricularia*, and *L. stagnalis*. *Neritina fluviatilis*: although this shell is by no means rare, it may be worth while to know that it is exceedingly abundant in the Thames near Hammersmith bridge. *Valvata piscinalis*, var. *subcylindrica* : I found a dead shell at Hammersmith which agrees fairly well with the description of this var. *Planorbis lineatus*: I have taken this not uncommonly in a pond on Barnes Common, in company with *Limnæa palustris*, monst. *decollatum* (one specimen only), *Valvata cristata*, *Physa fontinalis*, *Planorbis complanatus*, *P. vortex*, &c. *Planorbis nitidus*: Fulham. *P. nautileus*, var. *crista*: Acton, in a pond. *Limnæa truncatula*: Hyde Park. *Limnæa stagnalis*, var. *roseolabiata*: in a pond at Grove Park, with *Ancylus lacustris*. I have also found *Ancylus lacustris* at Bromley ; I once saw a specimen "spinning a thread." *Arion hortensis*: I have found amongst nettles at Bedford Park two varieties of this species ; one is orange brown, with the bands on either side just visible, the margin of the foot orange, and the foot grey beneath ; the other form is light yellowish-grey, with the bands also lightly marked out in a somewhat darker colour ; the orange-brown variety seems to be the more common of the two. *Limax lævis*: I found a little slug near Southall which I sent to Mr. J. W. Taylor, and which is thought to belong to this species. *Zonites glaber* (apparently the *Hyalina alliaria* of continental authors, but certainly the *Z. glaber* of Jeffreys) : abundant on a bank near Bromley ; I obtained four specimens of a variety having the shell of a greenish-white colour, corresponding to similar and well-known varieties of other species of Zonites. This species, when alive, has a slight odour of garlic ; they live amongst moss. *Z. nitidulus*, var. *Helmii*: one specimen near Chislehurst. *Helix aculeata*: Sevenoaks. *Helix aspersa* var. *exalbida*: not uncommon amongst *Clematis vitalba* near Dartford ; my brother has found it on *Pteris aquilina* on Chislehurst Common, but it seems to be very rare in the latter locality. *H. nemoralis*: on a bank near Crayford I have found some curious varieties of this species, living amongst ivy ; one form is deep lilac, and corresponds nearly to the var. *lilacina* of *hortensis*, except that the colour is somewhat darker. Another form has the shell of an orange colour, quite different from the ordinary pink form, and a third variety when alive appears of a delicate green colour, but this is due to the animal, for when it is extracted the shell is seen to be yellow, like the ordinary Libellula, but rather thinner than usual. On the same bank I found the beautiful larvæ of *Hadena pisi*. Close by I found specimens of *Cyclostoma elegans*. *Helix hortensis*, var. *albina*: Sidcup, in Kent, and Acton in Middlesex. Var. *lilacina*: Eltham, Sidcup, St. Mary Cray, and Chislehurst in Kent, one specimen at Gunnersbury in Middlesex. *H. arbustorum* and var. *flavescens*: St. Mary Cray. *H. rotundata* var. *alba*: Otford, on a tree, one specimen. *H. lapicida*: Bickley, with *Clausilia laminata* and *Bulimus obscurus*, var. *albinos*. *Vertigo antivertigo*: two specimens on Barnes Common. *Balea perversa*: Bickley ; not common. *Achatina acicula*: my brother found one specimen at Chislehurst. Other species will be found mentioned in former notes. —*T. D. A. Cockerell*.

NATTERJACKS AT WIMBLEDON.—It may interest some of your readers to hear that three specimens of the natterjack toad (*Bufo calamita*) have been found by two of my boys this summer on Wimbledon Common. I send you this communication because I think its existence here has been denied.—*Franklin J. Sonnenschein*.

NATURAL HISTORY TRANSACTIONS OF NORTHUMBERLAND, DURHAM, AND NEWCASTLE-UPON-TYNE.—Part I. vol. viii., has just been issued, and contains the following papers : Presidential Address, by the Rev. Canon Tristram, describing the Proceedings of the members at the various meetings held between the spring of 1879, up to April 1880. Notes on a hitherto undescribed Roman Camp near Foulplay Head, Rochester, Rudwater, by G. T. Clough, of H. M. Geol. Survey, with two plates. Miscellanea, consisting of notes on various subjects, by Mr. John Hancock and others. Memoir of the late Thomas J. Bold, by Mr. Joseph Wright, acting curator of the Newcastle Museum. Mr. Bold was an extensive contributor to these and other natural-history publications; the list of his papers, as given by Mr. Wright, amounting to about 160, which are chiefly on Entomology. Notes on the Vertebral column and other remains of Loxomma Allmanni, Huxley, by the late Thomas Atthey. Mr. Atthey left behind him one of the largest and best collections of coal-measure fossils in the kingdom, which fortunately have been secured for the Newcastle Museum. The Yorkshire Caves; a three days' trip with the Tyneside Field Club, 1882, by Thomas T. Clarke; a very interesting article, and especially to those who have not visited these natural curiosities. Address to the members of the Tyneside Naturalists' Field Club, by the President, the Rev. A. M. Norman, D.C.L., F.L.S., May 1881. This is a very important paper and is divided into two parts—one describing the Proceedings of the members at the various meetings, held during the year, and the other describing the faunæ met with at various dredging expeditions, in which he himself took part. He also gives an extensive catalogue of the Fauna, as far as yet known, which lives in the North Atlantic Ocean at greater depths than 1,000 fathoms. A voyage to Spitzbergen and the Arctic Seas, by Abel Chapman, Esq., Silksworth Hall, Sunderland, with four beautifully executed plates. Mr. Chapman describes the ornithology of Spitzbergen, so far as was witnessed by himself. Presidential Address to the members by E. T. J. Browell Esq., in May 1882. Mr. Browell gives a very interesting account of the field meetings held during his year of office, and concludes with a reference to the death of Charles Darwin and his works. The concluding section of this part is headed "Miscellanea," and consists of several notes contributed by Earl Percy and others.—*Dipton Burn.*

NATURAL HISTORY OF JERSEY.—In consequence of the author's not receiving proof of his paper last month, the following errata occur:—"Graps" pools, instead of "grass pools;" *Trochus magnus,* for *T. magus, T. pallastra* for *T. pallustra,* and suckers *of* the skin, instead of *on* the skin.

ZOOLOGICAL CHARACTER OF THE DUCK-BILLED PLATYPUS.—At the recent meeting of the British Association at Montreal, the president of the Biological Section, Professor Moseley, read a telegram, stating that Mr. Caldwell finds the Monotremes to be viviparous with mesoblastic ovum. Professor Moseley said that this contained the most important scientific news that had been communicated to the meeting of the Association. Briefly its significance was thus explained. The lowest known mammal, the ornithorhynchus or duck-billed platypus, and the echidna or so called spiny anteater of the Australian region, although like other mammals they suckle their young, lay eggs like birds. Further, the early stages of development of the ovum are unlike all other mammals, including, as Mr. Caldwell has shown, marsupials, but are identical with those of birds and reptiles. The segmentation of the ovum is mesoblastic, not holoblastic. These important results, Professor Moseley continued, point to the origin of the monotremata, and thus of all mammalia, and incidentally of man himself, from the reptilia rather than from amphibia, which latter origin has been lately advocated by several naturalists of the highest distinction.

GIGANTIC EARTHWORMS.—A gigantic earthworm has been sent from Cape Colony for Mr. Frank Biddard, the prosector of the Royal Zoological Society. The Rev. G. Fisk, F.Z.S., with whom Mr. Biddard has corresponded on the subject, received the worm from Mr. H. W. Bidwell, who found it in the Botanic Garden at Uitenhage. The longest measurement of the creature yet taken reaches six feet five inches. The surface of the upper portion of the body shows a bright green colour of variable intensity, but otherwise it is a loathly animal. *Lumbricus microchæta* is the name by which it will be known.

THE POLYZOA AND THE ROTIFERÆ.—At the Biological Section of the British Association Meeting, Mr. Sidney F. Harmer's paper on the Development of Polyzoa gives the results of his work on Loxosoma, conducted while occupying the Cambridge table at the Naples Zoological station. They indicate that the polyzoa are allied to the rotiferæ, have little connection with the brachiopoda. The excretory organ of the adult belongs entirely to the head; the dorsal organ of the larva takes no part in budding. Its development and structure exhibit the characteristics of nervous tissue; it is in fact a brain. During the degeneration of the alimentary canal the free larva produces a pair of buds, probably never becoming itself adult.

FOX CUBS REARED BY A CAT.—I send you a copy of the "East Sussex News," which gives a short account of the curious fact in natural history, viz., the rearing of two fox cubs by a cat. It appears the mother of the cubs was killed in the hunting-field and the cubs were dug out of a hole some feet deep. The cat took readily to the cubs, and seems very fond of them. I have seen them to-day, about the twelfth of their existence, and they appear to be strong and healthy with their eyes still closed.—*J. G. Braden.*

BOTANY.

RARE PLANTS.—I wish to record the re-discovery of *Saxifraga hirculus* on July 31, growing in marshy ground near Loughnaroon, townland of Glenbuck, parish of Rasharkin, co. Antrim. It was found in 1837 by the late Mr. David Moore, curator of Glasnevin. When on the Ordnance Survey, I found it growing sparingly. I found the parsley-fern and beech on Slievananee, co..Antrim, and both the filmy ferns in Glendun, in the same county.—*S. A. Brenan.*

MIMICRY OF MINT BY DOG'S MERCURY.—In answer to your correspondent P. S. Taylor, respecting "mimicry of mint by dog's mercury," I think he has jumped at conclusions too quickly. The same occurrence of *Mercurialis perennis* took place in my own mint bed, but it also occurred all over the garden where the ground was bare and the sown crops had not come up. This was also the case with the mint bed which had been newly planted and failed, and hence the so-called mimicry. In former years when the mint has done well, although dog's mercury has occurred in other parts of the garden, where no mimicry could be supposed, the mint has been singularly free.—*Collis W.*

SAGITTARIA SAGITTIFOLIA.—In answer to your correspondent B. B., respecting *Sagittaria sagittifolia*, there is nothing at all unusual in the appearance of the two small hard tubers about the size of snowdrop bulbs, although I do not think the fact is generally known; but anyone in taking up one of these plants with care, even in early summer, and when in full foliage, will be sure to find these bulbs on stalks varying in length according to the nature of the soil through which they have to pass; these stalks being very brittle, will break, and the bulb be lost, unless great care is taken, and this may account for them not being noticed. It appears that the plant reproduces itself by these as offsets as well as by flower and seed. Early this summer, accompanied by an eminent botanist, I removed some of these plants from the river Roden, Essex, with these stalks and tubers intact, and he then mentioned that the fact was new to him, although I, a perfect novice, had often noticed them.—*C. Willmott.*

TO SKELETONISE LEAVES.—The leaves should be perfect. Place them in a vessel containing water and allow them to remain for three weeks, or longer, according to size; then examine, and if the colouring matter appears soft, place the leaf on a china plate, and with a camel-hair brush dipped in water carefully remove the thin part of the leaf from between the veins. Should it not be possible to do this at one operation without injury to the veins, &c., allow the leaf to remain in water a short time longer, then repeat the operation until a perfect skeleton is obtained. *Richard A. Crombleholme.*

THE LATE GEORGE BENTHAM, F.R.S.—Every English botanist will hear with profound regret of the death of this veteran botanist at the ripe age of 83 years. His name is associated with the development of the natural system almost more than any other man's, and his intellectual vigour and elasticity was shown by his quick perception of the value of the evolution theory to practical botany. He continued a vigorous worker almost up to the last.

THE CHEMISTRY OF PLANTS.—Messrs. Berthdot and André, two French scientists, have given an account of their recent researches in organic botany. They have attempted a complete analysis of a vegetable organism, with a view to determining the chemical equation during its development from the fertilised germ to its fructification and reproduction.

GEOLOGY, &C.

GEOLOGY AT THE BRITISH ASSOCIATION.—Geology this year was extremely local, though the locality was a great continent. The address of the president, Dr. Blanford, has an important general bearing on the characteristics of the *locale* of this year's meeting. He attacked a fallacy which has exercised a pernicious influence in geology, and which it will take some time to thoroughly eradicate. It has been very rashly assumed by some geologists that because we find the remains of similar animals in certain strata in regions far distant from each other, as Europe and America, Africa and Australia, therefore these strata must necessarily be of the same geological age. Dr. Blanford entered with great minuteness into the question, showing, by the production of numerous actual examples, how absurd the hypothesis often is. The subject is one having important bearings on the question of the development of life in various parts of the globe. It was specially appropriate in the president to take it as the topic of his address at the first meeting on the American continent; for naturally American and indeed English geologists attempt to co-ordinate American with European formations and fossils. Indeed, one of the most prolonged discussions turned to a large extent on the relations of the one to the other. Professor Bonney introduced it in his paper on the Archæan Rocks of Great Britain, which was supplemented to a large extent by papers by Professor Sterry Hunt, the Rev. J. F. Blake, and others. This was the introduction to American soil of a controversy that has been waging among English geologists for some years. Canada is the home of *Eozoon Canadense*, to which there was allusion in one or two of the papers; but its true nature, whether organic or inorganic, was not seriously discussed. Section C abounded with papers of interest, and not the least interesting were those by the Rev. E. Hill and others on Ice-age theories. Perhaps the most appropriate and interesting paper in this section was that by Professor Newberry, on the development of the North American Continent ; and the fact that such a momentous subject could be seriously discussed in an assembly of distinguished scientific men is a testimony to the completeness and value of the work of American geologists and American geological surveys.

THE MINERAL VEINS OF THE LAKE DISTRICT.— No geologist has so thoroughly examined the probable formation of the mineral veins of the district he lives in as Mr. J. D. Kendall, F.G.S., whose paper, entitled as above, appears in the last number of the Proceedings of the Manchester Geological Society. Mr. Kendall's profession as a mining engineer has given him unusual facilities for observation ; which he has turned to good purpose. His paper is a lengthy one, well illustrated, and his summary is as follows :—" Veins are not filled fissures. The variations in breadth are not due to the sliding of the walls upon one another, but to variations in the solubility of the rock. Veinstone is part of the rock which originally existed where the veins now are, and is a result of metamorphism. The metallic minerals, Hæmatite excepted, were deposited in cavities of the veinstone from chemical solutions. Hæmatite veins are substitutional deposits."

NOTES AND QUERIES

THE LAW OF TRESPASS.—Presuming that your note on the law of trespass refers only to Scotland, I venture to suggest that it would prove a valuable piece of information to many English botanists if the opinion of some competent legal authority could be obtained and published in SCIENCE-GOSSIP accurately defining what in point of law is a trespass in England, and clearly stating the penalties attaching thereto, and whether they can be enforced or recovered in the police court or only by county court process.—*F. J. George.*

NAMES ON TREES.—A long time ago, a gentleman now living in Dorset cut his name on a tree in this town. A schoolfellow of his did the like. Thirty years after, the tree was cut down. The second name was found easily enough, but, at first, the owner of the other could not discover his ; but, at last, he discovered at thirty feet towards the top of the tree. Both names were cut at four or five feet from the ground. I have the strongest possible assurance that this is a fact. How do you account for it?—*H. J. Moule.*

PARIS QUADRIFOLIA.—In reply to K. D.'s question, I beg to state that I found two plants of *Paris quadrifolia* in this neighbourhood (Colwall, Malvern), last June, and have a dried specimen with six leaves, found at Malvern Wells. It will be interesting to know if these variations are frequently met with, as Mr. Johns in his " Flowers of the Field," derives the name from the Latin *Par-paris* on account of the unvarying number of the leaves. —*A. D. Colwall.*

NOTICES TO CORRESPONDENTS.

TO CORRESPONDENTS AND EXCHANGERS.—As we now publish SCIENCE-GOSSIP earlier than heretofore, we cannot possibly insert in the following number any communications which reach us later than the 8th of the previous month.

TO ANONYMOUS QUERISTS.—We receive so many queries which do not bear the writers' names that we are forced to adhere to our rule of not noticing them.

TO DEALERS AND OTHERS.—We are always glad to treat dealers in natural history objects on the same fair and general ground as amateurs, in so far as the " exchanges " offered are fair exchanges. But it is evident that, when their offers are simply disguised advertisements, for the purpose of evading the cost of advertising, an advantage is taken of our *gratuitous* insertion of " exchanges " which cannot be tolerated.

WE request that all exchanges may be signed with name (or initials) and full address at the end.

W. DUCKWORTH.—There are two kinds of galls on your oak leaves. The round hairy ones are those made by *Spathegaster tricolor;* the others are the galls of *Spathegaster baccarum.* Thanks for the curious specimen of *Tragopogon pratense,* showing prolification.

R. H. W.—The black spots and patches on the leaf of sycamore have been caused by drops of rain or dew acting as sunburners by condensing the solar rays.

A. S. MACKIE.—Your fossil appears, from your verbal description, to be a species of Orthoceras or Gomphoceras ; or it may even be nothing but a septarian nodule. Send us a sketch of it.

MARSHALL.—Get some stout brass wire, bend it into a ring, and fit on it a fine muslin bag. This makes a capital pond net for rotifers, minute crustacea, &c. Have the ring soldered on to a conical ring to fit a rod or stick. You will find it best to make an aquarium of your own. See Taylor's " Aquarium : Its Management, &c.," price 6*s.* (London : W. H. Allen & Co.), or advertise for a secondhand one in our columns.

C.—You will get living Hippocampi from Mr. King, Sea Horse House, Portland Road, London. See Taylor's " Aquarium : Its Principles, Structure, and Management," for rest of query. How to make artificial sea water is there described on page 157.

A. K. P.—If you turn to the volumes of SCIENCE-GOSSIP for 1879, 1880, and 1881, you will find under the title of " Assisting Naturalists " the names and addresses of numerous gentlemen, skilled in every department of natural science, who have kindly volunteered to assist students and readers of this magazine in their various difficulties.

T. W. HOLSTEAD.—If you turn to the past vols. of SCIENCE-GOSSIP, you will find, in the index, reference to articles and notes respecting the preservation of Crustacea, star fish, &c.

W. S. W.—The " Fleurs du Lac," found as a yellow scum on the surface of the Lake of Thun at the end of May, is the pollen from the Pine forests.

F. R. T.—Tethya is a genus of living sponges. Dichorisandra is not a fossil tree, but a genus of Commelynaceæ, with the habit of Tradescantia. They are Brazilian herbs. *Peristeria elata* is a genus of Orchidaceæ, familiarly known as the Holy Ghost, or Dove plant. It grows in Panama. The other name you mention we have not heard of.

G. WARD.—Get Huxley and Martin's " Practical Instruction in Biology " (London : Macmillan), price 6*s.*

J. R. R.—Get Spencer Thomson's " Walks and Wild Flowers," price 2*s.* We don't know of any similar book for Ireland.

G. A. SIMMONS.—The specimen of Carp with both ventral fins on same side is very curious and interesting.

R. H. B. (King's Lynn).—You are quite right in supposing the species to be *Stellaria media;* it is a rare form of it, called *S. umbrosa,* Opitz.

G. B. L. (Cupar Angus).—It is *Plantago maritima,* L. You must not however imagine it is confined to maritime situations, it is often, as in your case, found far inland.

G. M. (Brechin).—No. 1. *Cynoglossum officinale.* No. 2. *Atriplex erecta,* this is very unlike the Pelitory. No. 3. *Euphorbia helioscopia,* a very common weed. No. 4. *Ononis arvensis.* No. 5. Tansy (*Tanacetum vulgare*).

H. L. (Maidstone).—We are sorry, but the small specimen you send is difficult to decide. It is, however, not a British species.

A. E. P. (Wolverhampton).—a. *Honkeneja peploides;* b. Thyme (*Thymus Chamædrys*) ; c. *Parmelia;* d. the pretty bellflower (*Wahlenbergia hederacea*).

W. F. H.—From the sketch you sent, your fungus appears to be a malformed agaric.

EXCHANGES.

SPLENDIDLY-preserved and correctly-named Swiss Alpine plants. Price 6d. each.—Address, Dr. B., care of Editor of SCIENCE-GOSSIP, Piccadilly.

WANTED, first ten parts of "English Illustrated Magazine." Will give London discount price for them. — L. Francis, 4 Somerville Road, Queen's Road, Peckham, S.E.

VIOLIN and buw offered for paraboloid or small achromatic condenser; approval both sides.—S. C. L., 276 Middleton Road, Oldham.

CONCHOLOGY: *H. pisana, arbustorum, ericetorum, obvoluta,* and other land and freshwater duplicates. Wants marine, need *not* be named (loose), also starfish.—Prudential Agent, Lees Street, Lodge Road, Birmingham.

WANTED, in exchange for Schwann and Schleiden's "Microscopical Researches into the Accordance in Structure and Growth of Animals and Plants," Carpenter's, Huxley's, or Tyndall's works, or Darwin's "Descent of Man."—C. Craxton, Higham Ferrers, Northants.

A BEAUTIFUL slide of New Zealand ruby sand in exchange for grouped diatoms or Lagena forams.—W. Aldridge, Upper Norwood, London, S.E.

WANTED, pieces of horns and hoofs, fit for cutting sections, from rhinoceros, bison, antelope, &c.; also skin from rhinoceros, alligator, or any interesting specimens, English or foreign. Well-mounted microscopic objects offered in exchange, anatomical, botanical, or insects, or material for mounting; list sent if required. Also zoophytes, correctly named, British or foreign.—R. M., 59 Hind Street, Poplar, London, E.

To exchange: "Nature," from Nov. 2nd, 1882, to April 19th, 1883 (inclusive), and from Nov. 1st, 1883, to May 8th, 1884 (inclusive, but one number awanting), for back volumes of SCIENCE-GOSSIP (unbound) or works on zoology or botany.— J. H. W. Laing, Downie Mount, Tayport, N.B.

WANTED, *H. sericea, H. concinna, H. revelata, C. Rolphii,* and *Z. alliarius.* Will give good specimens of *H. ericetorum, C. aubia, C. caminata, H. lapicida,* and *H. rupestris.*—Wm. Webster, National School, Lofthouse, Wakefield.

TEN years' back numbers of SCIENCE-GOSSIP, also thirteen parts of Yarrell's " British Birds," by Newton, for. sale or exchange.—J. Bracewell, 39 Hood Street, Accrington.

STAMP album, in good condition, in two volumes, cost 17s., containing 850 stamps, lately valued at £3, would form a good basis for a large collection. Wanted, micro cabinet to contain not less than 500, or mounting instruments, or 2½ in. objectives by R. J. Beck, Crouch, or Swift, or any good maker.—H. J. Parry, 10 Windsor Terrace, Newcastle-on-Tyne.

WANTED, 197, 243, 529, 701, 792, 939, 1200, 1219, 1277, 1437, 1552. In exchange, 171, 172, 771, 812, 944, 958, 1141, 1293.— Rev. F. H. Arnold, Hermitage, Emsworth.

FOSSIL shells wanted from all formations; good exchange in land and freshwater shells.—C. T. Musson, 1 Clinton Terrace, Derby Road, Nottingham.

WANTED, to exchange mosses, hepatica, and lichens to complete collections. Send numbers wanted from Lond. Cat.— J. McAndrew, New Galloway, N.B.

SHELLS for exchange: *Bulla hydatis, Trochus lineatus, Cardium aculeatum, Littorina neritoides, Helix Cartusiana, Helix pomatia, Zonites glaber, Bythinia Leachii,* and numbers of others. Send lists of duplicates and desiderata to—S. C. Cockerell, 51 Woodstock Road, Bedford Park, Chiswick, W.

WANTED, Pupæ of Cecropia, Pernyi, Cynthia, Polyphemus, Luna, Pyri, Io, and other exotic species. Will exchange British Lepidoptera, choice flower seeds, &c.—Robert Laddiman, Helleston Road, Norwich.

SOME well-mounted slides of various kinds to exchange, including butterfly scales, palates of snails, insects, spicules, infusoria, and others. Wanted, botanical specimens in exchange, histological mounts greatly preferred. List on application.— F. R. Tennant, Port Hill, Stoke-on-Trent.

FOR exchange, a 15s. stamp album, containing about 800 stamps; some few of the stamps are slightly damaged; many rare. Wanted, fossils, micro slides, or science books.—F. R. Tennant, Port Hill, Stoke-on-Trent.

STAVELEY'S "British Spiders" (coloured plates), Pye Smith's "Scripture and Geology," Wilson's "Chapters on Evolution," Hindley's "Roxburgh Ballads " (2 vols.), for books on natural history.—George Roberts, Lofthouse, Wakefield.

WILL send young plants of Valisneria for unmounted cleaned diatoms.—John Sim, West Cramlington.

OFFERED, L. C., 7th or 8th edit., Nos. 121, 192, 327, 326, 367, 398, 406, 628, 642, 859, 924, 1008, 1040, 1124, 1333, 1349, 1412, 1422, 1448, 1458, 1501, 1515, 1577, &c. Send lists.—H. J. Wilkinson, 17 Ogleforth, York.

WANTED, a few live British snakes, lizards, &c., for preserving in spirits, good specimens; will exchange birds' eggs— S. E. W. Duvall, Ranelaph Road, Ipswich.

MANGE mites and allied genera wanted in exchange for rare and well-mounted Acari, including *Glyciphagus plumiger,* Myobia, *Cheyletus flabellifer,* &c.—H. E. Freeman, 60 Plimsoll Road, Finsbury Park, N.

WANTED, rare foreign shells and British fossils, also first-class shell cabinets; state requirements.—J. E. Linter, Arragon Close, Richmond Road, Twickenham.

WANTED, Berkeley's "Cryptogamic Botany," SCIENCE-GOSSIP, No. 1 to 1879 inclusive, bound or unbound, with other works on Cryptogams. Offered, micro slides and material.— W. E. Green, 24 Triangle, Bristol.

EGGS of *Branta antarctica,* Gm., *Larus dominicanus,* V., *Graculus Brasiliensis,* Gm., and *Graculus Gaimardii,* Less., from Patagonia, single and in clutches, in exchange for other good eggs, British or exotic, former preferred.—J. M. Campbell, Kelvingrove Park, Glasgow.

WANTED, Marshall's "Anatomy" and Liddell and Scott's "Greek Dictionary," 4to edition, in exchange for British land, freshwater, and some foreign shells, lepidoptera, greenhouse plants (not budding), and some of the finest varieties of the Cactus tribe. Lists sent for selection.—E. R. F., 82 Abbey Street, Faversham.

REMARKABLY fine specimens of *Limnea stagnalis* and *Planorbis corneus,* also *L. peregra, L. glabra, P. hypnorum, P. xiveiorbis, N. fluviatilis, S. lacustre, D. polymorpha, P. viviparu.* Desiderata very numerous; other land, freshwater, and marine shells.—W. Hewett, 26 Clarence Street, York.

DUPLICATES: Corydon, *S. populi, Z. trifolii,* Potatoria, Sambucata, Cratægata, Betularia, Rhomboidaria, Atomaria, Piniaria, Ruberata, Pyraliata,. Dotata, Bucephala, Perla, Phragmitidis, Elymi, Cubicularis, Instabilis, Verbasci, Jota, Typica, Hybridalis, Alvearjella, Lafauryana, &c. Desiderata: numerous Macro- and Micro-Lepidoptera.—George Balding, Ruby Street, Wisbech.

TWO ounces guano from Peru, containing large and fine diatoms, free, for three good slides.—Tylar, 20 Geach Street, Birmingham.

WANTED, geological specimens or books for old "Boys' Own Papers," or "Chambers's Journal," or emu's eggs.—John T. Millie, Clarence House, Inverkeithing, Fifeshire, Scotland.

OFFERED, L.C., 7th ed., Nos. 65, 159, 164b, 192, 204d, 246, 510b, 542, 779, 1076, 1265, 1374b, 1385, 1391, 1431, 1432, 1435, 1445, 1569c, 1621, 1668. Lists exchanged.—Rev. W. R. Linton, 180 Upper Street, Islington, London.

FOR Tubes of *Campanularia angulata* and several species of diatoms *in situ,* send stamp to—J. Sinel, David Place, Jersey.

Hyalonema mirabilis (glass rope sponge) in exchange for other good sponges or Gorgonias for Spiculæ; also fine mounted slides of Multihamate spicula of Hyalonema for large spines of *Echinus cedaris, Spatangus,* &c.—W. White, 17 York Street, Nottingham.

FOR prepared leaves of *Hippophae rhamnoides* (scales splendid polarizer), send stamped envelope to—W. White, 17 York Street, Nottingham.

OFFERED, L. C., 7th ed., Nos. 40, 41, 133, 136, 161, 173, 185, 207d, 287, 313, 315, 528a, 634, 809, 835, 914, 998, 999, 1361, 1426, 1440, 1447, 1465, 1567, 1641, &c. Send lists.—H, Fisher, 52 R. Y. C., Clifton, Bristol.

DUPLICATES. *Anodonta cygnea, A. anatina, Unio tumidus, U. pictorum, Paludina vivipara, Limnea stagnalis, L. glabra, Planorbis corneus, P. contortus, P. nautileus, &c.*—Desiderata other land and freshwater shells. —R. Dutton, 13 St. Saviourgate, York.

BOOKS, ETC., RECEIVED.

"Synopsis of British Mosses," 2nd edition, by Chas. P. Hobkirk, London: Van Voorst. — "Summer," by H. D. Thoreau, London: T. U. Gill.—"Report of Penzance Nat. Hist. & Antiquarian Society," 1883-84.—"Proceedings of the Liverpool Naturalists' Field Club for 1883-4."—"Thirteenth Annual Report of South London Microscopical & Nat. Hist. Club."—"Studies in Microscopical Science," edited by A. C. Cole.—"The Methods of Microscopical Research," edited by A. C. Cole.—"Popular Microscopical Studies," By A. C. Cole. —"The Gentleman's Magazine."—"Belgravia."—"Journal of Conchology."—"Journal of Microscopy."—"The Illustrated Science Monthly."—"Midland Naturalist."—"Ben Brierley's Journal."—"Science."—"American Naturalist."—"American Monthly Microscopical Journal."—"Popular Science News."— "The Botanical Gazette."—"La Feuille des Jeunes Naturalistes."—"Le Monde de la Science." &c. &c. &c.

COMMUNICATIONS RECEIVED UP TO 12TH ULT. FROM :— L. F.—G. R.—J. A. C.—T. D. A. C.—F. H. A.—T. G.— H. J. W.—R. H. W.—C. T. M.—S. C. C.—J. M.—J. S.— H. G. P.—W. H. B.—G. R.—G. W.—E. L.—J. P. G.—F. R. T. —T. M. R.—M. J.—J. B.—R. L.—C. B. M.—M. R.—G. M. —J. R. R.—F. S.—F. W. C.—J. S.—W.—A. S. M.—E. H. —J. E. L.—E. R. F.—W. H.—R. A. R. B.—G. B.—A. H. W. —W. T.—J. T. M.—W. R. L.—F. M.—H. E. U. B.—F. K.— A. E. P.—W. G. A. S.—J. M. C.—H. E. F.—S. E. D.—S. C. L. —E. J. E.—G. A. S.—F. R. T.—L. C. M.—J. W. H.—C. C. —W. A.—R. A. C.—A. K.—P. S. C.—W. W.—H. W. G. W.— R. M.—W. D.—S. A. B.—Miss Y.—W. H. C.—T. H. S.— E. A. S.—S. A. S.—H. J.—C. G.—S. R. M. —W. J. S.—W.— J. H. W. L.—W. M. W.—H. F.—F. J. S.—G. B. S.—J. T. N. —H. C.—W. S. W.—T. M. R.—D. H.—J. C. S.—J. Y.—J. H. —H. V. F.—A. Mc P.—R. D.—J. S.—&c.

GRAPHIC MICROSCOPY.

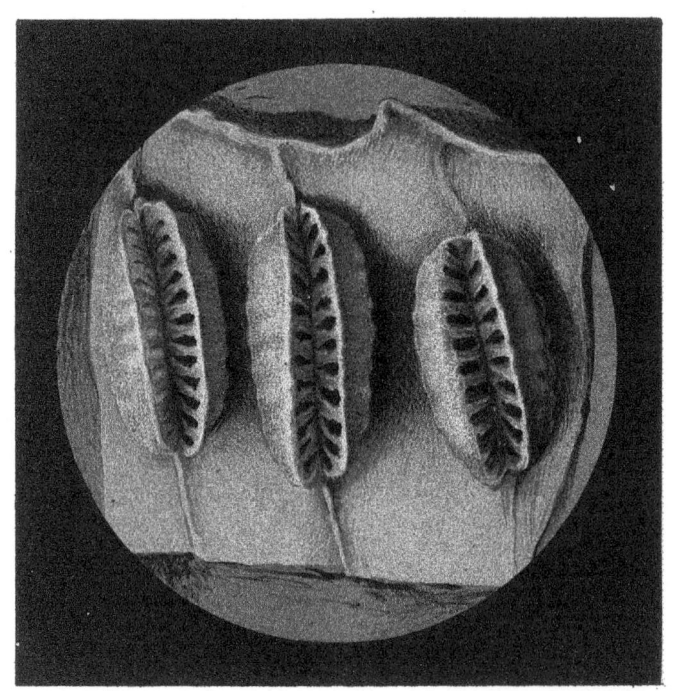

E.T.D del, ad nat. Vincent Brooks, Day & Son lith.

SORI OF FERN-MARATTIA.
× 50.

GRAPHIC MICROSCOPY.

By E. T. D.

No. XI.—Sori of Fern : Marattia alata.

THE outward appearance of the fructification of ferns has always attracted observation, groups and species affording great variety of forms variously situated, united with perfect symmetry, and in many cases exhibiting beauty and interest of an extraordinary character.

The under-side of a frond, as well known, reveals yellow, dust-like spots, lines, or patches, made up of groups, or masses of receptacles (the sori) containing, or composed of, a number of minute cases (the sporangia). Their positions greatly differ, and they are scattered over the under-side of the frond, localised in various parts, or confined to the very edge; they originate and are developed in the tissues, in contact with a vein, beneath the cuticle of the leaf, a portion of which is forced up as the receptacle is developed and presses for room, and often remains as a delicate membrane which eventually constitutes a protecting cover, (the indusium). As maturity increases, the indusium becomes partly detached, sometimes entirely, and eventually it either shrivels up, or falls off. In some ferns it opens in the centre and surrounds the sori, like a frill, forming a cup; in other species, the epidermis of the leaf is raised into ridges, including and protecting the sori beneath, thus fulfilling the functions of an indusium. The form and position of these receptacles and the presence or absence of the indusium are peculiarities bearing upon the distinction of genera.

No. 239.—November 1884.

The spore cases may be popularly described as globular boxes, which, in some cases, are encompassed by an elastic ring (the annulus). The tension of this band, breaks open and fractures the little case, releasing and scattering its contents. The presence or absence of this annulus, has led to the division of ferns into two groups—the *Annulate*, and the *Exannulate*. The Marattiaceæ are included in the latter, being destitute of this peculiarity; the dispersion of the spores is effected by the separation of the two halves of a compound and united sporange, which splits longitudinally, not unlike the pod of a pea, as seen in the plate.

Marattia alata is a native of the West Indies, but specimens of the fronds in fructification may be easily procured, as the plants are found in almost every collection of tropical ferns.

Beautiful as the general elegance of the little clusters of sporangia on any fern may be, when seen with a low-power and good reflected light, deeper interest is excited in tracing the future of an individual spore in countless thousands packed in their little cases.

These mere specks or particles, which in their minuteness might be scattered into comparative invisibility, exhibit a most marvellous process, only revealed by the aid of fine and high magnifying powers.

In flowering plants, fertilisation and the mature seed is a final result; in the ferns this process is reversed, fertilisation curiously following the commencement of a cell development, but preceding the establishment of the ultimate growth; and this wonderful power is potential in each individual spore, always microscopic, however large or grand in dimensions the perfected plant may be, and many tropical ferns assume tree-like dimensions.

To reach this exhibition of innate power, a well-ripened frond of an English hardy species should be dusted on the surface of a porous brick, placed in a shallow pan of water, the whole covered with a bell-glass, and, placed in a warm and shady corner, the spore begins to grow, projecting from the cell wall a delicate tubular prolongation, increasing by subdivisions, forming a minute flattened expansion, of a rather bright green colour, the "prothallus." For microscopical investigation this minute structure may be carefully raised and placed under-side uppermost, on a glass slip, in water, dilute glycerine, or (for continuous observation) in a solution of chloride of calcium, covered with the usual thin glass. Amongst irregular cells may be seen a few with peculiarly distinctive features; some, the "antheridia," containing a number of free granules, each enclosing a motile filament; others, the "archegonia," containing a germ cell. By the fracture of an "antheridium" contact is effected with contents of the "archegonia," through an intercellular passage, the primordial cell of the young fern then sends forth rudimentary rolled up leaves and roots. The expanded pro-embryo or prothallus, from which this result emanated, fades and the future plant becomes established; in various species the prothallus somewhat differs in appearance and character, but its essential functions are similar. This development has been graphically described in detail in the higher botanical books (Henfrey, Sachs, and others), and its bare outline is only repeated here, as suggesting to young microscopists a practical "observation" of the deepest interest, likely to encourage future research in the inexhaustible field of structural and morphological botany, a study greatly facilitated in having the object itself under inspection, either as a "preparation" or "living condition."

Crouch End.

A CHAPTER ON PHEASANTS.

THIS beautiful and well-known bird of our woods and plains belongs to the same class in natural history as our barn-door fowl (Gallinæ), but differs from them in many points. The head is destitute of a comb, the tail long, more or less drooping, composed of long, gently arching feathers, of which the middle exceeds the rest; the legs of the male are armed with spurs. The pheasant has nothing in his port of the upright, gallant bearing of the game cock, his attitude is more crouching, and the whole figure lower and more elongated. Their wings being short, they are ill-adapted for long flights, and therefore do not often wander from the preserves where they have been brought up. From this circumstance, it is most probable that in every country in Europe and America, where they are found, they have been introduced by the agency of man and not by chance. In Blain's Encyclopædia of Rural Sports, we are told that the pheasant is completely imprisoned on the island of Madie in Lake Maggiore, Italy; for should they ever attempt to escape by flight, they are sure to fall into the lake and be drowned, unless picked up by the boatmen. The male pheasant is not at all given to domestic affection, but passes an independent existence during part of the year, associating with others of its own sex during the rest of the season. The pheasant breeds in April, the young being hatched at the end of May or the beginning of June. The nest is a very rude attempt at building, being merely a heap of leaves and grasses collected together upon the ground, with a slight depression, caused apparently quite as much by the weight of the eggs, as by the art of the bird. The eggs are generally about eleven or twelve in number; the colour is of an uniform olive brown, and their surface perfectly smooth.

It is now generally admitted, says Sir William Jardine, that the pheasant was originally introduced into Europe from the banks of the river Phasis, now known as the Rion, in the country between the Black and Caspian Seas, where, it is said, this splendid bird is still to be found wild and unequalled in beauty.

That these magnificent birds were held in high estimation by the ancients for the beauty of their plumage, and the delicacy of their flesh, the following notes from the Greek and Latin authors abundantly testify.

According to the traditions sung by Aristophanes, a celebrated comic poet of Athens, B.C. 434, it was Jason, the leader of the Argonauts, when sailing up the river Phasis to Colchis, B.C. 1263, who discovered, and subsequently introduced this bird into Greece, where it soon became highly appreciated. It is related that when Crœsus, King of Lydia, seated on his throne in all the pomp of eastern splendour, asked Solon (the great Athenian lawgiver), then his guest, whether he had ever seen such magnificence before, the philosopher replied that he had seen the beautiful plumage of the pheasant, which he thought far superior. Athenæus, one of the early Greek authors, says that these birds were carried in cages, composed of precious woods, to adorn the triumphant march of Ptolemy II. King of Egypt into Alexandria. The same author tells us, that when the rich and luxurious inhabitants of Athens gave their magnificent feasts, such was their foolish prodigality and love of ostentation, that they caused a whole pheasant to be served to each guest, and live ones, placed in cages, ornamented the tables. The great naturalist, Pliny (A.D. 29), is the first Roman author who mentions the pheasant; he calls them "Phasianæ aves," birds of the Phasis, to the banks of which river the Romans in his day went in quest of them, which proves they were not common in Italy at that period. A few years later they must have become more plentiful, as Suetonius states in his "Lives of the Cæsars," that Vitellius, that imperial glutton, used to enjoy, among other viands, pheasants' brains mixed with many unheard-of delicacies in an immense dish, called by him the Shield of Minerva.

Heliogabalus, another Roman Emperor, in his ostentation and extravagance, is said to have fed the lions of his menagerie with these birds. Caligula, that monster of every crime, had placed in his temple a statue of gold, the exact image of himself, which was daily dressed in garments corresponding to those he wore; hen pheasants and other birds were offered in sacrifice to it, while a vile troop of his courtiers prostrated themselves at its foot at the very time when Caligula was wishing that the Roman people had but one neck, that he might sever it at a blow. Among other dishes that used to grace a grand supper-table was one containing pheasant sausages, which were made of the fat of this bird, chopped very small, mixed with pepper, gravy, and sweet sun-made wine, to which was added a small quantity of hydrogarum. Garum was a sauce made with brine, and mixed with the blood of mackerel or other fish. It appears to have been largely used in making savoury dishes by the Roman cooks of that period.—(See Soyer's "History of Foods.")

There are no records to afford a clue as to when and by whom the pheasant was first introduced into Britain. The Romans perhaps imported it with other imperial luxuries, but that it was protected by the laws of the country at a very early period is shown by the following extract from Dugdale's Monasticon Anglicanum. "That in the first year of Henry I. (1100) the Abbot of Amesbury obtained a license to kill pheasants."

In the Life of Thomas à Becket, by Canon Morris, it is mentioned that the archbishop, on the day of his martyrdom (1170), dined at three o'clock, and that his dinner consisted of a pheasant.—"One of his monks said to him, Thank God, I see you dine more heartily and cheerfully to-day than usual. His answer was, A man must be cheerful who is going to his Master."

The price of a pheasant in 1299 (being the 27th of Edward I.) was fourpence, a mallard three halfpence, and a plover one penny. In the early days of the Tudors, pheasants, and other game, used to be found in the woods and fields on the outskirts of the metropolis of that period, now covered with streets of houses and a teeming population. We find an Act of Parliament in the reign of Henry VII. entitled the Forferture for taking of Feasants and Partridges or the eggs of Hawkes and Swans; and a proclamation dated 7th July, in the 27th of Henry VIII. in which the king recites his great desire to preserve the pheasants, partridges, and hares from his palace at Westminster to St. Gyles's-in-the-Fields, as well as from thence to Islington, Hampstead, Highgate and Hornsey Park, and that of any person of any rank or quality presumed to kill any of these birds, they were to be imprisoned as also to suffer such other punishment, as to His Highness should seem meet.

Sir T. Elyot, a writer in the days of Henry VIII. states in his "Castle of Health," that "the pheasant excedeth all fowls in sweetness and holsomness, and is equal to capon in nourishment, but the partridges, of all fowls, is most soonest digested, and have in them more nutriment."

Jeffreson, in his amusing work, "A Book about the Table," tells us that the epicures and physicians of Elizabethan days spoke handsomely of the pheasant. Logan calls it "meat for princes and great estates, and for poor scholars when they can get it." One would think more highly of the epicurean discernment of the eulogists of a noble bird, had they not recommended us to stew it with celery— a miserable way of spoiling fine fare that cannot be denounced too warmly, though boiled pheasants and celery are still seen on the tables of intelligent, albeit whimsical gourmands.

Mr. Stevenson, in his "Birds of Norfolk," mentions that the earliest notice of the pheasant in that county occurs in the Household Book of the L'Estranges, of Hunstanton, dating back to 1519, and containing some curious entries, in which this bird is specially mentioned, both as a "quary" for hawks and occasional article for the luxury of the table. Thus, in the 11th year of Henry VIII. (1519), appears, amongst other "rewardes for bryngyng of psents—" Item: to Mr. Asheley svnt for bryngyng of a fesant cocke an iiij woodcocks ye xviijth daye of Octobre reward iiij, also Item : a fesant of gyste (articles received in lieu of rent) ; and twice in the same year we find the following record. Item : a fesant kylled wt ye goshawke, and again, in 1533, ij fesands and ij ptrychyes (partridges) kylled wt the hawke.

The following extract is from the "Magazine of Natural History," on the habits of the pheasant, by Mr. C. Waterton, of Walton Hall, whose Essays on Natural History, with that of White's Selborne, ought to be on the bookshelf of every intelligent gamekeeper, and in every village library in these days of education and school-boards.

Mr. Waterton, speaking of the destruction of the eggs of this bird by vermin, &c., attributes it, in many cases, to the custom of looking up their nests. A track is made through the grass, which is sure to be followed up by the cat or weasel, often to the direful cost of the setting bird, or the hen pheasant flies precipitately from the nest on being disturbed ; the eggs are left uncovered, and become cold, or fall an easy prey to the carrion crow or other destroyer of eggs. In the wild state, when wearied nature calls for relaxation, the pheasant first covers her eggs, and then takes wing directly without running from the nest. Waterton tells us he once witnessed this fact from the sitting-room window in the attic story of his house. He saw a pheasant fly from her nest in the grass, and on her return she kept on the wing till she dropped down upon it. By this instinctive precaution of rising immediately from the nest on the bird's departure, and its dropping on it at its return, there

M 2

is neither scent produced, nor track made in the immediate neighbourhood by which an enemy might have a clew to find it out and rob it of its treasures.

Waterton goes on to say these little wiles are the very safety of the nest, and he suspected that they are put in practice by most birds that have their nests on the ground; and to this circumstance he attributes the great increase of his pheasants by not allowing them to be disturbed, though they were surrounded by hawks, jays, crows and magpies, which had large families to maintain and bring up in the immediate neighbourhood.

(*To be continued.*)

THE NERVOUS SYSTEM OF THE COCKROACH.*

By Professor L. C. Miall and Alfred Denny.

THE nervous system of the cockroach comprises ganglia and connectives,† which extend throughout the body. We have, first, a supra-œsophageal ganglion, or brain, a sub-œsophageal ganglion and connectives which complete the œsophageal ring. All these lie in the head; behind them, and extending through the thorax and abdomen, is a gangliated cord, with double connectives. The normal arrangement of the ganglia in Annulosa, one to each somite, becomes more or less modified in insects by coalescence or suppression, and we find only eleven ganglia in the cockroach, viz., two cephalic, three thoracic, and six abdominal.

The nervous centres of the head form a thick, irregular ring, which swells above and below into ganglionic enlargements, and leaves only a small central opening, occupied by the œsophagus. The tentorium separates the brain or supra-œsophageal ganglion from the sub-œsophageal, while the connectives traverse its central plate. Since the œsophagus passes above the plate, the investing nervous ring also lies almost wholly above the tentorium.

The brain is small in comparison with the whole head; it consists of two rounded lateral masses or hemispheres, incompletely divided by a deep and narrow median fissure. Large optic nerves are given off laterally from the upper part of each hemisphere; lower down, and on the front of the brain are the two gently rounded antennary lobes, from each of which proceeds an antennary nerve; while from the front and upper part of each hemisphere a small nerve passes to the so-called "ocellus,"

* For the "Natural History, Outer Skeleton, Alimentary Canal, and Organs of Respiration and Circulation of the Cockroach," see this Journal, March, May, July, and September, 1884.
† Mons. Yung ("Syst. nerveux des Crustacées décapodes, Arch. de Zool. exp. et gén.," tom. vii. 1878) proposes to name *connectives* the longitudinal bundles of nerve-fibres which unite the ganglia, and to reserve the term *commissures* for the transverse communicating branches.

a transparent spot lying internal to the antennary socket on each side in the suture between the clypeus and the epicranium. The sub-œsophageal ganglion gives off branches to the mandibles, maxillæ and labrum. While, therefore, the supra-œsophageal is

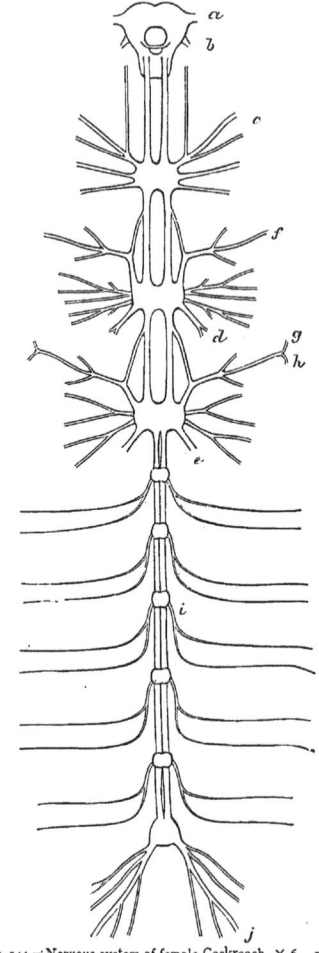

Fig. 144.—Nervous system of female Cockroach, × 6. *a*, optic nerve; *b*, antennary nerve; *c*, *d*, *e*, nerves to first, second, and third legs; *f*, to wing-case; *g*, to second thoracic spiracle; *h*, to wing; *i*, abdominal nerve; *j*, to cerci.

largely sensory, the sub-œsophageal ganglion is the masticatory centre.

The œsophageal ring is double below, being completed by the connectives and the sub-œsophageal ganglion; also by a smaller transverse commissure, which unites the connectives, and applies itself closely to the under surface of the œsophagus.*

Two long connectives issue from the top of the sub-œsophageal ganglion, and pass between the tentorium and the submentum on their way to the neck and thorax. The three thoracic ganglia are large, in correspondence with the important appendages of this part of the body, and united by double connectives. The six abdominal ganglia have also double connectives, which are bent in the male, as if to avoid stretching during forcible elongation of the abdomen. The sixth abdominal ganglion is larger than the rest, and is no doubt a complex, representing several coalesced posterior ganglia; it supplies large branches to the reproductive organs and rectum.

In the cockroach the stomato-gastric nerves found in so many of the higher invertebrates are conspicuously developed. From the front of each œsophageal connective, a nerve passes forwards upon the œsophagus, outside the chitinous crura of the tentorium. Each nerve sends a branch downwards to the labrum, and the remaining fibres, collected into two bundles, join above the œsophagus to form a triangular enlargement, the frontal ganglion. From this ganglion a recurrent nerve passes backwards through the œsophageal ring, and ends on the dorsal surface of the crop (·3 inch from the ring), in a triangular ganglion, from which a nerve is given off outwards and backwards on either side. Each nerve bifurcates, and then breaks up into branches which are distributed to the crop and gizzard.* Just behind the œsophageal ring, the recurrent nerve forms a plexus with a pair of nerves which proceed from the back of the brain. Each nerve forms two ganglia, one behind the other, and every ganglion sends a branch inwards to join the recurrent nerve.

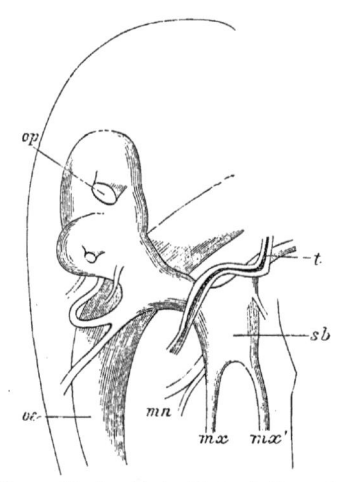

Fig. 145.—Side-view of brain of Cockroach, × 25. *op*, optic nerve; *oe*, œsophagus; *t*, tentorium; *sb*, sub-œsophageal ganglion; *mn*, *mx*, *mx'*, nerves to mandible and maxillæ. [Copied from E. T. Newton.]

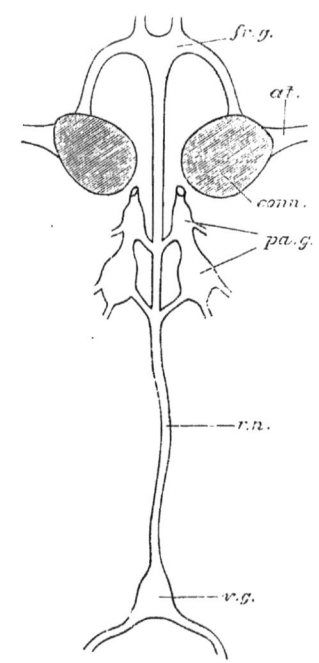

Fig. 146.—Stomato-gastric nerves of Cockroach. *fr. g.*, frontal ganglion; *at.*, antennary nerve; *conn.*, connective; *pa. g.*, paired ganglia; *r. n.*, recurrent nerve; *v. g.*, ventricular ganglion.

Fine branches proceed from the paired nerves of the œsophageal plexus to the salivary glands.

The stomato-gastric nerves differ a good deal in different insects; Brandt † considers that the paired and unpaired nerves are complementary to each other, the one being more elaborate, according as the other is less developed. A similar system is found in mollusca, crustacea, and some vermes (*e.g.*, Nemerteans). When highly developed, it contains unpaired ganglia and nerves, but may be represented only by

---

* This commissure, which has been erroneously regarded as characteristic of Crustacea, was found by Lyonnet in the larva of Cossus, by Strauss-Durckheim in Locusta and Ruprestis, by Blanchard in Dytiscus and Otiorhynchus, by Leydig in Glomeris and Telephorus, by Dietl in Gryllotalpa, and by Liénard in a large number of other Insects and Myriapods, including Periplaneta. See Liénard, "Const. de l'anneau œsophagien," Bull. Acad. roy. de Belgique, 2me sér. t. xlix. 1880.

* The stomato-gastric nerves of the cockroach have been carefully described by Koestler ("Zeitsch. f. wiss. Zool.," Bd. xxxix. p. 592).
† "Mem. Acad. Petersb.," 1835.

an indefinite plexus (earthworm). It always joins the œsophageal ring, and sends branches to the œsophagus and fore-part of the alimentary canal. The system has been identified with the sympathetic, and also with the vagus of vertebrates, but such correlations are hazardous; the first indeed may be considered as disproved.

Between and above each pair of connectives of the ventral cord runs a transparent nerve, composed largely of ganglionic cells, which is often termed the sympathetic, though its physiological relations are uninvestigated. These branches proceed from the middle of the connective, on the right and left side alternately, and pass backwards to the neighbourhood of the succeeding ganglion. Here the nerve forks, each half forming a spindle-shaped enlargement, and joining the lateral branches from the ventral cord.

For our knowledge of the internal structure of the ventral nerve-cord of insects we are chiefly indebted to Leydig. The connectives consist of nerve-fibres only, which, as in invertebrates generally, are non-medullated. The ganglia include (1) rounded, usually unipolar nerve-cells; (2) tortuous and extremely delicate fibres collected into intricate skeins (*punkt-substanz*); (3) commissural fibres, and (4) connectives. The chief fibrous tracts are internal, the cellular masses outside them. A double investment protects the cord. Closely embracing the nervous structures is a transparent, structureless, chitinous sheath, within which a matrix of branched mother-cells may be here and there distinguished. Outside the proper sheath is a peritoneal layer of loose and irregular cells. Tracheal trunks pass to each ganglion, and break up upon and within it into a multitude of fine branches.

Two bundles of commissural fibres connect the ganglia of the same pair.* Of the peripheral fibres some pass direct to their place of distribution, others traverse at least one complete segment and the corresponding ganglion before separating from the cord.

Many familiar observations show that the ganglia of an insect possess great physiological independence. The limbs of decapitated insects, and even isolated segments, provided that they contain uninjured ganglia, exhibit unmistakable signs of life. Yersin and Baudelot's experiments imply that the ventral cord is divisible into an upper motor tract and a lower sensory, the centres of both motion and sensation lying in the ganglia exclusively.

The minute structure of the brain has been investigated by Leydig, Dietl, Flögel and others, and exhibits an unexpected complexity. It is as yet impossible to reduce the many curious details which have been described to a completely intelligible account. The physiological significance, and the homologies of many parts are as yet altogether obscure. The comparative study of new types will however, in time, bridge over the wide interval between the insect-brain and the more familiar vertebrate brain, which is partially illuminated by physiological experiment. Mr. E. T. Newton has published a clear and useful description * of the internal and external structure of the brain of the cockroach, which incorporates what had previously been ascertained with the results of his own investigations. He has also described † an ingenious method of combining a number of successive sections into a dissected model of the brain. Having had the advantage of comparing the model with the original sections, we offer a short abstract of Mr. Newton's memoir as the best introduction to the subject. He

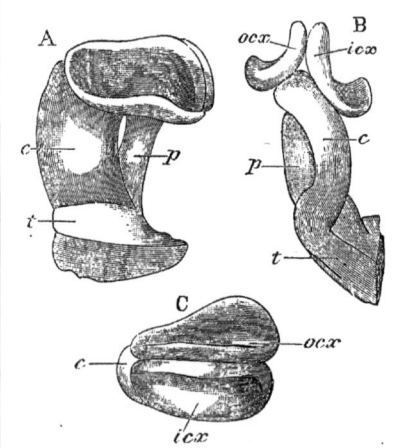

Fig. 147.—A, Lobes of the brain of the Cockroach, seen from within; *c*, cauliculus; *p*, peduncle; *t*, trabecula. B, ditto, from the front; *ocx*, outer calyx; *icx*, inner calyx. C, ditto, from above. [Copied from E. T. Newton.]

describes the central framework of the cockroach brain as consisting of two solid and largely fibrous trabeculæ, which lie side by side along the base of the brain, becoming smaller at their hinder ends; they meet in the middle line, but apparently without fusion or exchange of their fibres. Each trabecula is continued upwards by two fibrous columns, the cauliculus in front, and the peduncle behind; the latter carries a pair of cellular disks, the calices (the cauliculus, though closely applied to the calices, is not connected with them); these disks resemble two soft cakes pressed together above, and bent one inwards, and the other outwards below. The peduncle divides above, and each branch joins one of the calices of the same hemisphere.

---

* Michels finds in Oryctes that such transverse commissures do not exist, but that there are numerous transverse fibres connecting the ganglionic cells of one side with the peripheral nerves of the other ("Zeits. f. wiss. Zool.," Bd. xxxiv., p. 696, 1881).

---

* "Q. J. Micr. Sci.," 1879, pp. 340-356, pl. xv. xvi.
† "Journ. Quekett Micr. Club," 1879.

This central framework is invested by cortical ganglionic cells, which possess distinct nuclei and nucleoli. A special cellular mass forms a cap to each pair of calices, and this consists of smaller cells without nucleoli. Above the meeting-place of the trabeculæ is a peculiar laminated mass, the *central body*, which consists of a network of fibres continuous with the neighbouring ganglion-cells, and enclosing a granular substance. The trabeculæ, peduncles and cauliculi appear to be a peculiar modification of that constituent of the ganglia, which Leydig has termed "dotted substance" (*punkt-substanz*). It can be resolved by the highest powers of the microscope into densely interwoven fibrillæ, which Krieger[*] has traced into both ganglion-cells and peripheral nerves. In the antennary lobes the endings, and by peculiar sensory rods or filaments upon the antennæ. These are taken to be the organs respectively of sight, hearing, and smell. Insect eyes vary conspicuously in structure; the auditory organ is found in places so dissimilar as the tibia of the fore-leg and the first abdominal segment. Other sense-organs, not as yet fully elucidated, may co-exist with these. The maxillary palps of the cockroach, for example, are continually used in exploring movements, and may assist the animal to select its food; the cerci, where these are well-developed, and the halteres of Diptera, have been also regarded as sense-organs of some undetermined kind, but this is at present wholly conjecture.[*]

Our scanty remaining space forbids us to describe

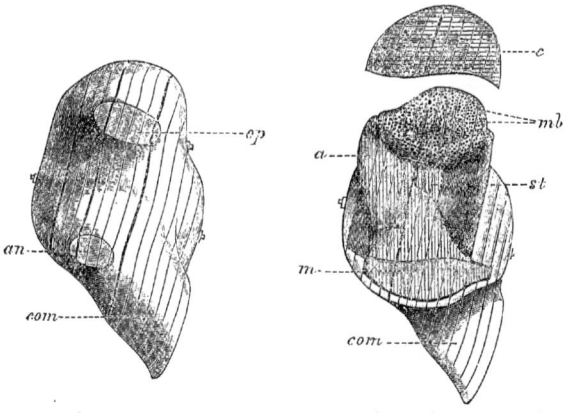

Fig. 148.—View from the outer side of the left half of model of upper part of brain of Cockroach. The oblique lines in this and Fig. 149 indicate the successive slices of which the model is composed, their direction being that of the sections in Fig. 150. *op*, cut end of optic nerve; *an*, cut end of antennary nerve.
Fig. 149.—Right half of model-brain seen from the inner side, with the parts dissected away, so as to show the anterior nervous mass, *a*; the median mass, *m*; the mushroom-bodies, *mb*; and their stems. *st*. The cellular cap, *c*, has been raised, so as to display the parts below: *com*, is a part of the connective uniting the brain and infra-œsophageal ganglia. [Figs. 148–150 are taken from Mr. E. T. Newton's paper in "Journ. Quekett Club," 1879.]

punkt-substanz appears in a more ordinary form as a network of fine fibres enclosing ganglion-cells, and surrounded by a layer of the same. It is remarkable that no fibrous communications can be made out between the calices and the cauliculi, or between the trabeculæ and the œsophageal connectives.

Each optic ganglion contains two lenticular bodies, as in various insects and crustacea; the nerve-fibres cross before entering the first of these, again between the two, and a third time between the outer lenticular body and the eye.

The sense-organs of insects are very variable, both in position and structure. Three special senses are indicated by transparent and refractive parts of the cuticle, by tense membranes with modified nerve-what has been made out respecting these various organs. We can only notice one, and that briefly. The most interesting and the best understood is certainly the organ of sight.

The compound eyes of the cockroach occupy a large, irregularly oval space (see fig. 151) on each side of the head. The total number of facets may be estimated at about 1800. The number is very variable in insects, and may either greatly exceed that found in the cockroach, or be reduced to a very small one indeed. According to Burmeister, the Coleopterous genus Mordella possesses more than 25,000 facets. Where the facets are very numerous, the compound eyes may occupy nearly the whole surface of the head, as in the house-fly, dragon-fly, or gad-fly.

[*] In the crayfish, "Zelts. f. wiss. Zool.," Bd. xxxiii., p. 541 (1880).

[*] It is to be remarked that unusually large nerves supply the cerci of the cockroach.

Together with compound eyes, many insects are furnished also with simple eyes, usually three in number, and disposed in a triangle on the forehead. The white fenestræ, which in the cockroach lie internal to the antennary sockets, may represent two simple eyes which have lost their dioptric apparatus. In many larvæ only simple eyes are found, and the compound eye is restricted to the adult form; in larval cockroaches, however, the compound eye is large and functional.

Each facet of the compound eye is the outermost element of a series of parts, some dioptric and some sensory, which forms one of a mass of radiating rods or fibres. The facets are transparent, biconvex and polygonal, often, but not quite regularly, hexagonal. In many insects the deep layer of each facet is separable, and forms a concavo-convex layer of different texture from the superficial and biconvex lens. The facets taken together, are often described as the cornea; they represent the chitinous cuticle of the integument. The subdivision of the cornea into two layers of slightly different texture suggests an achromatic correction, and it is quite possible, though unproved, that the two sets of prisms have different dispersive powers. Beneath the cornea we find a layer of crystalline cones, each of which rests by its base upon the inner surface of a facet, while its apex is directed inwards towards the brain. The crystalline cones are transparent, refractive, and coated with dark pigment; in the cockroach they are compara-

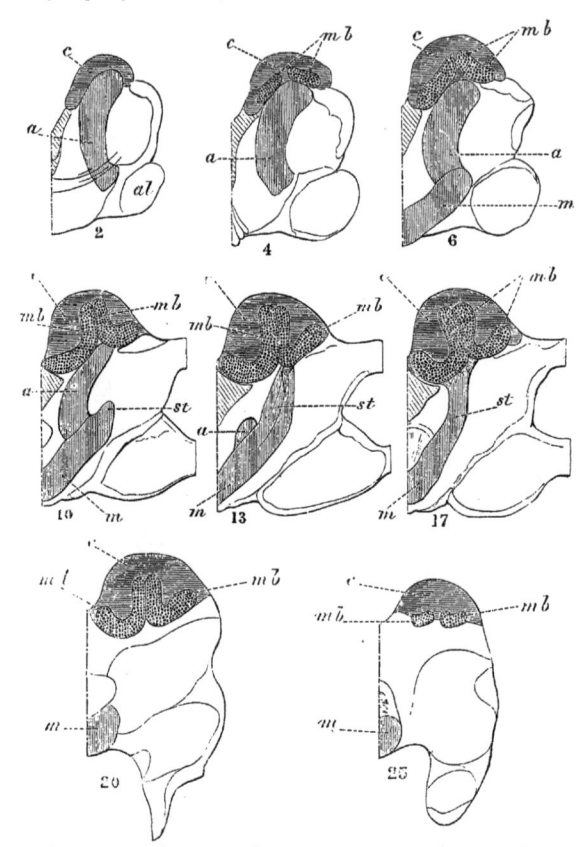

Fig. 150.—Diagrammatic outlines of sections of the brain of a Cockroach. Only one side of the brain is here represented. The numbers indicate the position in the series of thirty-four sections into which this brain was cut. *al*, antennary lobe; *mb*, mushroom bodies (*calices*), with their cellular covering, *c*, and their stems (*peduncles*); *st*; *a*, anterior nervous mass (*cauliculus*); *m*, median nervous mass (*trabecula*).

tively short and blunt. Behind each cone is a nerve-rod, which though outwardly single for the greater part of its length, is found on cross-section to consist of four components,* these diverge in front, and receive the tip of a cone, which is wedged in between them; the nerve-rods are densely pigmented. To their hinder ends are attached the fibres of the optic nerve, and the internal boundary of the eye is defined by a "fenestrated membrane."

The compound eye is thus divisible into three strata, viz. (1) the facetted cornea; (2) the crystalline cones; (3) the retinula of nerve-rods. In the simple eye the non-facetted cornea and the retinula are readily made out, but the crystalline cones are not developed as such. The morphological key to both structures is found in the integument, of which the whole eye, simple or compound, is a modification. A defined tract of the chitinous

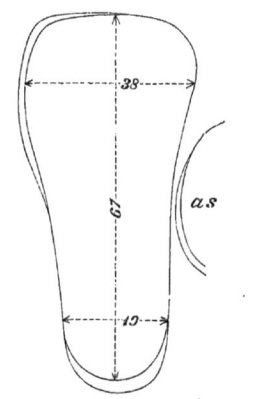

Fig. 151.—Plan of eye of Cockroach, showing the number of facets along the principal diameters.

cuticle becomes transparent, and either swells into a lens (fig. 154), or becomes regularly divided into facets (fig. 156), which are merely the elaboration of imperfectly separated polygonal areas, easily recognized in the young cuticle of all parts of the body. Next, the chitinogenous layer is folded inwards, so as to form a cup, and this by the narrowing of the mouth is transformed into a flask, and ultimately into a solid two-layered cellular mass (fig. 154). The deep layer undergoes conversion into a retinula, its chitinogenous cells developing the nerve-rods as interstitial structures, while the superficial layer, which loses its functional importance in the simple eye, gives rise by a similar process of interstitial growth to the crystalline cones of the compound eye (fig. 156). The basement-membrane underlying the chitinogenous cells, is transformed into the fenes-

* The number in insects varies from eight to four, but seven is usual; four is the usual number in Crustacea.

trated membrane, which marks the internal limit of the eye, and into whose perforations the nerve-rods are inserted, like organ pipes into the soundboard. The mother-cells of the crystalline cones and nerve-rods are largely replaced by the interstitial substances they produce, to which they form a sheath; they are often loaded with pigment, and the nuclei of the primitive cells can only be distinguished, after the colouring-matter has been discharged by acids or alkalis.

As to the way in which the compound eye renders

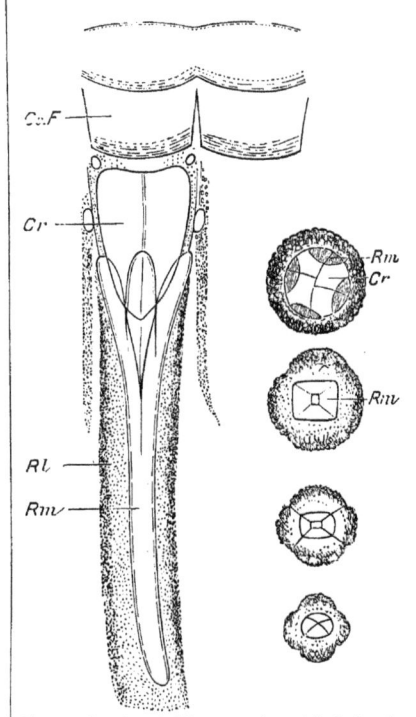

Fig. 152.—One element of the compound eye of the Cockroach, × 700. *Co. F*, corneal facets; *Cr*, crystalline cones; *Rm*, nerve-rod (rhabdom); *Rl*, sheath of ditto. To the right are transverse sections at various levels. [Copied from Grenacher.]

distinct vision possible, there is still much difference of opinion. A short review of the discussion which has occupied some of the most eminent physiologists and histologists for many years past will introduce the reader to the principal facts which have to be reconciled.

The investigation, like so many other trains of biological inquiry, begins with Leeuwenhoeck (Ep. ad Soc. Reg. Angl. iii.), who ascertained that the cornea of a shard-borne beetle, placed in the field of a microscope, gives images of surrounding objects, and that these images are inverted. When the cornea is flattened out for microscopic examination, the images (*e.g.* of a window or candle-flame) are similar, and it has been too hastily assumed that a multitude of identical images are perceived by the insect. The cornea of the living animal is, however, convex, and the images formed by different facets cannot be precisely identical. No combined or collective image is formed by the cornea. When the structure of the compound eye had been very inadequately studied, as was the case even in Cuvier's time (Leçons d'Anat. Comp. xii. 14), it was natural to suppose that all the fibres internal to the cornea were sensory, that they formed a kind of retina upon which the images produced by the facets were received, and that these images were transmitted to the brain, to be united, either by optical or mental combination, into a single picture.

great number of inverted partial images. How then can insects and crustaceans see with their compound eyes? Müller answered that each facet transmits a small pencil of rays travelling in the direction of its axis, but intercepts all others. The refractive lens collects the rays, and the pigmented as well as refractive crystalline cone further concentrates the pencil, while it stops out all rays which diverge appreciably from the axis. Each element of the compound eye transmits a single impression of greater or less brightness, and the brain combines these impressions into some kind of picture, a picture like that which could be produced by stippling. It may be added that the movements of the insect's head or body would render the distance and form of every object in view much readier of appreciation. No accommodation for distance would be necessary, and the absence of all means of accom-

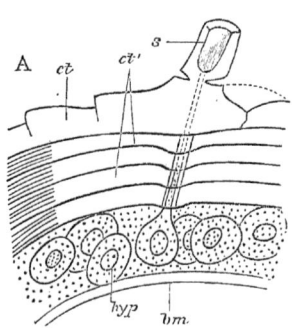

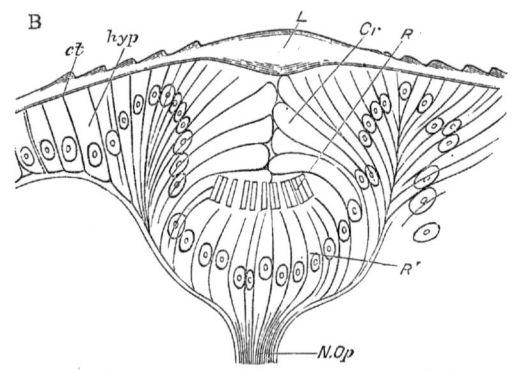

Fig. 153.—Diagram of Insect integument, in section. *bm*, basement-membrane; *hyp*, hypodermis, or chitinogenous layer; *ct*, *ct'*, chitinous cuticle; *s*, a seta.

Fig. 154.—Section through eye of Dytiscus-larva, showing the derivation of the parts from modified hypodermic cells. *L*, lens; *Cr*, crystalline cones; *R*, nerve-rods; *N. Op*, optic nerve. [From Grenacher.]

Müller† in 1826 pointed out that so simple an explanation was inadmissible. He granted that the simple eye, with its lens and concave retina, produces a single inverted image, which is able to affect the nerve-endings in the same manner as in vertebrates. But the compound eye is not optically constructed so as to render possible the formation of continuous images. The refractive and elongate crystalline cones, with their pointed apices and densely pigmented sides, must destroy any images formed by the lenses of the cornea. Even if the dioptric arrangement permitted the formation of images, there is no screen to receive them.* Lastly, if this difficulty were removed, Müller thought it impossible for the nervous centres to combine a

modation ceases to be perplexing. Such is Müller's theory of what he termed "mosaic vision." Many important researches, some contradictory, some confirmatory of Müller's doctrine,* have since been placed on record, with the general result that some modification of Müller's theory tends to prevail. The most important of the new facts and considerations which demand attention are these :—

Reasons have been given for supposing that images are formed by the cornea and crystalline cones together. This was first pointed out by Gottsche (1852), who used the compound eyes of flies for demonstration. Grenacher has since ascertained that the crystalline cones of flies are so fluid that they can hardly be removed, and he believes that Gottsche's images were formed by the corneal facets

---

* "Zur vergl. Phys. des Gesichtsinnes."
† Exner has since determined by measurement and calculation the optical properties of the eye of Hydrophilus. He finds that the focus of a corneal lens is about 3mm. away, and altogether behind the eye.

* A critical history of the whole discussion is to be found in Grenacher's "Seh-organ der Arthropoden" (1879), from which we take many historical and structural details.

alone. He finds, however, that the experiment may be successfully performed with eyes not liable to this objection, *e.g.*, the eyes of nocturnal lepidoptera. A bit of a moth's eye is cut out, treated with nitric acid to remove the pigment, and placed on a glass slip in the field of the microscope. The crystalline cones, still attached to the cornea, are turned towards the observer, and one is selected whose axis coincides with that of the microscope. No image is visible when the tip of the cone is in focus, but as the cornea approaches the focus, a bristle, moved about between the mirror and the stage, becomes visible. This experiment is far from decisive. No image is formed where sensory elements are present to receive and transmit it. Moreover, the image is that of an object very near to the cornea, whereas all observations of living insects show that the compound eye is used for far sight, and the simple eye for near sight. Lastly, the treatment with acid, though unavoidable, may conceivably affect the result. It is not certain that the cones really assist in the production of the image, which may be due to the corneal facets alone, though modified by the decolorised cones.

Grenacher has pointed out, that the composition of the nerve-rod furnishes a test of the mosaic theory. According as the percipient rod is simple or complex, we may infer that its physiological action will be simple or complex too. The adequate perception of a continuous picture, though of small extent, will require many retinal rods; on the other hand, a single rod will suffice for the discrimination of a bright point. What then are the facts of structure? Grenacher has ascertained that the retinal rods in each element of the compound eye rarely exceed seven, and often fall as low as four —further, that the rods in each group are often more or less completely fused so as to resemble simple structures, and that this is especially the case with insects of keen sight.*

Certain facts described by Schultze tell on the other side. Coming to the Arthropod eye, fresh from his investigation of the vertebrate retina, Schultze found in the retinal rods of insects the same lamellar structure which he had discovered in Vertebrata. He found also that in certain moths, beetles and crustacea a bundle of extremely fine fibrils formed the outer extremity of each retinal or nerve-rod. This led him to reject the mosaic theory of vision, and to conclude that a partial image was formed behind every crystalline cone, and projected upon a multitude of fine nerve-endings. Such a retinula of delicate fibrils has received no physiological explanation, but it is now known to be of comparatively rare occurrence; it has no pigment to localise the stimulus of light; and there is no reason to suppose that an image can be formed within its limits.

The optical possibility of such an eye as that interpreted to us by Müller has been conceded by physicists and physiologists so eminent as Helmholz and Du Bois Reymond. Nevertheless, the competence of any sort of mosaic vision to explain the precise and accurate perception of insects comes again and again into question whenever we watch the movements of a housefly as it avoids the hand, of a bee flying from flower to flower, or of a dragon-fly in pursuit of its

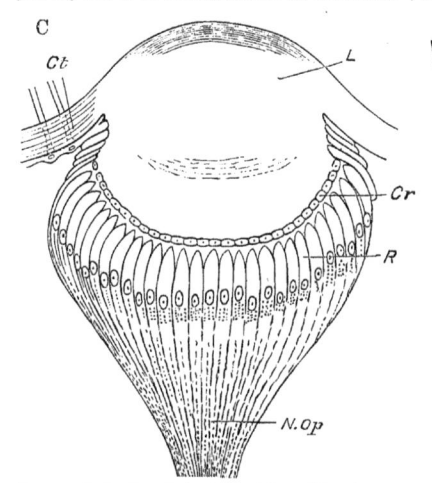

Fig. 155.—Section through simple eye of Vespa. The references as above. [Simplified from Grenacher.]

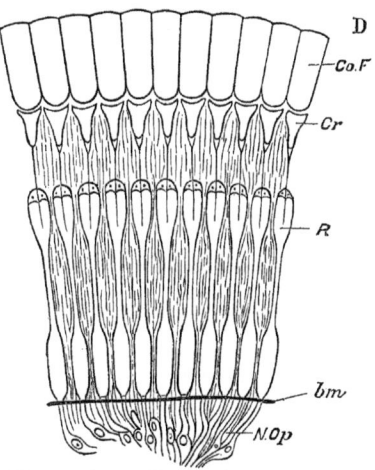

Fig. 156.—Diagrammatic section of compound eye. The references as above.

* Flies, whose eyes are in several respects exceptional, have almost completely separated rods, notwithstanding their quick sight.

prey. The sight of such insects as these must range over several feet at least, and within this field they must be supposed to distinguish small objects with rapidity and certainty. How can we suppose that an eye without retinal screen, or accommodation for distance, is compatible with sight so keen and discriminating? The answer is neither ready nor complete, but our own eyesight shows how much may be accomplished by means of instruments far from optically perfect. According to Aubert, objects, to be perceived as distinct by the human eye, must have an angular distance of from 50″ to 70″, corresponding to several retinal rods. Our vision is therefore mosaic too, and the retinal rods which can be simultaneously affected comprise only a fraction of those contained within the not very extensive area of the effective retina. Still we are not conscious of any break in the continuity of the field of vision. The incessant and involuntary movements of the eyeball, and the appreciable duration of the light-stimulus partly explain the continuity of the image received upon a discontinuous organ. Even more important is the action of the judgment and imagination, which complete the blanks in the sensorial picture, and translate the shorthand of the retina into a full-length description. That much of what we see is seen by the mind only is attested by the inadequate impression made upon us by a sudden glimpse of unfamiliar objects. We need time and reflection to interpret the hints flashed upon our eyes, and without time and reflection we see nothing in its true relations. The insect-eye may be far from optical perfection, and yet as it ranges over known objects, the insect-mind, trained to interpret colour, and varying brightness, and parallax, may gain minute and accurate information. Grant that the compound eye is imperfect and even rude, if regarded as a camera; this is not its true character. It is intended to receive and interpret flashing signals; it is an optical telegraph.

In closing this series of sketches we would say, that what has been here set down comprises but a small fraction of what has been made out by many good observers, while all that has been seen and studied bears no ratio whatever to the inexhaustible detail presented to us in the simplest insect. Whether in a busy age it is worth while to dwell minutely upon the structure and mode of life of so small a creature is a question which we and our readers would probably answer in a different sense from the majority of mankind. It seems to us that of all subjects of study, Life is the most deeply interesting to the living man, and that to know some little of the possibilities of life under various conditions is worth any expenditure of pains and time.

[ERRATUM.—Fig. 125, p. 207 should have been placed with the arrow horizontal, and pointing to the left. Fig. 121, p. 205 is magnified 15 times.]

BIFURCATION OF THE ELM LEAF.

AMONGST the many notices of the morphology and teratology of plants and leaves which have

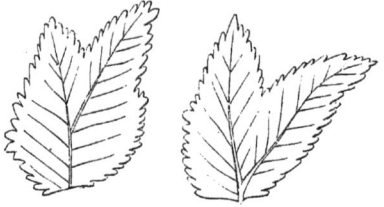

Fig. 158.—Partial Divisions.

Fig. 159.—Completed Division.

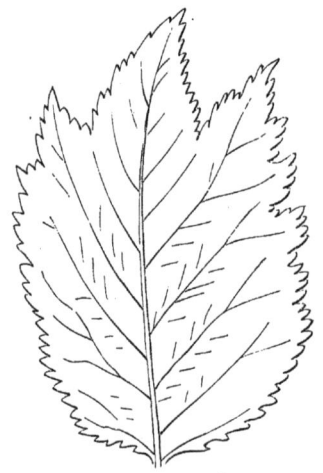

Fig. 160.—Leaf of *Ulmus montana*, with trifid tip, but no division of central vein.

appeared in SCIENCE-GOSSIP, I have not met with any allusion to the peculiar tendency of the leaves o the common elm (*Ulmus campestris*) to bifurcation of the central vein, often to the extent of complete division of the leaf into two leaves. When the division is slight, the tip of the leaf is only slightly bifid, but we often meet with all intermediate degrees of bifurcation between that and complete duplication of both leaf and leaf-stalk. I have been observing this for seven or eight years, and have, thus far, only found it in *U. campestris*; I have never seen it in *Ulmus suberosa*.

In the Wych elm (*U. montana*) I have sometimes found an apparent division of the tip of the leaf into three; but, upon careful examination, it has proved to be only an elongation of two lateral veins, and not a bifurcation or trifurcation of the central.

Although this trifid condition is more common in young trees of *U. campestris*, I have also found it in the terminal leaves of branches on old trees: in some young and vigorous shoots, nearly all the leaves are sometimes partly, or completely divided. I shall be glad to learn through the SCIENCE-GOSSIP columns, whether other observers have noticed the same thing, as soil and situation may possibly make it more or less a local peculiarity. I have looked in vain for similar forms in hazel, hornbeam, beech, &c.

NORRIS F. DAVEY, M.R.C.S.L., &c.

GOSSIP ON CURRENT TOPICS

By W. MATTIEU WILLIAMS, F.R.A.S.

THE natural law of demand and supply is well illustrated by Microbia. The public mouth is wide open and eager to swallow any and every bugbear concerning these creatures, and they are discovered accordingly. A scientific contemporary tells us that "The dangers to public health which lurk in out-of-the-way places appear inexhaustible," and then describes the researches of M. Reinsch of Erlangen, who "has devoted much study to the matter," and thereby discovered that old and recent coins of all metals from all the European states are infested with "micro organisms of algæ and bacteria." Reinsch obtains these by industriously scraping away the matter which accumulates in the interstices of the relief with a needle, placing the scrapings in a drop of distilled water, and examining this under a microscope. We are further told that "a recent writer in 'Science et Nature' refers to this discovery as of great importance from a hygienic point of view." This has been repeated in many newspapers and magazines,—has in fact "gone the round." If any of my readers have been alarmed thereby they may send me by parcels post, addressed to Stonebridge Park, N.W., all the questionable coins in their possession, as I am not frightened at all, having discovered long ago that, in spite of tooth brushes and dentifrice, the interstices of my own teeth and tongue and palate are in the same condition as those of the coins, and yet I have lingered on for threescore years and still remain alive. If these readers (after despatching the coins as above) will use a needle and scrape between their own teeth, as Reinsch did between the relievi of the coins, and examine the result with a high power, ($\frac{1}{8}$-inch or upwards) of a good microscope, they will find abundant microbia all alive and wriggling. Some of them may possibly be "comma shaped." A good haul of these may be readily captured by scraping the tongue with the edge of a knife. I doubt whether we can scrape anything that has ever been moistened and exposed to the air without repeating Reinsch's discoveries, if we put the scrapings in water and magnify sufficiently.

Another bugbear has been set on foot by the newspapers. We are to be infected by means of maccaroni, as the following quotation will show: "A correspondent writes to suggest the connection there may be between cholera and maccaroni. Any one who has followed the road along the coast from Naples to Pompeii must have noticed the numberless rows of maccaroni tubes along the wide margin of the roadway." This is followed by a picture of the consequences of the wind depositing "the seeds of pestilence" on the adhesive surface of the paste and the paste accordingly carrying cholera microbia throughout Europe.

"Our correspondent" has, however, omitted two facts which completely refute his alarming suggestions. The first is that the maccaroni which he saw hanging out to dry on the roadside is not exported at all. It was manufactured exclusively for domestic use, and no more likely to reach England than the pancakes made on Shrove Tuesday in English domestic kitchens are likely to be exported to Naples. Maccaroni is the staple food of Neapolitans, and the Neapolitan housewife makes it as familiarly as the English housewife makes the paste for pies and puddings. That which comes here is produced in large factories, where the dangers described are no greater than in ordinary bakeries. The second fact, unseen by the writer, is, that before we eat maccaroni we stew it in boiling water for about half an hour. This would effectually wash away and destroy any infection germs that might possibly be adherent to it.

These alarmists say nothing concerning another possible carrier of infection that is not cooked, not washed, nor cleansed by any other process, but which in the course of its manufacture is handled freely by questionable fingers and in very hot fever-smitten climates, and not unfrequently moistened by questionable saliva. The consumers of this very questionable product simply suck it in its raw uncleansed condition. I refer to the cigar. Further details concerning the picking of the leaves, their exposure like the maccaroni for drying, the rolling, and the final finishing twisting and moistening of the pointed end are unnecessary. I leave the cigar smoker to meditate thereon while this end is between his lips, merely adding that my own reflections suggest the desirability of an amber mouthpiece.

In my own house I insist upon the washing of all fruit that is eaten in its raw state, and suspect that plums have been falsely accused—as plums—of propagating cholera. Most of them in our markets are imported from the South, after much handling by dirty people. This is also the case with figs, dates, and raisins. Fortunately for us, the dried dwarf grapes of Greece ("currants") are usually washed by dealers to improve their appearance, and are cooked. Those who have travelled in the interior of Greece will understand why I mention these with special emphasis.

A communication from B. Sc., in the September number of SCIENCE-GOSSIP, page 212, reminds me of some experiments I made many years ago in mounting microscopic objects. I bought specimens of all the gums and gum resins commercially obtainable, treated them with various solvents, for the purpose of obtaining a substitute for Canada balsam, something that should hold down the thin glass cover, envelope and hermetrically seal organic preparations, especially sections, and yet be sufficiently limpid when in solution to permit the free removal of air bubbles, which as all microscopists know, are so obstinately retained by the viscosity of Canada balsam. I succeeded, as usual, in securing a large crop of failures, but obtained one success by dissolving *gum Thus* (the basis of frankincense, a resin of the "*arbor thurifera*" or thuja) in bisulphide of carbon. I found that this solution is very limpid and transparent, and rapidly solidifies. The vegetable sections that I mounted stood well, but having almost forsaken the microscope, I had nearly forgotten it. Probably some of the readers of SCIENCE-GOSSIP who are faithful to microscopic work will try it, and report results in SCIENCE-GOSSIP for the benefit of brother amateurs.

The drosera controversy has not by any means subsided. Darwin, Kellermann, Raumer and Rees, contend that the insects captured serve as food, that these plants are truly carnivorous; while Regel and others deny this. H. Büsgen in a recent paper, of which an abstract is published in the September number of the Journal of the Chemical Society, attributes the different conclusions to the varying conditions under which the experiments were made, with plants possibly in different stages of development. The comparison should commence with the weighing of the seeds, and finish with the estimation of the total dry matter of the plant.

As the seeds of the *Drosera rotundifolia* are so very minute, Büsgen commenced with the seedlings. The comparisons were made between plants growing in similar soil—peat previously boiled in a nutritive solution and placed in saucers covered with bell-glasses; one set fed with lice from vine leaves, the other without animal food. The unfed plants were less strong and healthy than the others; the fed plants had seventeen flower branches on fourteen plants against nine buds on sixteen unfed; and ninety seed capsules on fourteen plants against twenty on the sixteen; the total dry weight of the fed plants was 0·352 gramme against 0·119 of the unfed. Other trials gave similar results.

Cut flowers wither more slowly than leaf twigs from the same plant. This is due to a greater transpiration from the leaves than from the flowers. If the transpiration of the leaves is arrested the flower will remain fresh for a much longer time than if the leaves are allowed to remain. In like manner the terminal leaf bud lasts longer on cut twigs if all the other leaves are removed.

Some interesting researches on the self-purification of rivers have been recently made by F. Hulwa. He finds that the water of the Oder on entering the city of Breslau, although slightly contaminated, yielded an excellent drinking-water when filtered. In the course of its progress through the city it showed continuously increasing pollution, the maximum at its exit. A short distance lower down, the self-purification of the river by the combined action of the oxygen of the air and of vegetable and animal life became very marked, impurities diminishing so rapidly that at a distance of 8¾ miles from the city the water was in about the same condition as when it entered the city, neither chemical nor microscopic tests showing serious impurity. Complaints have been made against the penny steamboats on the Thames, because they disturb the bottom. So far as the bridges are concerned they may be mischievous, but this mischief is not difficult to repair and prevent by the simple device of surrounding the foundations of the threatened arches with heavy stones, or fundamentally improving the foundations. From a sanitary point of view we may regard these and other steamboats (including even the much abused steam launches) as valuable agents, seeing that agitation of water polluted with sewage is the most effectual mode of purification. The steam tugs on canals which are otherwise stagnant are great benefactors.

The action of frost upon plants has been recently studied by H. R. Goppert (*Annales Agronomiques*, vol. x. p. 41). It had been previously stated that plants, bulbs and roots are killed by sudden thawing, not by the preceding freezing. Goppert submitted potatoes, and the bulbs of hyacinths, narcissus, &c., to a temperature of about 3 degrees below freezing ($-1\cdot4°$ Centigrade) and then suddenly to 15 degrees ($-8°$ C.) below freezing. This killed all the bulbs, whether they were afterwards thawed either slowly or rapidly, but none of the bulbs cooled to $-1\cdot4°$ C. were damaged, the potatoes only being frozen. Flowers of the exotic orchids *Pajus* and *Calanthe* which develop indigo when they die, and thus turn blue, were killed directly when frozen, and no amount of gradual thawing revived them. The buds of some ligneous plants bore an exposure to a temperature of $-16°$ to $-26°$ Centigrade, and subsequent thawing at 25° C. (77° Fahr.) without injury.

All who are able to understand the importance of sound meteorological data, must be gratified with the progress of the Ben Nevis Observatory, placed at an elevation of 4406 feet, only four miles distant from the sea, and in the track of the great storms that come to us from the wide Atlantic. Ordinary observatories tell us the story of local atmospheric currents as written by themselves with the anemometer, but the Ben Nevis anemometer is supplying us with autographs of main currents, and just those which most directly affect us. The observations already recorded concerning variations of temperature due to elevation are very curious. The mean difference between Fort William at sea level and the observatory is at the rate of 1° Fahr. for every 270 feet of ascent, or 16°·3 for the whole height. The greatest average difference is in May 18°, the smallest in December when it sinks to 14°·9. At times the difference has fallen to nothing, and it has even been higher at the observatory than at 4406 feet lower down. This was the case on December 31, 1883, when at 11 A.M. the temperature at Fort William was 27°·5, while that on Ben Nevis was 30°·0. This was accompanied with excessive dryness of the air. Relative rainfall, and relative barometric fluctuating are also extremely interesting, and the collations of all these promise to supply us with valuable inductions displaying intelligible law and order in the apparently capricious movement of our proverbially variable climate.

NOTES ON NEW BOOKS.

MANUAL OF THE MOSSES OF NORTH AMERICA, by Leo Lesquereux and Thomas P. James. (Boston : S. E. Cassino & Co.) This excellent work will be warmly welcomed by all botanical students in the United States, and it will be scarcely less acceptable to bryologists in Europe. No two better men could have been selected for the task than the authors, and the microscopical analyses of the species by Mr. James are admirably done and artistically depicted. The manual includes descriptions of all the mosses known to occur in the United States (about 900 species). Of course a vast number of them are common to this country, and bryologists will be interested in noting varietal differences, &c. The work is supplemented by six plates, exquisitely engraved, which illustrate the genera.

Synopsis of British Mosses, by Charles P. Hobkirk, F.L.S. (London : Van Voorst. Second edition.) Nearly ten years have elapsed since we first had the pleasure of reviewing Mr. Hobkirk's much-needed work, and we are glad to welcome this second edition. It is much improved and enlarged, and bears a more attractive appearance than its predecessor. As the first edition has long been out of print, and copies of it much in demand, this second edition ought to be very successful.

The Life of the Fields, by Richard Jefferies. (London : Chatto & Windus.) The publication of every new book by Mr. Jefferies confers a fresh pleasure. No man living has a keener eye for all kinds of natural phenomena, or a more catholic sympathy with every living object. Add to this the possession of a felicitously simple style of writing, and a stranger to his books may form some idea of their matter and style. The present volume (like some others) is a collection of papers written for various magazines and journals. One reads some of them, as we should drink, old wine, slowly, conscious of every word and idea, so as not to allow anything to escape attention. Among such we particularise, " The Pageant of Summer," " Clematis Lane," " January in the Sussex Woods," " Mind under Water," " Field Play," and "Notes on Landscape Painting ;" although to specially refer to any appears invidious.

Summer, from the journal of Henry D. Thoreau, edited by H. G. O. Blake. (London : T. Fisher Unwin.) A second volume of selections from the journal of this well-known American writer. It is a very pleasant, thoughtful, peaceful book—just the volume to calm perturbation of spirits, or to be read as an antidote to the chafing and fret of a business life. Thank God for such sweet resting-places ! Thoreau was a kind of Gilbert White, strongly flavoured with Emersonianism. He was not such a keen or accurate observer as the former, nor did he possess such an attractive style of delineation. But his books are particularly pleasant, and none more than his " Summer."

Petland Revisited, by the Rev. J. G. Wood, F.L.S., &c. (London : Longmans & Co.) This is an attractively got up volume in every way—paper, type, and especially the woodcuts. Mr. Wood has a large range of practical sympathy with animals, and in this book we have a collection of pleasantly-related anecdotes, mostly personal (some of which fairly try one's faith) concerning cats and dogs (but cats especially, for which Mr. Wood owns a fondness which ought to make this volume very valuable), chameleons, hedgehogs, coatamundis, monkeys, rabbits, rats, snakes, &c. &c. Mr. Wood's story-telling powers are well known, and in this book he shows no falling off in either style or spirit.

The Honey-Bee; its Nature, Home, and Products, by W. H. Harris, B.Sc. (London : Religious Tract Society.) In spite of the numerous well-written and valuable books already in existence concerning bees and bee-keeping, there was room for another on the lines Mr. Harris has adopted. There are many original points of treatment, and the style is clear and attractive. The illustrations are numerous, many are quite new, and all are good. To a novice in bee-keeping particularly, this book will be very welcome; and many whose experience leads them to think they know all about the bee, will still be able to pick up a few hints from Mr. Harris's attractively got up little volume.

A Season among Wild Flowers, by the Rev. Henry Wood. (London: W. Swan Sonnenschein & Co.) An acceptable book to put into the hands of a young student of British botany. There is a very clearly written description of the Linnean and natural systems of classification, and a good account of all the leading kinds of English flowers. The woodcuts are for the most part excellent, and very numerous.

The Wonders of Plant Life under the Microscope, by Sophie Bledsoe Herrick. (London: W. H. Allen & Co.) This is a volume of Studies in Plant Life which has achieved such a success in America that an English publisher has introduced it to British readers. We think he has done rightly. It is beautifully printed on good paper, and the illustrations are almost novel for their artistic beauty. A very large portion of the book is devoted to Cryptogamic plants, especially microscopical kinds; and there is an extensive summary of all that has been written concerning insectivorous plants. As might be expected, the illustrations deal almost entirely with microscopic structures, so that they are all very useful, both to the student and the general reader.

The Blowpipe in Chemistry, Mineralogy, and Geology, by Col. W. A. Ross. (London: Crosby Lockwood & Co.) The author makes out a capital reason for using the blowpipe in anhydrous analysis. He shows both how to make one and how to use it, and enters fully, and even enthusiastically, into all the auxiliary details. Indeed, the book is in everyway a good guide and manual, and will prove especially helpful to young geologists and mineralogists, whom (and especially the latter) the author appears to have borne particularly in mind. Not the least valuable chapter is the last, giving full details of analysis of minerals by the blowpipe, &c.

Photography for Amateurs, by T. C. Hepworth. (London: Cassell & Co.) A capital, easily understood, and non-technical manual for all who either dabble or work in the photographic art. The writer is well skilled in the art of conveying to the minds of the most uninitiated the facts he has himself well mastered.

The Dynamo: How made and How used, by S. R. Bottone (London: W. Swan Sonnenschein & Co.) Although this is a book especially written for amateurs, it will prove serviceable to students of electricity generally. The chapters originally appeared in the "English Mechanic," where they excited so much interest that we are pleased to see them republished in this cheap and attractive form. To those who think of constructing small dynamos it is the best book they could study.

A Dictionary of Miracles, by the Rev. Dr. Brewster. (London: Chatto & Windus.) This is a curious volume, full of attractive bits. It is a cheap and unusually well bound book withal, giving nearly 600 pages for 7s. 6d. It is so many-sided that its abundance of strange stories touch the naturalist as well as others, although the chief aim of the veteran author (this is the 50th year of his authorship) is to show a mode of thought which prevailed in Christendom for many centuries, and which has not yet died out.

Diseases of Field and Garden Crops, by Worthington G. Smith. (London: Macmillan & Co.) Few manuals have been more required than the one before us. Both to agriculturists and horticulturists the subject is one of all-absorbing interest and importance, whilst to workers with the microscope who have taken up the subject of parasitic fungi this book will be scarcely less welcome. The name of the author is a sufficient guarantee both for accuracy and fulness. Perhaps no other English author was so fit to deal with the subject. Mr. Worthington Smith's descriptions are remarkable for their lucidity. They are chiefly based upon the reports of addresses delivered at the request of the Institute of Agriculture at the British Museum, South Kensington. The work is illustrated by 143 capital woodcuts, all drawn and engraved by the author. A very copious index adds to the value of the book, which is one we cordially recommend.

Practical Taxidermy, by Montagne Brown. (London: L. Upcott Gill.) We are pleased to see that this capital manual has already passed into a second edition. The author has taken the opportunity to revise and considerably enlarge it; not the least valuable part of the new matter are further instructions in modelling and artistic taxidermy. To amateurs this book is peculiarly valuable.

SCIENCE-GOSSIP.

AT the recent Annual Meeting of the American Society of Microscopists, Mr. F. M. Hamlin read a paper on an "Ideal slide," whose construction was that the cell was sunk below the surface. Mr. Hamlin is evidently not aware that Mr. B. Piffard, of Hemel Hempstead, has patented such a slide, which we described some months ago.

IT is stated that the segmentation of ovum in the monotremata is me*r*oblastic; the "s" should be "r" thus me*r*oblastic as in birds, as distinguished from holoblastic as in mammals, where the segmentation involves the whole ovum. Mesoblast has of course an entirely different signification and is not applied to segmentation at all.—*Geo. D. Brown.*

AT a recent meeting of the Haggerston Entomologist Society, a discussion was held on Mr. South's new list of British Lepidoptera, and the universal opinion was, "That many of the alterations were uncalled for, and that a re-issue of the Doubleday list, with the addition of the new species discovered since the date of its publication, would have been far more acceptable to the great body of British Entomologists."

MICROSCOPISTS will hear with regret of the death of Col. J. J. Woodward, of the United States Army, whose name frequently occurred in the discussions reported in the "Monthly Journal of Microscopy," when Dr. Lawson was editor. One of the most important of the many papers he published was "Applications of the Photograph to Micrometry," which had special reference to the micrometry of blood in criminal cases.

MR. F. ENOCK, of Woking, has just issued an exquisite slide of the gall cynips, the smallest of its kind. With spot-lens the delicate structure of the wings comes out exquisitely. It is a most interesting object, beautifully mounted.

A COMMITTEE of the American Association for the Advancement of Science has been appointed to secure Government aid in the investigation of fungoid diseases.

WE beg to call attention to the Diatomescope, invented by Lord S. G. Osborne, made and sold by Mr. Ernest Hinton, of Upper Holloway, as a most useful instrument for out-of-door work, and particularly for identification of diatoms, &c.

THE "Naturalists' World" for October publishes a sheet of autographs of the following naturalists :—Dr. Alfred R. Wallace, W. F. Kirby, Professor W. H. Flower, Sir John Lubbock, Grant Allen, Dr. J. E. Taylor, the Rev. J. G. Wood, Richard Jefferies, Dr. M. C. Cooke, W. Saville Kent, Professor Huxley, W. Mathieu Williams, R. Bowdler Sharpe, Dr. F. Buchanan White, and Worthington G. Smith.

MR. B. PIFFARD, of Hemel Hempstead, has forwarded us some beautifully mounted slides, of much value to science teachers and students. Among them are sections of the fasciculated stems of the sweet pea, vertical section of *Lecanora tartarea* (a lichen) showing asci, &c., section of leaf of coltsfoot, showing cluster-cups (æcidium) *in situ*, and a slide of the male perichætum of *Atrichum undulatum*, showing antherozoids.

MR. CHARLES COLLINS, jun., micro-naturalist, has sent us three admirably mounted slides. One of the head of the cockroach, another of the soldier beetle (these two illustrating the structure of biting mouths), and the last of the head of the water measurer (illustrating sucking-mouth). The names of the subkingdom, class, order, family, genus, and species are on each slide.

M. BALBIANI, professor at the Collège de France, was commissioned a short time ago by the Minister of Agriculture to report upon the best mode of destroying the winter eggs of the phylloxera, as it has been found that it is in this way the progress of the parasite is very materially checked. M. Balbiani tried several fresh experiments, among others a mixture of oil, naphtha, quicklime and water. This mixture has been tried upon a very large scale in the vineyards of the Lot-et-Garonne and the Loir-et-Cher, and it possesses, according to M. Balbiani, the double recommendation of being effectual and cheap, as the cost is under a franc for a hundred stocks.

THE best coral grounds yielding the most and best red coral are still those on the Algerian coast, fished for that purpose from the middle of the sixteenth century, the others being the coasts of Sicily, Corsica, Sardinia, Spain, the Balearic Isles, Provence. Over 500 Italian boats manned by 4200 men, are employed in the coral fishery, 300 of these boats being from Torre del Greco in the Bay of Naples. The quantity gathered by these 500 boats amounts in all to about 56,000 kilogrammes annually, valued at 4,200,000 lire; that by other boats, Spanish, French, &c., to 22,000 kilogrammes, at 1,500,000 lire—total for the year, 78,000 kilogrammes at 5,750,000 lire. The gross gains per boat may be set down at 8,000 lire for the season, and the expenses at 6033 lire, leaving only 1967 lire net profit. In Italy are sixty coral workshops, of which forty are in Torre del Greco alone, employing about 9200 hands, mostly women and children.

THE sunset glories of last autumn are being repeated. In the earlier part of October over the Yorkshire Wolds a most brilliant effect was observable for about two hours. The sky above the horizon was a mass of gorgeous colours, orange tints predominating.

A CAPITAL and highly readable paper by Mr. Worthington G. Smith, F.L.S., was lately read at the meeting of the Essex Field Club on "The Politics of the Potato Fungus—a Retrospect." It appeared in full in the "Journal of Horticulture" for October 9.

THE new salt-field in Western New York is expected to produce six hundred thousand bushels of "the finest salt in the world" this year, and expects to double the production next year. A wide belt of country, the extent of which is not yet determined, is underlaid at a depth of from twelve hundred to sixteen hundred feet, with a deposit of pure rock-salt from sixty to ninety feet in thickness.

THE various phases in the eclipse of the moon were viewed with much interest on the night of October 4th. In most cases the sky was quite clear, so that stars of the twelfth magnitude were plainly visible close to the moon.

M. A. RICCO, of Palermo, has invented a very original electro-magnet. He rolls a long strip of sheet-iron around a nut of soft iron, placing oiled paper between the different layers of the strip to isolate them. A pole is connected with the nut, to which the inner end of the iron strip is soldered, while the other is connected with the outer end. The current passing through the strip magnetises not only the nut, but also each layer of the strip of iron, which plays the double *rôle* of conductor and magnetic substance, so that the lines of power produced by the conductor are condensed. It is stated that the power of such a magnet is considerably greater than that of an ordinary electro-magnet.

A SAD accident has occurred at the Healtheries. Henry Pink, an attendant employed in the dynamo-shed, had one hand on one of the brushes of a 25-arc light Hockhausen machine. By some means he got his other hand on the other brush, or on another part of the machine, with the result that a part of the current passed through his body. His death was not instantaneous, but occurred a few minutes afterwards.

LARGE bluish-green topazes, weighing several pounds, each, have been found at Mudgee, New South Wales.

IN its application to carious teeth, creosote is often inconvenient in consequence of its fluidity producing ill effects upon the mucous membrane of the mouth. This may be obviated by giving to it a gelatinous solidity by adding ten parts of collodion to fifteen of creosote. This, besides being more manageable than liquid creosote, also closes up the orifice in the tooth, preventing the access of the air to the dental nerve.

STERN paddle-wheel gun-boats have been lately built by the French Government for service on the rivers Tonkin and Gaboon.

A NEW small motor, actuated by explosions of small charges of gun-cotton, has been brought out. It is said to be useful whenever small powers are required.

ON the 22nd of October, Wolf's Comet was observed as follows: right ascension 21 h. 50 m. 31 s.; north declination 6° 24'·2.

THE last part (XX.) of Dr. Lang's "Butterflies of Europe," has just appeared. It deals with the genera Cœnonymphea, Triphysa, Erebia, Œneis, Ypthmia, Pararge, Epinephele, and those of the Hesperidæ, &c., giving detailed accounts of each species. There is also a list of addenda. Altogether this is the completest work of the kind yet published, whilst the high finish of the coloured plates raises its artistic merits to the highest rank. A systematic list of European butterflies is given in the present part, and a full index of species, varieties, synonyms, &c.

THE French Commission appointed to inquire into the nature, &c., of cholera, reject Dr. Koch's "Comma bacillus," and maintain that the blood contains the cholera poison, and that the initial lesion of cholera takes place in the blood. They state that by the hourly examination of the blood of cholera patients, the progress of the malady can be mathematically followed.

M. PERREY has found a solution of sulphate of copper of the greatest benefit, when applied to vines not more than six years old, in preventing and over-coming mildew.

A PATENT has been brought out for coating the surfaces of other metals with oxide of copper, which can be varied in their colour.

SULPHIDE of carbon is stated to be an antiseptic, and a sure destroyer of all living germs.

WE recommend all those who are about to get up conversazioni in connection with scientific societies, to procure a short handbook drawn up for the Chester Society of Natural Science by Mr. C. F. Fish. It sets forth in an admirable manner what to show, and how to show objects illustrating almost every depart-ment of scientific inquiry.

THE volume of the "Smithsonian Report for 1882" has just been published. It contains, besides the Report, records of recent scientific progress in astronomy, geology, geography, physics, chemistry, botany, mineralogy, zoology, anthropology, &c.

ZOOLOGY.

THE MOLLUSCA OF KENT, SURREY, AND MIDDLE-SEX.—I am now working out the Mollusca of these counties, and am very anxious to obtain all possible information on the subject, and local lists from those districts which I have not been able to visit myself. Can any of the readers of SCIENCE-GOSSIP help me in the matter? Any information whatever relating to the Mollusca of these three counties would be most welcome.—*T. D. A. Cockerell*, 51 *Woodstock Road, Bedford Park, Chiswick, W.*

THE BOOK-WORM.—In your issue September last there are some remarks about the book-worm. Having had some experience of this excavator in a volume dated 1484, thus exactly 400 years old, I beg to state my conviction that it is identical with the Scolytus which perforates beech-wood. Early books had covers of wood, chiefly of beech-wood, though sometimes of oak. Our very term "book" is derived from the Saxon "boc," beech. Now it is the wood, not the leaves, which is specially the object of the larvæ. In my volume the covers were entirely riddled, and only held together by the leather envelope. Some twenty-five years ago I discovered that a volume, recently bound, which stood by its side, showed signs of attack. Thus I became informed that the creature was living in the covers. To eject it, or try to do so, I beat them well with a hammer, and so ultimately I removed twenty larvæ and five of the beetle or perfect insect, and so got rid of all further mischief. The worm had perforated the volume in many ways, but it was the covers they preferred. Oaken covers are rarely attacked and only under special conditions. The millboard now used they do not like, as I found they retired after having pierced the leather envelope. In fact, banish the beech-wood and you banish the worm. It is only an accident when it appears in other bound books. I am quite convinced that, properly so-called, there is no book-worm, but it is the larvæ of the Scolytus, common to the beech-wood, and which plays such havoc with furniture made of that material.—*J. G. Waller.*

MARINE BORERS.—An interesting lecture was recently delivered at the Edinburgh Forestry Exhibition by Professor M'Intosh, in which he called attention to the serious damage inflicted upon submarine wood-work by marine borers. Among the most destructive of this class are the crabs known as the *Cheluria terebrans* and the *Limnoria lignorum*, or Scotch gribble, of which the former is the most mischievous, as being able to make larger and more oblique excavations. It was thought that the gribble paid attention only to timber, but it is now known that it is equally unremitting in its attentions to the sheaths of gutta-percha and other materials which protect submarine cables. The ravages of the gribble, great as they are, are surpassed by those of the xylophaga, a very small bivalve occupying a position between the stone and rock boring pholas and the wood boring teredo. The tunnels which the latter made into timber were of astonishing length, varying from one to two feet in the common teredo to three feet in the case of the great teredo. Up to the present time, no wood has been found capable of resisting the attacks of these little creatures; and although various remedies have been tried in the shape of immersion of the wood in silicated lime, bitumen, and creosote, by forcing them under great pressure into the tissue, the latter material was the only one which had been found to be efficacious, while mechanically nothing short of metallic sheathing protects the timber. On the other hand, the Professor pointed out, that the borers were frequently useful in their proper place, and particularly in the case of drifted timber, and old wrecks, which would be very dangerous to navigation were they not rapidly disintegrated by the action of the teredo.

ARVICOLA AMPHIBIUS.—I have given much attention of late to the habits of this vole, and feel assured it is entirely vegetarian in its diet. I know we do sometimes find the shells of water snails in its burrow, but I suspect they are often dragged in with weeds by accident and not carried there intentionally. The common brown rat often dispossesses the aquatic rodent of its burrow and appropriates it to its own use, and as this gentleman does relish a molluscous diet the water vole is apt to be charged with the brown rat's sins. Beautifully dark coloured specimens of *Arvicola amphibius* may be seen, in the fens near Cambridge, by those who know how to watch warily for such timid animals — timid with good reason, for not unfrequently they, unwillingly, serve as a living target for a pistol bullet.—*Albert H. Waters, B.A., Cambridge.*

LOCAL SCIENTIFIC SOCIETIES.—The Annual Report of the Penzance Natural History and Antiquarian Society is to hand, containing, among other matter, papers of considerable interest on "The Marine Algæ of West Cornwall," by John Ralfs; "The Marine Polyzoa of the Land's End District," by J. B. Magor; "The Sphagnums or Bog Mosses of West Cornwall," by W. Curnow; "The Ichneumonidæ of the Land's End District," by Ernest D. Marquand; and an account of a "Lichen Supper," describing some culinary experiments on certain of the common lichens, including *Sticta pulmonata, Peltigera canina* and *Parmelia perlata*, of which the sticta is described as being horribly tough and bitter, whilst the peltigera is compared to half-boiled cabbage. The results of these experiments will hardly commend lichens to the epicurean palate as special luxuries. The Proceedings of the Liverpool Naturalists' Field Club is mainly occupied with descriptions of various excursions made by the club during the past year : a list is also given of the most interesting botanical "finds" during those excursions. In the Report of the South London Natural History Club, we note the address of the President, Mr. B. Daydon Jackson, and abstracts of various papers read before the Club, including: "Insectivorous Plants," by Mr. Henry Groves ; and "British Sea Anemones," by Mr. W. T. Suffolk. "The Ear," by Mr. H. Belham Robinson ; "*Aconitum Napellus*," by Mr. D. G. Simpson, and "Endemic Species and their Lessons," by Mr. G. C. Chisholm. The Fourth Annual Report of the Walthamstow Natural History Society contains an account of the year's work, and an interesting description by the curator, Mr. A. H. Hinton, of the principal objects of natural history &c., in the possession of the Club. In the Transactions of the Yorkshire Philosophical Society are to be found reports by the respective curators of the departments of Botany, Comparative Anatomy, Mineralogy, &c., of these collections under their care, a list of various donations to the Museum and Library, and a valuable communical article by the Rev. W. C. Hey to the Society on "The Forms of Pond Snails in Yorkshire."

MICROSCOPY.

QUERY.—After staining sections in piero-carmine, what is the best method of fixing the yellow stain before clearing and mounting?—J. J. A.

THE DIATOMACEÆ OF THE AMERICAN WATERS.—I thought that the subjoined list might prove of some little interest to students of the Diatomaceæ, as it proves conclusively the great variety and versatility of form to be found along the American seaboard:—The list consists of some 268 different forms (according to Dr. Engels of Virginia), and is compiled from a gathering made in Mobile Bay. That I may not weary the reader, I have simply given the various families with their varieties, which are as follows:—Amphora, 30 varieties; Amphiprora, 4; Achnanthes, 3; Anaulus, 2; Amphitetras, 1; Auliscus, 2; Actinoptychus, 3; Actinocyclus, 3; Asteromphalus, 1; Biddulphia, 7; Bacteriastrium, 2; Campylodiscus, 5; Cerataulus, 3; Coscinodiscus, 14; Cocconeis, 3; Cymatosira, 1; Dimmeregramma, 4; Enyonema, 5; Euonologramma, 2; Eupodiscus, 2; Epithemia, 1; Euonotia, 3; Fragillaria, 2; Grammatophora, 1; Gomphonema, 2; Hyalodiscus, 2; Mastogloia, 4; Navicula, 76; Nitzschia, 14; Orthosira, 3; Odontiscus, 1; Pseudo-auliscus, 1; Pleurodesmium, 1; Pleurosigma, 3; Plagiogramma, 4; Rhaphoneis, 6; Sceptroneis, 1; Synedra, 3; Striatella, 1; Scoliopleura, 1; Systephania, 1; Stauroneis, 6; Surirella, 7; Terpsinoe, 1; Triceratium, 8; Triphyllopelta, 1. —*Alfred W. Griffin.*

THE NORFOLK DIATOMACEÆ.—Mr. F. Kitton, Hon. F.R.M.S., has issued the first series of the Norfolk Diatomaceæ in a strong, neat, and elegant case. The slides are all named, and accompanied by a catalogue. The mounting, ringing, and general turn-out of the series are remarkable for carefulness and good taste. This series ought to be very successful, for to possess the actual objects themselves named by so high an authority as Mr. Kitton, is surely better than mere illustrations of them.

A CHEAP MICROSCOPE HOLDER.—I hit upon a microscope holder the other day, which I dare say many would like to try. It costs about a penny, and works as well as a guinea one with universal brass hinge. It consists of a turned American clothes peg, held between two upright strips of wood, and these are bound at the top with an elastic band, which is passed three times round them. The bottom end of the strip is held by one screw to a block of wood. The clothes peg thus has every motion, up and down between the strips of wood, round upon its own axis, and sideways on a hinge. My boys find it very useful.—*William Linton Wilson.*

THE JOURNAL OF MICROSCOPY.—This excellent serial, edited by Mr. Alfred Allen, will be better known to our readers as "The Journal of the Postal Microscopical Society." Part 12, vol. iii., has just been published, containing, besides a good deal of various notes appertaining to microscopy and natural history, the following papers—"On the Peronosporæ," by George Norman; "The Organisms in Yeast," by Henry C. A. Vine; "On the collection and preparation of the Diatomaceæ," by Alfred W. Griffin; "Senecio vulgaris," by R. H. Moore; "Half an Hour at the Microscope with Mr. Tuffen West," &c. The papers on Peronospora, Yeast, and Senecio are illustrated by beautifully executed lithographed plates.

MICROSCOPIC SLIDE CENTRERER.—We have just seen a very compact little instrument, registered by Mr. A. B. Chapman, of Ipswich, for mounting objects accurately in the centre of the glass slips, and for applying the thin cover glass concentrically with the object. It has two revolving backgrounds to contrast with the colour of the object, one being black with white circles, the other white with black circles, and so arranged that, by simply turning a little knob, either can be used or both removed as desired without touching the slip, which can be finished entirely (except the ringing) before it is taken off the instrument. It is so simple that there is nothing to prevent any manipulation required in mounting the object, and we recommend it to our microscopic friends.

TOLU AS A MOUNTING MEDIUM.—Mr. C. Henry Kain calls attention to tolu as a mounting medium, as it has a higher index of refraction than styrax. For mounting, it should be dissolved in alcohol or chloroform, preferably the latter. The colour is a disadvantage, but this does not seriously affect thin mounts like diatoms.

THE OBSERVATION OF THE LOWER ORGANISMS.—"Science Record" says, M. Léo Ewera has successfully used a solution of India ink in studying the physiology of lower organisms. The ink is rubbed up in water and added to the fluid in which the animals and plants are living. The carbon of which it is composed is so fine that it readily stays in suspension, while by the absence of all noxious qualities it does not interfere with the life of the organism. It will be found very useful by those studying Protozoa and unicellular algæ.

GEOLOGY OF TENBY.—Reply to M. L. C.'s questions in July number: 1. Tenby is a good centre for a collector of fossils to stay at. The coal-measures and mountain limestone in the immediate neighbourhood are fairly fossiliferous, and fossiliferous Cambrian and Silurian rocks are within easy access. 2. There is no book on the Geology of Tenby, but a sketch of the geology of that district will be found in a new edition of "A Guide to Tenby," published at Mason's Library, Tenby.—*T. R.*

BOTANY.

PRIMROSES.—During the early part of this spring, whilst plucking primroses, I observed what I consider an apparent contradiction to the two distinctly defined varieties. Off one root I obtained flowers of both kinds. There were two whose andrœcium was above the gynœcium to three of the opposite arrangement. Subsequently, in the hundreds of plants I examined, I was unable to find a repetition. Has this before been noticed? I have been unable to find any note of it elsewhere?—*G. F. G.*

PARIS QUADRIFOLIA.—In answer to Mr. Colwall's inquiry as to the variation of this plant, I can inform him that the number of the leaves is by no means constant. Herb Paris is frequent in this neighbourhood, and has five leaves almost as often as four, and I have gathered it this year with six, or even seven leaves.—*C. W. Greenwood, Froxfield, Petersfield.*

PARIS QUADRIFOLIA.—It may interest K. D. and others to know, that my sister has this season collected a large number of specimens of this plant with varying numbers of leaves. In a large clump of it growing at Rudden Brow, near Goosnargh, the number of specimens with three, five, six, and, in one instance seven leaves, considerably exceeded those with the usual four. I may state that this particular group of plants was remarkably luxuriant in growth, the situation seeming to agree well with them.—*R. Standen.*

SAXIFRAGA HIRCULUS.—The note at p. 239 of the re-discovery of the rare *Saxifraga hirculus* in co. Antrim is not the first record of that interesting fact, and S. A. Brennan is only second among the happy finders. J. L. Praeger found it on the 8th of July on the headlands north of Cairnlough in co. Antrim in abundance and in beautiful bloom, and the fact was mentioned in the account of the "long summer excursion of the Belfast Naturalists' Field Club" which appeared in three Belfast newspapers on the 17th July. I could particularise the exact spot in which this treasure flourishes, but neither Mr. Praeger, nor any other botanist would wish to have it visited by the plant exterminators whose ravages are unfortunately extending even to Ireland. In Mackay's "Flora Hibernica" is the remark—"It is singular that this plant, which Dr. Hooker found in Iceland, should not be found in the north of Ireland."—*H. W. Lett, M.A.*

TO SKELETONISE LEAVES.—May I add a postscript to Mr. Crombleholme's note on this subject on page 238. The best time to gather the leaves is from the third week in June to the fourth week in July—just when the foliage is in perfection, and before it begins to get too dry and woody. When the leaves have been skeletonised they may be bleached to an almost pure white by being immersed in a bath of dilute chlorine; the older the leaf, the stronger the solution required. They must be watched in the bath, and carefully removed as soon as bleached, then washed in clean cold water and laid out to dry on clean blotting-paper. They may, when dry, be tastefully arranged on a stand. This will be found a pleasant occupation for the winter evenings. It may also be mentioned that photographic views of these groups form some of the most charming stereoscopic pictures.—*C. Beale, Rowley Regis.*

GEOLOGY, &C.

INCRUSTATION IN A WATER-PIPE.—At the Inkerman coal-pit number one, Renfrewshire, a wooden water-pipe, which had been in use for upwards of twenty-four years for carrying water, gave occasion for examination, which, on being opened, was found to be nearly filled with an incrustation which the miners called an "incrustation of salt." The incrustation when removed from the pipe was dark gray or slate coloured, hard, and rock-like, and the pipe in which it was formed being square the incrustation measured outside 2⅛ × 2⅛ in., and which is the original size of the pipe, but by it was reduced to a square hole measuring only 13/20 × 13/20 in., which is in the centre of the incrustation, while the walls of the incrustation measure in thickness 14/20, 15/20, 14/20, 15/20 in. From each of the corners to the corners of hole in the centre are lines of cleavage. This form of incrustation is common in limestone districts, and affords a good example of rock formation. I analysed a sample of the above incrustation and found:

Carbonate of lime	77·50
Gangue (i.e., mud or clay)	20·00
Hydrated peroxide of iron	2·50
	100·00

from which it will be seen that carbonate of lime is chief in the deposit. The rock thus formed is called Travertine, as it is compact, hard and semi-crystalline which distinguishes it from a similar deposit, viz., calc-tuff or tufa, which is loose and porous.—*Taylor, Sub-curator, Museum, Paisley.*

LARGE UNIOS.—Last July, the artificial lake, near the hall at Ossington, having been drained for the purpose of removing a deep deposit of mud, I found a number of the finest specimens of *Unio pictorum* I have ever seen. The largest measured 4⅞ inches in length. Has this been exceeded? I obtained as many as the soft mud allowed me to reach, the remainder have since been taken away with the material removed by the workmen.—*W. Gain Tuxford.*

NOTES AND QUERIES.

BIRDS IN AUGUST.—As Gilbert White pronounced August "by much the most mute of months," and one of his editors, Mr. Jesse, considered the robin its only songster, it may be worth noting that throughout great part of the recent August, not only robins, but wrens, goldcrests, white-throats, chiff-chaffs, willow-wrens, sedge-warblers, grasshopper-warblers, water-ousels, gold-finches, green-finches, linnets, yellow-hammers, buntings, and wood-pigeons were frequently in song. The coo of the ring-dove might be heard almost any day of the month; and on fine dry nights I saw the grasshopper-warbler, as White saw it in the early morning, steal from its covert, mount the spray of a furze-bush, and trill gaily away "on the top of a twig, gaping and shivering with its wings." On the 11th, I found a goldfinch's nest half finished, and on the 13th another, containing three eggs; both now contain young birds. It is remarkable that these two nests should have been built in the hottest week of the hottest summer we have had for many years: at a time when nearly all aquatic birds had forsaken the dried-up marshes, and teals, water rails, and groups of herons formed a quite unwonted spectacle along the shores of the rivers. Only the airiest of birds could dream of building in such weather. A congregation of golden plover appeared as early as the 8th of August, and by the 19th chaffinches and missel-thrushes were in flocks.—*C. B. Moffat.*

HYDROGEN GAS GENERATED, BUT NOT CONSUMED.—The enquiry of S. C., on this subject, in the September number, revives the remembrance of a curious pamphlet, published many years ago, propounding precisely the same question; a pamphlet, that, had Professor de Morgan expanded his criticisms beyond the examination of mathematical crazes, might have found a place in his most entertaining "Budget of Paradoxes." The theory of the pamphleteer was, that as (to him, it appeared) the rainfall was in excess of evaporation, the increase could only be explained by a production of water, through electric force, from an accumulated ocean of hydrogen above meeting the excess of oxygen also continually pouring into the atmosphere from the surface of the earth; an obvious difficulty (which the writer evaded) would be the ultimate result of such an evolution, and the prevention of an accumulating deluge. As alcohol is "diffused" when mixed with water, gases are "diffused" in the ocean of the atmosphere, free hydrogen, oxygen, carbonic acid, and the many sulphuretted mixtures locally produced, rapidly combine with the atmosphere, and eventually enter into the maintenance of the equilibrium of existences. The idea of a surplus of hydrogen filtering its way to, and floating upon the surface of our sea of atmosphere, is alarming, as, by a parity of reasoning, a denser gas (say) carbonic anhydride, would be lurking at the bottom. The law of "diffusion" is a mercy, and the only explanation of the disposal of ponderous, etherial, or contaminating gases. Any class book on Chemistry explains how hydrogen accepts combinations long before, as a distinct element, it can "work its sinuous course upward and form a stratum above the atmosphere."—*E. T. D.*

MOTION IN A SPIDER'S SEVERED LEG.—On the afternoon of September 2, as I was sweeping for larva among some heath, I noticed in my net something wriggling about, which, on examination, proved to be a spider's leg. Its late owner was on the other side of the net, so I had no doubt as to its identity. On taking it out and putting it on my hand I was astonished to see it writhing and hopping about just as if it had been still connected with the spider, and it continued in a state of gradually decreasing motion for some minutes, when it ceased altogether. Has such an occurrence been noticed before? I should be much obliged if some reader of SCIENCE-GOSSIP would enlighten me on the subject. The spider was a small-bodied, long-legged species, but I am quite ignorant of its name.—*H. E. U. Bull, Foundry Lane, W. Southampton.*

CAN any of the readers of SCIENCE-GOSSIP please inform me what a common land tortoise will eat? I have given it milk and lettuce, but it never appears to drink or eat either, and it will not eat raw meat.—*K. H. J.*

SKELETON LEAVES.—The maceration of leaves in cold water, even for months, is, as far as my experience goes, utterly useless; others may have been more successful, but, to say the least, it is a tedious process. In SCIENCE-GOSSIP, 1867, will be found practical directions for the preparation, bleaching, &c., of leaves (see pp. 22, 141, 246). I have tried them, and can speak to their value.—*F. K.*

BIRDS AND THE ARUM.—Can any reader of SCIENCE-GOSSIP inform me whether anything is known of birds or ground game eating off the young spathe and spadix of *Arum maculatum* before fully expanded? I am led to make the inquiry in consequence of having had my attention drawn to the matter by a friend whilst botanising in an out of the way wood near Preston; where nearly every specimen was thus mutilated. The leaves were seldom injured; but the floral portion of the plant was rarely perfect. Are pheasants fond of these things?—*F. J. George.*

A GENERAL INDEX TO SCIENCE-GOSSIP.—With the completion of the present volume, SCIENCE-GOSSIP will have appeared continuously for twenty years. It has occurred to me that, to those subscribers who are in possession of the complete set, a general index would be exceedingly useful as a means of ready reference. The work contains many excellent papers on various subjects, and where it is desired to collate them on any particular one, it involves no little labour and loss of time. An extra charge should be made for this addition, and doubtless many, like myself, would gladly subscribe for it: if the present proprietors can see their way to afford the accommodation. If those who feel disposed to do so would kindly intimate their desire to the publishers it would enable the latter to form an opinion as to the possibility of carrying out the scheme.—*W. H. Harris.*

CURIOUS ACT OF A NEWT.—There is a newt in my aquarium who, a short time ago, after casting his skin, swallowed it, just as would a small worm. He is a young one of the larger kind.—*G. A. Simmons.*

PHOSPHOROUS INSECTS.—The following fact in natural history may prove interesting to some of your readers. On the night of the 1st of May, returning home at about 9 P.M. the night being somewhat cold and damp, and the sky heavily clouded, my attention was attracted by a brilliant light on the side of the gravel-drive. Thinking for a moment that it was a centipede showing an unusual degree of luminosity, I

stooped down to examine, and saw directly that the light proceeded from some insect, neither centipede nor glow-worm. I think the light extended to the under side of the body, as the creature left very brilliant streaks on the pebbles as it ran over them. The upper part of the thorax was extremely brilliant, and I only regret that I cannot remember whether the wing cases stood in the luminosity. The light was really beautiful, being more brilliant and crystalline in its character than I have ever seen on a centipede. After a minute, I caught the insect and brought it in, when I was astonished to see a common-looking black-beetle. Another insect very faintly luminous was close by the one described when first seen, but the light of this second insect disappeared almost immediately, and I could see no more of it in the darkness. I imagine the luminosity of the hard case of a black-beetle at least to this extent to be no very common occurrence, but it is a well-known fact that many insects are occasionally, though rarely, luminous. Mr. Gosse, in the "Romance of Natural History," gives instances of strong luminosity in a mole-cricket, and in a crane-fly. This and other evidence seems to show that insects may be occasionally luminous of which thousands may be taken without a trace of the phenomenon. I have sent the insect to a scientific relation of mine, C. F. George, Esq., of Kirton-in-Lindsey, for examination. It will be interesting to know if your readers have had any similar experiences.—*John C. Scudamore, Norfolk.*
P.S.—The beetle is *Steropus madidus*.—*C. F. George.*

PRAYING (?) MANTIS.—On the 15th of April, I found a mantis which differs so much from those I have found here during the last thirteen years, that I should like some information from you or your readers. Colour: ash grey, with five white spots on fore legs (or arms), and two white bands on the other legs. Head: back of head elongated into a kind of tower standing above the eyes, about twice as long as the lower part of the face; the top flattened in front and white at the tip; antenna with double curve and black tips. Abdomen: divided into seven segments; the joints very distinct on the upper side, but carries it curved over the back, showing the under side of abdomen with three projections on each segment, the centre one very large and forming a kind of pouch. As this has been found much earlier in the year than usual, is it a survivor from last year or a young form? Can some reader give me any information as to the life-history of the mantis?—*W. Harvey.*

NOTICES TO CORRESPONDENTS.

TO CORRESPONDENTS AND EXCHANGERS.—As we now publish SCIENCE-GOSSIP earlier than heretofore, we cannot possibly insert in the following number any communications which reach us later than the 8th of the previous month.

TO ANONYMOUS QUERISTS.—We receive so many queries which do not bear the writers' names that we are forced to adhere to our rule of not noticing them.

TO DEALERS AND OTHERS.—We are always glad to treat dealers in natural history objects on the same fair and general ground as amateurs, in so far as the "exchanges" offered are fair exchanges. But it is evident that, when their offers are simply disguised advertisements, for the purpose of evading the cost of advertising, an advantage is taken of our *gratuitous* insertion of "exchanges" which cannot be tolerated.

WE request that all exchanges may be signed with name (or initials) and full address at the end.

H. E. FOUNTAINS.—Wood's "Common British Beetles," price 1s., and Cooke's "Ponds and Ditches," price 2s. 6d. The latter is published by the Christian Knowledge Society.

G. D. BROWN.—Thanks for your friendly hint. We are aware of the leaf-fungus you mention; but in the case to which the answer was given, the black spots were not due to fungoid growths, but to the cause usually assigned.

A. S. MACKIE.—From your sketches, we have no doubt the fossil is a Belemnite, probably from the Oolitic formation, and perhaps re-deposited in the Drift beds. It is a very common species in the Oolite and Lias. See Taylor's "Geological Stories."

J. H. D.—You would do best to address your query to the editor of "The English Mechanic."

J. BORING.—No doubt the sparrow you saw was partly an Albino. They are not uncommon; you may generally find one or more in every village.

E. H. W.—The phosphorescent millipede was no doubt *Geophilus electricus*, a not uncommon species, remarkable for its glowworm-like power of emitting light. This insect has long been known to naturalists. Its phosphorescence seems most powerful in the autumn.

W. JEFFERY (Chichester).—We hope to publish your interesting paper very shortly.

A. W. WEYMAN.—Get Lankester's "Half Hours with the Microscope," edited by F. Kitton, published by Messrs. Allen & Co., Waterloo Place, at 2s. 6d.

T. JOY.—We conclude the insect sent us had escaped, for none was in the smashed box when it reached us.

G. D. B.—Accept our best thanks for your kindly remarks and criticisms.

F. R. T.—We have not heard the name of the plant you mention, nor been able to find it. Of course, we cannot answer when you are not sure you have spelled the name correctly.

C. BURTON.—Either the Hampstead Naturalists or the Western Club (of which see notice in present SCIENCE-GOSSIP) would be glad to welcome you, and both would be well situated for you. Write to the secretaries.

J. M.—See Lecture by Professor Huxley, delivered at the Fisheries Exhibition, Norwich, on "The Herring;" also paper (price 6d.) published by the International Fisheries Exhibition last year, by R. W. Duff, M.P., on "The Herring Fisheries of Scotland."

G. A. H.—It is difficult to judge of a bird from a slight portion of wing like that you sent, but we have little doubt the bird is the female greenfinch. For an account of the death-watch beetle and its sounds, see an article in SCIENCE-GOSSIP for November, 1880, called "A Wood-Carver's Experience of the Death-Watch Beetle."

EXCHANGES.

SPLENDIDLY-preserved and correctly-named Swiss Alpine plants. Price 6d. each.—Address, Dr. B., care of Editor of SCIENCE-GOSSIP, Piccadilly.

DUPLICATES: *Bythinia tentaculata* (Sarno, Rio, Naples), *Helix calcarata* (Malta), *Mytilus minimus* (Sicily), *Dentalium stragulatum* (Messina), *Helix elata* (Sicily), and few other Sicilian shells. Desiderata: *Balea lucifuga, Vitrina Draparnaldi*, and other British land and marine shells. Send lists.—John Platania-Platania, Via S. Giuseppe No. 14, Acireale, Sicily.

To exchange, a microscope, 1 in., ¼ in., and ⅛ in. combined powers, in cabinet, also books and microscopical slides, for other microscopic slides or good watch. Lists to—S. Harrison, Dalmain Road, Forest Hill, London, S.E.

WILL exchange twelve dozen microscopic slides in cabinet, or three dozen physiological ones, for good magic lantern.—G. Brocklehurst, Roundhay, Leeds.

EGGS of shrike, nightingale, spotted flycatcher, black-headed bunting, blackcap, redstart, golden-crested regulus, titmouse, chiffchaff, grebes, coots, rooks, jays, and many others, for eggs of black and red grouse, gulls, terns, &c.—H. Medley, Palmerston Square, Romsey, Hampshire.

OFFERED, L.C., 7th ed., Nos. 150, 174, 197, 280, 313, 375, 406, 528a, 564, 587, 594, 813, 835, 859, 999, 1008, 1039, 1971, 1208, 1322, 1333, 1363, 1422, 1447, 1448, 1504, &c. Send lists.—H. Fisher, 52 K. Y. C., Clifton, Bristol.

DUPLICATES: *Mitra cinctella, Conus tessellatus, Pyrula vespertilio, Clanculus Pharaonis, Helix hæmastoma, Cypræa Arabica, C. argus, C. canrica, C. lynx, C. stercoraria, C. vitellus*, &c. Wanted, foreign shells, and British fossils. Exchange lists.—J. E. Lustér, Arragon Close, Richmond Road, Twickenham.

GOOD microscopic slides, various, in exchange for other slides, materials, or apparatus.—J. J. Andrew, 2 Belgravia, Belfast.

GOOD microscopic slides for exchange. What offers?—Samuel M. Malcolmson, M.D., Union Hospital, Belfast.

FRESH specimens of *Vanessa atalanta*, also healthy pupæ of *Smerinthus populi*.—D. Fergusson, 77 Skene Street, Aberdeen.

WANTED, diatom sides in exchange for slides of *Arachnoidiscus ornata* ("in situ") on coralline, a splendid object for binocular and dark ground illumination; also other slides, chiefly diatoms. Send list.—H. Morland, Cranford, near Hounslow.

WANTED, foreign butterflies; will exchange British lepidoptera, coleoptera, and foreign and English shells.—C. M. D. Dods, 47 Chepstow Place, Westbourne Grove, Bayswater, W.

"PHONETIC JOURNAL," 204 weekly parts, 1876, 1880, 1881, and 1882, in good condition, unbound; will exchange for geological works.—W. L. Atkinson, 205 Humberstone Road, Leicester.

WANTED, snakes, lizards, &c.; exchange "Knowledge," vol. i., unbound (50 numbers), one number missing; or money; or birds' eggs, side blown, and poisoned against mildew, &c.; state wants.—G. Simmons, 102 Ladbroke Grove Road, Notting Hill, London.

PUPÆ of Menyanthidis, S. populi, and imagos of Davus; desiderata numerous.—John Mearns, 48 Jasamine Terrace, Aberdeen.

A COLLECTION of over 700 foreign stamps, many rare and unused, in Stafford Smith's Permanent Album (6s. 6d.), all different, catalogue, value over £3. What offers? Wanted, entomological cabinet and apparatus, also dissecting implements: no reasonable offer refused.—H. H., 4 Albert Terrace, Old Trafford, Manchester.

DUPLICATE micro slides for exchange, including many good fish scales; insect preparations specially wanted, also small flat air pump. Lists exchanged.—H. Moulton, 37 Chancery Lane, London, W.C.

WANTED, Rimmer's "Land and Freshwater Shells" in exchange for books on natural history, vols. i.-iii, of SCIENCE-GOSSIP, &c.—A. Alletsee, 7 Glendale, Clifton, Bristol.

ON receipt of specimen of any unmounted material, I will send unmounted specimen of Penicillium glaucum fungi on germinating grain.—J. W. Horton, Brayford Wharf, Lincoln.

WANTED, British and foreign Hepaticæ. Offered, mounted slides of the same or other species in fructus, or micro slides, various. Address, with lists.—W. E. Green, 24 Triangle, Bristol.

WANTED, cryptogamic micro slides; will exchange books or well-mounted slides of cocoon of silkworm for polariscope.—W. S. P., Birstall, near Leeds.

VALENTINE'S section knife, in case, never been used; will exchange for well-mounted slides, or what offers?—H. J. Parry, 10 Windsor Terrace, Newcastle-on-Tyne.

LINDSAY'S "Lichens," good copy, what offers?—W. P. Quelch, 8 Eccleston Road, Ealing Dean, London, W.

DUPLICATES: remarkably fine L. stagnalis and P. corneus; also L. glabra, peregra, var. fragilis of stagnalis, L. palustris, H. arbustorum, and var. flavescens, H. rufescens, Lapicida ericetorum, vars. alba and submaritima of virgata, C. rugosa, &c. &c. Desiderata very numerous.—W. Hewett, 26 Clarence Street, York.

MICROSCOPIC slides; a number of good miscellaneous mounts in exchange for others recently mounted; also 12 tubes cleaned diatoms, and some miscellaneous unmounted objects for disposal.—Mathie, 42 McKinlay Street, Glasgow.

WANTED, well-mounted slides, and a slide cabinet; have a large assortment of British shells in duplicate.—S. C. Cockerell, 51 Woodstock Road, Bedford Park, Chiswick, W.

STARCHES, duplicates of eighteen kinds, dry, mounted, offered for like number of good sections of British trees on type forms of native desmids.—J. H. Morgan, St. Arvan's Lodge, Chepstow, Monmouthshire.

OFFERED, 150, 940, 1142, 1212, and many others, for 830 (if Irish), 1042, 1347. Lists exchanged; note change of address.—C. A. Oakeshott, Marlborough Avenue, Torquay.

THREE hundred dried plants, including many rare species, all British, for shells.—C. A. Oakeshott, Marlborough Avenue, Torquay.

FOR exchange, about four or five hundred Portuguese wild flowers and plants, dried and named; for mosses or fresh or salt water algæ, duly classified.—Isaac Newton, Oporto.

MICRO slides and material, or a large number of herbarium examples of British and foreign phanerogams for named and localised rocks, minerals, fossils, or what offers?—J. H. Lewis, 145 Windsor Street, Liverpool.

FIFTY varieties of micro-fungi, also diatoms, histological and entomological, and other well-mounted micro slides to exchange for other good preparations. For lists apply to—Dr. Moorhead, Errigle, Cootehill, Ireland.

DUPLICATES: Anodon cygnea, Planorbis corneus, Paludina Listeri, Bythinia tentaculata, and Limnea stagnalis. Desiderata: Unio pictorum, Unio margaritifera, Cyclostoma elegans, Succinea oblonga, Helix lapicida, and Helix arbustorum.—C. O., Syrian House, Sale, near Manchester.

WANTED, Pisidium nitidum, P. roseum, P. Henslowanum, P. pulchellum, P. cinereum, Vertigo Lilljeborgi (Westerlung), V. moulinsianum (dup.), V. tumida, V. alpestris, V. pusilla, V. angustior, V. minutissima, Planorbis glaber. Will exchange for either of the above a 4¼ inch specimen of Unio pictorum from my recent find of these large shells in Ossington Lake, or other Unios, Anodons, &c.; state wants.—W. Gain, Tuxford, Newark.

WANTED, Scott's "Weather Charts and Storm Warnings" in exchange for other book or micro slides, if cheap.—B., 36 Windsor Terrace, Glasgow.

FOR exchange, Lon. Cat. 7, 125, 130b, 147, 286, 361, 368, 593, 739, 913, 1382, and very many other plants. Send lists to—J. R. Neve, Campden, S.O., Gloucestershire.

FOR exchange, "European Butterflies and Moths," by W. F. Kirby, neatly bound in half morocco, gilt edges; What offers?—J. M. Mackay, 15 Gordon Street, Aberdeen.

WANTED, geological specimens for books, Cornish specimens preferred.—John T. Millie, Clarence House, Inverkeithing.

WANTED, Devonian corals or Silurian trilobites for Cooke's "Rust, Smut, Mildew, and Mould" (coloured illustrations) and Cooke's "British Fungi" (coloured illustrations).—Cairns, 111 Princess Street, Hurst, Ashton-under-Lyne

WANTED, geological specimens and minerals, Cornish preferred, in exchange for books, foreign coins, and eggs.—John B. Douglas, High Street, Inverkeithing.

OFFERED, L.C., 7th edit., 180, 273c, 317, 490, 584, 715, 1130, 1210, 1422, and others. Wanted, 101, 119, 377, 665, 762, 997, 1035, 1240, 1302, and others. Send lists.—A. W. Preston, Thorpe Hamlet, Norwich.

FIRST-CLASS anatomical and pathological sections from the human subject, and carmine injected sections from kitten and rabbit, in exchange for well-mounted slides. Desiderata, insect preparations, grouped diatoms, and botanical objects.—H. Vial, Crediton, Devon.

MICRO slides: fifty varieties of micro fungi, diatoms, histological sections (animal and vegetable), to exchange for other well-mounted slides. Apply for lists to—Dr. Moorhead, Errigle, Cootehill, Ireland.

MACGILLIVRAY'S "Conchologists' Text-book," 21 coloured plates, 9th edition, offered for micro slides.—F. Adams, 92 Upper Alma Street, Newport, Monmouth.

WANTED, a few good specimens of Dytiscus marginalis, fresh (this season's), either living or dead; also specimens of rhinoceros horn, whalebone, and various hoofs for section-cutting. Good exchange in well-mounted objects or material.—R. M., 59 Hind Street, Stanisby Road, Poplar, London, E.

L. stagnalis, P. corneus, from Strensell Common; Paludina vivipara, Unio tumidus, from Riverfoss; in exchange for land and freshwater shells or marine.—Robert Barker, 11 Towend Street, Groves, York.

WANTED, SCIENCE-GOSSIP for 1869, 71, 73, 74, and 75, unbound or bound in recent blue covers, in exchange for coins, fossils, loose geological papers, or books.—C. Beale, Lime Tree House, Rowley Regis, Dudley.

SCIENCE-GOSSIP for 1865, 1874, 1875, bound in cloth case, for 1876 in numbers with case, and for 1877, 1878, in numbers; a micro lamp by Horne and Thornthwaite, with porcelain shade, in polished pine-wood case; a powerful home-made induction coil of good workmanship, made of over three miles of wire, with carbon-zinc bichromate battery; also a £2 2s. Carter's reading and writing desk, quite new, adjustable to any angle, and can be used as a small table for invalids in bed. Micro slide cabinet or micro accessories to value.—B. B. W., 23 Batoum Gardens, West Kensington Park, W.

WANTED, to exchange living specimens of rare and cutical species of British plants. List of duplicates will be sent.—Arthur Bennett, High Street, Croydon.

WANTED, SCIENCE-GOSSIP, Nos. 229, 230, 231; also botanical slides for vols. iv. and v. "Oracle," first vol. "Knowledge," and Balfour's "Outlines" (1862).—F. Marshall, Benwick, March.

WILL send tube of fine budding specimens of Hydra viridis for good mounted object or unmounted physiological sections.—T. W. Lockwood, Lobley Street, Heckmondwike, Yorkshire.

OFFERED, L. C., 7th ed., 148, 197, 406, 859, 924, 1008, 1319, 1333, 1349, 1412, 1422, 1446, 1447, 1458, 1494, 1497, 1501, 1504, 1577. Send lists.—H. J. Wilkinson, 6 Alexandra Terrace, York.

BOOKS, ETC., RECEIVED.

"Smithsonian Report, 1882."—"Diseases of Farm and Garden Crops," by Worthington G. Smith. London: Macmillan & Co.—"The Honey Bee," by W. H. Harris. London: Religious Tract Society.—"The Amount of the Atmospheric Absorption," by S. P. Langley.—"The Scottish Naturalist."—"The Gentleman's Magazine."—"Belgravia."—"The Journal of Microscopy."—"The Science Monthly."—"Midland Naturalist."—"Ben Brierley's Journal."—"Science Record."—"Science."—"American Naturalist."—"The Electrician and Electrical Engineer."—"Canadian Entomologist."—"American Monthly Microscopical Journal."—"Popular Science News."—"The Botanical Gazette."—"Revue de Botanique."—"La Feuille des Jeunes Naturalistes."—"Le Monde de la Science."—"Cosmos, les Mondes."—"Revista," &c. &c. &c.

COMMUNICATIONS RECEIVED UP TO 12TH ULT. FROM:—N. F. D.—J. F.—A. S. W.—F. K.—J. H. M.—W. H. J.—J. T. L.—G. A. H.—J. G. W.—J. C. S.—Dr. J. A. O.—Dr. H. W. C.—H. F. F.—G. P.—C. A. O.—F. K.—W. H. H.—F. J. G.—C. O.—G. A. G.—F. R. T.—H. M.—H. E. F.—W. G.—J. T. M.—G. R. G.—C. B.—J. B.—J. R. N.—J. C. M.—R. S.—R. C.—S. C. C.—W. M.—J. B. D.—A. W. P.—G. D. B.—H. V.—T. H. M.—F. A. A.—R. M.—R. B.—C. B.—J. P. P.—F. W.—W. H.—J. W. H.—E. D. T.—W. P. Q.—A. W. S.—H. T. P.—W. S. P.—W. E. G.—C. C., jun.—A. M.—D. F.—A. W. G.—F. J. G.—G. A. S.—E. H. W.—J. E. L.—A. H. W.—H. W.—W. L. A.—J. M.—H. F.—J. B.—A. S. M.—A. W. W.—S. H.—H. M.—C. M. D. D.—H. F. M.—A. A.—H. W. L.—T. D. A. C.—S. M. M.—J. J. A.—G. B.—F. M.—H. L. B.—J. M.—A. B. H. W. S. W. B.—&c.

GRAPHIC MICROSCOPY.

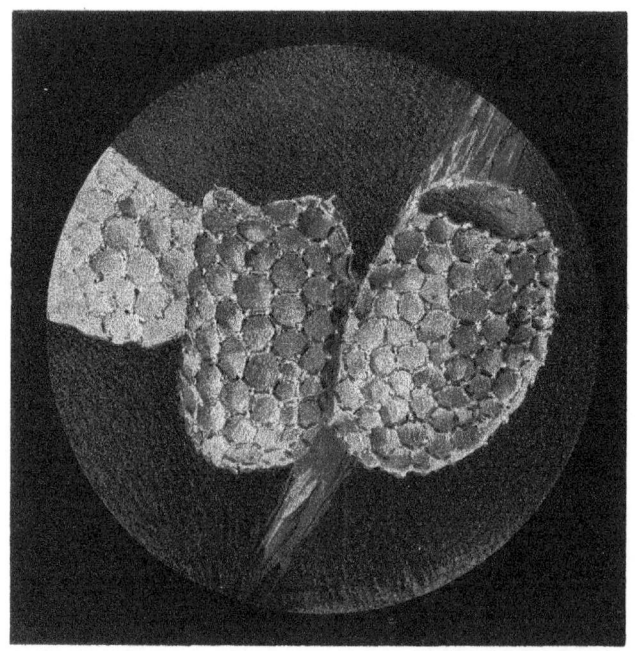

E:T.D. del, ad nat. Vincent Brooks Day & Son lith

EGGS OF MOTTLED UMBER MOTH
×75

GRAPHIC MICROSCOPY.

By E. T D.

No. XII.—Eggs of Mottled Umber Moth (Hybernia defoliaria).

BEYOND beauty of configuration, surprisingly varied in character, the eggs of insects exhibit under magnification the most delicate combinations of colour; including a gamut of white, yellow, gray, pink, and browns, blending with extreme daintiness, and, in many instances, aided by iridescent interferences.

The eggs of the mottled umber moth involve both qualities, play of colour and elegance of design, disclosing reticulated hexagons, the angles studded with white nodules, the depressions flashed with opalescence. They are found on the stems of the buck-thorn and white-thorn.

Early December is a favourable time to procure eggs of insects on and beneath the bark, and in the interstices of those trees or plants which serve the larvæ with food; decaying wood and old palings near the source of future supply are promising spots.

Want of space contracts a list; but of elegant and typical eggs may be mentioned: puss moth (*Lerura vinula*), in colour and shape like a ripe Seville orange, found on willow and poplar. Magpie moth (*Abraxis grossulariaria*), egg shell oval, of a delicate puce, scintillating with iridescence; currant, gooseberry, sloe. Vapourer moth (*Orgyia antiqua*), egg round, flattened, creamy yellow, bordered with a brown ring, thickening at the summit, found on hardy shrubs, even in smoky London squares. Thorn moth (*Ennomos crosaria*), a beautiful egg, elongated, square, the top frilled, bottom perfectly flat; oak, birch. Dingy shears (*Orthosia ypsilon*), subconical, reticulated, with raised ribs running in regular order from the base to the top; firmly attached to the slender stems of willow and poplar. Lappet moth (*Gastropacha quercifolia*), spherical, blue, with circular brown bands, delicately blending; willow and blackthorn. Chocolate tip (*Clostera curtula*), globular, colour like antique Roman glass; aspen, sallow, and poplar. Bordered rustic (*Caradrina morpheus*), subconical, showing a ribbed structure leading to a curiously-formed lid, found on teazle.

Many eggs are sufficiently transparent to reveal the young larvæ within; notably, the buff tiger moth (*Diacrisia russula*), appearing like a globule of glass, covered with a delicate net-work of hexagons; and the small emerald (*Iodis Venaria*), an oval silvery egg, singularly clear, showing the contents, found on stems of clematis.

Eggs change in colour as they mature; those of the Kentish glory (*Eudromis versicolor*), on birch, beech, and lime, are at first, brimstone, changing to deep green, red, and finally purple; silkworm eggs, as well known, pass through successive tones, from lemon yellow to dingy slate.

Eggs should be empty and seen as mere shells; the difficulty is to get them intact, when hatched naturally, the larvæ, on emerging split the structure generally at the most interesting point, the apex. Beautiful as are the eggs of the Lepidoptera, for quaint device, in lids, caps, fringes, markings, and for colour, they are surpassed by those of parasitic insects; under fine reflected light few objects excel in beauty the eggs of an Anopluran, packed with singular regularity in the shaft and barbs of the feather of a bird.

Crouch End.

THE experiments first broached by an American naturalist, that dying fish can be restored by brandy, have been confirmed by Mr. W. Chambers.

A CHAPTER ON PHEASANTS.

(Continued from page 244.)

BOTH pheasants and partridges are partial to nesting in clover; when this is cut, a large number of eggs are often destroyed by the scythe or mowing machine.

The pheasant is more than a half reclaimed bird, while the partridge wanders in wildest freedom through the land, heedless of the fostering care of man. The bird in question when hatched under a domestic fowl will come to be fed at all hours of the day; it will sometimes associate with the poultry on the farm, and, where it is not disturbed, will roost in trees close to our habitation; and notwithstanding the proximity of this bird to the nature of the barn-door-fowl, still its innate timidity baffles every attempt on the part of man to render its domestication complete. Waterton says he spent some months in trying to overcome this timorous propensity, but he failed completely in the attempt. The young birds which he had hatched under a domestic hen, soon became very tame, and received food from the hand when it was offered cautiously to them. They would fly up to the windows and feed in company with the common poultry. But if anybody approached them unawares, off they went to the nearest cover with surprising velocity, where they would remain till all was quiet, and then return with their usual confidence. Two of them lost their lives in water by the unexpected appearance of a pointer, while the barn-door-fowl seemed scarcely to notice the presence of the intruder; the rest of the young birds finally took to the woods at the commencement of the breeding season.

Hybrids between the pheasant and common hen are by no means uncommon, and the peculiar form and colour of plumage, together with the wild and suspicious mien, are handed down through several generations. A hybrid was shot in a wild state in the woods at Wolterton, West Norf., December, 1854, apparently a cross between a pheasant and a Cochin China fowl; and in November, 1848, was killed at Snettisham, in the same neighbourhood, a hybrid between the pheasant and black-grouse. — (See "Birds of Norfolk.")

The Rev. J. Wood, in his "Natural History," mentions that the turkey and guinea fowl have been known to mate with the pheasant. The cock pheasant is a very pugnacious bird, and the author just quoted tells us that it can maintain a stout fight with the barn-door cock, and often comes off victorious by his irregular mode of proceeding, for, after making two or three strokes at his enemy, up goes the pheasant into a tree to breathe, leaving the cock looking about for his antagonist. Presently, while his opponent is still bewildered, down comes the pheasant again, makes another stroke or two, and retires to his branch. The cock gets so puzzled at this mode of fighting that he often yields the point.

There is another variety of pheasant found in our woods and preserves, introduced many years ago from China (*P. torquatus*), chiefly distinguished by a white ring round its neck, which has so intermingled with the common sort that it is difficult at the present time to find a specimen of the old English type without some traces, however slight, of the ring-neck and other marked features of this Chinese bird. The author of the "Birds of Norfolk" says he has been informed, that no little difficulty is sometimes experienced by gamekeepers, from the fact of the eggs of the ring-neck pheasant hatching more quickly than those of the common pheasant, and hence should a mixed "clutch" of eggs be placed under a hen, which is very likely to happen when supplies are purchased from different places, she comes off with her first hatched young, leaving perhaps a majority of good eggs still unincubated in the nest.

In olden time the pheasant, like the peacock, was considered a royal bird, and the great ones of the earth used to swear by it. Gibbon, the historian, gives the following account of the origin of this custom. Shortly after the taking of Constantinople by the Turks, a chivalrous meeting was convened at Lille, by Philip, Duke of Burgundy, to concert measures for the defence of Christendom. In the midst of the banquet, a gigantic Saracen entered the hall, leading a fictitious elephant with a castle on his back. A matron in a mourning robe, the symbol of religion, was seen to issue from the castle; she deplored her oppression, and accused the slowness of her champions. The principal herald advanced, bearing on his fist a live pheasant, which, according to the rites of chivalry, he presented to the duke. At this extraordinary summons, Philip, a wise and aged prince, engaged his person and powers in a holy war against the Turks. His example was imitated by the barons and knights of the assembly; they swore to God, the Virgin, the ladies, and the *pheasant*.

The food of the pheasant in its wild state appears to be of a very varied character, consisting of grain of all kinds, seeds, green leaves, insects, slugs, &c. Yarrel says he has several times seen them pulling down ripe blackberries from a hedge-side, and later in the year, has seen them fly up into high bushes to pick sloes and haws. The roots of ranunculus, bulbosus and ficaria (the buttercup, pilewort), and crow's-foot, form a large portion of its food in the spring. The quantities of noxious insects and grubs destroyed by these birds is something extraordinary. Some years since, Mr. Milton, of Great Marlborough Street, found in the crop of a cock pheasant 852 larvæ of tipulæ, or crane flies, those long-legged insects popularly known as daddy-long-legs. A correspondent of the "Sporting Magazine" writes, that no fewer than 1225 of these destructive larvæ were taken from the crop of a hen pheasant. No doubt, adds Mr. Curtis, in his work on

Farm Insects, these birds pick out the larvæ in corn and turnip fields; and when it is remembered that the almost incredible number contained at one time in the stomach only makes a single meal, the extent of their services may in some measure be estimated.

During the last few years various artificial foods for pheasants have been introduced, and consequently birds of larger size and finer flavour have been reared. The quantity of pheasants reared and killed during the season at battues, &c. in various parts of the kingdom is something enormous. Dr. Wynter in his "Curiosities of Civilization," has a paper on the London Commissariat, in which he states that 70,000 pheasants and 125,000 partridges were annually sent to the London market. As Dr. Wynter's book appeared in 1860, no doubt the number sent to the metropolis has greatly increased since then. To these must be added the large quantity sold by poulterers in other cities and towns all over the kingdom. This will give us some idea of the number of "head" reared for the gratification of the sportsmen and the luxury of the table.

Blair, in his Encyclopædia, says, avoid killing a hen pheasant, except on some very particular occasions. The principal one is the increase of the hen birds to such a degree as to outnumber the cocks: as this does happen occasionally, then to thin them assists the general stock; but we fear that this excuse is sometimes made when it ought not to be. It was a very excellent conventional understanding, that a fine of half-a-guinea should be paid to the keeper of the manor whenever a hen bird was killed. It is, however, often evaded, and, from the novice it is not exacted sometimes, when we think it ought to be. The size of the cock, the length of his tail, and his occasional call to his mates, when all of them are absent, are sufficient guides for the most inexperienced, therefore never *excuse* the fine. What says the poet Pye?

> But when the *hen*, to thy discerning view,
> Her sable pinion spreads, of duskier hue,
> The attendant keeper's prudent warning hear
> And spare the offspring of the future year:
> Else shall the *fine* which custom laid of old,
> Avenge her slaughter by the forfeit—gold!

Our limits will not allow us to gossip further, but much interesting matter may be found respecting this bird in the works already mentioned in this paper, and books on natural history by various authors.

HAMPDEN G. GLASSPOOLE.

THE "Albertian" is the title of a magazine issued at Framlingham College, Suffolk. It is better than the usual run of school magazines, although we are pleased to see that in all these praiseworthy literary productions, science is made one of the most prominent features. Such a fact is significant of the future. "The Butterflies of the District" is a very creditable paper in the "Albertian."

CHAPTERS ON FOSSIL SHARKS AND RAYS.

By ARTHUR SMITH WOODWARD.

III.

HYBODONTIDÆ.

THE Hybodonts form a large Selachian family of very great palæontological interest and importance, and appear to have come into existence in early Carboniferous times, attaining their maximum development in the Jurassic, and dying out again at the close of the Cretaceous epoch. While, however, their remains are conspicuous among the fossil faunas of these different periods, individuals and species being both abundant, the number of genera is comparatively few, and the type-genus, *Hybodus*, has an exceedingly wide range. *Ctenacanthus* (*Cladodus*) and *Tristychius* are found throughout the Carboniferous formations; *Pristicladodus* and *Carcharopsis*, probably belonging to the same family, are characteristic of the Lower Carboniferous; and species of *Hybodus* occur in most marine and estuarine strata from the Muschelkalk to the Upper Chalk.

Although long suspected[*] to be parts of the same fish, it is only recently[†] that the dorsal fin-spine known as *Ctenacanthus* has been definitely proved to be generically identical with the teeth termed *Cladodus*. A specimen from the well-known Calciferous Sandstone of Eskdale has decided the question, and affords interesting information regarding the sharks that have left so many of their spines and teeth in nearly all regions where the Carboniferous rocks are developed. The Eskdale fossil exhibits two dorsal fins, each armed with a spine in front, the first of these being slightly longer and more arched than the second; it shows, further, that the fish was covered with shagreen granules, that the notochord was persistent, and that there were no anterior spines to either pectoral or ventral fins; and, though not well seen, there are distinct indications of the dentition being of the Cladodont type. The length of the specimen is about 30 inches, and the anterior spine measures nearly 5; consequently, if other species of the same genus had approximately the same proportions, the largest individuals must have attained a length of about nine or ten feet, the spines known as *C. major*, from the Carboniferous Limestone of Armagh and Bristol, occasionally measuring 18 inches from base to summit. The spine of *Ctenacanthus* is easily recognised, being somewhat laterally compressed, having the sides of the exposed portion ornamented with longitudinal, more or less denticulated ribs and furrows, and possessing a double series

---

[*] James Thomson, "On *Ctenacanthus hybodoides*, Egerton," Trans. Geol. Soc., Glasgow, Vol. IV, pp. 59-62: Hancock & Athey, "Ann. & Mag. Nat. Hist.," Vol. IX, Ser. 4, 1872, p. 260.
[†] "On a New Fossil Shark," by Dr. Traquair, F.R.S., in "Geol. Mag.," Jan. 1884.

of recurved hooklets on the hinder border. It much resembles the spine of *Hybodus* (fig. 165), and differs chiefly in the denticulation of the longitudinal ridges. The teeth included in Agassiz' genus, *Cladodus* (fig. 162), consist of a central principal cone, with several lateral cones, and are particularly remarkable from the circumstance, that the outermost of the lateral cones are the largest.

The Carboniferous spines known as *Tristychius* have been found in association with teeth that cannot be distinguished from those of *Hybodus*,* and these detached teeth have, indeed, been previously referred to the familiar Mesozoic genus. The more our knowledge progresses, however, the more difficult does it appear to define these ancient Selachians from the characters of isolated spines and teeth, for modern researches seem to show that spines alone are of very little value in determining affinities, and that some types of dentition are common to several distinct genera.†

Carcharopsis and *Pristicladodus* (fig. 161) are two very imperfectly known genera, represented in the Lower Carboniferous strata by solitary detached teeth, and may possibly belong to Hybodonts, although some palæontologists are inclined to regard them as indicating sharks of the existing family of Carchariidæ. Agassiz, who founded the first genus, pointed out the resemblance of the tooth to that of *Carcharodon*, and suspected that *Carcharopsis* and *Carcharodon* might be closely related to each other, but, after considering all that is at present known concerning the obscure fossils, Mr. J. W. Davis (of Halifax) has lately‡ decided that they may be more correctly placed in the family now under consideration.

The general characteristics of *Hybodus* are comparatively well-known, owing chiefly to the excellent preservation of many examples in the Lower Lias of Lyme Regis; but we are still ignorant of its exact shape, and the peculiarities of the paired, anal, and caudal fins. The body is covered with shagreen, and, as in *Ctenacanthus*, there are two dorsal fins, each armed with an anterior spine (fig. 165), the first of these powerful weapons of defence being longer, more slender, and more arched than the second. The teeth vary much in form in the different species, but generally consist of a central principal cone, with one or more secondary cones on each side; sometimes (*H. minor*) they are remarkably acute, and at other times (*H. Delabechei*) blunt; sometimes (*H. grossiconus*) the cones are well marked off one from another, while at other times (*H. raricostatus*) only slight furrows separate them; and, in a few instances (*H. medius*, &c.), none but a single cone with gently sloping sides can usually be recognised. The first complete mouth of a *Hybodus* was described by Sir Philip Egerton, in 1845,* when the nearly perfect skull of a new species (*H. basanus*) was discovered in the Wealden strata of Pevensey Bay, Sussex. The specimen proved that the teeth of the upper jaw scarcely differed at all from those of the lower, that the dentition varied much less in different parts of the mouth than is the case in recent sharks, and that, therefore, detached teeth almost always admitted of specific determination. These conclusions have been confirmed to a great extent by more recent discoveries in other formations, and there are very few species in which all the teeth are not readily recognisable as variations of a single type-form; the small *H. Dubrisiensis*, of the Chalk, appears to have the most complex dentition of any of the British Hybodonts, the front teeth being markedly acute and prehensile, while the hinder ones are much flattened and elongated.

One of the most singular features of *Hybodus*, consists in its possession of curious curved spines on the upper part of the head.† These cephalic spines (fig. 164) were only known to Agassiz in an isolated condition, and were originally described by that eminent ichthyologist as teeth, under the name of *Sphenonchus*. They have been found in association with remains of most of the English species, and it seems probable that they were common to all: it is uncertain at present how their number and arrangement varied in each, but *H. Delabechei*,‡ from the Lower Lias of Lyme Regis had four of these dermal appendages,—a pair on each side of the head, a little above the orbit,—and there are indications of some others possessing the same number. They are generally regarded as being of the same nature as the dermal spinous tubercles so characteristic of some of the Rays.

As already stated, the earliest remains of *Hybodus* occur in the Continental Muschelkalk, and the first appearance of the genus in Britain is in the Keuper Series. Teeth and spines (dorsal and cephalic) of *H. Keuperinus* have been found both in Warwickshire and Somersetshire. The English Rhætics yield *H. minor*, *H. raricostatus*, and some others, and a formation of similar age in Moray, Scotland, contains spines and teeth of *H. Lawsoni*. The Lias, particularly the lower division of Lyme Regis, is exceedingly rich in remains of *Hybodus*, and the beautiful specimens obtained from it have afforded more information concerning the genus than examples from any other stratum. The principal species are as follows:—*H. reticulatus* (of which the posterior dorsal spine was originally described as *H. curtus*) with teeth (fig. 166) consisting of sharply-pointed cones, well marked off from each other, and the principal cone much the largest; *H. Delabechei*, with teeth (fig. 167) somewhat

---

* T. Stock, "Ann. & Mag. Nat. Hist.," Sept. 1883.
† T. Stock, "Nature," Vol. 27 (1882) p. 22.
‡ "Trans. Roy. Dublin Soc.," Ser. 2, Vol. I. (1883), pp. 381–383.

* "Quart. Journ. Geol. Soc.," 1845, pp. 197–199, Pl. IV.
† First pointed out by E. Charlesworth, "Mag. Nat. Hist.," 1839, p. 245, Pl. IV.
‡ E. C. H. Day, "Geol. Mag.," 1865, p. 565.

resembling the latter, but readily recognised by the bluntness of the cones ; *H. raricostatus*, having teeth much elongated,—cones not often well defined, but pointed, and the central cone usually not very prominent ; and *H. medius*, of which the tooth (fig. 168) consists of a single cone, with rather indistinct indications of secondary or lateral cones. The Oolites have afforded several species, and also the Purbeck and Wealden ; *H. grossiconus* (fig. 163) and *H. polyprion* range from the Stonesfield Slate to the Wealden ; spines described as *H. dorsalis* characterise the Purbeck and Wealden beds ; and numerous very perfect skulls of *H. basanus* have been found in the latter strata on the shore of Pevensey Bay. *Hybodus* spines are occasionally met with in the Cambridge Greensand ; the remains of *H. Dubrisiensis* are somewhat rare fossils of the Chalk ;* and from this formation, also, Agassiz has described fragments of a spine under the name of *H. sulcatus*.

ORODONTIDÆ.

The characteristics of the members of this extinct family are very imperfectly known ; nothing beyond their dentition and dermal tubercles has hitherto been discovered, and there is no definite evidence at present as to the external form, or as to whether they possessed dorsal fin-spines or not. It may be questioned, in fact, whether the basis of the "family" itself is secure, and whether De Koninck is not nearer the truth in considering the group as a subdivision of the Cestraciontidæ.

Orodus is regarded as the typical genus, and appears to be exclusively confined to the Lower Carboniferous formations, both in the Old and New World, although *Agassizodus*, of the American Coal Measures, is closely allied and probably its representative in that higher horizon. The best known British species is *O. ramosus* (fig. 169), from the Carboniferous Limestone. This tooth is elongated in shape, sometimes as much as four inches in length, and has the crown slightly raised into a blunt cone with sides gradually sloping to the extremities ; it is also characterised by a more or less prominent longitudinal ridge, and its surface is embellished with a number of ramifying lateral ridges, which are so arranged as to constitute a most elaborate ornamentation. The larger specimens from Oreton (Salop) and Bristol are, indeed, among the most beautiful of fossils, but, in a palæontologist's estimation, the remains are all of a very unsatisfactory nature. Never more than three or four teeth in consecutive series have yet been discovered, and the same remark applies to the other genera from British strata referred to the same family. It is interesting to note, however, that a nearly perfect jaw of *Agassizodus* has been described* from the American Coal Measures and upon it alone we must depend at present for our knowledge of the arrangement of the teeth in the mouth of Orodontidæ. This remarkable specimen exhibits nearly 500 teeth in their natural order, and indicates that their disposition was very similar to that observed in the living *Cestracion*, but that the rami of the jaw formed a much more obtuse angle at the symphysis than is the case in the latter genus. The majority of the teeth themselves are distinctly Hybodont in shape, like our *Orodus*, while those occupying the symphysial area are very similar to (or possibly identical with) the little conical bodies known under the name of *Petrodus* (fig. 172) when found detached.

Some of the species of Agassiz' genus, *Helodus*, are also included in this family, and the well-known names of *Helodus mammillaris*, *H. didymus*, *H. lævissimus*, &c., must henceforth be replaced by *Lophodus mammillaris*, *L. didymus*, &c., if the most recent ideas on the subject are to be accepted as correct. It has been shown† that *Helodus* originally comprised two very different types of teeth,—one distinctly conical, with a concavity of the base of the crown corresponding to the convexity of its surface, the other with a nearly even coronal base-line, whatever the contour of the tooth. Those of the former type are transferred to the Orodont *Lophodus* ; and those of the latter are provisionally retained under the old generic name, although it is not unlikely that future researches will prove them to be symphysial teeth of *Cochliodus*, *Psephodus*, and other shark of the Cochliodont family.‡

COCHLIODONTIDÆ.

Perhaps the most typical Selachians of the Carboniferous period are those referred to Owen's family of Cochliodontidæ.§ Their remains seem to be exclusively confined to Carboniferous strata, and are most abundant in the lowest divisions, although a few—such as *Pleurodus*—occur also in the Coal Measures, and attain their greatest development in that formation. The group comprises a large number of forms, no less than twelve genera having been described from British rocks alone, and hence it will only be possible to notice here very few of them selecting those that are most prominent and characteristic.

Cochliodus gives its name to the family, and is at present known solely by the dentition. There is no evidence yet forthcoming by which any of the

---

* See "The Geologist," Vol. VI, p. 241.

* "Geol. Surv. of Illinois (Palæontology)," Vol. VI, pp. 311-318.
† H. Romanowsky, "Bull. d. l. Soc. Imp. des Nat. de Moscou," 1864, p. 160 ; L. G. de Koninck, "Fauna du Calc. Carb. d. l. Belgique," Pt. 1, p. 42 ; J. W. Davis, "Trans. Roy. Dublin Soc.," 1883, p. 403
‡ "Geol. Surv. of Illinois (Palæontology)," Vol. II, p. 88.
§ "Geol. Mag.," 1867, pp. 59-63, Pls. III, IV.

numerous ichthyodorulites of the Carboniferous Limestone may be safely associated with the teeth, and very perfect jaws from American strata show that even our admirable specimens from the Bristol quarries are exceedingly incomplete examples. Everyone acquainted with geological text-books is familiar with the figure of the mandible of *Cochliodus contortus* (fig. 173), which exhibits three inrolled teeth on each ramus, and it is not difficult to perceive how these few *single* teeth represent the successive transverse *series* of teeth in the living *Cestracion*,—the type of the Cestraciontidæ. As in the latter family, the new dental substance is added at the inner side (Y), and growth thus proceeds from the inside, outwards ; but while, in *Cestracion*, the worn-out teeth of the outer border fall away as soon as they are no longer required, such cannot take place in *Cochliodus* without fracture, and to avoid this the border becomes inrolled,

possess median or symphysial teeth that were once regarded as species of *Helodus*. *Psephodus* is one of the most interesting, and, although no specimens exhibiting the actual arrangement of the teeth has hitherto been described; a study of the large series in the collection of the Earl of Enniskillen has led Mr. J. W. Davis to suggest* that each side of the jaw was provided with a row of three The teeth themselves are much (fig. 175) flatter than those of *Cochliodus* and very variable in shape ; some exhibit a slightly concave surface, while others, equally worn, are distinctly convex, and this appears to show that the dentition of the upper jaw was very similar to that of the lower, the only difference being in the convexity or concavity of the crushing surface. *Helodus planus*, Agass., is definitely proved to belong to *Psephodus magnus*, but it is uncertain at present what part of the mouth the teeth of this form occupied.

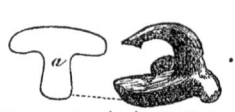

Fig. 161.—*Pristicladodus dentatus*.

Fig. 162.—*Cladodus mirabilis*.

Fig. 163.—*Hybodus grossiconus*.

Fig. 164.—Cephalic spine of Hybodus. *a*, outline of base.

Fig. 165.—Hybodus (dorsal fin-spine). One-third natural size.

Fig. 166.—*Hybodus reticulatus*.

Fig. 167.—*Hybodus Delabechei*.

Fig. 168.—*Hybodus medius*.

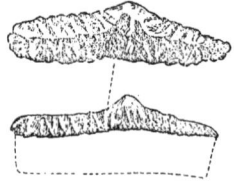

Fig. 169.—*Orodus ramosus*.

forming quite a small spiral in the older individuals, as shown in the diagrammatic section, (fig. 173 *a*). These are two of the most remarkable peculiarities of the genus now under consideration, and there is also one other to which Sir Richard Owen attached great importance when he established the Cochliodont family, namely, that the symphysial area (x) appears totally destitute of teeth, whereas the corresponding part of *Cestracion* is armed with dentition of a prehensile type. This conclusion, however, has been invalidated by more recent researches in America, and some beautiful specimens from the Carboniferous Limestone of the United States, leave no doubt that the symphysis was occupied with teeth of the kind commonly known as *Helodus* when they happen to be found detached.

The other genera of Cochliodontidæ differ from the Cestracionts in the same way as *Cochliodus*,—in the large transverse teeth, and in the inrollment by growth,—and some have been proved, likewise, to

Another characteristic genus is *Deltodus*, in which the upper teeth are known to differ very much from the lower. The imperfect specimens hitherto met with, seem to show that the upper jaw was only armed with two dental plates, of the form represented in fig. 176, while the lower possessed at least two, and probably three, in each ramus. The latter (fig. 174) are not so "deltoid" as the former, being much rounder and convex, and exhibit very distinctly the inrollment of the outer border. *Sandalodus* is a closely allied genus, represented in Britain by the large species *S. Morrisii* (Davis), from the Carboniferous limestone of Oreton and Bristol, and particularly remarkable because the lower jaw appears to resemble the upper in not being provided with more than two of the inrolled teeth.

Pleurodus is interesting in consequence of the structural details revealed by a specimen discovered

---

* "Trans. Roy. Dub. Soc." Ser. 2, Vol. I, pp. 416, 417.

in the Northumberland Coal Measures.* Usually the teeth (fig. 170) are only met with in an isolated state, but the very complete fossil just referred to is said to show that there were three of these on each side of the jaw, that the body of the fish was covered with shagreen, and that at least one dorsal spine was present. The spine is described as being short, stout, and marked with prominent longitudinal ridges.

Other genera of Cochliodonts are *Streblodus*, *Deltoptychius*, *Pœcilodus*, *Tomodus*, and *Xystrodus*, and the species of *Helodus* (restricted) are also associated with these in the most recent monographs on the subject.

Lastly, considering the Cochliodontidæ as a whole, it is evident that they form a family quite distinct from any inhabiting recent seas, and there can be little doubt that they are sharks somewhat approach-

appears to be exclusively confined to the Lower Carboniferous. The teeth are more or less flat, with the root deeper than the crown, and the surface either smooth and punctated, or covered with fine rugose markings. In form, they vary considerably, being occasionally much elongated (fig. 177 *b*), and sometimes triangular (ib., *c*) ; but generally they are irregular parallelograms (ib., *a*), with the long axis transverse, the hinder border somewhat concave, the anterior slightly convex, one lateral margin at right-angles, and the other obliquely inclined to the anterior border. Though the diagram is rather hypothetical in some respects, we may assume that they were arranged in the mouth in the manner shown in fig. 177, as suggested by Professor De Koninck* and Mr. Davis,† and many palæontologists are now of opinion that the teeth with a smooth

Fig. 170.—*Pleurodus affinis*.

Fig. 171.—*Lophodus mammillaris*.

Fig. 172.—*Petrodus patelliformis*.

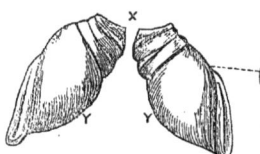

Fig. 173.—*Cochliodus contortus*. *a*, transverse section of tooth.

Fig. 174.—*Deltodus sublævis* (lower tooth).

Fig. 175.—*Psephodus magnus*.

Fig. 178.—*Copodus cornutus*.

Fig. 179.—*Pleurogomphus auriculatus*.

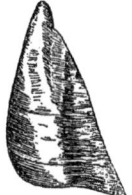

Fig. 176.—*Deltodus sublævis* (upper tooth).

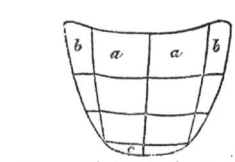

Fig. 177.—Diagram showing the probable arrangement of teeth of Psammodus. One-sixth nat. size. (After J. W. Davis.)

ing the living *Cestracion*. It is noteworthy, however, that the dentition bears some slight resemblances to that of the Dipnoan *Ceratodus*, as was first noted by Agassiz, and as has recently been discussed by Mr. Davis,† and palæontologists will anxiously look forward to the discovery of more satisfactory specimens to ascertain how far these surmises are correct and how far *Cochliodus* and its allies may be deemed "missing links."

PSAMMODONTIDÆ.

This group, or family, is only represented in British rocks by a single genus, *Psammodus*, which

surface belong to the same species as those with the rugose markings,—retaining only the name *P. rugosus*, and regarding *P. porosus* as a synonym.

COPODONTIDÆ.

The Copodonts are a most peculiar and problematical group of flat, crushing teeth, and sufficient is not yet known about them to decide whether or not they constitute a natural family. Their relationships are even more obscure than those of the Orodontidæ and Psammodontidæ, but they have been provisionally placed in a distinct subdivision of the Plagiostomi by the author of the elaborate monograph,‡ already often

---

* Quoted by John Ward, "Essays North Staffs. Nat. Hist. Field Club," 1875, p. 223. See also Hancock & Atthey, "Ann. & Mag. Nat. Hist." Ser. 4, Vol. IX., 1872, pp. 249–252, and J. W. Davis, "Quart. Journ. Geol. Soc.," 1879, pp. 181–183.
† "Trans. Roy. Dub. Soc.," Ser. 2, Vol. I, pp. 417–420.

* "Fauna du Calc. Carb. d. l. Belgique," p. 41.
† "Trans. Roy. Dub. Soc.," 1883, p. 462.
‡ "On the Fossil Fishes of the Carb. Limst. Series of Gt. Britain," by James W. Davis, F.G.S., in "Trans. Roy. Dub. Soc.," 1883.

referred to, which must form the basis of all future work upon the Selachian fossils of the Carboniferous strata. Most of the genera and species are founded upon single detached teeth, which have been met with in considerable numbers in the Carboniferous Limestone of Armagh, and are occasionally found in the Lower Carboniferous of England. Figs. 178 and 179 are sketches of the forms known respectively as *Copodus cornutus* and *Pleurogomphus auriculatus*, and these will give an idea of the most prominent characteristics of the group.

PLANT-NOTES.

By the REV. H. FRIEND, AUTHOR OF "FLOWERS AND FLOWER-LORE."

THE interesting jottings on this subject by Mr. Robinson suggest a few reflections, and demand some comment. It is interesting to hear of the courage-cup still made in Cheshire, but is it true that the word borage " is merely a corruption of the word. corage or courage (cor and ago)"? If so, where can the form be found? The word occurs in the Romance languages, thus : French, bourrache ; old French, borrace (see Brachet); Italian, borraggine; Spanish, borraja ; Portuguese, borragem ; Low Latin, borrago. Some suggest that the root of the word is Oriental, but seeing that the family bears the general name of bugloss (see Britten's and Holland's Dict. of Plant Names), from the leaf being rough like an ox-tongue, we may, I think, fairly accept the most general derivation of the word from borra, rough hair, whence several words denoting roughness (see Skeat and Diez, also Pritzel and Jessen, "Volknamen der Pflanzen," i. 60).

The elder is largely employed for hedges in Bucks and Northants, and I have already referred to the idea that the plant may have been employed for a protection against witchcraft (see "Flowers and Flower-lore," p. 543, and Index s. v. Elder). If Mr. Robinson will consult the valuable Dictionary of Plant Names now being issued by the English Dialect Society, I think he will find the synonymous expressions used in various parts of the country for the elder support the idea that it was called bourtree or bore-tree, because of its bore, or hollow stem.

The Clote of Drayton can scarcely be any other than the yellow water-lily. The burdock is the only important rival. Earle gives ("English Plant Names," page 46) "Lappa, bardane, clote," from a trilingual vocabulary of plants written in the thirteenth century. But this does not bear a yellow flower (see Dr. Printzel, i. 201-2). Moreover the *Nuphar lutea* is still called Clot or Clote in the south-west of England. Marshall tells us that in East Norfolk it is applied to the coltsfoot (*Tussilago farfara*). In any case there appears to be reference to the large globular flower-head, from clot, "a ball," hence particularly inapplicable to Galium or Galeobdolon, the former of which could scarcely claim attention, except on the ground that some varieties, especially Aparine or Cleavers, produced round seeds. Moreover, the words clate, cleat, clite, &c., which are employed in many parts for cleavers, coltsfoot, and other plants have quite distinct histories and etymologies, and refer either to the shape of the leaf, the habit of the seed, or some other similar peculiarity, so that the water-lily is in undisputed possession of the field.

Mr. Robinson does not seem to be aware that the lines he quotes from Hogg have been more than once discussed, and that Dr. Johnston has pretty clearly proved that the line "When the blewart bears a pearl," has reference to the beautiful little speedwell (*Veronica Chamædrys*), the "pale, glaucous under-side" of the corolla being remarkably like a pearl when the blossom is closed. It is, however, true that the corn bluebottle is known in the North as blewart or blawart. Th hairbell question was fully discussed in SCIENCE-GOSSIP for 1881. The cockle of St. Bernard was no doubt the *Lychnis Githago*, still known under that name in the Midlands. One of the German names of the plant is kuckel.

A great deal of plant-lore yet remains uncollected. I have found some interesting names and customs in the Midlands quite recently. Thus, at sheep-shearing, it used to be the fashion here to decorate the shearers with posies, and the small white rose so common in cottage gardens being in bloom at this season, it became a favourite flower for the purpose, and bore the name of sheep-shearing rose. This name is still in use in certain parts of Bucks.

Brackley.

GOSSIP ON CURRENT TOPICS.

By W. MATTIEU WILLIAMS, F.R.A.S., F.C.S.

THE researches of M. Miquel, recorded in the "Semaine Médicale," on the distribution of bacteria in the atmosphere, are very interesting, whether regarded from a physical or physiological point of view. The standard quantity of air examined was ten cubic metres. At heights of 2000 to 4000 metres on the Alps (6561 to 13,123 feet), none were found. Thus bacteria cannot cross the Simplon or Mount Cenis, the heights of these passes being respectively 6578 and 6773 feet, therefore Dr. Koch's cholera germs cannot pass directly from Italy to Switzerland, unless it travels through the tunnels.

On the Lake of Thun, 560 metres above sea level, eight were found ; at 500 metres, near the Hotel Bellerne, twenty-one. In a room in the hotel, 600 ; in the Parc de Montsouris, 7600 ; and in the Rue de Rivoli, Paris, 55,000.

Why is this? It is not the low temperature, as

M. Miquel found 750,000 living bacteria in a block of ice from Lac de Joux, which had been kept for eleven months, and that atmospheric microbia resisted thirty-six hours' exposure to a temperature of −100° C. and revived in three days. The difference at the different elevations must be due to the rarity of the atmosphere. It is evident that these microscopic creatures fall through thin air as the feather falls through that in the air pump receiver in the well-known guinea and feather experiment. Those who still cling to the desperate theory of the suspension of the dust of Krakatoa should ponder on this.

The editor of the "Journal of Science" contends that the failure of the attempts to promote scientific education in China is mainly attributable to the system of literary examination there prevailing.

Official employment and social rank are made dependent on these examinations, and such rank being the sole and universal ambition of the Chinese people, Chinamen should be superlatively intellectual if examinations have any educational value. Chinese examinations are no sham, but are so severe that "only a minority of the candidates pass, the plucked ones coming up year after year, even to old age." Mere memory of the lowest order, *i.e.* memory of words, is so extravagantly cultivated that the reasoning powers of the men are as effectually crushed by this wretched system of cerebral strangulation, as the locomotive powers of the women are suppressed by the crippling of their feet.

Among the books presented during the current year to the Royal Astronomical Society is a translation of Sir John Herschel's "Outlines of Astronomy" into Chinese, by A. Wylie. The fact that this is the second edition published at Shang-hai, appears to contradict the above, but the probable explanation is that its circulation is limited to the outlying regions of China, where British commerce and British influence is beginning to break up the ancient conservation of this typically conservative people.

A paper by Professor Sachs, published in the early part of the present year, in the "Arbeiten des botanischen Instituts," vol. iii. of Würtzburg, describes some researches that must be interesting to most of the readers of SCIENCE-GOSSIP. If fresh green leaves are immersed in boiling water for about ten minutes, and afterwards in alcohol, their chlorophyll is extracted without rupture of the cells, and the leaves become etiolated. In some instances more or less of starch remains. This is readily detected by immersing the blanched leaf in a solution of iodine in alcohol. If there is much starch, the cellular tissue of the leaf becomes blue-black, the venation being displayed as a pale network on a dark ground. With less starch, the colour is paler; with none, only a yellow stain of iodine is shown. Any amateur may thus repeat the interesting experiments of Sachs, which show that the quantity of starch contained in leaves varies with their exposure to light. If a part of a leaf is covered with tinfoil, and the rest is exposed to sunlight, the darkened patch is displayed as a light patch, when the leaf has been treated as above.

The differences between the condition of the leaves in the day and in the night is very strikingly shown, and variations during the day may be thus demonstrated, a very short time effecting the formation or disappearance of the starch. Leaves full of starch in the evening may be found quite empty in the morning. This is best shown by halving a given leaf in the morning, and another in the evening, removing the one half, and comparing the halves on the following evening and morning. The half leaf separated before sunrise and tested shows no starch, the remaining half tested in the evening displays much starch. The difference is explained by the conversion of the starch into soluble glucose which is carried away to nourish the plant. This occurs both during day and night, but in the day more starch is formed than is carried away. During the night no starch is formed. Thus it happens that the composition of a given leaf varies according to the hour of the day on which it is plucked, and this may account for the discrepancies observable in different analyses of the leaves of the same plant.

The further quantitative researches of Sachs have led him to the conclusion that the formation of as much as 20 to 25 grammes of starch is commonly formed daily, per square metre of leaf surface (1 oz. avoirdupois = 28¼ grammes). The quantity varies with different plants.

In the Proceedings of the Royal Society of Tasmania for 1883, is a paper by E. T. Higgins, and W. F. Petterd on a new cave spider (*Theridion troglodites*), the female of which measures six and a half inches from the claw of the anterior, to the claw of the posterior leg. It is found in a cave in the Chudleigh district. We are told that no insects have been discovered in this cave, and that mammalian remains are agglutinated in the rock by stalactitic incrustations. My own experience leads me to ask whether the explorers looked for fleas as the possible food of these spiders. I have found fleas in limestone caverns, or, rather, they have found me, where no other supplies of food existed, excepting the animal matter that may have remained in the fossils of which the limestone was chiefly composed. The most impressive instance of this kind was in a very extensive cavern, near Syracuse, that had long been closed, but which, for want of better occupation, I attempted to explore, in company with a guide and fellow tourist. We were driven out of the cavern by an overwhelming force of these creatures; compelled to strip in the open fields, shake each garment separately, then run twenty or thirty yards away, and deposit it at this distance from the shaking place and when all had been thus removed go there to dress. Our guide "Jack Robinson," an old Anglo-Syracusan sailor, estimated the quantity we brought to the surface as "three hats fulls." My own estimate was very much smaller, but I never before or since witnessed any approach to such a display of that form of animal life. When we emerged we appeared to be wearing brown gaiters, the enemy was marching from our boots upwards. The limestone there is simply a conglomeration of fossils. There was a legitimate natural demand for very large cave spiders. It would be a curious experiment to take a brood of the Tasmanian cave spiders, and deposit them in the Syracusan flea cavern.

There is a large harmless lizard in Tasmania (*Cyclodus nigroluteus*), which ignorant prejudice is treating as our harmless lizard, the "blind worm," is treated here. The Tasmanian victim is named the "death adder," and hunted down as a poisonous reptile, while, according to Mr. Dyer, it should be jealously protected, as it kills young poisonous serpents. Our blind worm feeds voraciously on garden slugs. I have supplied those I have tamed, and still keep as domestic pets, with three or four slugs daily, and all have been swallowed. A walled garden might be cleared of slugs, by the aid of these pretty little creatures. After swallowing the last slug, they would still have earthworms to fall back upon, besides the larvæ of beetles, and other underground grubs, which I find they greatly enjoy as a change of diet.

The report on "The Mineral Resources of the United States," published at Washington, by the United States Geological Survey, is very interesting, not only to United States citizens, but also to ourselves. Mere politicians may be alarmed at this display of the magnitude of the mineral resources in the hands of those of our own race, who are regarded by such politicians as our rivals.

Those who look a little deeper and from a truly scientific point of view, will see therein so much prospective increase of the world's wealth, of which, in the natural course of events, we shall, next to the Americans themselves, obtain the largest share. There is no mistake about the fact that they have vastly more coal, more iron, more copper, more lead, more of all the raw materials of the earth than we have; that we shall cease to send them coal, pig iron, iron rails, copper ingots, &c., or otherwise we shall cease to be their coal-diggers, their iron miners and smelters, &c.; but shall we cease to be Englishmen, because we cease to do such dirty work? I think not. We must look the facts in the face, and prepare for the inevitable change by elevating the character, and enriching the intellect of our artizans; this must begin with education in science, whereby they shall be fitted to do higher and better paid work than coal-digging, furnace feeding, puddling, casting, forging, and thus, instead of sending out crude iron at 40s. or 50s. per ton, we shall work it up with elaborate machinery, into other useful and artistic products of which every ton weight shall be raised by skilled labour to the value of a hundred or a thousand tons of pig-iron or mere rails. Our "black country" will be but whity-brown compared to that of Lake Superior, with its inexhaustible deposits of rich hæmatite, suitable for Bessemer pigs, and its wonderful deposits of native copper. As an example of growth, Colorado only produced 56 tons of lead in 1873; now the production of the state has risen to 58,642 tons, chiefly from the carbonate deposits of Leadville in the Rocky Mountains, where the lead is regarded as a secondary product, and named "base bullion," the ore being smelted for the gold and silver it contains. The volume above named should be studied, not only by mineralogists, but by manufacturers, especially by those who have capital invested in blast furnaces; and by all concerned in the education of the productive classes.

SCIENCE-GOSSIP.

SMALL-POX was accurately described by Rhazes, an Arabian physician, about the year 900 A.D. The disease is supposed to have been introduced into Europe by the Saracens.

DR. WILSON, of Louisville, Kentucky, states he has made some of his most successful skin-grafts from the inner membrane of a perfectly fresh hen's egg. He prefers it, for this purpose, to human skin.

A HUNGARIAN scientist has beaten M. Reinsch. The latter only professed to find bacteria, &c., in dirty old money; the former states he finds them in bank-notes. Both money and bank-notes will have no attraction after this!

A SCOTTISH Geographical Society has been formed, with Lord Rosebery as its first President.

THE opening meeting of the session for 1884-5, of the Geologists' Association took place at University College on November 7th, when the president, Dr. Hicks, gave an address on "Bone caves."

ON Friday evening, October 17th, the Society of Amateur Geologists held their first meeting at 31, King William Street, E.C., when Professor Boulger, one of the vice-presidents of the Society, delivered the inaugural address, taking for his subject the "History of Geology." The society, which is receiving the support of many eminent geologists, has secured excellent rooms at 31 King William Street, E.C., where its meetings will be held on the third Friday in each month.

AN International Electrical Exhibition has lately been held at Turin. A notable feature was a Siemens' Dynamo-electric machine of thirty-horse power, which generated alternative currents, simultaneously utilised by the Swan, Siemens, and Bernstein systems, distributed over a circuit of twenty-four miles.

PASTEUR declares sulphuret of carbon to be the most effectual, as well as the cheapest, antidote against Phylloxera.

WE are very sorry to hear Prof. Huxley's health is so feeble that he has been ordered not only to leave off all work, but also to go abroad for change and rest.

A VERY thoughtful and suggestive Essay on "Domestic Water Supply, in its Relation to Sanitary Reform," has been published by Mr. J. Tertius Wood, Civil Engineer (Rochdale: James Clegg).

THE King's College Science Society (President, Mr. G. G. Hodgson, Hon. Sec. Mr. J. H. Leonard) have commenced their session for 1884-5, and their programme includes a fortnightly series of papers to be read, extending over a large area of scientific investigation.

DR. RICHARDSON, in the last number of his "Asclepiad," says that a simple disinfecting lamp may be easily made by burning bisulphide of carbon in a lamp, after the manner of an oil or a spirit lamp.

THE fourth part of Dr. Richardson's "Asclepiad" has appeared. It is brighter and more readable than ever. The article entitled "A Great Medical Reformer, Thomas Wakley, M.P.," is particularly pleasant reading.

MR. GARDINER has been investigating the function of tannin in vegetable cells. He finds it very abundant in the motor tissue of the sensitive plant, and in similar tissues, and he thinks there may be a possible connection between the presence of tannin and the irritability of the tissues.

A SUGGESTIVE paragraph, headed "Is the full moon red hot?" forms one of the chief of the "Science Notes" contributed by Mr. W. Mattieu Williams in the last number of "The Gentleman's Magazine."

MESSRS. SINEL & CO., have, amongst other numerous issues, just sent forth a slide of unusual zoological interest—the Zoea of *Galathea*. It is most remarkable for the long spine borne in front.

THE only gold medal for maps gained by any British Exhibitor at the Healtheries, was awarded to Mr. Edward Stanford, 55 Charing Cross.

THE Birmingham Natural History Society held a successful conversazione in the Town Hall on Nov. 4th.

THE Ipswich Scientific Society had a very instructive and entertaining conversazione on Nov. 27th.

IT was ascertained from observation at the Observatory of Bordeaux, during the recent total eclipse of the moon, that none of the stars disappeared at the exact moment of its occultation, almost implying that the edge of the lunar disk is transparent.

IT is gratifying to the many friends of Mr. C. P. Hobkirk, F.L.S., author of "A Synopsis of British Mosses," to hear he has been presented with a splendid testimonial, consisting of a silver tea and coffee service, and an illuminated address, on the occasion of his removing from Huddersfield to Dewsbury. The testimonial was presented "in recognition of his eminent service in promoting scientific work and education for a period of nearly thirty years," &c.

PROFESSOR HERDMANN, of Liverpool, writes to say he has found the feather-star (*Antedon rosaceus*), in the pentacrinoid stage in Lamlash Bay up to the end of September last.

DURING November, Dr. J. E. Taylor, F.G.S., &c., Editor of SCIENCE-GOSSIP, lectured before the Hemel Hempstead Natural History Society, on "Mountains and Valleys: How they were formed" (illustrated with the lantern); at Bowness Literary and Scientific Society, on "Dust;" at Ambleside Literary and Scientific Society, on "Flowers and Fruit, and their Relation to Insects and Birds;" and at the Chelmsford Museum, on "The Natural History of the Oyster."

MICROSCOPY.

FREEZING MICROTOME.—Requiring to cut some fine sections of soft tissues (animal), and having nothing but the old-fashioned section cutters, I shall be obliged by some information as to the construction of a freezing microtome, as being the most likely machine to serve my purpose. I have seen very fine sections (as regards the cutting) made with a home-made freezing microtome, and shall be glad to receive particulars of the construction of such an instrument, and the methods of using it.—*E. Lamplough, Hull.*

ON MOUNTING SECTIONS STAINED WITH PICRO-CARMINE.—On removing the section from the staining solution, do not wash it; but absorb the superfluous piero-carmine with blotting-paper, and then mount in glycerine containing 1 per cent. of formic acid. It is not necessary to remove all the piero-carmine from the section; in fact it is advisable to leave a little adhering to the section, for, within a few days after mounting, the trace of dye left will be absorbed by the section. This method I have always found most effective; and the preparations improve by keeping, for after several days nuclei become stained, that at first were unaffected.—*Dunley Owen, B.Sc.*

"STUDIES IN MICROSCOPICAL SCIENCE," &c.—Many of our readers will be glad to hear that Mr. Cole will issue the monthly parts of the third vol. of these important "Studies" very shortly. The Botanical Histology will be illustrated by twelve slides, and descriptive letter-press will be contributed by Mr. D. Houston, F.L.S.; Animal Histology will be similarly illustrated; the essay will be written by Mr. F. Greening; Pathological Histology by Mr. W. Fearnley; and the Popular Microscopical Studies will be written by the Editor, Mr. Cole.

MICROSCOPIC SLIDE CENTERING.—We now and again see in one or other of the many journals devoted to microscopy ingenious devices for centering objects on the slide, but the simplest way is, I think, a method I have adopted for many years, and which requires no other apparatus than the turn-table. I turn one or more rings in ink with a fine steel pen on the back of the slide, and by this means have a test for the accuracy of position of both object and cover-glass so long as required, and which is readily seen over any colour; the slide may be laid upon for mounting. I keep a number of slides so marked ready for use; should the ink from age not be readily washed off, a little liquor ammoniæ will dislodge it immediately. I once, under pressure for time, mounted an object on the inked side of the slide, the rings came out gray, and the effect was not objectionable. In mounting small objects, such rings would facilitate the finding of them, for which purpose, the rings might be turned in coloured ink according to the taste of the mounter.—*W. D. Saunders.*

STAINING SECTIONS.—After having stained the sections as required, place them in water acidulated with a few drops of acetic or picric acid, and leave for an hour. The former is, I think, the best of the two acids. When making experiments in treble staining a number of sections, stain in P. C., and then place in methylated spirit; there they may remain until required, as the spirit does not affect the stain, which forms a very good ground colour on which to try combinations of different anilines. The second method is as follows:—lay the section out flat on the glass slip, drain off the superfluous water, and run several drops of the staining fluid (not diluted) over it; allow to stand for from three to five minutes exposed to sunlight, covered with a watch glass to keep off the dust. (In winter it is well to warm gently over a spirit lamp the slide on which the section is being stained, as slight heat causes the tissues to stain both more rapidly and more brilliantly.) Do not wash the section, but simply run off the superfluous fluid by tilting the slide and then wiping round the section with the thumb, or a very soft clean cloth; but be careful not to remove the whole of the staining fluid, as any slight excess is gradually taken up by the tissues after the section has been mounted in either Farrant's sol. glycerine to which from 1 to 5 per cent. of formic acid has been added. (C. balsam or dammar may be used, though some say that it spoils the sections.) The full effects of the stain are not seen at first, but after the section has been mounted for two or three days, especially if a small quantity of the stain fluid has been left on the section, a beautiful selective double staining is found. There are many other methods, which are but slight modifications of the two methods given.—*V. A. Latham.*

NAVICULA CUSPIDATA AS A TEST-OBJECT.—Permit me to call the attention of those of your microscopical readers who are interested in the Diatomaceæ to *Navicula cuspidata*, as a beautiful and delicate test-object. In Smith's "Synopsis of the British Diatomaceæ," and also in Mr. Ralfs' article on the same subject in Pritchard's "Infusoria," it is represented as having only transverse striæ, and although these transverse striæ can be shown by a good ¼ inch object-glass, a good ⅛th power fails to show any longitudinal striæ. But with a good ¹⁄₁₀th and careful illumination, both sets of striæ can be beautifully seen, and then what a charming object it is! It is like the most delicate gauze that it is possible to conceive; even fine specimens of *Nav. rhomboides* appear coarse beside it. My first view of these extremely delicate lines was got from a slide mounted in balsam. I caught a glimpse of them, but all my subsequent efforts during that evening failed to procure me another view of them, although I succeeded the next night in seeing them well, and also in shewing them to an enthusiastic microscopical friend who had come about twelve miles to visit me, and who declared he would willingly have walked all the distance to have seen the sight. The double set of striæ is much easier shown when the frustules are mounted dry, or in media less transparent than balsam, such, for instance, as "Styrax," and will well repay the care and trouble that is necessary to exhibit it, besides the advantage of having another good test for the higher powers of the microscope.—*Joseph Davison.*

STAINING VEGETABLE TISSUES IN PICRO CARMINE —Place the sections in alcohol for one hour; immerse them in the (recently filtered) staining solution for from half an hour to three hours, i.e. until they are sufficiently stained. Wash them in alcohol, immerse them in an alcoholic solution of picrate of ammonia for one hour, and for a second hour in a like solution; in other words, change the solution once during the two hours. Place them in alcohol for a few minutes, and then place in oil of cloves for a short time and mount.—*William F. Pratt.*

THE DIATOMESCOPE.—We are sorry that another and a quite different instrument was last month noticed under this name. The Diatomescope is constructed by Mr. E. Hinton, and our readers will find a full account of it in the "English Mechanic" for August 22nd. It is a beautifully finished and ingenious adjunct to the microscope, being a kind of supplementary stage, which can be instantly placed on the ordinary stage. In the centre is inserted a powerful lens, so that the reflected light from the mirror beneath passes through it, and produces an exquisite side illumination of the objects viewed. In this manner the dots or striæ of diatom, &c. are brought out with remarkable distinctness. Mr. Hinton has conferred a favour on microscopists by bringing out this cheap and effective auxiliary.

ZOOLOGY.

THE MOLLUSCA OF DERRY.—During the last six months I have been busily hunting for shells in the district immediately round the mouth of the river Baun, co. Derry, and I think some of your correspondents might be interested to know that there is an excellent field for conchological exploration in that locality. Of course fresh-water shells were in the mud during most of the time, and the only capture worth mentioning was *L. palustris*, var. *tincta*, which swarms on the river banks. With land shells I was much more fortunate, securing in all forty-four different species. I will mention a few of the most noteworthy. All the Zonites except *Z. glaber* and *Z. excavatus* figure in my list, *Z. cellarius* being very fine and variable. The type of *Z. alliarius* I rarely found, but the variety is very common. The varieties of *Z. purus* and *Z. radiatus* I also found—the former commonly. *H. concinna*, *H. hispida*, *H. rotundata*, and *H. aculeata*, all afforded me one or more albino varieties. *C. lubrica* is extremely variable in size and is frequently clouded, I secured one specimen of the var. *hyalina*. The sandhills at the Baun mouth are rich in some of the less common Vertigoes; *V. pygmæa* (com.); *V. substriata* (m.c.); *V. pusilla* (m.r.); *V. angustior* (c.); all of which I found among the drift in the hollows dead; but as the season advances, live specimens ought to be met with. I secured one live specimen of *V. alpestris* under a stone by the river, and as this last was new to Ireland, I was doubtful of its identity and sent it to Dr. Jeffreys who confirmed my opinion. *V. edentula* and its var. *columella* are common in the district among leaves and in woods, but not among the sand hills. One captive has particularly pleased me, viz. a specimen of *V. pellucida*, with two beautiful milk-white bands running spirally from apex to mouth. *B. acutus* and its varieties are particularly large and handsome on the Portrush sandhills. On these sandhills I have observed a phenomenon which has puzzled me much, and should be glad of an explanation. At the mouths of the countless rabbit burrows there were, during this autumn, piles of *H. nemoralis*, with *H. aspersa* and a few *H. ericetorum*, empty and apparently gnawed—not broken in the way that we find them on "thrush stones." The only conclusion I can come to is that the rabbits eat them, but I do not know that anyone has observed this habit in rabbits—though rats are said to eat snails. I will not encroach further on your space, but should any one, interested in the subject, wish for more minute details of the district, with a view to further search, I should be glad to furnish him with them. A great deal might be said about the marine shells on this part of the coast.—*Lionel E. Adams*.

THE MOLLUSCA OF MAIDENHEAD.—As far as I am aware, there is no published list of the land shells of the Maidenhead district, and so the following list of shells, although by no means complete, as they were all taken on the evening of July 6th, may be of some use to collectors subsequently visiting the locality. They were all taken in the immediate neighbourhood of Castle Hill, and are as follows:— *Arion ater:* I found three full-grown specimens, two were entirely black, and the third belonged to the var. *rufa*, being of a dark brown colour. I found one of the black specimens in the act of devouring a dead worm, thus proving conclusively that this species is not altogether a vegetarian. *Arion hortensis*, one specimen; *Limax agrestis*, *Zonites cellarius*, *Z. nitidulus*, *Z. alliarius*, var. *viridula*—in shape this specimen is not unlike a young Cellarius, but from the dark colour of the animal, the strong garlic odour, and the fact that no greenish-white form of cellarius has hitherto been recorded, I think I am right in referring it to viridula; I only got one specimen. *Helix aspersa*, *H. rufescens*, *H. Cantiana*, *H. hortensis*, and vars. *arenicola*, *lutea*, and *albina*, *arenicola* being very common. *H. nemoralis*, amongst ivy, *H. lapicida*, *H. hispida*, *Cochlicopa lubrica*, *Clausilia laminata*, and *C. rugosa*. On the same day I found *H. hortensis* and var. *incarnata*, and *H. aspersa* near Dropmore, on the Bucks side of the river.—*T. D. A. Cockerell*.

HYDROZOA AND POLYZOA.— The name of the illustration in your September number should be *Lepralia variolosa*. The writer of the paper speaks of the polyzoa and hydrozoa as being closely related, whereas they belong severally to distinct divisions of the animal kingdom; viz. the hydroida to the cœlenterata, while the polyzoa, as lepralia, flustra and others, are allied to the mollusca, a much higher group, and have been placed accordingly in the molluscoida (a subdivision of mollusca), including these (polyzoa), and the tunicata (ascidians, &c.). The author (p. 193) quotes Johnston as an authority in placing lepralia in ascidioida. More recent writers, as Busk and Hinks (British Polyzoa, 1870), and many others, separate lepralia and allied genera from the ascidians, although they admit a relationship between them which I may illustrate thus:

MOLLUSCOIDA.
1.—*Polyzoa.*   2.—*Ascidioida.*

The polyzoa being members of the suborder molluscoida, and only allied laterally to ascidioida. Referring to the mode of growth of these organisms, the author does not distinguish between polyzoa and other classes. Passing over the description of flustra, I come to sertularia, arborescent and plant-like (page 193, column 2, line 23). It would have been better to state that the resemblance between flustra and sertularia is only superficial, and that in zoological classification they are widely separated. Third, " stony and calcareous (madrepores and coral-lines)." These are stony and calcareous respectively, but neither belong to the group we are discussing. The madrepores are corals which belong to the cœlenterata, a sub-kingdom which includes classes, sub-classes, and orders, one class being hydrozoa, and another actinozoa. The latter includes the madrepores, which are thus widely separated from hydrozoa, and still more widely from all the polyzoa. As regards corallines, which are grouped with madrepores as being stony and calcareous, it must be stated that they are not animals at all, but are algæ with calcareous matter deposited in their cellular substance.—*Geo. D. Brown, F.L.S.*

NATURAL HISTORY LISTS.—I should like to suggest that accurately-compiled lists of the various groups of Natural History would be of great use to many readers not possessed of standard works. Thus, of marine shells, I know of no list, except an old one (not following the arrangement and nomenclature of Dr. Jeffreys, which is generally accepted now), and the want of one is much felt by conchologists. No doubt it is the same with many other things as well.—*Sydney C. Cockerell.*

INSECTS' EGGS.—I would suggest to our numerous entomological readers that a series of sketches, drawn to scale, of the eggs of our British butterflies and moths would be of the greatest interest. A few were given in the earlier volumes of SCIENCE GOSSIP. Whilst we possess beautiful figures both of butterflies and moths in the caterpillar, chrysalid, and imago states, it is surprising what few figures of the eggs we have. Will some reader take the hint?—*J. E. Taylor.*

THE FAUNA AND FLORA OF WORCESTERSHIRE.—As I am engaged in a work on this subject, may I ask the kind co-operation and assistance of workers in enabling me to obtain a clear and precise account of the distribution of plants and animals in this interesting county? All names of species recorded should bear the locality where taken, and the time of the year when captured. Varieties are earnestly solicited.—*J. W. Williams, B.Sc., Mitton, Stourport, Worcestershire, and Middlesex Hospital, W.*

A FOUR-FOOTED BIRD.—An announcement is made of the discovery of a living species of four-footed bird, inhabiting the island of Marajo, in the Lower Amazons. It is only four-footed, however, during infancy. The discovery has been announced to the Chicago Academy of Sciences, by Mr. E. M. Brigham, who has been for some time back engaged in studying the bird's embryology. The Academy will shortly publish the paper, with plates, &c. Mr. Brigham finds that in the development of this remarkable bird (whose zoological name is *Opisthocoma cristata*), from what corresponds to about the embryonic development of our common fowl at the tenth day of incubation, the fore-feet showed their characters unmistakably through the egg development. Thence to a period of several days after hatching, the fore-feet, toes, claws, &c., were as well marked as similar parts on the hind or true legs. Later on, the digits, claws, &c., exfoliated, and the true bird asserted itself. This example, however, roughs out for the palæontologist, as it were, the course which may have been taken in the development of birds from four-footed ancestors.

BOTANY.

DOUBLE DAHLIAS.—In raising the stem of a dahlia from an oblique to a vertical position, I observed a double flower, that is, two complete full-sized flowers on one peduncle. Not having observed this before, although I have grown the same kind for many years—(old-fashioned double, dark red)—I have inquired of several florists, but cannot find that such a thing has ever been observed by them. Perhaps some of your correspondents will be able to throw some light on the subject. It struck me at first that the "freak" arose from the want of light, the stem being very much bent down, but then how am I to account for the two flowers being of full size and almost exactly alike?—*J. Wallis, Deal.*

PARIS QUADRIFOLIA.—In June, 1878, I gathered, at Bishop Frome, Herefordshire, a very large specimen of *Paris quadrifolia*, having four leaves, one leaf having an inclination to divide, and six larger perianth segments. One segment is considerably longer, and about four times broader than the others. At the same time and place I found the variety with five leaves and five perianth segments, but I have never seen any other specimen in which the number of perianth segments did not correspond with the number of the leaves.—*Henry L. Graham, Buildwas, Salop.*

FERNS OF THE ARLBERG PASS.—Now that the recent opening of the Arlberg Railway, with its remarkable tunnel eight miles in length, has brought that district so prominently into public notice, a list of the ferns to be found by the side, or within a short distance of the road leading from Bludenz to Landeck, the two places between which the newly opened portion of the railway lies, may be not without some interest to your readers. The list numbers eighteen in all, and comprises the following species :—*Poly-*

podium phegopteris, P. dryopteris, P. Robertianum, Lastræa filix-mas, L. oreopteris, L. dilatata, Athyrium filix-fœmina, Pteris aquilina, Polystichum aculeatum, P. Lonchitis, Cystopteris fragilis, C. montana (near the summit of the pass), *C. alpina* (some of the specimens very fine), *Asplenium trichomanes, A. viride, A. ruta-muraria, A. septentrionale* (between Flirsch and Landeck) and *Blechnum spicant.*— T. W. B.

FLEURS DU LAC.—Le numéro d'Octobre de ce journal, page 239, répond à Mr. "W. S. W." que "les fleurs du lac" trouvées en écume jaune sur la surface du lac de Thun, fin Mai, sont le pollen des forêts de pins. Cette réponse a-t-elle été vérifiée, et n'avait-on point affaire à des algues postochinées dont parlent Mr. W. Phillips dans son travail sur "The Breaking of the Shropshire Meres," tiré des "Transactions of the Shropshire Archæological and Natural History Society," de Février 1884, et MM. Bornet et Flahault dans leur article sur les "Rivulaires," paru dans le "Bulletin de la Société Botanique de France," tome 31*e*, 1884, page 76? Ces articles pourraient, peut-être, être utilement signalés à votre correspondant.—*C. C., Doullens, Somme.*
[In answer to our able correspondent, we beg to say that in the case in question there is no doubt of the "fleurs" being pine-pollen.—ED. S.-G.]

THE DESTRUCTION OF TREES AROUND LONDON. —The growth of our modern Babylon may be matter for the wondering pride of its citizens; but, in one aspect at least, it is calculated to excite feelings of a very different character. I allude to the encroachments of speculative builders and their reckless and increasing demolition of our choicest trees, if they stand in the way of the advancing tide of bricks and stucco. Highgate, Muswell Hill, Hornsey, Crouch End—I restrict my references to this side of London —all are being ravaged and disfigured without hindrance, except from the unavailing protests of a few lovers of natural scenery. That the naturalist and the suburban rambler are being deprived of their favourite haunts is but a minor part of the evil in progress. Readers of SCIENCE-GOSSIP do not need to be reminded of the hygienic value of foliage, or of the effects produced on rainfall and climate by the destruction of forest trees. We have lately been reminded, by the threatened water famine in the Manchester district, that the supply of rain, even in this humid island, is but limited; and we have only to cross over to France to learn how fertile provinces may be rendered desolate by the clearing of their timber trees. Can anything be done to check this growing evil? Neither the sale nor the purchase of land can be prevented, while mere reasoning would be thrown away on the stolid perpetrators of the mischief. Legislative interference is called for; and why should not the power which has already protected the wild birds throw little protection around our woodland trees? The law has stepped in to regulate the width of roads and the sanitary arrangements of houses. Cannot our naturalists and other men of science secure a little regulation of the laying out of "building plots"? In the course of a few days, as I have recently witnessed, the growth of more than half a century is destroyed, while all the suburban householder can do is to plant a few saplings in his garden, from which he can scarcely hope to derive any benefit in a lifetime! Unless public interest be wakened to check this selfish greed and reckless destructiveness, we shall find eventually, in an altered climate, that a Nemesis waits on those who set at naught Nature's laws for the sake of expediency.— *W. H. G., Crouch End, N.*

NEW DISCOVERIES IN GEOGRAPHICAL BOTANY. —Mr. J. Thomson, the African traveller, has arrived home in safety, and he received a brilliant reception at the Royal Geographical Society, on November 3rd, when he delivered an address descriptive of his travels and discoveries. One of Mr. Thomson's objects was to collect plants from the mountainous regions of Central Africa. These have been placed in the hands of Sir Joseph Hooker, who has lost no time in making them known, or in pointing out their geographical significance. Thirty-five species were collected from Kilmanjaro at from 9000 to 10,000 feet above the sea-level. A few came from Lake Nairasba, at from 7000 to 8000 feet elevation. Thirty-four species were gathered on the Kápté plateau, 5000 to 6000 elevation; and fifty-eight came from Lykipia, where they were found at an elevation of from 6000 to 8000 feet. These species bear out the idea of a mixture of northern and southern forms, due, perhaps, to alternate glacial periods in each hemisphere, driving them to this common meeting-ground. Sir Joseph Hooker shows that Mr. Thomson's specimens are of this character. Among the northern forms are an anemone, delphinium, and cerastium. The most striking of the southern forms is the wild chestnut of Natal (*Calodendron capensis*). One northern form is a juniper, growing to the height of a hundred feet, and forming groves at a height of 6000 feet above the sea.

BATS.—I have often noticed bats flying about the streets of our town at night in the autumn and winter months in mild weather. They look very strange when they swoop down in the light from shop windows and lamps—a pale brown object which suddenly disappears again in the dark above. I have seen them in the town in October, November, December, January, and February, according to notes made at the time. Do they find insect food at this time of the year in such situations, or are they merely attracted by the light? I once saw a bat flitting about in broad daylight here.—*Henry Lamb, Maidstone.*

GEOLOGY, &c.

THE DENUDATION IN AMERICA.—Mr. T. Mellard-Read, C.E., F.G.S., in the annual address to the Liverpool Geological Society, chose for his subject "The Denudation of the Two Americas." After an introduction dwelling on the importance to geology of an accurate determination of the magnitude of the agencies engaged in fashioning the earth, it was shown from careful analyses and calculations that the river Mississippi brought down and delivered into the Gulf of Mexico matter in solution in its waters equal to 150 million tons per annum. Analyses of the waters of the La Plata, the St. Lawrence, and the Amazon were then given, and the amount denuded by chemical agencies from each river basin was calculated, the author arriving at the conclusion that a mean of 100 tons per square mile per annum of matter in solution is removed from the American continent by rivers draining into the Atlantic Ocean. The author had on a previous occasion calculated that 100 tons per square mile per annum was the mean chemical denudation of the whole of the land of the globe. Taking this determination as a unit of measure, the astounding result was arrived at, that all the rivers draining into the Atlantic Ocean from America, Europe, Africa, and Asia, an area equal to 21 million square miles, delivered into it, each six years, one cubic mile of mineral salts in solution, estimating them as reduced to rock at two tons to the cubic yard.

THE IRISH DRIFT.—An important paper appears in the Scientific Proceedings of the Royal Dublin Society by G. H. Kinahan, M.R.I.A., on "The Classification of the Boulder Clays and their Associated Gravels." He thinks that some of the gravels, &c., under glacial drift may be younger than the overlying deposits, owing to the ice-sheet melting above and below, so as to leave cakes and patches on differing horizons. The waters due to such cakes when they finally melted away, would wash portions of the glacial drift into sands and gravels formed and arranged subsequent to the overlying drift.

NOTES AND QUERIES.

BIRDS AND ARUM.—If F. J. George, who, in the November number, p. 262, asks if "any thing is known of birds or ground game eating off the spathe and spadix of *Arum maculatum*," will look in Gilbert White's "Natural History of Selborne," Letter XV., dated March 30th, 1768, he will there see an answer to his question.—*N. A. D.*

LARGE UNIOS.—Mr. Tuxford's note of *U. pictorum* measuring $4\frac{13}{16}$ in. just surpasses the $4\frac{7}{10}$ in. Leicestershire specimen recorded by Norman. Mr. Tuxford, I presume, is sure of his species? Such measurements are normal for *Anodon cygnæus*, a very similar-looking shell, of which I hold a $6\frac{1}{4}$ in. specimen, taken under identical circumstances.—*Ernest G. Harmer.*

REPTILES EATING THEIR SLOUGHS.—With respect to newts eating their cast-off skins, this, I think, they invariably do, as I have noticed it frequently. The toad, it is well known, has the same habit. A tame toad that I kept for a number of years used to retire into a deep hole, and first proceed to take off his old clothes, by rubbing his hands down his sides, till the skin became wrinkled and loose; he then seized the skin of his breast with his mouth and stript it up, drawing it over his head with his hands. The skin of his right arm and hand he drew off like a glove with his left hand; then, changing hands, served the other the same. His stockings he drew off, by grasping the skin of his toes in his hand and pulling and hauling it off, always ending by making a meal of it. Though I had him from his babyhood to full-grown toadhood, and he was always very tame and familiar, I only saw him change his skin three or four times, as he would always retire out of sight in the darkest hole he could find. His companions in the same case were a fine slowworm (which, by the way, was the largest I ever saw, being 18 inches long), and a salamander (*Salamandra maculosa*); the toad and the salamander were very good friends, generally occupying the same hole; but Ben (the toad) and the slowworm never could agree, Ben being of a waggish disposition, often sallying out, seizing it by the tail, and making a great show of swallowing it alive, probably mistaking it for an extra fine earthworm. Occasionally Ben would seize one end of a worm, and the slowworm, the other; and then would commence a struggle fierce and long, generally ending in a victory for the slowworm, when Ben would retire in disgust to sulk, and, though offered fresh worms, would refuse them for an hour or so till he had forgotten his disappointment, and his temper had become smooth again. One morning I found the slowworm dead, having lived with me but ten or eleven months. Ben got very stout and very restless, so I gave him his liberty, though sorry to part with him. The salamander died this summer, having lived with me seven years.—*G. Currie.*

THE CAMEL. (*Camelus bactrianus*).—There was a statement in the "Field," a few months back, as to the probability of the camel (the above species, I suppose) still existing in a wild state in Asia, according to the testimony of a recent traveller—I forget his name. Has this been confirmed? The camel used in Egypt, and that which is commonly known by that name, is the dromedary (*C. dromedrius*), is it not? and is, also, I suppose, the one mentioned in the Scriptures. Will some reader kindly reply?—*Henry Lamb, Maidstone.*

THE HEDGEHOG (*Erinaceus Europæus*).—This animal evidently has not any religious sentiment, or it would understand that cleanliness is next to godliness, which would make it as agreeable as it is useful in a kitchen as a beetle-trap. We had one for a short time, and it used to clear up the cat's food, bread and milk, or anything else; they will eat nearly anything, slugs, insects, &c., but my father says, and another confirms his opinion, that they could not manage a live rabbit. It is nocturnal in its habits, and my father knows a man who says that when he has been out of a night sugaring it has frequently come up quite close to him to watch for any insects which might fall. It hybernates during the winter, and my father and others say the same, that they have frequently found it in the middle of a wood stack, and thickly covered with leaves like bills on a file. No doubt this is done for a double purpose: warmth, and to hide itself from being seen. My father has often seen it simulate death; when thrown into the river it would immediately sink, but as soon as his back was turned it would rise and swim away. Some years ago my grandmother tasted soup made from the hedgehog, which is also good eating, cooked gipsy fashion, with a paste of clay all over it, and baked in a bonfire. This was told me by a lady who went among the gipsies to learn their language. In "Museum of Animated Nature" I find the following. The hedgehog's food consists of insects, slugs, frogs, toads, mice, and even snakes, eggs, young nestlings, and various kinds of vegetables, as the roots of grass and plantain, and ripe orchard-fruits which fall from the trees. As regards the destruction of snakes by the hedgehog, it appears that the cunning quadruped makes a sudden attack on the reptile, and giving it a hard bite, instantly rolls itself up for safety, then cautiously unfolds, and inflicts another wound, repeating the attack until the snake is "scotched." It next passes the body of the snake gradually through its jaws, cracking its bones at short intervals; which done, it proceeds to eat its victim as one would eat a radish, beginning with the tip of the tail. The female breeds early in the summer, forming an artful nest, roofed so as to throw off the rain; within, it is well lined with leaves and moss. The young, from two to four in number, are blind at their birth, about two inches long, perfectly white, and naked, though the rudiments of the prickles are perceptible. These soon develope themselves, and harden even before the eyes are opened, but it is not till a later period that the young are able to draw down the skin over the muzzle, and fold themselves into a complete ball. The mother is devoted to her offspring.—*Clara Kingsford, Barton House, Canterbury.*

PECULIAR HAILSTONES.—A heavy shower of hail fell at Polmont, N.B., on the 6th of last May. I examined carefully about thirty of the hailstones which I picked up at different parts of my garden. The longest diameter of each was about one-fourth of an inch in length. I do not remember previously to have seen such structural appearances as I will now, with your leave, very briefly describe. Each hailstone was in shape an almost perfect pyramid; hitherto those which I have observed have been all more or less spheroidal, but every one of these was distinctly pyramidal in form. I made rough transverse sections of about a dozen and found that most of them exhibited an internal radiated structure which could be seen by the naked eye, but could be much more clearly made out by means of the addition of a good magnifying glass. The radii were all focused along the longest axis of the pyramid. Would any of your readers, with your permission, kindly inform me if such appearances have been seen before?—*Alex. Johnstone, F.R.S.S.A.*

WATER VOLES.—In answer to Mr. F. H. Parrott's query: Are water voles entirely vegetable feeders? I wish to quote my observations of this little animal, as it has long been my favourite of the *feræ naturæ* of North Yorkshire. Many times have I stolen at eventide to the banks of the Wiske, and watched them gambolling amongst the grass, swimming and diving in the river, or nibbling at the herbage. They delight in gnawing the base of the leaves of the iris. Later in the evening they may be found a little way from the banks of the stream, luxuriating amongst the grass or corn. Of the hundreds I have watched, I never saw them partaking of anything but a vegetable diet. The finding of shells in their runs is hardly sufficient proof of their carnivorous nature, as these runs may not be inhabited by the vole, or the shells might be deposited there by some other agency. It will be a sorry day for the poor little creature when flesh-eating proclivities are assigned to it, as the angler will without doubt banish them from his preserves. I must repeat a fact I mentioned in this journal some time ago, viz. that the brown rat is a great lover of water, as witness his delight in drains, dockyards, &c. In some streams in Yorkshire he is frequently seen, and in one near Scarborough he is so abundant, that the more innocuous vole is entirely supplanted. As the brown rat swims and dives with ease, he is liable to be confounded with a vole, although a closer observation at once reveals the wide difference between them. I believe I have seen the small black variety of the vole by the Wiske, but being unable to get a close view, I cannot be certain of its identity. Could any of your correspondents inform me if the black rat is still to be procured in the London dockyards?—*J. A. Wheldon, Northallerton.*

EARLY EMERGENCE OF INSECTS.—In SCIENCE-GOSSIP lately there was a paragraph on the early emergence of insects. The writer did not think that a second brood of *Attacus Peryni* can be reared in this country. His experience is, eggs laid 25th August, hatched 12th September, and the larvæ after passing through several changes all died by 10th November. I have much pleasure in giving you my experience with dates. Second brood of *Attacus Peryni*: eggs laid 5th September 1882, and hatched on the 18th and 19th of the same month. First moulting on 26th and 27th, and on the 3rd and 4th October they threw off their "Old Clo." Their third on 12th and 13th, their fourth between 22nd and 24th October, and went into their cocoons between 5th and 13th November, eleven in all. They were reared in cages made up of four panes of glass cemented together at the corners, with stout paper bottoms, and perforated zinc covers, and the temperature ranged from 65 to 72 degrees of heat. I unfortunately put them away into a colder room before they had changed to the pupæ and consequently no insects came out. In 1883 I was more fortunate, although later in the season the eggs did not hatch till about the end of September,— the particular dates of moulting have not been preserved, but, if anything different from those of the previous year, they were, perhaps, a little longer between the moultings. The nearly full fed larvæ were exhibited at the South of Scotland Entomological Society's exhibition of Insects, held here on 1st December, and about the middle of the month the whole of them, 22 in number, had "spun up." Owing to the advanced season two or three of the larvæ before spinning had to be content with eating dry and withered leaves, as every favoured spot and sheltered nook had been sought in vain to find them fresh food. The cocoons are thinner in the shell, and the silk much whiter, than those of the first brood, and a number of them were hung up in the same temperature in which the larvæ were reared, and up to the 18th of February last, ten fine specimens had emerged.—*A. Litster, 67 Channel Street, Galashiels.*

OSPREYS.—It may interest some of your readers, to hear that the osprey eagles (water-eagles) have this year again paid their visit to Loch-an-Eilan, and, like good parents, have nursed their young ones. Loch-an-Eilan stands in the midst of the Grampians, and almost at the foot of the mighty Ben Macdhui, from which the snow never disappears. It is surrounded by the famous woods of Rothiemurchus, and is fitting summer home for the osprey. It is within easy reach of both boat of Garten, and Aviemore stations on the main-line of the Highland railway. In the middle of the loch stands an old castle of the Wolf of Badenoch, and on one of the exposed corners of this castle is built the huge nest of the ospreys. On the 15th of July last, I, in company with my brother, visited Loch-an-Eilan, and we were fortunate to see one of the birds on the water. After some time it flew on to the nest, and we got a splendid view of it as it fed the two young ones. When it had finished feeding the young pair, it stood by the side of the nest watching its charge, and every now and then turning its head to the side whence any sound proceeded. It was indeed a fine sight to see the noble bird standing and defying, so to speak, the approach of anything to harm the young ospreys. A glimpse could be got of the young as they playfully rose to feed the old bird. The mother bird was a little larger than a pretty large hen, with a short black bill, white breast with sprinklings of grey, grey wings, and a white tuft of hair on the top of the head. Those two birds are said to be the only now remaining in the British Isles. They go off in the winter season—the young never return—and return in the following summer, and they have done so now for many years, and with such regularity that by the inhabitants a certain day of the year is called the "eagle day." Long may they make Loch-an-Eilan their summer home, and long may the peaceful dwellers of Rothiemurchus resist the tempting offers of some of our Cockney friends of large sums of money for the possession of either eggs or young birds.—*J. G. Sharp*.

STRANGE TENACITY OF LIFE.—A few days ago the body of a puss moth (*Cerura vinula*) was brought to me. It was found under a poplar terribly mutilated by ants or birds ; the head and greater part of the thorax had been eaten away, as also the entire wings, and there was a wound on the left side of the abdomen ; nevertheless, the insect remained alive for twenty-four hours, and during that time was continually thrusting out its ovipositor in search, as it seemed, for a convenient place to lay its eggs. But whether it was not satisfied with the lime leaf on which I had placed it, or whether in its singular state it was unable to produce any eggs, is to me a mystery. Is it possible that if I had given it the leaf of a willow or poplar, the natural food of the larvæ, it would have deposited its eggs? For insects generally show a wonderful instinct in laying their eggs where the larvæ when hatched will be at once supplied with food.—*R. A. F.*

THE MINNOW.—Has it ever occurred to any of your readers what a voracious little shark we have in our common minnow? I put one of them in a small aquarium containing pond life, and was surprised to find next day that nearly all my microscopical captures had disappeared. I always gave the palm for gluttony to Messrs. Dragon-Fly-larvæ, and Dytiscus, but this little minnow beat them. Thinking he looked hungry one day, I commenced the difficult task (as it turned out) of trying to appease his appetite. With a dipping tube I forced into the aquarium, one at a time, a hundred and fifty good sized daphniæ which the minnow took as fast as I could supply. I then gave him a few cyclops, three small red worms, and a small water louse. As I had been thirty minutes supplying this meal, I thought surely he ought to be satisfied ; not so, however. After watching for the dipping tube for a minute or so, off he goes in evident disgust, but, to my surprise, at once proceeded to assist a tadpole to get rid of its tail, and with one dash took the only remaining spiracle from the tail of Mr. Dragon-Fly-larvæ.—*Henry Burns*.

PET TOMTITS, &c.—I have been much interested in Mr. Robertson's account of his tame tomtits, as two pairs of those engaging little birds, one pair blue tits, the other the greater titmouse, have been my constant companions for some months, coming continually to pay me little visits, and to feed on a bench just inside my window. I first attracted them by beef suet hung outside the window, but soon coaxed them in by placing the delicate morsel to which they had got accustomed inside. They know my voice well, and come at my call, and answer me at a great distance, and on some occasions, when, on finding their way in through a very small opening they have got alarmed and confused in trying to find their way out, I have succeeded in quieting them by my voice ; one of them once ceasing all endeavours to free itself and hopping on the handle of a basket, where it composedly sharpened its beak till I opened the window wider for its flight. This little sensible individual was the greater tit, and the cock—and by far the boldest of the quartet. I knew him always by his pretty habit of raising the crest when answering me, and by the broader line down the chest. I never saw any indication of a crest on the hen, and would be glad to know if this is an invariable distinction of their sex. I think the best way of distinguishing their note from the chaffinch's is to observe that they frequently repeat their call three times, " Twink, twink, twink," while I never heard the chaffinch repeat more than the double " Twink, twink." I found the blue tits the quickest to tame, but the others bolder and more sympathetic, once their confidence was won. The blue tits liked bread, taking it regularly in turn with the suet, a few pecks of one and a few peeks of the other, while the larger tits scarcely touched it, and, if the fat was not provided for a few days, gave up coming. The staler the bread the better it was appreciated. Their manner of eating both it and the suet was very pretty ; the pieces were fastened to the bench by a long steel pin, and one little claw was always clasped firmly round this upright pin, which gave them a sideways attitude, and then they hacked away with their strong little bills like a pickaxe, making a surprising noise, and having evidently great enjoyment in the process.— *C. G. Grierson*.

NOTES FROM ROTTERDAM.—The "Nieuwe Rotterdams Courant" relates (12th August) :—"The gardener P. v. Leeuwen at Noordwijk, near Leiden, has obtained a double white crocus (*Crocus vernus*, All.). This is the first double crocus which ever was gained." That same day the above-mentioned newspaper writes :—From Groningen we are informed : The 10th, a swarm of small black-winged insects came over here from the east, and covered gardens, houses, &c. ; this insect is one stripe long, and as thick as a horsehair. What insect can this be ? In the end of July, the same newspaper mentioned that a scholar of the Hoogere Burger School at Arnhem had photographed, during a violent thunderstorm, some flashes of lightning.—*W. H. Croockewit*.

A BEECH WITH "GNAURS."—On the grounds of Hawkhead, Renfrewshire, by the storm of the 26th January last many trees were thrown, among which is a beech (*Fagus sylvatica*) on which there are many "gnaurs" or "burrs." On this beech I counted thirty-nine "gnaurs" or "burrs," one of these, which had grown at a height of thirty feet from the root of the tree, was a foot in diameter and four inches deep, from which grew several small branches, many not exceeding an inch in length. This tree is about two feet nine inches in diameter, and has lived to the age of ninety-five years. "Gnaurs" or "burrs" are common on the elm (*Ulmus montana*), and are often polished and put in lobbies for ornaments, but I have never observed them as common on the beech.—*Taylor, Sub-Curator, Museum, Paisley.*

THE "CLOTE."—"This is the clote bearing a yellow flower," Drayton (SCIENCE-GOSSIP, No. 235, July 1884, p. 149). The clote is a common Dorset name for the yellow water-lily (*Nuphar lutea*). In "Poems in the Dorset Dialect," by Rev. W. Barnes, the name occurs many times. One poem is upon the clote itself; the first verse runs as follows:—

O zummer clote! when the brook's a gliden'.
So slow an' smooth down his zedgy bed,
Upon thy broad leaves so seäfe a-riden,
The water's top wi' thy yollow head.
By Alder's heads, O,
An' bulrush beds, O,
Thou then dost float,
Goolden zummer clote!

—*O. P. Cambridge, Bloxworth Rectory.*

PARIS QUADRIFOLIA.—My experience of this plant in Rhenish Prussia was that, wherever it occurred at all, by searching for a little while plants with five or six leaves could be found.—*J. C.*

THE NATURAL HISTORY CLAUSE IN THE RAILWAY BILL.—It can hardly be doubted that to sustain the moral of an engineering speculation social problems as well as those regarding individual rights have to be adjusted, and that various natural phenomena concerning which we seek and pray for enlightenment have to be considered, and which doubtless, if understood, would be no more incomprehensible than an error in calculation, or a fault in material. Indeed, strictly speaking, an error, fault, and unforeseen conjuncture, are all evils presided over by Providence, and naturalists who study the phenomena of the universe are as truly portion of the social life as mathematicians and chemists. Apart from all such momentous issues we yet seem to be in want of a right understanding in regard to the finding of objects of interest to scientific corporations by workmen in public employ. Of course, if a man's spade turned up a gold watch, our moral conviction would be that he should advertise the same, but scientific interest in the matter would not be awakened. Indeed, it is at the outset ridiculous to suppose that among the multiplicity of natural objects the untrained mind should determine what is esteemed rare and interesting to the learned, and it is almost impossible to suppose otherwise than if they should seek to sell their finds, the results would prove futile, on account of an inability to grasp even the trade valuation, and purchasers at fancy prices in these days of scientific supremacy prove few, especially when the objects lack colour, form, and other sensual attractions. Thus many objects of real interest are overlooked, simply because the workmen do not understand that the scientific public cares about them, and even when they find an object that recommends itself to their eyes and understanding, the public have no further guarantee than it will be procured with ordinary care; they cannot suppose that any technical resources will be resorted to, or memoranda of import to naturalists taken down. Let us suppose, for example, a huge tusk of an elephant is found in a sandy bed, crushed and impregnated with iron oxide, and it would be of interest to form a correct idea of its size. Two or three men insert their spades, and each digs up a handful of rotten stuff, causing the mirage of the tusk to vanish in a moment, gone with many a good find that will not return. Yet in this case a conviction might occur to the field naturalist, that had the large tusk been uncovered gently above with a trowel, and the fractures gone over with glue, and after the glue had hardened, the tusk had been gently moved on to a stretcher and thus transported into the laboratory, its chances of fair preservation would have been greater. Indeed, what has seemed most desirable is, that in such cases the railway company should employ a competent naturalist, make the workmen that acknowledgment which is proper, and claim the finds. The initiative to this system has been taken by the London and South-Western Railway in respect to the elephant fragments lately found at this place, in claiming "all future finds," and it is to be hoped this may pave the way to a greater recognition of the community of working naturalists, and bring interesting objects into the possession of those who appreciate them.—*A. H. Swinton, Guildford.*

NOTICES TO CORRESPONDENTS.

TO CORRESPONDENTS AND EXCHANGERS.—As we now publish SCIENCE-GOSSIP earlier than heretofore, we cannot possibly insert in the following number any communications which reach us later than the 8th of the previous month.

TO ANONYMOUS QUERISTS.—We receive so many queries which do not bear the writers' names that we are forced to adhere to our rule of not noticing them.

TO DEALERS AND OTHERS.—We are always glad to treat dealers in natural history objects on the same fair and general ground as amateurs, in so far as the "exchanges" offered are fair exchanges. But it is evident that, when their offers are simply disguised advertisements, for the purpose of evading the cost of advertising, an advantage is taken of our *gratuitous* insertion of "exchanges" which cannot be tolerated.

We request that all exchanges may be signed with name (or initials) and full address at the end.

B. H.—Every number of SCIENCE-GOSSIP contains the coloured plate. If your bookseller did not get one, let him apply to the publisher.

SYDNEY C. C.—Many thanks for your suggestion, which is a very valuable one.

INDEX TO SCIENCE-GOSSIP.—We have been much gratified by the flood of letters received in reply to Mr. Harris's suggestion about a General Index to the last twenty years' vols. of SCIENCE-GOSSIP. It proves the large *clientèle* we possess, and the active individual interest taken in the welfare of the dear old Fourpenny! Due announcement will be made of the Index.

J. B. J.—You are right. The word "viviparous" instead of "oviparous" in connection with the duck-billed platypus was a telegraphic error.

N. A. D.—"Diseases of Garden Crops," by W. G. Smith, is published by Macmillan at 4s. 6d. "The Honey Bee," by W. H. Harris, is published by the Religious Tract Society at 5s. "The Blow-Pipe in Chemistry," by Colonel Ross, is published by Crosby Lockwood & Co. at 3s. 6d. "The Botanical Gazette" is an American journal, published at Indianapolis by Carlon & Hollenbeck, at 10 cents per number.

H. G. W. A.—The insects sent are spring-tails (Podura). A weak solution of carbolic acid, or a weak emulsion of paraffin, will get rid of them.

E. B. L. B.—Thanks for the excellent hint.

A. PEARSON.—Get Houston's "Practical Botany," price 2s., or Vine & Prank's "Botany."

H. B. H.—The size of the dredge mentioned is the inside size. A rope of the thickness of an ordinary clothes line is what we always use.

A. P.—See Taylor's "Half Hours at the Sea-Side," p. 186, for remarks on the eyes of cuttlefish.

H. R. H. AND OTHERS.—You will find a tolerably full account of the earthquake in the eastern counties last April in the June number of SCIENCE-GOSSIP, entitled "A Genuine British Earthquake."

J. R. M.—You will find all your queries as to number and varieties, winter visitors, &c., of British birds in Morris's work. "Oology" stands for egg-collecting.

EXCHANGES.

FINE collection of minerals, crystals, and chalk fossils in exchange for English coins.—A Butt, 97 Burton Road, Brixton, S.W.

OFFERED, "Science Monthly," first twelve numbers, clean. —W. Berridge, F.R.Nat.Soc., Loughborough.

WANTED, unmounted parasites or fleas, in exchange for double-stained wood sections or whole insects.—G. Mackrill, 29 Tower Street, Hull.

WANTED, to exchange British birds' eggs: must be good specimens, with full data.—Address X., Post Office, Penpont, Dumfries-shire.

GOOD collection of fossils, over 700 species and 1000 specimens, including Silurian, Devonian, Carboniferous, Lias, Wealden, Greensand, Gault, Cretaceous, and Tertiaries, in exchange for stone implements, either Palæolithic or Neolithic (British or Irish).—G. F. Lawrence, 49 Beech Street, London, E.C.

WANTED, apparatus for photographing small objects or small engravings. Illuminating colour box, new books, curios. Tunley, Southsea.

"KNOWLEDGE," wanted, the back monthly numbers, from Nov. 1881 to Oct. 1882, inclusive, with the astronomical plates in good condition—F. W. Clark, The Middlesex Hospital, London, W.

FINE specimens of *Trifolium stellatum* from Shoreham Point, gathered in July this year. Wanted, 50, 59, 254, 274, 286, 287, 302, 321, 331, 332, 338, 350, 357, 361, 366, 376.—J. H. Bloom, Thornham, Wilbury Road, Brighton.

MICRO-SLIDES: a series of cotton, woollen, and linen fabrics, mounted for the microscope, in exchange for other good slides. —W. Henshall, Bredbury, near Stockport.

WANTED, Continental plants in exchange for Continental or English plants. Send lists.—A. R. Waller, 17, Low Ousegate, York.

L. glutinosa, Bulla hydatis, and a large number of other rare shells for exchange. Wanted, well-mounted slides and *P. roseum, L. involuta, Succinea oblonga, Acme lineata,* var. *Vertigos, Limnees,* &c.—S. C. Cockerell, 51 Woodstock Road, Bedford Park, Chiswick, W.

EGGS of pheasant, partridge, red-legged partridge, moorhen, wood pigeon, jackdaw, thrush, missel thrush, blackbird, chaffinch, bullfinch, greenfinch, blackcap, redstart, bunting, gold crest regulus, blue tit, long tail tit, willow wren, &c., in exchange for other eggs or British butterflies.—James L. Mott, 18 Grafton Place, Northampton.

WANTED, Hassall's "Adulterations in Food and Medicine, and How to Detect Them," for Stanley's "Dark Continent," in good condition.—William Brunton, Inverkeithing, Fifeshire.

EXCHANGE organic remains from the Cretaceous (chalk) formation for those from the Oxford clay, Oolite (great or inferior), or Lias.—G. E. East, jun., 10 Basinghall Street, London, E.C.

EXCHANGE for other micro-slide, *Polypodium vulgare,* showing sori, venation, &c., mounted in C. balsam, without pressure. —Rev. H. W. Lett, M.A., Lurgan, Ireland.

FEW copies of "British Birds" and "Varieties of British Lepidoptera" to exchange for other works on natural history.— S. L. Mosley, Beaumont Park Museum, Huddersfield.

FOR scales of perch, bream, &c., send stamped and addressed envelope to—H. Moulton, 37 Chancery Lane, W.C.

DUPLICATES: *Anodonta anatina, Unio tumidus, U. pictorum, Clausilia laminata, C. rugosa, Helix hispida, H. cantiana, H. arbustorum, H. rotundata, Limnæa glabra, Planorbis contortus, P. nautileus* and var. *cristata, Paludina vivipara,* &c. Desiderata: very numerous.—R. Dutton, 13 St. Saviourgate, York.

WELL-MOUNTED diatoms (*in situ*) on coralline from Portugal; also other slides of spread diatoms, in exchange for good mounts of diatoms or foraminifera, or unmounted material.— George Bailey, 1 South Vale, Upper Norwood.

OFFERED, L. C., 7th ed., Nos. 267, 564, 577, 858, 940, 944, 1015, 1059 a, 1280, 1330, 1334, 1412, 1485, 1486, 1577, 1636, 1665 b, &c. Desiderata, 562, 574, 580, 598, 600, 1479, 1484, 1487, 1545, 1597, &c.—H. Fisher, 52 R. Y. C., Clifton, Bristol.

WILL exchange birds' eggs for British moths and butterflies. —A. Milne, 23½ Hutcheon Street, Aberdeen.

WANTED, to exchange fossils from Silurian, Dudley, for specimens from Solenhofen, &c. (lithographic stone), or best sp̄cimens from chalk formations. Send list.—H. C. Lambert, 34 Talbot Road, Stafford.

"KNOWLEDGE," Nos. 35 to 87, minus No. 63, will exchange for Hayward's "Botanist's Pocket-book," or condenser for microscope.—A. Wright, 9 High Street, King's Lynn.

WANTED, rubbings of monumental brasses from old churches; exchange others or natural history objects, fossils, &c.— F. Stanley, Margate.

A FEW dozen physiological and pathological slides to exchange for others of interest; micro photographs and selected diatoms, &c., wanted.—W. Tutcher, 22 North Road, Bristol.

EXOTIC Lepidoptera: many duplicates to exchange for others, exotic only; also wings of *Urania rhypheus, Morpho menelaus,* and other brilliant species for microscopic purposes. Rare papilios much wanted for figuring; condition immaterial; 210 species already figured. These figures have all been drawn and coloured from specimens in my own collection, and as they now stand it is believed they form the most extensive series of illustrations of the genus papilio in existence.—Hudson, Railway Terrace, Cross Lane, near Manchester.

SCIENCE-GOSSIP for 1876, 77, and 78 (vols. xji. xiii. xiv.), bound, cloth; exchange for botanical micro slides.—F. R., 65 Devonport Road, Shepherd's Bush, London, W.

WANTED, "The Microscope," by Carpenter, microscope lamp, and micro photographs, in exchange for books and microscopic slides.—B. Berry, Dudfleet, Horbury, near Wakefield.

OFFERED, L. C., 7th ed., Nos. 31, 57, 59, 67, 588, 595, 620, 699, 809, 1012 b, 1921, 1024, 1043, 1149, 1369, 1287, 1288, 1477, and others. Send lists.—A. Sangster, Cattie, Oldmeldrum, N.B.

WANTED, Dana's "Geology" or "Mineralogy," Macaulay's "Lays of Ancient Rome," for B. O. P., eggs, or coins (foreign).—John T. Millie, Clarence House, Inverkeithing, Fifeshire.

WANTED, good material for mounting, more especially insects (in spirit); will give well-mounted slides in exchange.— C. Collins, 25 St. Mary's Road, Harlesden, N.W.

GENTLEMEN having for disposal a quantity of any one insect (providing that it is not common) will do well to send their requirements in micro slides at once to—C. Collins, 25 St. Mary's Road, Harlesden, N.W.

DUPLICATES, principally carboniferous, offered for mesozoic or other palæozoic fossils; crustacea and echinodermata preferred.—J. A. Hargreaves, Westfield Terrace, Baildon, Leeds.

HAIRS of any of the following in exchange for other objects of interest, or twelve varieties for good mounted object, lion, tiger, leopard, black and white bear, hyaena, racoon, jackal, white and blue fox, wolf, seacalf, red and grey squirrel, and others.—A. Draper, 275 Abbey Dale Road, Sheffield.

A COLLECTION of 785 postage stamps in Oppin's "Postage Stamp Album," many rare and all different, well worth £3 10s. Will exchange for standard work on British lepidoptera, or entomological apparatus, or offers.—W. B. Smith, Market Square, Potton.

FOR exchange, "Design and Work," nine volumes newly and strongly half bound. Wanted a good microscope lamp (Sear's preferred) and slides.—R. Ridings, 1, Hampton Terrace, Lisburn Road, Belfast.

MICRO-SLIDE in exchange for first class mounts only.—J. Harbord Lewis, F.L.S., 145, Windsor Street, Liverpool, S.

A BEAUTIFUL platyscopic lens by Browning, just new, for disposal, cost 18s. 6d.; offers to B. C. Hare, Polytechnic Institution, 309, Regent Street, W.

BOOKS, ETC., RECEIVED.

Phillip's "Manual of Geology," edited by R. Etheridge, F.R.S., & H. G. Seeley, F.R.S. London: Charles Griffin & Co.—"Text Book of Zoology," by Dr C. Claus, Translated and edited by Adam Sedgwick, M.A., & F.G. Heathcote, M.A., London: W. Swan Sonnenschein & Co.—"Plant Lore, Legends, and Lyrics," by Richard Folkard, jun., London: Sampson Low & Co.—"Origin of Cultivated Plants," by Alphonse de Candolle, London: Kegan, Paul & Co.—"Effie and Her Strange Acquaintances," by the Rev. John Crofts, M.A., Chester: Phillipson & Golder.—"The British Moss-Flora," by R. Braithwaite, M.D. Part viii.—"The Gentleman's Magazine." —"Belgravia."—"The Journal of Microscopy."—"The Science Monthly."—"Midland Naturalist."—"The Asclepiad.—"Ben Brierley's Journal."—"Science Record."— "Science."—"American Naturalist."—"The Electrician and Electrical Engineer."—"Canadian Entomologist."—"American Monthly Microscopical Journal."—"Popular Science News."—"The Botanical Gazette."—"Revue de Botanique." —"La Feuille des Jeunes Naturalistes."—"Le Monde de la Science."—"Cosmos, les Mondes." &c. &c. &c.

COMMUNICATIONS RECEIVED UP TO 11TH ULT. FROM :—
J. T. W.—E. T. D.—W. T. M.—T. M. R.—W. H. B.—
L. C. M.—W. W. D.—A. B.—A. K.—P.—H. F.—T. B.—
F. S. G.—T. D. A.—C. W.—W. W.—C. H. R.—J. W. D.—
W. H. T.—J. B. J.—H. L. C.—W. B.—F. W. C.—F. W. B.
Co.—J. S.—W. B.—B. H.—S. W. M.—E. L.—P. H.—T. W. B.
—G. F. L.—J. H. L.—C. P. H.—B. P.—W. H. H.—I. A. jun.
—J. H. M.—W. L. W. G.—J. L. M.—A. R. W.—W. R. H.—
C.—S.—S. C.—E. L.—C. J. W.—W. D. S.—T. E. D.—
W. H.—W. H. G.—J. W. W.—T. D. A. C.—J. H. B.—
G. O. H.—N. P.—D. O.—J. C.—A. L.—J. C. M. P.—F. T. M.
—W. B.—F. M. H.—M.—A. D. G.—C.—E. J. H.—G. E. E.
jun.—G. H. K.—V. A. L.—W. R. G.—G. B.—S. L. M.—
J. M. B. T.—R. D.—W. R.—H. M.—J. B.—R. M. C.—
H. W. L.—R. B.—R. B.—S.—A. D.—A. T. M.—Mrs. E. P.—
J. A. H.—L. J.—R. M.—A. E. P.—C. C.—S.—T. M.—
W. F.—A. F.—E. S.—A. B. D.—J. C. H.—F. J. R.—W. F.—
H. F.—F. S.—A.—W. H. C. L.—W. F.—F. H. F. F.—A. M.
—J. F.—H. B. H.—J. R. B. M.—J. H. L.—A. J.—B. C. H.
R. R.—Dr. W. T. G.—N. A. D.—H. G. W. A.—S. S.—F. F. F
—F. B.—W.—C. K.—A. P.—E. B. L. B.—R. C.—R. B. W
—H. B. H.—J. R. B. M., &c.

INDEX TO VOL. XX.

About Mosquitos, 85
Abundance of Star-fish, 142
Alcohol solidified, 113,
Algae, aerial, 88
Alps—The Gentians of, 128
American Diatomaceæ, 260
Ammoniaphone, The, 65, 86
Amphibian, new Labyrinthodont, 112
Anatomical Preparations, 119
Animals and Birds in Jersey, 95
Animals, Maternal Instinct in, 223
Animal Tissues to harden, 42, 89
Animal Tracks, supposed, 21
Annelida, British Marine, 180, 196, 231
Antelope in the Crag Beds, Fossil 117
Anthers of Flowers, Mounting, 186
Apanteles glomeratus, 19
Aphis, The Horned, 186
Aquarium, a New, 8
Aquaria, How to keep Marine, 74
Aquaria, Shrimps in, 115
Aregma mucronatum, 92
Arsenical Wall-Paper, 160
Arum and Birds, The, 262, 280
Arvicola amphibius, 259
Association of *Helix nemoralis* and *H. hortensis*, 18
Atoll, The Battle of the, 21
Australian Entomology, 31, 81

Bacillus of Cholera, 114
Bacillus, Stainer and Others, v., 198
Bacteria, destruction of, 234
Bats, 119, 279
Battle of the Atoll, The, 21
Beech with "Gnaurs," 282
Bee's Leg, 22
Behaviour of Plants, 116
Birds in August, 262
Birds and the Arum, 262, 280
Birds, Early, 95
Bird, Four-footed, 278
Birds in Jersey, Animals and, 95
Birds, Notes on, 196
Birds, Strange, 23, 98
Bivalves out of their Element, 91
Blood Prodigy, 46
Blowhole, Peculiar, 85
Books, Pamphlets, etc., Notices of, 18, 40, 41, 54, 67, 90, 92, 103, 114, 115, 135, 138, 160, 162, 164, 184, 186, 212, 236, 255, 258, 260, 275
Book-Worm, 210, 258
Borers, Marine, 259
Boring near London, Deep Well, 69
Boring Sponges, 187
Botany, 20, 44, 68, 92, 116, 139, 164, 188, 213, 238, 261, 278
Botanical Ramble round Weymouth and the Channel Islands, 36
"Botany, Topographical," Watson's 44, 68
Brazilian Diamonds, 184
British Earthquake, a Genuine, 122
British Fresh-water Mites, 80
British Mineral, a New, 141
British Shells, 163
Broom, White, fasciation in stem of, 189
Bubbles, Soap, 47
Bullfinch, Strange Habit of, 148, 191
Butterflies at Cambridge, 116
Butterflies, Scarcity of, 87, 138, 187

"Cain and Abel," 141
Camel, The, 280
Canker on Apple-trees, 113
Carbonic Acid and Cement, 114, 137
Carbonic Acid solidified, 113
Carex ovalis, var. *bracteata*, 213

Cartilage, Preserving, 70
Castles, Mole, 167
Cat, Wild, 163
Caterpillars, feeding, 94
Cells, glass, 41
Cerataphis lataniæ, 186
Changing the colours of violets, 92
"'Challenger' Musings," 164
Chapters on Fossil Rays and Sharks, 172, 227
Character of Duck-billed Platypus, Zoological, 237, 283
Chemistry of Plants, 238
Chitine mounting, 114
Chitonidæ, eyes of, 234
Cholera Bacillus, 89, 114, 177
Cholera, French Committee on, 212, 258
Cholera, Propagation of, 235
Christmas Flowers, 44
Christmas Notes, 35
City, a Pre-historic, 77
Classification of the Mollusca, 184
Climbing Mice, 190
"Close," 283
Clover in New Zealand, 20, 135
Coal, the Formation of, 165
Cockroach, The Natural History of, 59
Cockroach, Alimentary Canal of, 150
Cockroach, Nervous System of, 244
Cockroach, Organs of Respiration and Circulation in, 203
Cockroach, Outer Skeleton of, 106
Conchological Notes, 3, 47, 91, 115
Congo, Dr. Zintgraff's Expedition to, 88
Continuity of Protoplasm, 93, 139
Convallaria rubra, 140
Cormorant in Worcestershire, 9
Coronula diadema, 190
Correspondence, 23, 47, 71, 95, 119, 143, 167, 191, 215, 239, 263, 283
Crested Newt, The, 138
Crystal Palace Insectarium, The, 186
Cucumber Trees, 71, 164
Cultivation of the Herring, 19
Curious Bird-Trap, 214
Curious phenomena, 127
Cutting Sections in Ribbons, 89

Dahlias, Double, 278
Darwin at Home, 51
Darwin on Instinct, 10, 94
Denudation, a Chapter on, 126
Denudation in America, 250
Destruction of Trees around London, 279
Diamonds, origin of, 118
Diamonds, Indian, 87
Diatomaceæ of the American Waters, 260
Diatomaceæ of Norfolk, 90, 260
Diatomescope, The, 276
Diatoms, markings on Test, 136
Difficulties in Mounting, 185, 212
Discoveries in the Devonian Strata, 93
Discoveries in Geographical Botany, 279
Diseases of Fruit, 214
Disease, the Potato, 39
Disease of Salmon, 19
Distribution of Plants, 34, 164
Do Swallows Reason ?, 166
Does the Sparrow-hawk attack Toads ? 215
Domed Nests, Observations on, 26, 142
Dormice, Death of, 167
Dormouse in England, 167, 190, 215
Double Flowers, Origin of, 57, 99, 116, 140, 218
Dragon and Lace-Flies, Early, 215
Drawing from the Microscope, 17, 41
Dredging in the Frith of Clyde, 200, 221
Drift, Irish, 280
Droitwich Brine Springs, 45

Drying Flowers, 35, 71, 79
Dublin, Mineralogical Studies in the County of, 75, 219
Duck-billed Platypus, The, 237
Dust-free Spaces, 134

Early Birds, 95
Early Blooming of Wild Flowers, 190
Earthquake, A Genuine British, 122
Earthquakes, Prof. Milne on, 64
Earthworm, Gigantic, 237
Egg of Hedge Sparrow, Malformation of, 167
Eider Duck, 142
Electro-Magnet, A New, 258
Elephant, The Pedigree of, 130, 146
Elm Leaf, Bifurcation of, 252
Emeralds, New Locality for, 134
Entomological Slides, 67
Entomology, Recollections of Australian, 81
Entomology of Highgate, 101
Eruption at Krakatoa, Report on, 136
Evidences of Primeval Man, 50
Exchanges, 24, 47, 71, 96, 119, 143, 168, 192, 218, 240, 263, 284
Experiments with Virus of Hydrophobia, 77

Fasciation in stem of White Broom, 189
Fauna and Flora of Worcestershire, 278
Ferns of the Arlberg Pass, 278
Ferns of the Pyrenees, 170
Field Fare, The, 166
Field Mouse, White, 167
Fish and Air, 163
Fish, Curious habit of, 142
Fleurs du Lac, 279
Flora of Socotra, 20
Flow of Protoplasm, 44
Flowers, Christmas, 44
Flowers, Drying, 35, 71, 79
Flowers, Early Blooming of Wild, 190
Flowers, The Origin of Double, 57, 99, 116, 140, 218
Foliage, Unseasonable, 47
Fossil Antelope on the Crag Beds, 117
Fossil Pectens of the Isle of Wight, 52
Fossils, Sham, 94
Four-footed Bird, 278
Fox-cubs, Reared by a Cat, 237
Freckled Goby, The, 171
Free-swimming Rotifers, 124, 163, 187
Freezing Microtome, 275
Fruit, Diseases of, 214
Fungus, New British, 92

Galenite in Sandstone, 141
Gate of North Wales, At the, 156
Geese Migrating, 19, 74
Gentians of the Alps, 128
Geology, 21, 45, 69, 93, 117, 141, 164, 189, 214, 238, 261, 280
Geology at the British Association, 238
Geology of Lincolnshire, 105
Geology and Mineralogy of Madagascar, 189
Geology of Northampton, The Underground, 189
Geology of Tenby, 166, 260
Gloucestershire Slugs, 78, 163
Glycerine, Preserving in, 67
Gobius minutus, 171
Gold-fish, 22, 23, 46, 142
Gossip on Current Topics, 125, 155, 174, 194, 225, 253, 272
Graphic Microscopy, 1, 25, 49, 73, 97, 121, 145, 169, 193, 217, 241, 265
Greely, Lieut., at the British Association, 234

Growing Cell for Minute Organisms, 8
Guildford, Mammoths at, 21
Gum Styrax as a Mounting Medium, 66

HASCHISCH, THE EFFECTS OF, 63
Hail-stones, peculiar, 281
Hawk-moth, Striped, 118
Hedgehogs, 3, 45, 46, 70, 280
Helices, reversed, 213
Helix aspera, reversed, 141
Helix caperata, 3
Helix pisana, 161, 190
Herbarium, A Live, 133
Herring, Cultivation of, 19
Hibiscus, "Sport" in, 188
Highgate, The Entomology of, 101
Holly, The, 47,
Hoplophora ferruginea, 56, 166
Horse's leg, Swelling on, 118
House-fly, The 163
House-fly, as a Propagator of Disease, 211
House-fly, The Teeth of, 176, 225
House-Martin, 44, 70
Humming Bird, Taming, 191
Hyalina Draparnaldi, 115
Hydrogen, generated, but not consumed, 215, 262
Hydrogen, liquefaction of, 235
Hydrozoa and Polyzoa, 277

INCRUSTATION IN A WATER-PIPE, 261
Infernal Machines, Detection of, 183
Influence of High Pressure on Living Organisms, 112
Infusoria, Mounting, 185
Insectarium at Crystal Palace, 186
Insects' Eggs, 278
Insects, Early Emergence of, 45, 281
Insects, Mounting, 186
Instinct in Animals, 223
Instinct, Darwin on, 10, 94

JERSEY, ANIMALS AND BIRDS IN, 95
Jersey, Notes on Natural History of, 177, 202

KEEPING GOLD-FISH, 22
Keeping Marine Aquaria, 74
Keeping Serpulæ, 22

LADY-BIRD, 167. 215
Lake District, The Mineral Veins of, 239
Lapwing, Pied, 190, 215
Lathræa squamaria, 4, 44
Leaves, Skeleton, 166, 188, 238, 261, 262
Life, Strange Tenacity of, 282
Lily of the Valley, root action of, 100
Limnæa pereger, var. *picta*, 190
Lincolnshire Geology, an interesting bit of, 105
London, The Oolitic Rocks under, 189

MADAGASCAR, GEOLOGY AND MINERALOGY OF, 182
Magnetism, Prof. Hughes on the Theory of, 87
Malformation of Hedge-Sparrow's Egg, 167
Mammoths at Guildford, 21
Mantis, The Praying, 263, 283
Marine Aquaria, 74
Marine Borers, 259
Marine Denudation, 126
Marine Zoology, Slides Illustrative of, 42
Markings on Test Diatoms, 136
Medicinal Plants, 23
Medium for Hard Sections, 91
Metamorphic Rocks of South Devon, 69
Mice, Climbing, 190
Microscopy, 17, 41, 67, 89, 115, 136, 162, 185, 212, 235, 260, 275
Microscopy, Graphic, 1, 25, 49, 73, 97, 121, 145, 169, 193, 217, 241, 265
Microscope, Drawing from the, 17, 41
Microscopic Objects, Photographing, 17
Microscopic Slides, Improvement in, 136, 162
Microscopic Holder, a Cheap, 260
Microscopic Slide Centrerer, 260, 276
Microtome, a New, 137
Middlesbrough District, Shells in, 91
Migration of Geese, 19, 74
Mildness of Last Season, 186
Mimicry of Mint, by Dog's Mercury, 188, 238

Mind in the Lower Animals, 22
Mineral, A New British, 141
Mineralogical Studies in the County of Dublin, 75, 219
Mineral Veins of the Lake District, 239
Minnow, The, 282
Miscellaneous Notes, 215
Mites, British Fresh-Water, 80
Mole Castles, 167
Mollusca of Brentford District, 138
Mollusca, Classification of, 184
Mollusca of Derby, 277
Mollusca of Kent, Surrey and Middlesex, 258
Mollusca of Maidenhead, 277
Mollusca of Margate, 43, 67
Moon, Peculiar Appearance of, 140
Mosquitos, about, 85
Mosses, 83
Mounting the Anthers of Flowers, 186
Mounting Chitine, 114
Mounting, Difficulties in, 185, 212
Mounting Infusoria, 162, 185
Mounting Insects, 186
Mounting Medium, Gum Styrax as a, 66
Mounting Medium, Tolu as a, 260
Mounting Sections stained with Picro-carmine, 275, 276

NAME OF PLANT, ODD, 94
Names on Trees, 239
Narcissus, Peculiar, 140
Natterjacks at Wimbledon, 236
Natural History Clause in Railway Bill, 283
Natural History of Jersey, Notes on, 177, 237
Natural History Lists, 278
Natural History Transactions of Northumberland, &c., 236
Naturalists, Local, 23
Navicula cuspidata, 276
Nests, Domed, 26, 142
New Labyrinthodont Amphibian, 112
Newt, The Crested, 138
Newt, Curious action of, 262
Nightingales in York, 159
Northampton, Underground Geology of, 189
North Wales, At the Gate of, 156
Notes, Christmas, 35
Notes, Miscellaneous, 215
Notes from Rotterdam, 282
Notes and Queries, 22, 45, 70, 94, 118, 141, 166, 190, 214, 239, 262, 280
Notes for Science Classes : Mosses, 83
November Nosegay in Worcestershire, 20

OBITUARY:
Aitkin, Mr. John, 211
Bentham, Mr. George, 238
Brittain, Mr. Thos., 63
Le Conte, Dr., 17
Fullagar, Mr. J., 112
Guyot, M., 88
Kidd, Mr. H. W., 111
Merrifield, Mr. C. W., 40
Nilsson, Prof., 15
Parker, Mr. J, 112
Powell, Mr. H., 17
Thomson, Prof. Allen, 112
Todhunter, Dr., 88
Tolles, Mr. Robert, 39
Watts, Mr. Henry, 183
Wilson, Sir Erasmus, 212
Woodward, Col. J. J., 257
Observations on Domed Nests, 26, 142
Odd Name of Plant, 94, 117
Œcidium Ari, 140
Œcidium Jacobææ, 164
Oolitic Rocks under London, 189
Ophioglossum vulgatum, var. *ambiguum*, 148
Orchid Flowers, 20, 69
Organisms, The Lower, Influence of High Pressures on, 112,
Organisms, Observations on Living, 262
Origin of Diamonds, 118
Osprey, The, 281

PALATE-MYOGRAPH, 114
Paludina vivipara, 213
Parasites, Vegetarians and, 184
Paris quadrifolia, 190, 239, 261, 278, 283
Pectens, Fossil, in the Isle of Wight, 52

Peculiar Appearance of the Moon, 140
Peculiar Narcissus, 140
Pedigree of the Elephant, 130, 146
Peziza Sumneria, 116
Pheasants, A Chapter on, 242, 266
Phenomena, Curious, 127
Phosphorous Insects, 262
Phylloxera, M. Ba lbiani on the, 257
Phytological Record for March, 68
Plants, The Behaviour of, 116
Plants, Distribution of, 34, 164,
Plants, Fossil, in the Silurian Formation, 28
Plants, Medicinal, 23
Plant Names, 70, 117, 141
Plant Notes, 149, 272
Plants, Rare, 238
Pleasure and Pain, 67
Poison, Effect of, on the Hedgehog, 94
Polyzoa and Hydrozoa, 277
Polyzoa and Rotiferæ, 237
Post-Glacial Ravines in Chalk Wolds of Lincolnshire, 93
Potato Disease, 39
Potatoes, Species of, 68
Praying Mantis, 263
Pre-Historic City, 77:
Preparation, Anatomical, 119
Preservation of Protozoa and Small Larvæ, 66
Preservation of Soft Tissues, 42
Preserving Cartilage, 70
Preserving in Glycerine, 67
Pressures, High, Influence of, on Living Organisms, 112
Primeval Man, New Evidences of, 50
Primroses, 261
Protoplasm, Continuity of, 93, 139
Protoplasm, Facts concerning, 44
Puccinia, Early, 140

RAMBLE OVER THE SURREY DOWNS, A Midsummer, 95
Ramble round Weymouth, a Botanical, 36
Ravines in the Chalk-Wolds of Lincolnshire, Post-Glacial, 93
Rays and Sharks, Fossil, 172, 227, 267
Recollections of Australian Entomology, 31
Red Hill, 86
Redshank, Spotted, at Lynn, 19
Reproduction of the Zygnemaceæ, 20
Reptiles eating their Sloughs, 280
Rocks of South Devon, Metamorphic, 69
Rook, The, 119, 142
Root Action of the Lily of the Valley, 100.
Rotiferæ and Polyzoa, The, 237
Rotifers, Free-Swimming, 124, 163, 187

Sagittaria sagittifolia, 188, 238
Salmon Disease, 19
Salmon Ova for New Zealand, 67
Sample Microscopic Slides, 18
Saurian, New Fossil, 28
Saxifraga hirculus, 261
Scarcity of Butterflies, 138, 187
Science of Guide-Books, 70
Science-Gossip, 15, 39. 63, 86, 111, 134, 160, 183, 210, 233, 256, 274
Scilly, Ten Days in, 98
Season, The Mildness of, 186
Seasonable Notes from Cushenden, 94
Sections, Hard, a Medium for, 91
Sections in Ribbons, cutting, 89
Sections, stained, 275, 276
Seeds, M. Duchatre's Experiments with, 64
Selachians, Peculiar Japanese, 88
Senecio vulgaris, var. *radiatus*, 188
Serpulæ, Keeping, 22
Sham Fossils, 94
Sharks and Rays, Fossil, 172, 227, 267
Shells, British, 163
Shells in Middlesbrough District, 91
Shrimps in Aquaria, 115
Silurian Formation, Fossil Plants in, 28
Sinking of a Mountain, Reported, 112
Skeleton Leaves, 166, 188, 238, 261, 262
Skin-Grafts, 274
Slides Illustrative of Marine Zoology, 42
Slugs, Gloucestershire, 78
Soap-Bubbles, 23, 46
SOCIETIES ; meetings, transactions, &c., of :—
Belfast Naturalists' Club, 92
Belgian Microscopic Society, 90
Botanical Exchange Club, 140

SOCIETIES; meetings, transactions, &c., of (cont.):—
Bristol Naturalists' Society, 213
British Association, 87, 234
Carlisle Microscopic Society, 42
Chichester and West Sussex Natural History Society, 213
Cryptogamic Society of Scotland, 39
Essex Field Club, 188, 212
Geologists' Association, 41, 112, 118, 134, 182, 189
Greenwich Microscopic Society, 89
Hackney Microscopic Society, 185
Hampstead Naturalists' Club, 213
Huddersfield Naturalists' Society, 92
Lambeth Field Club, 41
Liverpool Geological Society, 45
Liverpool Literary Society, 103
Liverpool Microscopic Society, 137
Liverpool Naturalists' Field Club, 259
Marine Biological Association, 114, 101, 185, 212
Manchester Microscopic Society, 41, 213
Norwich Geologists' Association, 139
Nottingham Naturalists' Society, 89, 139
Ottawa Field-Naturalists' Society, 139
Penzance Natural History Society, 259
Postal Microscopic Society, 42, 113
Quekett Club, 90, 212, 236
Rochester Naturalists' Club, 113
Royal Society, 15, 16
Royal Society of Tasmania, 213
Royal Microscopic Society, 42, 90, 136, 162, 235
Society of Amateur Geologists, 274
South London Natural History Club, 259
Walthamstow Natural History Society, 259
Yorkshire Philosophical Society, 259
"Youths" Naturalist's Society, 18

Socotra, Flora of, 20
Solidified Acohol, 113
Solidified Carbonic Acid, 113
"Songs for the Times," 164, 214
South Essex, 91
Spaces, Dust-Free, 134
Spider's Severed Leg, Motion in, 262
Sponges, Boring, 187
"Sport" in Hibiscus, 188
Starling, White, 166
Star-Fishes, Abundance of, 142
Stoat in England, White, 46
Stoat in Jersey, 11
Storm-Glass, 22, 51
Strange Birds, 23
Strange Visitor, A, 33
Striped Hawk-Moth, 118
Strix brachyotus 43
Studies in Microscopic Science, 18
Sundews, 70, 117, 140
Sunsets, Blood-Red, 2, 118, 160
Swallows in Church, 191
Swallows, Do they Reason?, 166
Swelling on Horse's Leg, 118

TAMING WILD HUMMING-BIRDS, 191
Tannin in Vegetable cells, 275
Technology, Microscopical, 5
Tenacity of Ligo, Strange, 282
Tenby, Geology of, 166, 260
Teratology, Vegetable, 140
Thunderstorms, Prof. Tait on, 64
Tissues, Animal, to Harden, 89
Tissues, Soft, Preservation of, 42
Tolu, as a Mounting Medium, 260
Tomtits, 282
Tooth-Wort, Notes on (*Lathræa squamaria*), 4
"Topographical Botany," 44
Tortoise, Proper Food for, 262

Tracks, Supposed Animal, 21
Trap, for Bird, Curious, 214
Trees, Names on, 239
Trespass, Law of, 239
Trochus, Rare Species of, 115
Turbot, Scales of, 138

UNIOS, LARGE, 261, 280
Unseasonable Foliage, 47

VEGETABLE TERATOLOGY, 140
Vegetarians and Parasitism, 184
Vertebrate Remains in the Strata of Devonshire, 93
Violets, Changing the Colours of, 92
Visitor, A Strange, 33
Volcanic Group of St. David's, 69

WALES, NORTH, AT THE GATE OF, 156
Water-Voles as Vegetable Feeders, 118, 281
Weather-Glass, The Fitzroy, 203
Weymouth and the Channel Islands, a Botanical Ramble Round, 36
Whales and their Origin, 42
White Field-Mouse, 167
White Starling, 166
Wild Cat, in Kent, 163
Window Pets, My, 175
Worcestershire, The Fauna and Flora of, 278

YEAST-FUNGI, NOTES ON, 11

ZOOLOGY, 18, 42, 67, 91, 115, 138, 162, 186, 213, 236, 258, 277
Zoological Character of Duck-Billed Platypus, 237, 283
Zoological Slides, 67

LONDON :
PRINTED BY WILLIAM CLOWES AND SONS, LIMITED,
STAMFORD STREET AND CHARING CROSS.